Trees of Hyperbolic Spaces

Mathematical
Surveys
and
Monographs

Volume 282

Trees of Hyperbolic Spaces

Michael Kapovich
Pranab Sardar

AMERICAN
MATHEMATICAL
SOCIETY
Providence, Rhode Island

2020 *Mathematics Subject Classification.* Primary 20F67, 51F30.

For additional information and updates on this book, visit
www.ams.org/bookpages/surv-282

Library of Congress Cataloging-in-Publication Data

Names: Kapovich, Michael, 1963- author. | Sardar, Pranab, 1982- author.
Title: Trees of hyperbolic spaces / Michael Kapovich, Pranab Sardar.
Description: Providence, Rhode Island : American Mathematical Society, [2024] | Series: Mathematical surveys and monographs, 0076-5376 ; volume 282 | Includes bibliographical references and index.
Identifiers: LCCN 2024009568 | ISBN 9781470474256 (paperback) | ISBN 9781470477783 (ebook)
Subjects: LCSH: Hyperbolic spaces. | Geometry, Hyperbolic. | AMS: Group theory and generalizations – Special aspects of infinite or finite groups – Hyperbolic groups and nonpositively curved groups.
Classification: LCC QA685 .K363 2024 | DDC 516.9–dc23/eng/20240324
LC record available at https://lccn.loc.gov/2024009568

∞ The paper used in this book is acid-free and falls within the guidelines
established to ensure permanence and durability.
Visit the AMS home page at `https://www.ams.org/`

10 9 8 7 6 5 4 3 2 1 29 28 27 26 25 24

Contents

Preface

The goal of this book is to understand the geometry of metric spaces X which have the structure of trees of hyperbolic spaces. The subject originates in the papers [**BF92**, **BF96**] of Bestvina and Feighn, where they proved a *combination theorem*,[1] stating that under certain conditions such an X itself is hyperbolic:

THEOREM 0.1. *Suppose $\mathfrak{X} = (\pi : X \to T)$ is a tree of hyperbolic metric spaces, where vertex-spaces and edge-spaces are uniformly hyperbolic, incidence maps of edge spaces into vertex-spaces are uniformly quasi-isometric and which satisfies the hallway flaring condition. Then X is a hyperbolic metric space.*

In Chapter 2 we give definitions clarifying the result. Informally, the hallway flaring condition means that two K-quasi-isometric (K-qi) sections of $\mathfrak{X}$ over a geodesic interval I in T "diverge at a uniform exponential rate" as we move along I in one of the two directions. The original proof of this theorem was by verifying that X satisfies the linear isoperimetric inequality. In the book we give a new (and longer) proof under a weaker flaring assumption than the one made by Bestvina and Feighn; we name the weakened condition *uniform* (or, in another version, *proper*) flaring. Informally speaking, instead of requiring the exponential divergence of sections, we only require *some* rate of divergence, given by a uniform proper function of the arc-length parameter of I. We refer the reader to Theorem 2.62 and Corollary 2.63 for the precise statements.

The main benefit of our proof is that it is done by constructing a *slim combing* of X by paths $c = c_{xy}$ in X. These paths will be uniform quasigeodesics connecting points $x, y \in X$. The description of the paths c is a 6-step induction summarized in Chapter 7, starting with paths in the trees of spaces of the simplest kind that we call *narrow carpets*: These are metric interval-bundles over geodesics in T such that one of the interval-fibers has uniformly bounded length. We hope that our method of proof of Theorem 0.1 has a potential to generalize this theorem to complexes of hyperbolic spaces.

The combing paths c in X are mostly concatenations of K-quasi-isometric sections of $\mathfrak{X}$ over geodesics in T. Thus, we obtain (up to a uniformly bounded error) a description of geodesics in X in terms of its structure as a tree of spaces, i.e., vertex-spaces and sections.

We also refer the reader to the related work of Gautero describing uniform quasi-geodesics in groups obtained via the combination theorem in a special case in [**Gau03**] and in [**Gau16**] for trees of relatively hyperbolic spaces. Unfortunately, we were unable to follow Gautero's proof. We note, furthermore, that various forms of the Bestvina–Feighn combination theorem for relatively hyperbolic groups and spaces were proven by Dahmani [**Dah03**], Alibegović [**Ali05**], Gautero and Lustig [**GL04**, **GL07**], and, in greatest generality, by Mj and Reeves [**MR08**]. In the book we did not attempt to describe (quasi)geodesics

[1] Subsequently, alternative proofs of the group-theoretic version of this theorem were given by Kharlampovich and Myasnikov in [**KM98**] and Gautero in [**Gau03**], under certain extra assumptions.

in the trees of relatively hyperbolic spaces. However, we proved the existence of Cannon–Thurston maps for subtrees of relatively hyperbolic spaces $Y \subset X$ in Chapter 9 using techniques of proof of the existence of Cannon–Thurston maps for subtrees of spaces in trees of hyperbolic spaces.

Summary of the main results of the book. The central results of the book are the four theorems below. To state these results we will need some definitions. A *tree of spaces* $\mathfrak{X} = (\pi : X \to T)$ comes equipped with a simplicial tree T and a collection of *vertex- and edge-spaces* (X_v, d_{X_v}), (X_e, d_{X_e}) associated with the vertices and edges of T. These spaces are general geodesic metric spaces, it suffices to think of these as (connected) metric graphs equipped with graph-metrics. For every edge $e = [v, w]$ of T we have *incidence maps* $f_{ev} : X_e \to X_v, f_{ew} : X_e \to X_w$, which are uniformly proper Lipschitz maps. Again, in the context of graphs, it suffices to think of combinatorial incidence maps (they will be automatically 1-Lipschitz).

Remark 0.2. For most of the book, we will, in fact, assume more, namely, that the incidence maps are L_0-quasi-isometric embeddings and, moreover, that all vertex-spaces (and, hence, edge-spaces) are δ_0-hyperbolic (for some constants δ_0, L_0).

Given the above data, one defines a geodesic metric space X, the *total space* of the tree of spaces $\mathfrak{X}$, and a uniformly Lipschitz projection $\pi : X \to T$. In the case when vertex-spaces and edge-spaces are graphs, the space X can be taken to be a graph, obtained from the disjoint union of the vertex-spaces and edge-spaces by adding edges connecting each vertex $x \in X_e$ to its images $f_{ev}(x) \in X_v$ and $f_{ew}(x) \in X_w$. The 1-Lipschitz projection π then sends each X_e to the midpoint of the edge e and each vertex-space X_v to the vertex v. A *K-qi section* of $\mathfrak{X}$ over a subtree $S \subset T$ is a section σ of the restriction

$$\pi : \coprod_{v \in V(S)} X_v \to V(S)$$

such that $d(\sigma(v), \sigma(w)) \le K$ for every edge $e = [v, w]$ of S. (Here $V(S)$ is the vertex-set of the subtree S.) We will frequently conflate sections and their images.

We say that $\mathfrak{X}$ is a *tree of hyperbolic spaces* if all vertex- and edge-spaces of $\mathfrak{X}$ are δ_0-hyperbolic for a common constant δ_0. Next, we give simplified forms of the definitions of *hallways* and *carpets* in trees of spaces, which will suffice for the purpose of stating our main results.

Definition 0.3. A K-hallway $\mathfrak{A}$ in $\mathfrak{X}$ consists of a pair of K-qi sections (γ_1, γ_2) of $\mathfrak{X}$ over an (oriented) interval $[\![v, w]\!] \subset T$. A K-hallway is called a *(K, C)-carpet* if

$$d_{X_w}(\gamma_1(w), \gamma_2(w)) \le C.$$

A carpet is said to be *bounded* by a geodesic interval $\alpha = [\gamma_1(v)\gamma_2(v)] \subset X_v$, in which case we will also use the notation $\mathfrak{A} = \mathfrak{A}(\alpha)$.

Definition 0.4. We say that a tree of spaces $\mathfrak{X}$ *satisfies the thin bigon property for K-hallways* if there is a function $R = R(K, C)$ such that for every K-hallway $\mathfrak{A}$ as above, the inequality

$$\max\left(d_{X_v}(\gamma_1(v), \gamma_2(v)), d_{X_w}(\gamma_1(w), \gamma_2(w))\right) \le C$$

implies that for every vertex t of the interval $[\![v, w]\!]$,

$$d_{X_t}(\gamma_1(t), \gamma_2(t)) \le R.$$

Our proof of hyperbolicity and description of geodesics in hyperbolic trees of spaces hinges upon a construction of a *slim combing* and a theorem by Brian Bowditch that we review next.

Suppose that X is a metric space. A *combing* C of X is an assignment to each pair of points $x, y \in X$ of a rectifiable path c_{xy} in X (called a *combing path*) parameterized by its arc-length. A combing is called *slim and proper* if the following conditions are met:

(1) There exists a constant δ such that for any three combing paths c_{xy}, c_{yz}, c_{zx}, the image of c_{xy} is contained in the δ-neighborhood of the union of the images of c_{yz} and c_{zx}.

(2) There exists a Lipschitz function η such that each path c_{xy} is η-*proper*, i.e., for any two points p, q in the image of c_{xy}, the distance in X between p, q is at most $\eta(\ell(p, q))$, where $\ell(p, q)$ is the length of the subpath of c_{xy} between p and q.

Bowditch proved that whenever a path-metric space X admits a slim combing C, then X is δ'-hyperbolic and each combing path c_{xy} is K-quasigeodesic, where δ' and K depend only on δ and η. We refer the reader to Section 1.10 for details.

The following is the central result of this book:

THEOREM 0.5. *Suppose that $\mathfrak{X} = (\pi : X \to T)$ is a tree of δ_0-hyperbolic spaces, such that the incidence maps from the edge-spaces of $\mathfrak{X}$ to the incident vertex-spaces are L_0-quasi-isometric embeddings. Then there exists a constant K, depending only on δ_0, L_0, such that the following holds. Assume, that $\mathfrak{X}$ satisfies the* thin bigon property *for K-hallways. Then there exists an explicit (in terms of the structure of the tree of spaces $\mathfrak{X}$) slim combing of X.*

The detailed description of the combing in Theorem 0.5 is too long and technical to be reproduced here; it is summarized in Chapter 7. For now, we can say that there are two key ideas used repeatedly in this description:

(1) An easy description of geodesics in *quasiconvex chain-amalgamations* of hyperbolic spaces. For instance, if X is a hyperbolic space which is the union of two qi embedded geodesic subspaces A, B with intersection C (that is also a qi embedded geodesic subspace), then the geodesic in X connecting $a \in A, b \in B$ can be described (up to a uniformly bounded error) as the concatenation of three geodesics: Two of these are the geodesics in A and B connecting a, b to their nearest-point projections (taken, respectively, in A, B) $\bar{a}, \bar{b} \in C$, and a geodesic in C connecting $\bar{a}, \bar{b}$.

(2) Quasi-Isometric sections over intervals in T are quasigeodesics in X. In particular, suppose that $C \subset X$ is a quasiconvex subset (whose quasiconvexity is proven, for instance, by constructing a coarse Lipschitz projection $X \to C$) and x, y are points in X which can be connected to C via quasi-isometric sections γ_x, γ_y. Then a uniform quasigeodesic in X connecting x, y can be described (analogously to (1)) as a concatenation of three paths $c(x, \bar{x}), c(\bar{x}, \bar{y}), c(\bar{y}, y)$. Here $c(x, \bar{x})$, $c(\bar{y}, y)$ are subpaths in γ_x, γ_y between, respectively, $x, \bar{x}$ and $\bar{y}, y$, where $\bar{x}, \bar{y}$ are suitable points which are:
   (i) Either both in C (up to a uniform error), in which case $c(\bar{x}, \bar{y})$ is the geodesic in C connecting $\bar{x}, \bar{y}$.
   (ii) Or $\bar{x}, \bar{y}$ are points in γ_x, γ_y where these sections enter the same vertex-space X_t of X, such that the distance $d_{X_t}(\gamma_x, \gamma_y)$ is uniformly bounded (and the vertex-space X_t is "the first counting from x, y" where this happens). The path $c(\bar{x}, \bar{y})$ in this case is the geodesic $[\bar{x}\bar{y}]_{X_t}$ in X_t connecting $\bar{x}$ to $\bar{y}$.

By combining Theorem 0.5 with Bowditch's theorem, we obtain the second main result of the book, which is our version of the Bestvina–Feighn combination theorem.

THEOREM 0.6. *If $\mathfrak{X} = (\pi : X \to T)$ is a tree of spaces satisfying the hypothesis of Theorem* 0.5*, then X is hyperbolic.*

The description of combing paths in X simplifies greatly when the tree of spaces $\mathfrak{X}$ is (M, K, A)*-acylindrical* (for a suitable K depending on the parameters δ_0, L_0 of $\mathfrak{X}$ and some constants M, A), i.e., when any two K-qi sections γ_1, γ_2 over every interval $I \subset T$ of length $\geq A$ satisfy the inequality

$$d_{X_v}(\gamma_1(v), \gamma_2(v)) \leq M$$

for all vertices v in I. (In the setting of trees of spaces corresponding to graphs of groups, this notion is equivalent to the acylindricity of the associated group action on the Bass–Serre tree.) In this special case we also describe the Gromov–boundary of X in terms of the tree T and Gromov–boundaries of the vertex-spaces and edge-spaces of $\mathfrak{X}$, Theorem 7.13.

We now return to the discussion of general trees of hyperbolic spaces, $(\pi : X \to T)$. Suppose that $S \subset T$ is a subtree; one then obtains an induced subtree of spaces $(\pi : Y \to S)$ and a natural inclusion map $Y \to X$. If $(\pi : X \to T)$ satisfies the assumptions of Theorem 0.5, then so does $(\pi : Y \to S)$; in particular, Y is also hyperbolic. As an application of our description of geodesics in X (and in Y), we prove (see Theorem 8.48) the existence of *Cannon–Thurston maps* between the Gromov-boundaries $\partial_\infty Y \to \partial_\infty X$, extending an earlier result by Mitra [**Mit98**], who proved the existence of Cannon–Thurston maps for the inclusion maps of vertex-spaces into X. Mitra's proof (as well as the subsequent work of Mj and Sardar, [**MS09**]) was, in fact, a guideline for our description of geodesics in X. However, Mitra's description of geodesics stopped at geodesics connecting points in the same vertex-space (Step 3 of our 6-step description), leaving much of the work to be done in general.

The following existence theorem for Cannon–Thurston maps is the third main result of the book.

THEOREM 0.7. *Suppose that $(\pi : X \to T)$ is a tree of spaces satisfying the assumptions of Theorem* 0.5*. Then for every subtree $S \subset T$ and the corresponding induced tree of hyperbolic spaces $(\pi : Y \to S)$, the inclusion map $Y \to X$ admits a continuous extension to a map of Gromov-boundaries, $\partial_{Y,X} : \partial_\infty Y \to \partial_\infty X$.*

The map $\partial_{Y,X}$ is the Cannon–Thurston map between the ideal boundaries of Y and X. Furthermore, we prove a relative analogue of Theorem 0.7, for trees of relatively hyperbolic spaces, Theorem 9.47.

Our last main result is a detailed analysis of the *Cannon–Thurston laminations* of the inclusion maps $Y \to X$. Before stating this theorem, we will need more definitions.

DEFINITION 0.8. Suppose that X is a hyperbolic space and $Y \subset X$ is a hyperbolic subspace, such that the inclusion map $Y \to X$ admits a Cannon–Thurston extension $\partial_{Y,X}$: $\partial_\infty Y \to \partial_\infty X$. The *Cannon–Thurston lamination* (CT-lamination), $\Lambda(Y, X)$, of the inclusion map $Y \to X$ is the set of (unordered) pairs of distinct points ξ, η in $\partial_\infty Y$ such that $\partial_{Y,X}(\xi) = \partial_{Y,X}(\eta)$. A geodesic $\alpha \subset Y$ connecting points ξ, η with $\{\xi, \eta\} \in \Lambda(Y, X)$, is called a *leaf* of the CT-lamination $\Lambda(Y, X)$.

The next theorem gives a complete description of CT-laminations of the inclusion map $Y \to X$ in the setting of Theorem 0.7.

THEOREM 0.9. *Suppose that the total space X of a tree of spaces $\mathfrak{X}$ is a proper metric space. Then there are constants K, C depend only on the parameters δ_0, L_0 of $\mathfrak{X}$ and a function $D = D(k)$ such that the following hold.*

(1) *If $\{\xi^+, \xi^-\} \in \Lambda(Y, X)$, then there exists a vertex-space $X_u \subset Y$ and a complete geodesic $\alpha : \mathbb{R} \to X_u$, which is a uniform quasigeodesic in Y asymptotic to $\xi^\pm$, such that the intervals $\alpha_{-m,n} = [\alpha(-m)\alpha(n)]_{X_u} \subset \alpha$, bound (K, C)-narrow carpets $\mathfrak{A}(\alpha_{-m,n})$ in X for all $m > 0, n > 0$.*
(2) *Conversely, if X_u is a vertex-space of Y, $\alpha \subset X_u$ is a complete geodesic asymptotic to distinct points $\xi^\pm \in \partial_\infty Y$, such that each subinterval $\alpha_{-m,n}$ as above bounds a (K, C)-narrow carpet $\mathfrak{A}(\alpha_{-m,n})$ in X, then $\{\xi^+, \xi^-\} \in \Lambda(Y, X)$.*
(3) *Suppose that X_u is a vertex-space of Y, $\alpha \subset X_u$ is a complete geodesic. Then α is* not a leaf *of $\Lambda(Y, X)$ if and only if for some (equivalently, every) $k \geq 1$, there exist points $x, y \in \alpha$ and maximal k-qi sections Σ_x, Σ_y over, possibly different, subtrees T_x, T_y in T, such that $x \in \Sigma_x$, $y \in \Sigma_y$ and for every vertex $v \in T_x \cap T_y$, the distance in X_v between the points $\Sigma_x \cap X_v, \Sigma_y \cap X_v$ is at least D.*

We refer the reader to Theorem 8.63 for more detailed properties of ending laminations of the inclusion maps $X_v \to X$ of vertex-spaces.

Organization of the book. In Chapter 1 we review basic facts of coarse geometry and geometry of hyperbolic spaces. While most of the material of the chapter is standard and well-known, we included it for the ease of reference in the rest of the book.

In Chapter 2 we discuss definitions of the theory of trees of metric spaces, state and compare different flaring conditions in trees of spaces, formulate our main theorem (Theorem 0.5) and prove it in some easier cases, e.g., for *quasiconvex amalgamations* (Section 2.6.2).

In Chapter 3 we define a certain class of subspaces Y in a tree of spaces $\mathfrak{X}$, called *semicontinuous families*. Informally speaking, such a family is a subtree of spaces $\mathfrak{Y}$ in $\mathfrak{X}$ "centered" at a specific vertex $u \in T$. As one moves along edges $[v, w]$ away from u in T, the corresponding vertex-space $Y_w \subset X_w$ can get "much smaller" than Y_v but cannot get "much larger." This is the *semicontinuity* property. Furthermore, these subspaces Y have the property that their intersections $Y_v = Y \cap X_v$ with vertex-spaces of $\mathfrak{X}$ are uniformly quasiconvex in X_v and every point in Y is connected to the intersection $Y_u = Y \cap X_u$ of Y with the distinguished vertex-space X_u, by a K-quasi-isometric section of $\mathfrak{X}$ over the interval $[\![u, v]\!]$ in T. We prove that the subspaces Y are coarse Lipschitz retracts of X, which is a generalization of the horocyclic projections to a geodesic in the hyperbolic plane; its existence was first proven by Mitra, [**Mit98**, Theorem 3.8], in the case of semicontinuous families called *flow-spaces* $Fl_K(X_u)$. Flow-spaces and three other types of semicontinuous families (ladders, carpets and bundles) serve as key tools in our definition of the combing C of X. Ladders are certain (semicontinuous) families of intervals over subtrees in X. Bundles should be thought of as *continuous* families of quasiconvex subsets Q_v of vertex-spaces X_v of $\mathfrak{X}$ with two (nonempty) vertex-spaces Q_v, Q_w uniformly Hausdorff-close, whenever v and w span an edge of T. Carpets (more precisely, (K, C)-narrow carpets) form a special class of bundles and ladders; they are interval-bundles over intervals $[\![u, w]\!]$ in T where the fiber of the interval-bundle over w has length $\leq C$.

Chapter 4 primarily deals with Steps 1–3 of our description of geodesics in X: We describe combing paths in carpets, carpeted ladders and general ladders, and establish their hyperbolicity. Hyperbolicity of flow-spaces is proven in Chapter 5, which is technically the most difficult part of our work: We prove the slim triangle property for the combing paths

c by analyzing triples of ladders (with the common center u) in the flow-space $Fl_K(X_u)$ of a vertex-space X_u.

Our last challenge is to connect by combing paths points in different vertex-flow-spaces $Fl_K(X_u)$, $Fl_K(X_v)$. This is done in Chapter 6. The case of points in intersecting flow-spaces $Fl_K(X_u)$, $Fl_K(X_v)$ is handled in Section 6.1 where we primarily analyze the case of *special intervals* $[\![u, v]\!] \subset T$, i.e., when $Fl_K(X_u) \cap X_v \neq \emptyset$. This covers Step 4 of our description of geodesics in X and is quite technical. The main trick is to introduce a certain generalization of flow-spaces of vertex-spaces and appeal to a special (and easy) case of Theorem 2.62 proven earlier, the *quasiconvex amalgamation*, when the tree T contains a single edge (Corollary 2.68). Once the case of special intervals is done, we complete easily Step 5 of our description of geodesics by considering points in flow-spaces $Fl_K(X_J)$ for subintervals $J \subset T$ represented as unions of three special subintervals: For the proof we use the quasiconvex amalgamation again. (A good example of such an interval J is given by a *semispecial* interval $[\![u, v]\!]$, where the flow-spaces $Fl_K(X_u)$, $Fl_K(X_v)$ have nonempty intersection in X.) Lastly, we conclude the 6-step description of geodesics in X by appealing to the *horizontal subdivision* of geodesic intervals J in T, so that the consecutive subdivision vertices u_i, u_{i+1} define pairwise uniformly cobounded flow-spaces $Fl_K(X_{u_i})$, $Fl_K(X_{u_{i+1}})$ (their projections to the tree T are disjoint), while each interval $[\![u_i, u_{i+1}]\!]$ between u_i, u_{i+1} is a union of three special subintervals. This uniform coboundedness property allows us to reduce the problem of hyperbolicity of the flow-space $Fl_K(X_J) \subset X$ to the *pairwise cobounded quasiconvex chain-amalgamation* of hyperbolic spaces which is, again, a special and easy case of Theorem 2.62 proven earlier (Theorem 2.65). The combing paths c_{xy} in X are then defined as geodesics in flow-spaces $Fl_K(X_J)$, $x \in X_u, y \in X_v$, and $J = [\![u, v]\!]$. Lastly, we verify the slim combing property for these paths c_{xy} by considering flow-spaces $Fl_K(X_S)$ for geodesic tripods $S \subset T$ and appealing to Theorem 2.65 (or, more precisely, its consequence, Corollary 2.69) one last time.

In Chapter 7 we review the description of the combing paths c_{xy} by putting together different steps of the descriptions scattered in the earlier parts of the book. We also prove an easy application of this description by giving a characterizations of geodesics α in vertex-spaces of X which are quasigeodesics in X itself, in terms of carpets bounded by subintervals in α. Furthermore, assuming acylindricity, we give a simplified description of uniform quasigeodesics and quasigeodesic rays in X. We use this description to describe the ideal boundary of X in terms of ideal boundaries of vertex-spaces and of the tree T.

In Chapter 8 we apply our description of geodesics to prove Theorem 8.48, establishing the existence of Cannon–Thurston maps $\partial_{Y,X}$ for subtrees of spaces $\mathfrak{Y} = (\pi : Y \to S)$ in a hyperbolic tree of spaces $\mathfrak{X} = (\pi : X \to T)$. We begin the chapter by reviewing the definition of Cannon–Thurston extensions (also known as *Cannon–Thurston maps*, *CT-maps* or *CT-extensions*) of uniformly proper embeddings between Gromov-hyperbolic spaces. We also prove Mitra's criterion for the existence of CT-extensions in terms of Gromov-products, which is critical for our proofs. The main technical result of the chapter is Theorem 8.21 which relates quasigeodesics ϕ in X to quasigeodesics in Y via a certain *cut-and-replace* procedure, replacing *detour subpaths* in ϕ by geodesics in vertex-spaces of Y: If ϕ connects two points in Y but ϕ is not entirely contained in Y, then maximal subpaths of ϕ which are not in Y are *detour subpaths* ζ connecting points in the same vertex-space X_t (which is also a vertex-space of Y). The replacement procedure then amounts to replacing each ζ with a geodesic $\hat{\zeta}$ in X_t connecting the endpoints of ζ. Theorem 8.21 asserts that the resulting modification of ϕ is a uniform quasigeodesic in Y. Along the way, we relate nearest-point projections to flow-spaces $Fl_K(X_u)$ taken in X and in Y. We give a necessary and sufficient condition for ideal boundary points to have equal images under the Cannon–Thurston maps $\partial_{Y,X}$ (Theorem 8.52). In particular, such points have to belong to the Gromov boundary of the same vertex-space of $\mathfrak{Y}$. In the following four sections of the chapter we discuss *Cannon–Thurston laminations* for the inclusion maps $X_v \to X$ in more detail and relate these to *ending laminations*. In the last section of the chapter, we prove some group-theoretic applications of our results. In particular, we construct examples of non-Anosov undistorted surface subgroups of $PSL(2,\mathbb{C}) \times PSL(2,\mathbb{C})$ consisting entirely of semisimple elements.

In Chapter 9 we consider trees of relatively hyperbolic spaces and, generalizing the results of Chapter 8, prove the existence of Cannon–Thurston maps in this context and establish some properties of the associated Cannon–Thurston laminations.

There are many constants and functions used in the book. As a general rule, we label these using as the subscript the number of the theorem (or lemma, etc.) where these quantities are introduced.

Acknowledgments

During the work on this book the first author was partly supported by the NSF grant DMS-16-04241 as well as a Simons Fellowship, grant number 391602. The second author was partially supported by the DST-INSPIRE fellowship of the Government of India (DST/INSPIRE/04/2014/002236). Also the second author is extremely indebted to Mahan Mj who generously shared many ideas related to this book during the author's PhD under Mj. We are grateful to the referees of this book for numerous comments and suggestions.

CHAPTER 1

Preliminaries on metric geometry

1.1. Graphs and trees

Although we always work with unoriented metric graphs like Cayley graphs, we will also need oriented graphs, to describe graphs of groups. Definition 1.1 is taken from [**Ser03**].

DEFINITION 1.1. An *oriented graph* Γ is a pair of sets (V, E) together with two maps

$$E \to V \times V, \quad e \mapsto (o(e), t(e))$$

and

$$E \to E, \quad e \mapsto \bar{e}$$

such that $o(\bar{e}) = t(e), t(\bar{e}) = o(e)$ and $\bar{\bar{e}} = e$ for all $e \in E$.

We write $V(\Gamma)$ for V and $E(\Gamma)$ for E. We refer to $V(\Gamma)$ as the *set of vertices* of Γ and $E(\Gamma)$ as the *set of edges* of Γ. We will almost always conflate a graph Γ with its underlying space, i.e., its geometric realization as a 1-dimensional CW complex.

For an edge e of a graph we refer to $o(e)$ as the *origin* and $t(e)$ as the *terminus* of e; the edge $\bar{e}$ is the same edge e with opposite orientation. When $o(e) = v, t(e) = w$, we will use the notation $e = [v, w]$. While for general graphs this notation is ambiguous, for graphs which are trees (and this is the case we are mostly interested in), vertices v, w uniquely determine the oriented edge e.

We shall denote by $|e|$ the edge e of Γ without any orientation, regarded as a subset of the underlying space of Γ. For each edge e in a graph we define $\dot{e}$ to be $|e|$ with the endpoints removed.

Given a subset $W \subset V$, we define the *full subgraph* of Γ spanned by W as the maximal subgraph in Γ with the vertex-set W. The *valence* or *degree* of a vertex $v \in V$ is the cardinality of the set $o^{-1}(v) \subset E$ (equivalently, $t^{-1}(v) \subset E$).

If $\Lambda \subset \Gamma$ is a subgraph, then a vertex $v \in V(\Lambda)$ is a *boundary vertex* of Λ, if there is an edge $e = [v, w] \in V(\Gamma)$ such that $w \notin V(\Lambda)$. The edge e is then called a *boundary edge* of Λ in Γ. We will use this notion only when Λ is a subtree of a tree Γ.

A *graph-morphism*, or a *morphism of graphs* $\phi : \Gamma \to \Gamma'$ is a pair of maps $\phi_V : V(\Gamma) \to V(\Gamma'), \phi_E : E(\Gamma) \to E(\Gamma'), v \mapsto v', e \mapsto e'$ such that the following diagrams commute for all oriented edges $e = [v, w]$:

$$\begin{array}{ccccc} v & \longleftarrow & e & \longrightarrow & w \\ \downarrow{\scriptstyle \phi_V} & & \downarrow{\scriptstyle \phi_E} & & \downarrow{\scriptstyle \phi_V} \\ v' & \longleftarrow & e' & \longrightarrow & w' \end{array},$$

where the horizontal arrows are the origin/tail maps.

A *tree* is a simply connected graph.

We will use both the notation uv and $[\![u, v]\!]$ for the (geodesic) segment, or *an interval*, in Γ whose endpoints are the vertices u, v. (Since we will be mostly working with graphs which are trees, this notation is unambiguous.) Given a segment $[\![u, v]\!]$ in a tree, we define $]\!]u, v[\![$ as the maximal subtree of $[\![u, v]\!]$ containing all the vertices of $[\![u, v]\!]$ except for u and v. Similarly, we define subsegments $]\!]u, v]\!]$ and $[\![u, v[\![$.

CONVENTION 1.2. We will regard intervals $[\![u, v]\!]$ in simplicial trees as ordered sets with u the smallest element and v the largest. Accordingly, we will talk about supremums and infimums of subsets of $[\![u, v]\!]$ and $\sup(\emptyset) = u, \inf(\emptyset) = v$.

A *metric graph* is a connected graph Γ, every edge e of which is assigned a positive real number $\ell(e)$ (its *length*). The vertex-set of Γ then has a natural pseudometric d_ℓ, where the distance between vertices is defined to be the infimum of total lengths of edge-paths connecting these vertices. The metric d_ℓ extends to a pseudometric on the underlying space of Γ. Note, however, that the distance between the vertices of an edge e of Γ can be smaller than $\ell(e)$ even if the vertices are distinct. If $\ell : E(\Gamma) \to \mathbb{R}_+$ takes only finitely many values, then (Γ, d_ℓ) is a complete geodesic metric space. If the function ℓ is bounded away from 0, then d_ℓ is a metric, but, in general, metric graphs need not be complete nor geodesic and the pseudometric need not be a metric.

EXAMPLE 1.3. Let Γ be a graph with two vertices v, w and edges $\{e_i\}_{i\in\mathbb{N}}$ all of which connect v to w.

(1) Take the function $\ell(e_i) = \frac{1}{i}$. Then $d_\ell(v, w) = 0$. Hence, d_ℓ is not a metric in this example.
(2) Take the function $\ell(e_i) = 1 + \frac{1}{i}$. Then d_ℓ is a metric but Γ contains no geodesics between v and w.

EXAMPLE 1.4. Consider the graph Γ, which is the complete graph on the set $\{v_1, v_2, v_3\}$. Let $\ell([v_1, v_2]) = \ell([v_2, v_3]) = 1, \ell([v_3, v_1]) = 3$. Then $d_\ell(v_1, v_3)$ is 2 rather than $3 = \ell([v_3, v_1])$.

Most of the time, unless stated otherwise, we will metrize connected graphs Γ by declaring that every edge has unit length: The distance between vertices equals the minimal number of edges in an edge-path connecting these vertices. We refer to the resulting metric on Γ as the *graph-metric*.

1.2. Coarse geometric concepts

1.2.1. Metric notions. For a subset A of a topological space X, $cl(A)$ will denote the closure of A in X. By a *path* in a topological space X we will always mean a continuous map $I \to X$, where I is an interval in $\mathbb{R}$. Given a path $c : [a, b] \to X$, we denote by $\overleftarrow{c}$ the reverse path

$$c(t) = c(a + b - t).$$

A path c in a metric space (X, d) is *geodesic* if it is an isometric embedding $I \to (X, d)$. We will frequently conflate paths and their images. Since we are primarily interested in geodesic and quasigeodesic paths, this conflation is mostly harmless. Accordingly, if x, y are points in a path c in X, then $c(x, y)$ will denote the subpath of c between x and y. This notation is, of course, slightly ambiguous since c need not be injective, and a better notation would have been $c|_{[s,t]}$ where $c(s) = x, c(y) = t$. However, in practice, it will be always clear what the subpath $c(x, y)$ is. We will use the notation $c_1 \star c_2$ to denote the concatenation of two paths.

The *length* of a path $c : I \to (X, d)$, where $I = [a, b]$ is a finite closed interval in $\mathbb{R}$, is

$$\text{length}(c) = \sup \sum_{i=1}^{n} d(c(t_i), c(t_{i+1})),$$

where the supremum is taken over all subdivisions of interval I:

$$t_1 = a \leq t_2 \leq \cdots \leq t_n \leq t_{n+1} = b.$$

A metric space (X, d) is called *rectifiably connected* if every two points in it are connected by a path of finite length. A metric space (X, d) is called a *path-metric space* and the metric d a *path-metric* if for all points $x, y \in X$,

$$d(x, y) = \inf_{c} \text{length}(c),$$

where the infimum is taken over all paths in X connecting x and y. Examples of path-metric spaces are given by metric graphs (assuming, of course, that d_ℓ is a metric and not merely a pseudometric).

A metric space (X, d) is called *geodesic* if any two points $x, y \in X$ are connected by a geodesic path. We will use the notation xy, or $[xy]_X$, to denote a geodesic segment joining x to y in X. If X is a simplicial tree and u, v are vertices, then we will also use the notation $[\![u, v]\!]$ for this geodesic segment.

For $x, y, z \in X$ we shall denote by Δxyz a geodesic triangle with vertices x, y, z which is the union of three geodesic segments $xy \cup yz \cup zx$. Similarly, a *geodesic quadrilateral* in X with vertices x, y, z, w, denoted $\square xyzw$, is the union of four geodesics

$$xy \cup yz \cup zw \cup wx.$$

For any rectifiably connected subset Y in a metric space (X, d) we shall denote by $d_Y(\cdot, \cdot)$ the path-metric on Y induced from X: The distance between two points in Y is the infimum of lengths of paths in Y between these points, where the length of a path is computed using the restriction of the metric d.

For $R \geq 0$ and a subset $A \subset X$,

$$N_R^X(A) = N_R(A) := \{x \in X : d(x, a) \leq R \text{ for some } a \in A\}$$

will denote the (closed) R-neighborhood of A in X. A subset $A \subset X$ is said to be an *R-net* in X if

$$N_R(A) = X.$$

For subsets Y, Z in a metric space X, $\text{Hd}(Y, Z) \in [0, \infty]$ denotes the *Hausdorff distance* between Y and Z:

$$\text{Hd}(Y, Z) = \inf\{R : Y \subset N_R(Z), Z \subset N_R(Y)\}.$$

We will use the notation

$$d(Y, Z) = \inf\{d(y, z) : y \in Y, z \in Z\}$$

for the *minimal distance* between Y and Z. (Note that, unlike the Hausdorff distance, the minimal distance, in general, fails to satisfy the triangle inequality.) We will sometimes add the subscript X in this notation to emphasize that the distances and neighborhoods are taken in X.

For two maps $f, g : X \to Y$ between metric spaces, we define the distance between f, g as

$$d(f, g) = \sup\{d(f(x), g(x)) : x \in X\}.$$

1.3. Group actions

Throughout the book, we will be only considering *left group actions* of groups on sets. The notation for such an action is $G \times X \to X$ or $G \curvearrowright X$. For instance, if X is a group and G is a subgroup of X then the action of G on X via left-multiplication

$$L_g(x) = gx$$

is a left action of G on X.

Given a G-action on a set X and a point $x \in X$, one defines the *orbit map* for the action to be the map $o_x : G \to X$, by $o_x(g) = gx$. We will be primarily interested in isometric actions of discrete groups on metric spaces. Such an action is said to be *metrically*[1] *proper* if for each bounded subset $B \subset X$ the subset

$$\{g \in G : gB \cap B \neq \emptyset\}$$

is finite. In other words, pre-images of bounded subsets under orbit maps are finite. An isometric action is said to be *cobounded* if there exists a bounded subset $B \subset X$ such that $\bigcup_{g \in G} gB = X$. An isometric action is said to be *geometric* if it is both proper and cobounded.

REMARK 1.5. Geometric actions frequently refer to proper. cocompact, isometric actions in the literature. Clearly if the underlying space is proper then these two notions coincide. See [**Kap23**] for discussion on related issues.

Suppose that we are given an isometric action $G \curvearrowright X$ and a subset $Y \subset X$. The *stabilizer* of Y in G, denoted G_Y, is the subgroup of G consisting of elements preserving Y setwise. By the G-orbit of Y we will mean the collection of translates $\{gY : g \in G\}$ of Y under the G-action and we will denote it by GY.

DEFINITION 1.6. One says that the *G-orbit GY* of Y is *locally finite* if for each $x \in X$ and $r \in \mathbb{R}_+$, there exists a finite subset $\{g_1, \ldots, g_n\} \subset G$ such that

$$gY \cap B(x, r) \neq \emptyset \Rightarrow g \in g_i G_Y$$

for some $i = 1, \ldots, n$.

In order to see that this condition is natural, observe that for $h \in G_Y$,

$$gY \cap B(x, r) \neq \emptyset \iff ghY \cap B(x, r) \neq \emptyset.$$

LEMMA 1.7. *Suppose that X is a finitely generated group equipped with the word-metric and $Y < X$ is a subgroup. Then for each $G < X$, we have $G_Y = G \cap Y$ and the G-orbit GY is locally finite.*

PROOF. The first statement is obvious. For the second, without loss of generality, we may assume that $x = 1$. Since the ball $B(1, r)$ is finite, there exist a finite set of pairs (g_i, y_i), $i = 1, \ldots, n$, $g_i \in G, y_i \in Y$, such that whenever $y \in Y$, $g \in G$ satisfy $d_X(gy, 1) \leq r$, we have $gy = g_i y_i$ for some i. Then

$$h = g_i^{-1} g = y_i y^{-1} \in G \cap Y = G_Y$$

and, hence, $gY = g_i hY = g_i Y$ as required. □

COROLLARY 1.8. *Suppose that X is a geodesic metric space, $G' \curvearrowright X$ is a geometric action, $Y \subset X$ is a nonempty subspace whose G'-stabilizer G'_Y also acts geometrically on Y. Then for each subgroup $G < G'$, the G-orbit GY is locally finite.*

[1] We will omit the adjective "metrically" in what follows.

1.4. Length structures and spaces

Let X be a topological space. A *length structure* (see [**BBI01**, Chapter 2]) on X is a collection $\mathcal{P}$ of *admissible paths* (defined on closed intervals in $\mathbb{R}$) in X, together with a map

$$\ell : \mathcal{P} \to \mathbb{R}_+$$

(called a *length function*) satisfying the following axioms:

(1) $\mathcal{P}$ is closed under restrictions: The restriction of a path $c \in \mathcal{P}$ to a subinterval again belongs to $\mathcal{P}$.
(2) $\mathcal{P}$ is closed under concatenations.
(3) $\mathcal{P}$ is closed under linear reparameterizations.
(4) $\ell(c_1 \star c_2) = \ell(c_1) + \ell(c_2)$.
(5) ℓ is invariant under linear reparameterizations.
(6) For each $c : [a, b] \to X$, $c \in \mathcal{P}$, the length $\ell(c|_{[a,b]})$ depends continuously on a.
(7) The length function ℓ is consistent with the topology of X in the sense that for each $x \in X$ and each neighborhood U of x

$$\inf_{c \in \mathcal{P}(x, X-U)} \ell(c) > 0$$

where $\mathcal{P}(x, X - U)$ consists of all paths $c \in \mathcal{P}, c : [a, b] \to X$, $c(a) = x, c(b) \in X - U$.
(8) For each pair of points $x, y \in X$ the subset $\mathcal{P}_{x,y}$ consisting of paths $c \in \mathcal{P}$ connecting x to y is nonempty.

A *length space* is a topological space equipped with a length structure. Each length space $(X, \mathcal{P}, \ell)$ has a canonical metric $d = d_\ell$ defined by

$$d(x, y) = \inf_{c \in \mathcal{P}_{x,y}} \ell(c).$$

The topology defined by this metric is finer than the one of X; the metric d_ℓ is a path-metric (see [**BBI01**, Proposition 2.4.1]).

1.5. Coarse Lipschitz maps and quasi-isometries

Below, we let X, Y, Z denote metric spaces and let $L \geq 1$, $\epsilon \geq 0$.

(1) Suppose Z is a set. A map $f : Z \to Y$ is said to be *D-coarsely surjective* if $Y = N_D(f(Z))$, i.e., $f(Z)$ is a D-net in X.
(2) Suppose $\{Z_\alpha\}$ and $\{Y_\alpha\}$ are, respectively, a family of sets and a family of metric spaces. A family of maps $f_\alpha : Z_\alpha \to Y_\alpha$ is said to be *uniformly coarsely surjective* if there is a constant $D \geq 0$, such that for all α, $Y_\alpha = N_D(f_\alpha(Z_\alpha))$.
(3) A map $f : X \to Y$ is said to be *(L, ϵ)-coarsely Lipschitz* (or *coarse Lipschitz*) if $\forall x_1, x_2 \in X$ we have

$$d_Y(f(x_1), f(x_2)) \leq L d_X(x_1, x_2) + \epsilon.$$

A map f is coarsely Lipschitz if it is (L, ϵ)-coarsely Lipschitz for some $L \geq 1, \epsilon \geq 0$. When $\epsilon = L$, we say that f is *L-coarsely Lipschitz*.
(4) Let $\eta : \mathbb{R}_+ \to \mathbb{R}_+$ be an increasing function such that $\eta(1) \geq 1$. A map of metric spaces $f : X \to Y$ is called *(η, L)-proper* if f is L-coarsely Lipschitz and $d(f(x_1), f(x_2)) \leq R$ implies that $d(x_1, x_2) \leq \eta(R)$. The function η is a *distortion function* of f. We will frequently suppress the coarse Lipschitz constant L (it will be often equal to 1) and simply say that *f is η-proper*. For instance, if $Y \subset X$

is a rectifiably connected subset of a path-metric space (X, d_X), we say that Y is η-properly embedded in X if the inclusion map $(Y, d_Y) \to (X, d_X)$ is η-proper.

(5) Similarly, suppose that $f_\alpha : (X_\alpha, d_{X_\alpha}) \to (Y_\alpha, d_{Y_\alpha})$, is a family of maps between metric spaces. If these maps are (η, L)-proper for some η and L, then we will say that this family of maps is *uniformly proper*.

(6) A map $f : X \to Y$ is said to be an (L, ϵ)-*quasi-isometric embedding* if $\forall x_1, x_2 \in X$ one has

$$\frac{1}{L} d_X(x_1, x_2) - \epsilon \leq d_Y(f(x_1), f(x_2)) \leq L d_X(x_1, x_2) + \epsilon.$$

(7) A map $f : X \to Y$ will be simply referred to as a *quasi-isometric embedding* if it is an (L, ϵ)-quasiisometric embedding for some $L \geq 1$ and $\epsilon \geq 0$.

(8) An (L, L)-quasi-isometric embedding will be referred to as an *L-quasi-isometric embedding*.

(9) A map $f : X \to Y$ is said to be a (L, ϵ)-*quasi-isometry* if it is an (L, ϵ)-quasi-isometric embedding, which is also ϵ-coarsely surjective. If $\epsilon = L$ then we will refer to such f as an *L-quasi-isometry*.

(10) We will use the abbreviation *qi* for the word *quasi-isometric*.

(11) An (L, ϵ)-*quasigeodesic* (resp. an *L-quasigeodesic*) in a metric space X is a (L, ϵ)-quasi-isometric embedding (resp. a L-quasi-isometric embedding) $\gamma : I \to X$, where $I \subseteq \mathbb{R}$ is an interval.

(12) Given two maps $f : X \to Y$ and $g : Y \to X$, we say that g is an ϵ-*coarse left inverse* of f if $d(f \circ g, \mathrm{id}_Y) \leq \epsilon$. Similarly one defines an ϵ-*coarse right inverse*. If g is both ϵ-coarse left and right inverse then it is called an ϵ-*coarse inverse* of f.

(13) If $A \subset X$ and $i : A \to X$ is the inclusion map, then an (L, ϵ)-*coarse retraction* of X to A is a (L, ϵ)-coarsely Lipschitz map $g : X \to A$ such that $g|_A = \mathrm{id}_A$.

REMARK 1.9. More generally, one can define an (L, ϵ)-*coarse retraction* by requiring that

$$d(\mathrm{id}_A, g \circ i) \leq \epsilon.$$

However, in the book we only use the more restrictive definition.

EXAMPLE 1.10.

(1) Let G, H be finitely generated groups equipped with word metrics and $\phi : G \to H$ is a homomorphism. Then ϕ is a coarse Lipschitz map. If ϕ has finite kernel, then it is also uniformly proper.

(2) Suppose that G is a finitely-generated (discrete) group, $G \curvearrowright X$ is a proper isometric action on a metric space. Then for each $x \in X$, the orbit map $o_x : G \to X$ is uniformly proper, where G is equipped with a word metric.

DEFINITION 1.11. One says that a finitely-generated subgroup G of a finitely generated group H has distortion at most η if the inclusion map $G \to H$ is η-proper (when G, H are equipped with word metrics as above).

Thus, a subgroup G of a group H is at most *linearly distorted* if and only if it is qi embedded in H. In this case, one says that G is *undistorted* in H. One can make the notion of distortion independent of a generating set by working with a suitable equivalence relation on distortion functions. For instance, one can talk about polynomial distortion, exponential distortion, etc. We refer the reader to [**DK18**] for further details.

We next discuss quasi-isometries and qi embeddings of metric spaces. The following lemma is a direct calculation which we omit.

LEMMA 1.12.

(1) *Suppose that* $f_1 : X_1 \to X_2$ *and* $f_2 : X_2 \to X_3$ *are, respectively,* (L_1, ϵ_1) *and* (L_2, ϵ_2)*-coarse Lipschitz. Then their composition is* $(L_1L_2, L_2\epsilon_1 + \epsilon_2)$*-coarse Lipschitz.*
(2) *Suppose that* $f_1 : X_1 \to X_2$ *and* $f_2 : X_2 \to X_3$ *are, respectively,* (L_1, ϵ_1) *and* (L_2, ϵ_2)*–qi embeddings. Then their composition is an* $(L_1L_2, L_2\epsilon_1 + \epsilon_2)$*–qi embedding.*

LEMMA 1.13. *Let* $f : X \to Y$ *be an* L*-quasi-isometry. Then* f *admits a coarse* $3L^2$*-inverse which is a* $3L^2$*-quasi-isometry* $Y \to X$.

PROOF. For $y \in Y$ define $g(y) = x$ such that $d(y, f(x)) \le L$. Then

$$L^{-1}d(y, y') - 3 \le d(g(y), g(y')) \le Ld(y, y') + 3L^2$$

and

$$d(f \circ g(y), y) \le L, d(g \circ f(x), x) \le 2L^2. \qquad \square$$

LEMMA 1.14. *Let* $f_i : X_i \to Y$ *be* k*-qi embeddings such that*

$$\mathrm{Hd}(\mathrm{Im}(f_1), \mathrm{Im}(f_2)) \le r.$$

Define a map

$$g : X_1 \to X_2$$

sending $x_1 \in X_1$ *to a point* $x_2 \in X_2$ *such that* $d(f_1(x_1), f_2(x_2)) \le r$*. Then* g *is a* $K = K_{1.14}(r, k)$*-quasi-isometry.*

PROOF. Let $x_1, y_1 \in X_1$. Then

$$-k + \frac{1}{k}d_{X_1}(x_1, y_1) \le d_Y(f_1(x_1), f_1(y_1)) \le k + kd_{X_1}(x_1, y_1).$$

Setting $x_2 := g(x_1)$ and $y_2 := g(y_1) \in X_2$, we obtain

$$d_Y(f_1(x_1), f_2(x_2)) \le r \quad \text{and} \quad d_Y(f_1(y_1), f_2(y_2)) \le r.$$

It follows that

$$|d_Y(f_1(x_1), f_1(y_1)) - d_Y(f_2(x_2), f_2(y_2))| \le d_Y(f_1(x_1), f_2(x_2)) + d_Y(f_1(y_1), f_2(y_2)) \le 2r.$$

Hence, we get

$$-k - 2r + \frac{1}{k}d_{X_1}(x_1, y_1) \le d_Y(f_2(x_2), f_2(y_2)) \le 2r + k + kd_{X_1}(x_1, y_1).$$

Since f_2 is a k-qi embedding, we have

$$-k + \frac{1}{k}d_{X_2}(x_2, y_2) \le d_Y(f_2(x_2), f_2(y_2)) \le k + kd_{X_2}(x_2, y_2).$$

Using these two sets of inequalities we obtain

$$-\frac{2r + 2k}{k} + \frac{1}{k^2}d_{X_1}(x_1, y_1) \le d_{X_2}(x_2, y_2) \le 2k^2 + 2rk + k^2d_{X_1}(x_1, y_1).$$

Since $g(x_1) = x_2, g(y_1) = y_2$, it follows that g is a $(2rk + 2k^2)$-qi embedding.

Also, given any $x_2' \in X_2$, there is an $x_1' \in X_1$ such that $d_Y(f_1(x_1'), f_2(x_2')) \le r$. If $x_2'' = g(x_1')$. Then $d_Y(f_1(x_1'), f_2(x_2'')) \le r$. Hence,

$$d_Y(f_2(x_2''), f_2(x_2')) \le d_Y(f_1(x_1'), f_2(x_2')) + d_Y(f_1(x_1'), f_2(x_2'')) \le 2r.$$

Since f_2 is a k-qi embedding it follows that $d_{X_2}(x_2', x_2'') \le 2rk + k^2 < 2rk + 2k^2$. Hence g is a $K = (2rk + 2k^2)$-quasi-isometry. □

LEMMA 1.15. *Suppose $f : X \to Y$ and $g : Y \to X$ are (L, ϵ)-coarsely Lipschitz maps between metric spaces such that $d(\mathrm{id}_X, g \circ f) \le R$ and $d(\mathrm{id}_Y, f \circ g) \le R$. Then f as well as g is an $(L, \epsilon + 2R)$-quasi-isometry.*

PROOF. The proof of this lemma follows easily from definitions. We refer to [**MS09**, Lemma 1.1] for details. □

Lemma 1.16 follows immediately from definitions.

LEMMA 1.16. *Suppose that (Y, d_Y) is a metric space, $X \subset Y$ is a subset equipped with a metric d_X such that the inclusion map $(X, d_X) \to (Y, d_Y)$ is L-coarse Lipschitz and admits an L-coarse Lipschitz retraction $(Y, d_Y) \to (X, d_X)$. Then the inclusion map $X \to Y$ is an L-qi embedding.*

REMARK 1.17.

(1) A few categorical remarks are in order at this point. It is natural to consider the *coarse category* C, where objects are metric spaces and morphisms are equivalence classes of coarse Lipschitz maps (more generally, correspondences). Here two maps are declared to be equivalent if they are bounded distance apart. Isomorphisms in this setting are precisely quasi-isometries of metric spaces. *Monomorphisms* or *monic morphisms* in this category are *uniformly left-cancellative* morphisms, meaning that $f : X \to Y$ is monic if for each pair of (L, ϵ)-coarse Lipschitz maps $g_i : Z \to X, i = 1, 2$, if $d(f \circ g_1, f \circ g_2) \le D$ then $d(g_1, g_2) \le C(L, \epsilon, D)$. (Note the need for the uniform control on distances!) It is easy to verify that monic morphisms are precisely (equivalences classes of) uniformly proper maps; hence, monomorphisms are more general than qi embeddings. Epimorphisms are precisely the coarsely surjective maps. Most important examples of these, besides quasi-isometries, are given by coarse retractions to subsets of metric spaces. The coarse retractions frequently used in the book are *Mitra's projections* (in the setting of subtrees of hyperbolic spaces) and nearest-point projections to quasiconvex subsets of hyperbolic spaces.

(2) An even more general formalism of coarse structures is developed by John Roe in [**Roe03**].

(3) The above categorical notions, unfortunately, are not quite satisfactory for our purpose, since most of the time we have to keep track of various quantities such as distances between equivalent maps, coarse Lipschitz constants and distortion functions. For instance, when defining a graph (even a tree) of metric spaces, it is not quite enough to say that this is a functor from a graph (regarded as a category) to the category C, sending origin/terminus maps to monic morphisms of metric spaces: We will need uniform control of coarse Lipschitz constants and distortion functions. For a tree of hyperbolic spaces, we will need even more control, bounding hyperbolicity constants. For this reason, we will adopt a more pedestrian (and traditional) approach, and mostly refrain from using the categorical language.

LEMMA 1.18. *Suppose Y is a path-metric space, and let $X \subset Y$ be an η-properly embedded subset equipped with the induced path-metric from Y such that X is also an R-net in Y. Then the inclusion map $\iota : X \to Y$ is an L-qi embedding with $L = \eta(2R + 1)$. Consequently ι is a quasi-isometry.*

PROOF. Take $x, x' \in X$, let $c = c_\epsilon : I = [0, T] \to Y$ be an arc-length parameterized path in Y connecting x to x' whose length T is $\leq d(x, x') + \epsilon$. Subdivide the interval I into $n + 1$ subintervals $[t_i, t_{i+1}]$, $t_0 = 0$, such that $t_{i+1} - t_i = 1$ except for $i = n$, $0 \leq r = t_{n+1} - t_n < 1$. Let $P : Y \to X$ be a nearest-point projection map. We apply P to the sequence of points $y_i = c(t_i)$ and get a sequence $x_0 = x, x_1 = P(y_1), \ldots, x_n = p(y_n), x_{n+1} = x'$ such that

$$d_Y(x_i, x_{i+1}) \leq 2R + 1, i = 0, \ldots, n.$$

Hence, $d_X(x, x') \leq L(n + 1)$, $L = \eta(2R + 1)$. Therefore,

$$d_X(x, x') \leq L(d_Y(x, x') + 1),$$

i.e., ι is an L-qi embedding. □

Very similar arguments prove Lemma 1.19.

LEMMA 1.19. *Suppose Y is a path-metric space and X is any metric space and $f : X \to Y$ is an L-coarsely Lipschitz, η-proper embedding which is also R-coarsely surjective. Then there is a constant $K = K_{1.19}(L, R, \eta)$ such that f is a K-quasi-isometry.*

LEMMA 1.20. *Given a function $\eta : \mathbb{R}_+ \to \mathbb{R}_+$ and $R \geq 0$, there is a constant $K = K_{1.20}(\eta, R)$ such that the following holds.*

Suppose Z is a metric space and Z_1, Z_2 are two rectifiably connected subsets in Z such that both Z_1 and Z_2 are η-properly embedded in Z. Assume $\mathrm{Hd}(Z_1, Z_2) \leq R$ *and suppose $f : Z_1 \to Z_2$ is any map that such that $d_Z(z, f(z)) \leq R$ for all $z \in Z_1$. Then f is a K-quasi-isometry.*

PROOF. Since $\mathrm{Hd}(Z_1, Z_2) \leq R$ clearly there is a similar map $g : Z_2 \to Z_1$. We note that $d_Z(z, g \circ f(z)) \leq 2R$ for all $z \in Z_1$. Hence, $d_{Z_1}(z, g \circ f(z)) \leq \eta(2R)$ for all $z \in Z_1$, since Z_1 is η-properly embedded in Z, i.e., $d(\mathrm{id}_{Z_1}, g \circ f) \leq \eta(2R)$. Similarly, $d(\mathrm{id}_{Z_2}, f \circ g) \leq \eta(2R)$.

We claim that f, g are coarsely Lipschitz. Since the proofs are similar we shall show this only for f. Since (Z_1, d_{Z_1}) is a path-metric space it is enough to show that if $z, w \in Z_1$ and $d_{Z_1}(z, w) \leq 1$, then $d_{Z_2}(f(z), f(w))$ is bounded by a constant independent of z, w. However,

$$d_Z(f(z), f(w)) \leq d_Z(z, f(z)) + d_Z(w, f(w)) + d_Z(z, w) \leq 1 + 2R.$$

Hence $d_{Z_2}(f(z), f(w)) \leq \eta(2R + 1)$, since Z_2 is η-properly embedded in Z. Now the claim follows from Lemma 1.15. □

LEMMA 1.21. *Given $D \geq 0$, $k \geq 1$ and $\eta : \mathbb{R}_+ \to \mathbb{R}_+$, there is $K = K_{1.21}(D, k, \eta)$ with the following property:*

Suppose X is a metric space and $x, y \in X$ are arbitrary points. Suppose c is a 1-Lipschitz, η-properly embedded path in X and there is a k-quasigeodesic c_1 joining x, y in X such that $\mathrm{Hd}(c, c_1) \leq D$. *Then c is a K-quasigeodesic.*

PROOF. Let $c : [0, l] \to X$, $c_1 : [0, l_1] \to X$ be the given paths. Since c is 1-Lipschitz, for all $s, s' \in [0, l]$ we have $d_X(c(s), c(s')) \leq |s - s'|$. On the other hand, there are points $t, t' \in [0, l_1]$ such that $d_X(c_1(t), c(s)) \leq D$, $d_X(c_1(t'), c(s')) \leq D$. Without loss of generality we may assume that $t \leq t'$. Let

$$t_1 = t \leq t_2 \leq \cdots \leq t_n \leq t_{n+1} = t'$$

be points of $[0, l_1]$ such that $t_{i+1} - t_i = 1$, $1 \leq i \leq n - 1$, and $t_{n+1} - t_n \leq 1$. Then there are points

$$s_1 = s, s_2, \cdots, s_n, s_{n+1} = s'$$

in $[0, l]$ such that $d_X(c_1(t_i), c(s_i)) \le D$. It follows that

$$d_X(c(s_i), c(s_{i+1})) \le 2D + d_X(c_1(t_i), c_1(t_{i+1})) \le 2D + 2k, \quad 1 \le i \le n.$$

Hence, $|s_i - s_{i+1}| \le \eta(2D + 2k)$ and, therefore,

$$|s - s'| \le n\eta(2D + 2k) \le \eta(2D + 2k) + \eta(2D + 2k)|t - t'|.$$

However,

$$|t - t'| \le k^2 + kd_X(c_1(t), c_1(t')) \le k^2 + 2Dk + kd_X(c(s), c(s')).$$

It follows that

$$-\frac{1 + k^2 + 2Dk}{k} + \frac{1}{\max\{1, k\eta(2D + 2k)\}}|s - t| \le d_X(c(s), c(t)) \le |s - t|.$$

Thus, we may take $K = \max\{1, \eta(2D + 1), \frac{1+k^2+2Dk}{k}\}$. □

Lemma 1.22 follows immediately from the definition of a uniformly proper embedding.

LEMMA 1.22. *Suppose that (X, d_X) is a path-metric space, $Y \subset X$ is rectifiably connected and η-properly embedded in X, i.e., the inclusion map $(Y, d_Y) \to (X, d_X)$ is η-proper. Then for every subset $Z \subset Y$ and $R \ge 0$ we have*

$$N_R^Y(Z) \subset N_R^X(Z) \text{ and } N_R^X(Z) \cap Y \subset N_{\eta(R)}^Y(Z).$$

Here the R-neighborhood N_R^Y in Y is taken with respect to the induced path-metric d_Y.

Lemmas 1.23, 1.24, and 1.25 $I = [a, b]$, $I' = [a', b']$ denote nondegenerate intervals in $\mathbb{R}$ (equipped with the standard metric).

LEMMA 1.23 (Lipschitz approximation). *Let $f : I \to I'$ be a coarse (L, ϵ)-Lipschitz map. Then f is within distance $2(L + \epsilon)$ from a piecewise-linear $2(L + \epsilon)$-Lipschitz map $g : I \to I'$.*

PROOF.

(1) First, assume that $b - a \ge 1$. We then subdivide the interval I into subintervals

$$[a_0, a_1] = [a, a_1], [a_1, a_2], \dots, [a_n, b] = [a_n, a_{n+1}]$$

of length at least $1/2$ and at most 1. We replace the restriction of f to each subinterval $I_i = [a_i, a_{i+1}]$ with a linear function g_i such that $g_i(a_i) = f(a_i)$, $g_i(a_{i+1}) = f(a_{i+1})$. Since

$$|f(a_i) - f(a_{i+1})| = |g_i(a_i) - g_i(a_{i+1})| \le L|a_i - a_{i+1}| + \epsilon \le L + \epsilon, \tag{1.1}$$

it is easy to see that

$$d(f|_{I_i}, g_i) \le 2(L + \epsilon).$$

Combining the linear functions g_i we obtain a piecewise-linear function $g : I \to I'$ such that $d(f, g) \le 2(L + \epsilon)$. Since $a_{i+1} - a_i \ge 1/2$, the inequality (1.1) implies that the slope of each g_i is at most $2(L + \epsilon)$. Hence, g is $2(L + \epsilon)$-Lipschitz.

(2) Suppose that $b - a < 1$. Then we let g be the constant function, equal $f(b)$. Since $|f(s) - f(t)| \le L + \epsilon$ for all $s, t \in I$, $d(f, g) \le L + \epsilon$.

□

LEMMA 1.24 (Coarse monotonicity of quasi-isometries). *Set $D := L(5\epsilon + 4L)$. Suppose that $f : I \to I'$ is an (L, ϵ)-qi embedding. Then f is coarsely monotonic in the sense that if $r < s < t$ are in I and $\min(s - r, t - s) > D$ then $f(s)$ is between $f(r)$ and $f(t)$.*

PROOF. Let g be the Lipschitz approximation of f as in Lemma 1.23. Since f is an (L,ϵ)-qi embedding and $d(f,g) \le 2(L+\epsilon)$, we conclude that g satisfies

$$L^{-1}|t-t'| - (5\epsilon + 4L) \le |g(t) - g(t')|$$

for all $t, t' \in I$. Suppose that, say,

$$f(s) > \max(f(r), f(t)).$$

Then

$$\min(f(s) - f(r), f(s) - f(t)) \ge L^{-1}D - \epsilon.$$

Once again since $d(f,g) \le 2(L+\epsilon)$ we see that

$$\min(g(s) - g(r), g(s) - g(t)) \ge D' = L^{-1}D - (5\epsilon + 4L) = 0.$$

For concreteness, assume that $g(r) \le g(t)$. By the intermediate value theorem applied to the function g (restricted to the interval $[r,s]$) there exists $t' \in [r,s]$ such that $g(t') = g(t)$. We have $t - t' \ge t - s > D$. Thus,

$$L^{-1}D - (5\epsilon + 4L) < L^{-1}|t-t'| - (5\epsilon + 4L) \le 0 \Rightarrow D < L(5\epsilon + 4L),$$

which is a contradiction. If $f(s) < \min(f(r), f(t))$ then one can apply the same proof to the function $t \mapsto -f(t)$ to arrive at a contradiction. □

LEMMA 1.25 (Approximating quasi-isometries by homeomorphisms). *Let $I = [a,b]$, $I' = [a', b']$ be nondegenerate intervals in $\mathbb{R}$ (equipped with the standard metric). Suppose that $f : I \to I'$ is a k-quasi-isometry sending a to a' and b to b'. Then there exists a (piecewise-linear) homeomorphism $\tilde{f} : I \to I'$ within distance $D_{1.25} = D_{1.25}(k)$ from f which is also a $k_{1.25}(k)$-quasi-isometry.*

PROOF. The proof is similar to that of Lemma 1.23. Set $L = \epsilon = k$.

(1) Suppose first that $b - a \ge 2D$, where $D = L(5\epsilon + 4L) = 9k^2$ (as in Lemma 1.24). We subdivide the interval $I = [a,b]$ into subintervals

$$[a_0, a_1] = [a, a_1], [a_1, a_2], \ldots, [a_n, a_{n+1}] = [a_n, b],$$

each of length greater than D and at most $2D$. According to Lemma 1.24, f restricted to the subset

$$J = \{a_0, a_1, \ldots, a_n, a_{n+1}\}$$

is strictly monotonic. We then let $\tilde{f} = g : I \to I'$ be the piecewise-linear function equal to f on J and linear on the complementary intervals. In particular,

$$f(a) = a' = g(a), \quad f(b) = b' = g(b).$$

In view of monotonicity of $f|_J$, the function g is also (strictly) monotonic, hence, a homeomorphism. For $s, t \in [a_i, a_{i+1}]$ we have

$$|f(s) - f(t)| \le 2DL + \epsilon = 18k^3 + k.$$

Therefore, for all $t \in [a_i, a_{i+1}]$ we have

$$\begin{aligned} |f(t) - g(t)| &\le |f(t) - f(a_i)| + |f(a_i) - g(t)| \\ &\le |f(t) - f(a_i)| + |f(a_i) - f(a_{i+1})| \\ &\le 2(18k^3 + k), \end{aligned}$$

since $f(a_i) = g(a_i)$, $f(a_{i+1}) = g(a_{i+1})$ and g is monotonic.

(2) Suppose that $b - a \le 2D$. We then let $g = \tilde{f}$ be the linear function equal to f on $\{a, b\}$; $g(a) = a' < g(b) = b'$. As in Case 1, $d(f, g) \le 36k^3 + 2k$. Hence we can choose $D_{1.25} = 36k^3 + 2k$. Since f is a k-quasi-isometry it follows that $\tilde{f}$ is a $k_{1.25}(k) = (k + D_{1.25})$-quasi-isometry. □

Definition 1.26. Two subsets Y, Z of a metric space X are said to be *C-Lipschitz cobounded* if there exist (L, ϵ)-coarse Lipschitz retractions $X \to Z, X \to Y$ whose restrictions

$$r_{Y,Z} : Y \to Z, \quad r_{Z,Y} : Z \to Y,$$

satisfy:

(1) $r_{Y,Z}(Y)$ and $r_{Z,Y}(Z)$ have diameters D_Y, D_Z.
(2) $\max(L, \epsilon, D_Y, D_Z) \le C$.

In Section 1.21 we will relate this definition to the more standard notion of cobounded subsets in a Gromov-hyperbolic space.

Lemma 1.27. *If Y, Z are C-Lipschitz cobounded, then for every R there exists $D = D_{1.27}(R, C)$ such that if*

$$a_i \in Y, b_i \in Z, i = 1, 2$$

are points satisfying $d(a_i, b_i) \le R$, $i = 1, 2$, then $d(a_1, a_2) \le D, d(b_1, b_2) \le D$.

Proof. One can take $D = 2C(R + 1) + C + 2R$. □

1.6. Coproducts, cones, and cylinders

In this section we discuss several purely topological notions used elsewhere in the book. Let $\{Z_\alpha : \alpha \in A\}$ be an indexed collection of topological spaces. Then the *coproduct* topology on the disjoint union

$$Z := \coprod_{\alpha \in A} Z_\alpha$$

is the finest topology on the disjoint union such that all the natural inclusion maps $Z_\alpha \to Z$ are continuous. In particular, each $Z_\alpha \subset Z$ is a clopen subset homeomorphic to Z_α.

In the following two constructions, the unit intervals $[0, 1]$ are sometimes replaced by the half-intervals $[0, 1/2]$ with $1/2$ playing the role of 1.

Let X, Y be topological spaces, $f : X \to Y$ a continuous map. Then the *mapping cylinder $Cyl(f : X \to Y)$*, also denoted $X \cup_f Y$ and, sometimes, $Y \cup_f X$, is the quotient of $X \times [0, 1] \sqcup Y$ (with the coproduct topology) by the equivalence relation

$$(x, 1) \sim f(x), x \in X.$$

Lastly, the *cone* $C(a, X)$ over a topological space X is the quotient of the product $X \times [0, 1]$ by the subspace $X \times \{1\}$. The point a, the projection of $X \times \{1\}$ in $C(a, X)$, is called the *apex* of the cone. The projections of the intervals $\{x\} \times [0, 1]$ under the quotient map $q : X \times [0, 1] \to C(a, X)$ will be called the *radial line segments* in $C(a, X)$. We will identify X with the image of $X \times \{0\}$ in $C(a, X)$.

In the next section, we will metrize the cones $C(a, X)$.

1.7. Cones over metric spaces

Let (X, d) be a path-metric space. We equip $X \times [0, 1/2]$ with the product metric. The cone over the metric space X, denoted by $C(a, X)$, is defined to be the quotient of $X \times [0, 1/2]$ obtained by collapsing $X \times \{1/2\}$ to a point. We let $q : X \times [0, 1/2] \to C(a, X)$ still denote the quotient map and a to be the apex of $C(a, X)$. We metrize the cone $C(a, X)$ as follows. Mapping rectifiable paths

$$c : I = [0, 1] \to X \times [0, 1/2]$$

into $C(a, X)$ by q we obtain a family of *admissible* paths in $C(a, X)$. For a rectifiable path c as above, we define the length $L(q(c))$ to be equal to the length of c minus the total length of the intersection $c(I) \cap X \times \{1/2\}$. In other words, the length of the path $q \circ c$ is the total length of $c|_J$, where $J = I \setminus (q \circ c)^{-1}(\{a\})$. For instance, the radial line segment connecting any $x \in X \times \{0\}$ to a has length $1/2$. Using this notion of length we path-metrize the cone $C(a, X)$. We leave it to the reader to check that this metric, denoted $\hat{d}$, metrizes the topology of $C(a, X)$ and that for each $x \in X$,

$$\hat{d}(x, a) = 1/2.$$

1.8. Approximation of metric spaces by metric graphs

In this section we discuss a generalization of path-metric spaces, called *quasi-path metric spaces*.

Definition 1.28. A *finite r-path* in a metric space (X, d) is a map

$$c : [m, n] \subset \mathbb{Z} \to X$$

such that $d(c(i), c(i + 1)) \leq r$ for all $m \leq i \leq n - 1$. The *length* of such c is defined as

$$\text{length}(c) = \sum_{i=m}^{n-1} d(c(i), c(i + 1)).$$

The finite path c is said to *connect* the point $x = c(m)$ to the point $y = c(n)$.

A metric space (X, d) is called an *r-quasi-path metric space* for a constant $r > 0$ if for every pair of points $x, y \in X$ there exists a finite r-path c connecting x to y such that $\text{length}(c) \leq d(x, y) + r$.

For instance, every path-metric space is an r-quasi-path metric space for every $r > 0$.

Lemma 1.29. *Any r-quasi-path metric space is* $(1, 3r)$*-quasi-isometric to a path-metric space.*

Proof. Suppose X is an r-quasi-path metric space. We construct a metric graph Z with the vertex set $V(Z) = X$ such that $x, y \in X$ are connected by an edge e iff $x \neq y$ and $d_X(x, y) \leq r$, where the edge is assigned the length $\ell(e) = d_X(x, y)$.

Consider the inclusion map $\iota : X \to Z$. Suppose $x, y \in X$ are arbitrary points. Then there is an r-path $x = x_0, x_1, \cdots, x_n = y$ in X joining x to y in X. By the definition of the graph Z, x_i also form a sequence of vertices connected by edges in Z. Hence,

$$d_Z(x, y) \leq \sum_i d_Z(x_i, x_{i+1}) \leq \sum_i d_X(x_i, x_{i+1}) \leq d_X(x, y) + r.$$

Thus, ι is $(1, r)$-coarse Lipschitz. Let $\rho : Z \to X$ be the following map. The restriction of ρ on $V(Z)$ is simply the identity map and interiors of edges are mapped to one of the vertices. Let $\alpha : I \to Z$ be any piecewise-linear path (see [**BH99**, Chapter I.1, Section 1.9]). Then clearly, length(α) and the length of the r-path $\rho \circ \alpha$ differ by at most $2r$. Hence, ρ is $(1, 2r)$-coarsely Lipschitz. Moreover, it is clear that $d(\mathrm{id}_X, \rho \circ \iota) \le r$ and $d(\mathrm{id}_Z, \iota \circ \rho) \le r$. Hence, by Lemma 1.15, the maps ι, ρ are both $(1, 3r)$-quasi-isometries. □

DEFINITION 1.30 (Rips graph). Let (Y, d) be a metric space. For $R \ge 0$ the *R-Rips graph* of (Y, d) is the graph Z_R with the vertex-set Y and edges $[y_1, y_2]$ for all pairs of distinct points $y_1, y_2 \in Y$ such that $d(y_1, y_2) \le R$. We will equip Z with its graph-metric (each edge has unit length).

Note that for a general metric space Y, the graph Z_R is disconnected and the distance between points in different connected components is infinite. However, if Y is a path-metric space, then each graph Z_R is connected.

DEFINITION 1.31. A metric space (Y, d) is said to be *coarsely connected* if there exists $R < \infty$ such that the corresponding Rips graph Z_R is connected.

The following fundamental result of geometric group theory is usually stated for *proper* metric spaces Y and properly discontinuous actions, but, when the notion of metrically proper actions is used, properness of Y is not needed and the proof is the same as in the proper case, cf. **[DK18]**.

LEMMA 1.32 (Milnor–Schwarz lemma). *Suppose that (Y, d) is a (nonempty) metric space, G is a discrete group and $G \curvearrowright Y$ is a geometric*[2] *action. Then:*

(1) *If Y is coarsely connected, then G is finitely generated.*
(2) *If Y is a quasi-path metric space, then for one (equivalently, every) $y \in Y$ the orbit map $G \to Gy \subset Y$ is a quasi-isometry.*

LEMMA 1.33. *For a path-metric space X, let $Z = Z_1$ be the 1-Rips graph of X. Then the inclusion map $\iota : X \to Z$ is a $(1, 1)$-quasi-isometry with a $(1, 3)$-qi inverse $\rho : Z \to X$.*

PROOF. The proof is very similar to that of Lemma 1.32: We let the map $\rho : Z \to X$ be identity on the $V(Z) = X$, etc. Given $x, y \in X$, we join them in X by an arc-length parametrized path $\gamma : [0, l] \to X$ such that $l \le d_X(x, y) + \epsilon$, where $\epsilon > 0$ is chosen in such a way that $d_X(x, y) < m + 1$ where m is the nonnegative integer determined by $m \le d(x, y) < m+1$. Since length$(\gamma) < m+1$, it follows that $d_Z(x, y) \le m+1 \le d_X(x, y)+1$. Suppose $x, y \in X$ such that $d_Z(x, y) = n$. Let $x = x_0, x_1, \cdots, x_n = y$ the consecutive vertices on a geodesic in Z joining x, y. Then we know that $d_X(x_i, x_{i+1}) \le 1$. Thus,

$$d_X(x, y) \le \sum_{i=1}^{n} d_X(x_{i-1}, x_i) \le n$$

and we get $d_X(x, y) \le d_Z(f(x), f(y)) \le d_X(x, y) + 1$.

Finally, $\iota(X)$ is a 1-net in Z, and, hence, ι is coarsely 1-surjective. This proves the first statement of the lemma. The remaining parts of the proof follow from similar simple calculations and we leave details to the reader. □

COROLLARY 1.34. *Suppose X is an r-quasi-path metric space. Then there is a (connected) graph X' equipped with the graph-metric and a $(1, 3r + 1)$-quasi-isometry $\iota : X \to X'$ with a $(1, 3r + 3)$-qi inverse $\rho : X' \to X$ such that $\rho(X') = X$. In particular, each quasi-path metric space is $(1, \epsilon)$-quasi-isometric to a geodesic metric space.*

PROOF. This is a straightforward consequence of Lemma 1.29 and Lemma 1.33. □

[2] i.e., isometric, metrically proper, and cobounded

LEMMA 1.35 (Local-to-global principle for coarse Lipschitz maps from quasipath metric spaces). *Suppose that (X, d_X) is an r-quasi-path metric space, (Y, d_Y) is any metric space and $f : X \to Y$ is a map such that for all $x_1, x_2 \in X$, $d_X(x_1, x_2) \leq r$ implies that $d_Y(f(x_1), f(x_2)) \leq r'$ for some r' independent of x_1, x_2. Then f is $(\frac{2r'}{r}, 3r')$-coarse Lipschitz.*

PROOF. Take $x, y \in X$. Suppose $n \in \mathbb{N}$ is the smallest integer such that there is a finite sequence $x_0 = x, x_1, \cdots, x_n = y$ in X with $d_X(x_i, x_{i+1}) \leq r$ for all $i = 0, \ldots, n-1$ and $\sum_{i=0}^{n-1} d_X(x_i, x_{i+1}) \leq d(x, y) + r$. Then $d_X(x_i, x_{i+2}) > r$ for $0 \leq i \leq n-2$. It follows that $r(n-1)/2 < d_X(x, y) + r$. Hence, $n < 3 + \frac{2}{r} d_X(x, y)$. On the other hand,

$$d_Y(f(x), f(y)) \leq \sum_{i=0}^{n-1} d_Y(f(x_i), f(x_{i+1})) \leq nr' < 3r' + \frac{2r'}{r} d_X(x, y). \qquad \square$$

Let Y be a D-net in metric space (X, d) with the inclusion map $\rho : Y \to X$. Given $R > 0$ we let $Z = Z_R$ be the full subgraph of the R-Rips graph of (X, d_X) with the vertex set Y; we equip Z with its graph-metric. We extend the map ρ to the rest of Z, (the extension is still denoted by ρ), by taking an arbitrary orientation on Z and sending all points of any open directed edge $[v, w] \setminus \{w\} = [v, w)$ in Z to the point $v \in Y \subset X$. The next lemma (which is a form of the Milnor–Schwarz lemma for metric spaces, cf. [**DK18**, Theorem 8.52]) shows that, under some conditions, the graph Z is connected and the map ρ is a uniform quasi-isometry $Z \to X$. This result generalizes Lemma 1.33.

LEMMA 1.36. *Suppose (X, d_X) is an r-quasi-path metric space and $Y \subset X$ is a D-net in X. If $R \geq r + 2D$, then Z, as defined above, is a connected graph and the map $\rho : (Z, d_Z) \to X$ is a $(K_{1.36}(r, R), \epsilon_{1.36}(r, R))$-quasi-isometry.*

Moreover, there is a $(1 + r)$-coarse inverse $\iota : X \to Z$ to ρ such that $\iota(x) = x$ for all $x \in Y$ which is also a $(K_{1.36}(r, R), \epsilon_{1.36}(r, R))$-quasi-isometry.

PROOF. Consider vertices $y, y' \in Y$. Since (X, d) is a r-quasi-path metric space, there exists an r-path

$$y = x_0, x_1, \ldots, x_n x_{n+1} = y'$$

in X from y to y'. Since $Y \subset X$ is a D-net in X, there exist points $y_1, \ldots, y_n \in Y$ satisfying

$$d_X(x_i, y_i) \leq D, i = 1, \ldots, n.$$

By the triangle inequality, $d_X(y_i, y_{i+1}) \leq r + 2D \leq R$ for $i = 0, \ldots, n$ where $y_{n+1} = y'$, which implies that the vertices $x_0 = y, y_1, \ldots, y_n, y'$ are on an edge path in Z. This proves that Z is connected.

Suppose $z \neq z' \in Z$ are any two points. Let $\rho(z) = y, \rho(z') = y'$ and let $z, y_0, \cdots, y_n, z'$ be a geodesic in Z joining z, z' where $y_i \in Y$. We note that $d_Z(y, y_0) \leq 1, d_Z(y_n, y') \leq 1$ and $|d_Z(z, z') - n| \leq 2$. Now,

$$\begin{aligned} d_X(\rho(z), \rho(z')) &\leq d_X(y, y_0) + d_X(y_n, y') + \sum_{i=0}^{n} d_X(y_i, y_{i+1}) \\ &\leq 2R + nR \\ &\leq 2R + R(d_Z(z, z') + 2) \\ &= 4R + Rd_Z(z, z'). \end{aligned}$$

Hence the map $\rho : Z \to X$ is $(R, 4R)$-coarse Lipschitz.

Next, we define a coarse inverse map ι to ρ by defining it to be the identity map on Y and sending $x \in X \setminus Y$ to a point $\iota(x) = y \in Y$ such that $d_X(x, y) \leq D$. Then $d_X(x, x') \leq r$

implies that $d_X(\iota(x), \iota(x')) \le r + 2D \le R$, i.e., $d_Z(\iota(x), \iota(x')) \le 1$. Since (X, d) is a r-quasi-path metric space, the map ι is $(2/r, 3)$-coarse Lipschitz by Lemma 1.35.

By the construction,

$$d(\iota \circ \rho(t), t) \le 1, \quad d(\rho \circ \iota(x), x) \le r.$$

Hence, by Lemma 1.15 we can take

$$K_{1.36}(r, R) = \max\{R, 2/r\}, \quad \epsilon_{1.36}(r, R) = \max\{4R, 3\} + 2\max\{1, r\}. \qquad \square$$

Since Z constructed in this lemma is a complete geodesic metric space and since we are interested in coarse geometric properties of metric spaces, we can always replace quasi-path metric spaces with appropriate metric graphs. We will be using this in Section 2.2 when constructing total spaces of trees of spaces.

Similar to the approximation of coarse Lipschitz maps of line segments by piecewise-linear maps, one has a uniform approximation of coarse Lipschitz maps of metric (simplicial) graphs by simplicial maps, i.e., maps sending every edge linearly to an edge or a vertex.

LEMMA 1.37. *Fix numbers R, L, ϵ. Let X, Y be connected metric graphs with the same edge-length R. Then, after subdividing edges of X in at most $n(R, L, \epsilon)$ equal length sub-segments, every (L, ϵ)-coarse Lipschitz map $f : X \to Y$ is within distance $D_{1.37}(L, \epsilon, R)$ from a simplicial map $f_* : X \to Y$.*

PROOF. The proof is similar to that of Lemma 1.23 and we omit it. $\square$

1.9. Hyperbolic metric spaces

Hyperbolic metric spaces are coarsifications of the classical hyperbolic n-space $\mathbb{H}^n$ and are characterized by a form of *thin triangle condition*. The most common notions of hyperbolicity for metric spaces are the one due to Rips (for geodesic metric spaces) and one due to Gromov (for general metric spaces). One drawback of Gromov's definition is that his notion of hyperbolicity is not qi invariant, although it is invariant under $(1, \epsilon)$-quasi-isometries. One of the features (or bugs, depending on the perspective) of metric hyperbolicity is that it is stable under changes in metric below certain scale δ and that, accordingly, nothing can be said about general hyperbolic spaces below that scale. This also points to a limitation of Rips's notion of hyperbolicity since it applies only to geodesic metric spaces. This becomes somewhat important in the context of this book since metric spaces that we consider are frequently only *path-metric spaces*. One source where hyperbolicity along the lines of Rips definition is developed for path-metric spaces X is Väisälä's paper [**V05**]: Instead of geodesics he considers *h-short paths*, which are rectifiable paths between points $x, y \in X$ whose length is at most $d(x, y) + h$. A drawback is that one is forced to carry an extra constant. Another possible approach is to extend Rips's definition to the class of *quasi-path metric spaces*. We will give basic definitions in Section 1.9.3 but will not pursue this direction much further beyond proving that such metric spaces are $(1, \epsilon)$-quasi-isometric to geodesic metric spaces and, hence, Gromov's notion of hyperbolicity in the context of quasi-path metric spaces is preserved by general quasi-isometries, see Section 1.9.4. Yet, another possible approach is to work with path-metric spaces but instead of geodesics, work with sequences of paths whose lengths approximate distances between points. All the arguments appearing in the book will go through with constants unchanged comparing to the ones for geodesic metric spaces. A drawback is that this approach lengthens the proofs (which are already long and technical in Chapters 3, 4, 5, and

6). Thus, for most of the book, we work with geodesic metric spaces. In this section we present various notions of hyperbolicity starting with the most familiar ones.

We assume that the reader is familiar with the basic definitions and facts about hyperbolic metric spaces that can be found in [**BH99**], [**CDP90**], [**DK18**], [**Gro87**], [**Gd90**], [**ABC⁺91**], [**V05**]. In this section we collect some of these to fix the notions and for later use.

1.9.1. Hyperbolicity in the sense of Gromov.

Definition 1.38 (Gromov product). Let X be a metric space. Given points $x, y, z \in X$, the *Gromov product* of y, z with respect to x, denoted $(y.z)_x$, is defined as

$$\frac{1}{2}\left(d(x,y) + d(x,z) - d(y,z)\right).$$

Definition 1.39. A metric space X is said to be *δ-hyperbolic in the sense of Gromov* or simply *δ-Gromov-hyperbolic* if all $x, y, z, w \in X$ satisfy the inequality

$$(x.y)_w \geq \min\{(x.z)_w, (y.z)_w\} - \delta.$$

A metric space X is said to be *hyperbolic in the sense of Gromov* if it is *δ-Gromov-hyperbolic* for some $\delta \in [0, \infty)$.

A finitely generated group G is called *hyperbolic* if the metric space (G, d_G) is hyperbolic in the sense of Gromov, where d_G is the word metric on G for some finite generating set.

Example 1.40 ([**DK18**, Example 11.36]). Consider the graph $G \subset \mathbb{R}^2$ of the function $y = |x|$. We equip G with the restriction of the standard Euclidean metric on $\mathbb{R}^2$. For $n \in \mathbb{N}$ consider points

$$o = (0,0), p = (-n, n), q = (n, n), z = (2n, 2n).$$

Then $(p, q)_o = (\sqrt{2} - 1)n$, $(p, z)_o = 2\sqrt{2}n$, $(q, z)_o = \sqrt{2}n$. The difference

$$(p.q)_o - \min\{(p.z)_o, (q.z)_o\} = (\sqrt{2} - 1)n - \sqrt{2}n = -n$$

diverges to $-\infty$ as $n \to \infty$. Thus, G is not hyperbolic in the sense of Gromov. On the other hand, G is qi to the real line via the map $x \mapsto (x, |x|)$ and the real line is 0-hyperbolic.

Thus, Gromov-hyperbolicity is not preserved by quasi-isometries even for *quasigeodesic metric spaces*, i.e., metric spaces where all points are connected by uniform quasigeodesics. On the other hand, the following lemma is straightforward from the definition of Gromov-hyperbolicity:

Lemma 1.41. *Suppose X, Y are metric spaces and $f : X \to Y$ is a $(1, \epsilon)$-quasiisometry for some $\epsilon \geq 0$. Then X is Gromov-hyperbolic iff Y is. More precisely, if X δ-hyperbolic in the sense of Gromov, then Y is $\delta + 3\epsilon$-hyperbolic in Gromov's sense.*

1.9.2. Hyperbolicity in the sense of Rips. Suppose now that X is a geodesic metric space.

Definition 1.42. Consider a geodesic triangle $\Delta = \Delta x_1x_2x_3 \subset X$ with the vertices x_1, x_2, x_3, and let $\delta \geq 0$.

(1) The triangle Δ is said to be *δ-slim* if each side of Δ is contained in the δ-neighborhood of the union of the other two sides.

(2) For all $i \neq j \neq k \neq i$, let $c_k \in x_ix_j$ be such that $d(x_i, c_j) = d(x_i, c_k)$. The points c_i are called the *internal points* of Δ. Note that, for all $i \neq j \neq k \neq i$,

$$d(x_i, c_j) = \frac{1}{2}\Big(d(x_i, x_j) + d(x_i, x_k) - d(x_j, x_k)\Big) = (x_j.x_k)_{x_i}.$$

(3) If X is a tree, then $p = c_1 = c_2 = c_3$ and in this case we shall refer to the point p as the *center* of the Δ.
(4) The diameter of the set $\{c_1, c_2, c_3\}$ will be referred to as the *insize* of the triangle Δ.
(5) The triangle Δ is said to be δ-*thin* if for all $i \neq j \neq k \neq i$ and $p \in x_ic_j \subset x_ix_k$, $q \in x_ic_k \subset x_ix_j$ with $d(p, x_i) = d(q, x_i)$, one has

$$d(p, q) \leq \delta.$$

Lemma 1.43 is clear from the definitions:

LEMMA 1.43. *If Δ is δ-thin, then it is also δ-slim and its insize is $\leq \delta$.*

DEFINITION 1.44 (Rips hyperbolicity). A geodesic metric space X is said to be δ-hyperbolic in the sense of Rips if each geodesic triangle in X is δ-slim. A geodesic metric space is said to be *Rips-hyperbolic* if it is δ-hyperbolic in the sense of Rips for some $\delta < \infty$.

LEMMA 1.45 ([**ABC**+**91**, Proposition 2.1]). *Suppose X is a δ-hyperbolic metric space in the sense of Rips. Then the following hold:*

(1) *All the geodesic triangles in X have insize at most 4δ.*
(2) *All the geodesic triangles in X are 6δ-thin.*

It follows that a geodesic metric space is hyperbolic in the sense of Rips if and only if all geodesic triangles in X are uniformly thin.

The following lemmata are also very standard and follow easily from definitions, see [**DK18**], [**ABC**+**91**], or [**V05**].

LEMMA 1.46. *A geodesic metric space is Rips-hyperbolic if and only it is Gromov-hyperbolic. More precisely:*

(1) *If a metric space X is δ-hyperbolic in the sense of Rips, then X is 3δ-hyperbolic in Gromov's sense.*
(2) *If X is geodesic and δ-hyperbolic in Gromov's sense, then X is 2δ-hyperbolic in the sense of Rips.*

In view of this lemma, when talking about hyperbolicity for geodesic metric spaces, we will always mean hyperbolicity in the sense of Rips.

For geodesic hyperbolic spaces, the Gromov-product $(y.z)_x$ "almost equals" the distance from x to the geodesic yz.

LEMMA 1.47. *Suppose that X is a δ-hyperbolic space in the sense of Rips. Then for each triple $x, y, z \in X$,*

$$(y.z)_x - 2\delta \leq d(x, yz) \leq (y.z)_x.$$

LEMMA 1.48. *Every geodesic quadrilateral $\square = xyzw$ in a δ-hyperbolic metric space X is 2δ-slim, i.e., the side xy of $\square$ is contained in the 2δ-neighborhood of the union of the other three sides of $\square$. Similarly, each geodesic n-gon in X is $(n-2)\delta$-slim.*

1.9.3. Hyperbolicity for path-metric spaces. The form of Rips-hyperbolicity discussed in this section is a mild generalization of Rips-hyperbolicity for geodesic metric spaces. We refer the reader to [**V05**] for a more general discussion.

DEFINITION 1.49. A rectifiable path c connecting x to y in a metric space X is called *ϵ-short* if

$$\text{length}(c) \leq d(x, y) + \epsilon.$$

DEFINITION 1.50 (Triangles formed by paths). Suppose X is any metric space. Given any three points $x, y, z \in X$ and three (continuous or finite) paths $c(x, y), c(x, z), c(y, z)$ joining these points, the *triangle* formed by these paths is the set $\{c(x, y), c(x, z), c(y, z)\}$ and the members of this set will be called the *sides* of the triangle.

DEFINITION 1.51 (Slimness constant for a path-family).

(1) Suppose X is a metric space and $x, y, z \in X$. We shall say that a triangle formed by three paths $c(x, y), c(x, z), c(y, z)$ is *δ-slim* for some $\delta \geq 0$, if each side of the triangle is contained in the union of δ-neighborhoods of the remaining two sides.

(2) Suppose C is a family of paths in a metric space X such that for all $x, y \in X$ there is at least one path in C joining x, y. Then we define the *slimness constant* of C as

$$\delta_s(C) := \sup_{c(x_1,x_2),c(x_2,x_3),c(x_3,x_1)\in C} \inf\{r : \text{ the triangle } \{c(x_1, x_2), c(x_2, x_3), c(x_3, x_1)\} \text{ is } r\text{-slim}\},$$

where $x_i \in X$, $i = 1, 2, 3$ and $c(x_i, x_j)$ are paths in C joining x_i, x_j.

We are now ready to define a form of Rips-hyperbolicity for path-metric spaces.

DEFINITION 1.52 (Rips hyperbolicity of path-metric spaces). Suppose (X, d) is a path-metric space, and $\epsilon \geq 0$. For all $x, y \in X$ let $\mathcal{F}_\epsilon(x, y)$ be the family of all ϵ-short paths in X joining x to y. We say that X is *δ-hyperbolic in the sense of Rips* if (1) $\mathcal{F}_\epsilon(x, y) \neq \emptyset$ for all $x, y \in X$ and $\epsilon > 0$, and

$$(2) \limsup_{\epsilon\to 0+} \delta_s(\mathcal{F}_\epsilon) \leq \delta.$$

REMARK 1.53.

(1) (X, d) is a geodesic metric space if and only if $\mathcal{F}_0(x, y) \neq \emptyset$ for all $x, y \in X$. Elements of $\mathcal{F}_\epsilon$ are $(1, \epsilon)$-quasigeodesics in X.

(2) For a geodesic metric space, Definition 1.52 is equivalent to the standard notion of δ-hyperbolicity in the sense of Rips (with a slightly different hyperbolicity constant). There, a space is δ_0-hyperbolic (in the sense of Rips) if $\delta(\mathcal{F}_0) = \delta_0$.

(3) If for a path-metric space X we have $\delta(\mathcal{F}_\epsilon) < \infty$ for some $\epsilon \geq 0$, then X is $\delta(\mathcal{F}_\epsilon)$-hyperbolic in the sense of Rips.

1.9.4. Stability of quasigeodesics and qi invariance of hyperbolicity. One of the most fundamental facts about hyperbolic spaces is that quasigeodesics are uniformly close to geodesics. This fact is also known as the (hyperbolic) Morse lemma, as it first appeared in a work of Morse on geodesics in the hyperbolic plane, [**Mor24**]. Morse did not have the notion of quasigeodesics and he was interested in how geodesics on a surface change with a change of its hyperbolic metric. Since changing a Riemannian metric on a compact manifold results in a quasi-isometric change of the metric on its universal cover, Morse's result can be interpreted as stability of quasigeodesics. Morse's proof was quite general and most modern proofs of stability of quasigeodesics follow the same line of reasoning.

The next lemma is a converse to Lemma 1.21 in the setting of hyperbolic spaces.

LEMMA 1.54 (Morse Lemma or stability of quasigeodesics). *There is a function*

$$D_{1.54} = D_{1.54}(\delta, k)$$

defined for $\delta \geq 0$ and $k \geq 1$, such that the following holds:

Suppose X is a δ-hyperbolic geodesic metric space. Then for every k-quasigeodesic $\phi : [a, b] \to X$, the Hausdorff distance between the image of ϕ and that of any geodesic ϕ^* connecting the endpoints of ϕ, is $\leq D_{1.54}$.

More precisely, according to **[GS19]**, *for a (k, ϵ)-quasigeodesic ϕ in X,*

$$\mathrm{Hd}(\phi, \phi^*) \leq 92k^2(\epsilon + 3\delta).$$

Thus, for a k-quasigeodesic one can take

$$D_{1.54}(\delta, k) = 92k^2(k + 3\delta).$$

With minor modifications, the proofs go through for path-metric spaces, when geodesics are replaced with η-short paths ϕ^*_η (cf. **[V05]**). One obtains an estimate $D_{1.54}(\delta, k, \eta)$ and, hence,

$$D_{1.54}(\delta, k) = \lim_{\eta \to 0+} D_{1.54}(\delta, k, \eta).$$

As a consequence we get Lemma 1.55.

LEMMA 1.55. *There exists a function $D = D_{1.55}(\delta, k, r) \geq r$ such that the following holds. If X is a δ-hyperbolic geodesic space, and $\phi_i : I_i = [a_i, b_i] \to X$ are k-quasigeodesics satisfying*

$$d(x_1, x_2) \leq r, d(y_1, y_2) \leq r, x_i = \phi_i(a_i), y_i = \phi_i(b_i), i = 1, 2,$$

then the images $\phi_1(I_1), \phi_2(I_2)$ are D-Hausdorff close.

PROOF. Let ϕ^*_i be geodesics connecting the endpoints x_i, y_i of ϕ_i, $i = 1, 2$. Then, since quadrilaterals in X are 2δ-slim,

$$\mathrm{Hd}(\phi^*_1, \phi^*_2) \leq 2\delta + r.$$

Applying Lemma 1.54, we conclude that

$$\mathrm{Hd}(\phi_1, \phi_2) \leq D_{1.55}(\delta, k, r) = \max(2D_{1.54}(\delta, k) + \delta, D_{1.54}(\delta, k) + \delta + r).$$

More explicitly, since $D_{1.54}(\delta, k) = 92k^2(k + 3\delta)$, we get:

$$D_{1.55}(\delta, k, r) = \max(184k^2(k + 3\delta) + \delta, 92k^2(k + 3\delta) + \delta + r).$$ □

LEMMA 1.56. *Suppose that Y, X are path-metric spaces, X is δ-Rips-hyperbolic and $f : Y \to X$ is a (K, ϵ)-qi embedding. Then Y is also $\delta_{1.56}(\delta, K, \epsilon)$-hyperbolic.*

In particular, (Rips) hyperbolicity is qi invariant among path-metric spaces: If $f : Y \to X$ is a (K, ϵ)-quasi-isometry and Y is δ-hyperbolic in the sense of Rips, then X is $\delta'_{1.56}(\delta, K, \epsilon)$-hyperbolic in the sense of Rips.

PROOF. Consider a triple of points $x, y, z \in Y$ and let $\eta \geq 0$. Consider the triangle Δ_η in Y formed by η-short arc-length parameterized paths $c(x, y), c(y, z), c(z, x)$ connecting these points. These paths are $(1, \eta)$-quasigeodesics in Y.

Then $f(c(x, y)), f(c(x, z)), f(c(y, z))$ are (K, ϵ')-quasigeodesics in X, for

$$\epsilon' = K(1 + \eta) + \epsilon,$$

see Lemma 1.12. Hence, by Lemma 1.54, the quasigeodesic triangle formed by the quasigeodesic paths $f(xy), f(xz), f(yz)$ is $D = (2 \cdot 92K^2(\epsilon' + 3\delta) + \delta)$-slim. This implies that the triangle Δ_η is $K(D+\epsilon)$-slim. Sending η to 0, we conclude that Y is $\delta_{1.56}(\delta, K, \epsilon)$-hyperbolic for

$$\delta_{1.56}(\delta, K, \epsilon) = K(184K^2(K + \epsilon + 3\delta) + \delta + \epsilon).$$

The second statement of the lemma follows from the first, combined with Lemma 1.13. □

REMARK 1.57. For $\epsilon = K$ we will use then notation $\delta_{1.56}(\delta, K, \epsilon) = \delta_{1.56}(\delta, K)$ and $\delta'_{1.56}(\delta, K, \epsilon) = \delta'_{1.56}(\delta, K)$. Thus, for $\delta_{1.56}(\delta, K)$ we can take the number

$$K(184K^2(2K + 3\delta) + \delta + K).$$

Lemma 1.56 combined with Lemma 1.46 immediately imply:

COROLLARY 1.58. *Hyperbolicity is qi invariant among path-metric spaces.*

COROLLARY 1.59. *Gromov-hyperbolicity is quasi-isometry invariant among quasi-path metric spaces.*

PROOF. Suppose that X, Y are quasi-isometric quasi-path metric spaces and X is a Gromov–hyperbolic space. By Corollary 1.34, there are metric graphs X', Y' which are geodesic metric spaces, and are $(1, \epsilon)$-quasi-isometric to X and Y respectively for a suitable $\epsilon \geq 0$. Thus:

(1) by Lemma 1.41 X' is Gromov hyperbolic.
(2) X', Y' are quasi-isometric geodesic metric spaces.

However, then by Corollary 1.46, X' is Rips-hyperbolic since it is a geodesic metric space. Then, by Lemma 1.56, Y' is also Rips hyperbolic. Since Y' is a geodesic metric space, again by Lemma 1.46, it is also Gromov-hyperbolic. Finally, Y is also Gromov-hyperbolic by Lemma 1.41. □

Then we also get Corollary 1.60.

COROLLARY 1.60. *Suppose that Y is a path-metric space, $X \subset Y$ is a rectifiably connected R-net in Y, equipped with a path-metric such that the inclusion map $X \to Y$ is η-uniformly proper, and X is δ-hyperbolic. Then Y is $\delta_{1.60}(\delta, \eta(2R+1), R) = \delta_{1.56}(\delta, \eta(2R+1))$-hyperbolic.*

PROOF. According to Lemma 1.18, the inclusion map $X \to Y$ is an L-qi embedding, $L = \eta(2R + 1)$, hence, an L-quasi-isometry. Now the corollary follows immediately from Lemma 1.56. □

1.10. Combings and a characterization of hyperbolic spaces

One of the key tools in our work is a characterization of hyperbolicity in terms of *slim combings* due to Bowditch (see [**Bow14**, Proposition 3.1]). See [**MS13**, Theorem 3.15], [**Ham05**, Lemma 3.3] and [**DDLS23**, Proposition 2.2] for similar characterizations obtained by other authors. The idea is that if X is a δ-hyperbolic geodesic metric space, then for each pair of points we have a (typically nonunique) geodesic path xy between these points and these paths satisfy the δ-slim triangle property. Bowditch's characterization reverses this definition.

Let $\mathcal{P}(X)$ denote the space of paths in a topological space X.

DEFINITION 1.61.

(1) Two paths in a metric space are said to *Hausdorff D-fellow-travel* if their images are D-Hausdorff-close.

(2) A *combing* $\mathscr{C}$ of a metric space X is a map

$$c : X_0 \times X_0 \to \mathcal{P}(X)$$

sending each pair $(x, y) \in X_0^2$ to a path $c_{x,y}$ in X (also frequently denoted $c(x, y)$) connecting x to y, where $X_0 \subset X$ is a D-net for some D.

(3) For a function $C = C(r)$, a combing $\mathscr{C}$ is said to satisfy the *$C(r)$-Hausdorff fellow-traveling property* if for every triple of points $x, y, z \in X_0$ with $d(y, z) \le r$,

$$\mathrm{Hd}(c_{x,y}, c_{x,z}) \le C(r)$$

and $\mathrm{diam}(c_{x,x}) \le C(0)$.

(4) A combing $\mathscr{C}$ is said to be an *η-proper combing* for a function $\eta : \mathbb{R}_{\ge 0} \to \mathbb{R}_{\ge 0}$ if each path $c_{x,y} \in \mathscr{C}$ is an η-proper embedding into X.

A combing $\mathscr{C}$ will be called a *proper combing* if it is an η-proper combing for some function $\eta : \mathbb{R}_{\ge 0} \to \mathbb{R}_{\ge 0}$.

More generally, if $\mathscr{D} \subset \mathscr{C}$ then we say that $\mathscr{D}$ is *proper* if there is a function $\eta : \mathbb{R}_{\ge 0} \to \mathbb{R}_{\ge 0}$ such that each path in $\mathscr{D}$ is an η-proper embedding.

(5) A combing $\mathscr{C}$ will be called a *D-slim combing* for some constant $D \ge 0$ if $\delta_S(\mathscr{C}) \le D$ as per Definition 1.51.

A combing $\mathscr{C}$ will be called a *slim combing* if $\delta_S(\mathscr{C}) < \infty$.

While we define combings as maps, we will think of each combing as a subset $\mathscr{C} = c(X^2) \subset \mathcal{P}(X)$.

The following characterization of hyperbolicity is due to Bowditch, [**Bow14**, Proposition 3.1].

THEOREM 1.62. *Given $h \ge 0$, there is $k = k_{1.62}(h) < \infty$ with the following property. Suppose that X is a connected graph, and that for all $x, y \in V(X)$, we have associated a connected subgraph, $Y_{xy} \subset X$ containing both x and y, satisfying the following properties.*

(1) *For all $x, y \in V(X)$ with $d(x, y) \le 1$, the diameter of Y_{xy} in X is at most h.*
(2) *For all $x, y, z \in V(X)$, $Y_{xy} \subset N_h(Y_{xz} \cup Y_{zy})$.*

Then X is k-hyperbolic. In fact, we can take any $k \ge (3m - 10h)/2$, where m is any positive real number satisfying $2h(6 + \log_2(m + 2)) \le m$. Moreover, for all $x, y \in V(X)$, the Hausdorff distance between Y_{xy} and any geodesic from x to y is bounded above by $m - 4h$.

In the book we will be using the following corollary of Bowditch's characterization. Note that Property (2) amounts to the condition that the family $\mathscr{C}$ is D_2-slim, i.e., $\delta_s(\mathscr{C}) \le D_2$, see Definition 1.51. However, a similar characterization can also be found in [**DDLS23**, Proposition 2.2].

COROLLARY 1.63. *Given $D_0 \ge 0$, $D_1 > 0$, $D_2 > 0$ and a Lipschitz function $\eta : \mathbb{R}_{\ge 0} \to \mathbb{R}_{\ge 0}$ there are $\delta = \delta_{1.63}(D_0, D_1, D_2) \ge 0$, and $R = R_{1.63}(D_0, D_1, D_2) \ge 0$ such that:*

Suppose X is a path-metric space and $X_0 \subset X$ is a D_0-net such that for each pair of points $x, y \in X_0$ we are given a rectifiable path $c(x, y)$ in X joining x, y with the following properties.

Property (1) *For all $x, y \in X_0$ with $d(x, y) \le 1 + 2D_0$, the length of $c(x, y)$ in X is at most D_1.*

Property (2) *For all $x, y, z \in X_0$, $c(x, y) \subset N_{D_2}(c(x, z) \cup c(z, y))$.*

Then X is δ-hyperbolic in the sense of Rips. Moreover, for all $x, y \in X_0$ and all 1-quasigeodesics γ_{xy} in X joining x, y, $\mathrm{Hd}(\gamma_{xy}, c(x, y)) \le R$.

PROOF. Let $R = 1 + 2D_0$. Since path metric spaces are 1-quasi-path metric spaces, by Lemma 1.36 there is a $(K_{1.36}(1, R), \epsilon_{1.36}(1, R))$-quasi-isometry $\iota : Z \to X_0 \subset X$ which is the identity on the vertex set of Z, where Z is the full subgraph of the R-Rips graph of X with the vertex set X_0. Also there is coarse inverse $\rho : X \to Z$ to ι which is the identity map when restricted to X_0 and and which is also a $(K_{1.36}(1, R), \epsilon_{1.36}(1, R))$-quasi-isometry. Let $k_0 = K_{1.36}(1, R) + \epsilon_{1.36}(1, R)$. Then ι and ρ are k_0-quasi-isometries.

Now we shall verify that Z satisfies the conditions of Proposition 1.62 by finding a suitable set of edge-paths joining each pair of vertices. Since hyperbolicity of path-metric spaces is invariant under quasi-isometry it will then follow that X itself is uniformly hyperbolic. Suppose $x, y \in V(Z)$ and $\alpha : [0, l] \to X$ is the arc-length parametrization of the path $c(x, y)$. Let $n = \lfloor l \rfloor$. Consider the points $y_i = \alpha(i)$, $0 \leq i \leq n$ and $y_{n+1} = \alpha(l) = y$ if $l > nD_0$. For each i, there is a point $x_i \in X_0 = V(Z)$ such that $d_X(y_i, x_i) \leq D_0$. Then

$$d_X(x_i, x_{i+1}) \leq 1 + 2D_0.$$

In particular, there is an edge in Z joining x_i, x_{i+1} for $0 \leq i \leq n$. This defines an edge-path $\beta = \beta(x, y)$ in Z joining $x, y \in V(Z)$ and the length of $\beta(x, y)$ is at most $l + 1$.

We take the subgraph $Y_{xy} \subset Z$ to be the image of the path $\beta(x, y)$. This is clearly a connected subgraph in Z. We now verify the conditions of Proposition 1.62 for this family of subgraphs.

(1) Suppose $x, y \in V(Z)$ with $d_Z(x, y) \leq 1$. This implies $d_X(x, y) \leq R$. Hence the length of $c(x, y)$ is at most D_1 by property (1). Thus the diameter of Y_{xy} is at most $D_1 + 1$.
(2) If $V(\beta)$ denotes the vertex set $\{x_i\}$ of β then we note that

$$\mathrm{Hd}(c(x, y), V(\beta)) \leq 1 + D_0.$$

Thus, clearly, for all $x, y, z \in V(Z)$, we have

$$\beta_{xy} \subset N_{2+D_0+D_2}(\beta_{xz} \cup \beta_{zy}).$$

Thus if we take $h = \max\{D_1+1, 2+D_0+D_2\}$, then Z is $k_{1.62}(h)$-hyperbolic in the sense of Rips. Since $\rho : X \to Z$ is a k_0-quasi-isometry, X is $\delta = \delta_{1.56}(k_{1.62}(h), k_0)$-hyperbolic in the sense of Rips by Lemma 1.56.

For the second part of the corollary we set m to be the least positive number satisfying $2h(6 + \log_2(m + 2)) \leq m$. Let $x \neq y \in X_0$ be arbitrary points. Then by Theorem 1.62 any geodesic γ_{xy} in Z joining x, y we have $\mathrm{Hd}(\gamma_{xy}, \beta_{xy}) \leq m - 4h$. Suppose α_{xy} is a 1-quasigeodesic joining x, y in X. Then clearly $\rho(\alpha_{xy})$ is a $2k_0$-quasigeodesic in Z. Thus, $\mathrm{Hd}(\gamma_{xy}, \rho(\alpha_{xy})) \leq D_{1.54}(k_{1.62}(h), 2k_0)$.

Hence, $\mathrm{Hd}(\beta_{xy}, \rho(\alpha_{xy})) \leq m - 4h + D_{1.54}(k_{1.62}(h), 2k_0)$ and, therefore,

$$\mathrm{Hd}(V(\beta_{xy}), \rho(\alpha_{xy})) \leq 1 + m - 4h + D_{1.54}(k_{1.62}(h), 2k_0) = R_0.$$

But $V(\beta_{xy}) = \rho(V(\beta_{xy}))$ since ρ is the identity map when restricted to X_0. Hence,

$$\mathrm{Hd}(\rho(V(\beta_{xy})), \rho(\alpha_{xy})) \leq R_0.$$

Since ρ is a k_0-qi embedding we have $\mathrm{Hd}(V(\beta_{xy}), \alpha_{xy}) \leq k_0(R_0 + k_0)$. The inequality $\mathrm{Hd}(c(x, y), V(\beta_{xy})) \leq D_0 + 1$ implies $\mathrm{Hd}(c(x, y), \alpha_{xy}) \leq 1 + D_0 + k_0(R_0 + k_0)$. Hence we may set $R = 1 + D_0 + k_0(R_0 + k_0)$. □

COROLLARY 1.64. *Given $D_0 \geq 0$, $D > 0$, and a coarse Lipschitz function $\eta : \mathbb{R}_{\geq 0} \to \mathbb{R}_{\geq 0}$, there are $\delta = \delta_{1.64}(D_0, D, \eta) \geq 0$, and $K = K_{1.64}(D_0, D, \eta) < \infty$ such that:*

Suppose X is a path-metric space and $X_0 \subset X$ is a D_0-net such that for each pair of points $x, y \in X_0$ we are given a rectifiable arc-length parametrized path $c(x, y)$ in X joining x, y with the following properties:

Property (1) *For all $x, y \in X_0$ the path $c(x, y)$ is η-proper.*

Property (2) *For all $x, y, z \in X_0$, $c(x, y) \subset N_D(c(x, z) \cup c(z, y))$.*

Then X is δ-hyperbolic in the sense of Rips and moreover, the paths $c(x, y)$ are K-quasigeodesics in X.

PROOF. This follows from Corollary 1.63. Property 1 of Corollary 1.63 is verified with $D_1 = \eta(2D_0 + 1)$ and property 2 is verified with $D_2 = D$. Thus we can take $\delta = \delta_{1.64}(D_0, D, \eta) = \delta_{1.63}(D_0, \eta(2D_0 + 1), D)$.

For the second part of the corollary suppose γ_{xy} is an arc-length parametrized path in X joining $x, y \in X_0$ such that $\text{length}(\gamma_{xy}) \le 1 + d(x, y)$. Then clearly it is a 1-quasigeodesic. By the second part of Corollary 1.63 we have

$$\text{Hd}(\gamma_{xy}, c(x, y)) \le R_{1.63}(D_0, \eta(2D_0 + 1), D) = R.$$

Then, by Lemma 1.21, the paths $c(x, y)$ are $K_{1.21}(R, 1, \eta)$-quasigeodesics. □

REMARK 1.65.

(1) In the proofs of hyperbolicity of various spaces given in this books, based on Corollary 1.64, we first verify the uniform properness of the paths $c(x, y)$ and, frequently, also verify that they satisfy the *Hausdorff fellow-traveling condition*, before proving that (2) holds.

(2) In our proofs, instead of using arc-length parameterizations of the paths $c(x, y)$ we will be using some uniformly quasi-isometric reparameterizations of these paths. Clearly, uniform properness of one implies uniform properness of the other.

(3) We will refer to a family of paths c satisfying the assumptions of the corollary as a *proper slim combing* of X.

Lastly, we generalize this corollary to the case of discrete paths $c : [m, n] \cap \mathbb{Z} \to X$. We will be using the extension $\tilde{c}$ of maps c to the real interval $[m, n]$ defined by sending the open interval $(i, i + 1)$ to $c(i)$.

COROLLARY 1.66. *Given $r > 0$, $D_0 \ge r$, $D_1 > 0$, and a function $\eta : \mathbb{R}_{\ge 0} \to \mathbb{R}_{\ge 0}$ there are $\delta = \delta(D_0, D_1, \eta) \ge 0$, and $K = K(D_0, D_1, \eta) \ge 1$ such that:*

(a) *Suppose X is a r-quasi-path metric space and $X_0 \subset X$ is a D_0-net such that for each pair of points $x, y \in X_0$ we are given a discrete r-path $c(x, y)$ in X joining x, y with the following properties:*

Property (a1) *For all $D \ge 0$ and $x, y \in X_0$ with $d(x, y) \le D$, the length of $c(x, y)$ in X is at most $\eta(D)$.*

Property (a2) *For all $x, y, z \in X_0$, $c(x, y) \subset N_{D_1}(c(x, z) \cup c(z, y))$.*

Then X is δ-hyperbolic in the sense of Gromov.

(b) *Moreover, if paths $c(x, y)$ are η-proper, then each map $\tilde{c}$ is a K-quasigeodesic in X.*

PROOF. By Corollary 1.34, there is a $(1, \epsilon)$-quasi-isometry $\iota : X \to X'$, where X' is a metric graph all whose each edges are of length r and $\epsilon = \epsilon_{1.34}(r)$. Therefore, X' is a geodesic metric space. Now, define $X'_0 = \iota(X_0)$. We take any map $g : X'_0 \to X_0$ such that $x = \iota(g(x))$ for all $x \in X'_0$. Then for all $x, y \in X'_0$ we define a path $c'(x, y)$ as follows. Let $c_{xy} : I_{xy} \cap \mathbb{Z} \to X$ be the parametrization of $c(g(x), g(y))$. For any two consecutive points

$s, t \in I_{xy} \cap \mathbb{Z}$ we join $\iota \circ c_{xy}(s), \iota \circ c_{xy}(t)$ by a geodesic in X'. Concatenation of these forms a path $c'(x, y)$. We leave it to the reader to verify that the two properties of Corollary 1.63 hold for this family of paths in X'. This implies that X' is uniformly Rips-hyperbolic and, hence, uniformly Gromov-hyperbolic by Lemma 1.46. Since X is $(1, \epsilon)$-qi to X', it is also uniformly Gromov-hyperbolic by Lemma 1.41. The last part of the corollary follows from Lemma 1.21. □

We note that if the assumption in Part (b) of the above corollary holds. Then it is clear that condition (a1) necessarily holds as well.

1.11. Hyperbolic cones

Suppose that (Z, d_Z) is a path-metric space. In this section we define the *hyperbolic cone* or a *horoball* Z^h over Z. This definition will be used in Chapter 9 when discussing relatively hyperbolic spaces

As a topological space, Z^h is the product $Z \times [1, \infty)$, where we identify Z with $Z \times \{1\}$. We equip Z^h with the length structure (imitating the description of the hyperbolic metric on horoballs in the real-hyperbolic space): Paths in this length structure are concatenation of vertical and horizontal path, with respect to the product decomposition of Z^h. Given two points $y_1, y_2 \in [1, \infty)$, and $z \in Z$, we let the length of the interval between $(z, y_1), (z, y_2)$ in $\{z\} \times [1, \infty)$ equal $|\log(y_2/y_1)|$. We let the length of each horizontal path, contained in the "horosphere" $Z \times \{y\}$ equal y^{-1} times the length of the corresponding path in Z. This length structure defines a path-metric on Z^h. It is easy to verify that for all $z \in Z$ the map $t \mapsto (z, e^t)$ is a geodesic in this space.

Remark 1.67. We refer the reader to the book of Roe [**Roe03**, 2.5] and the paper by Bowditch [**Bow12**] for alternative definitions of hyperbolic cones. For instance, Roe's construction works for general metric spaces (Z, d_Z).

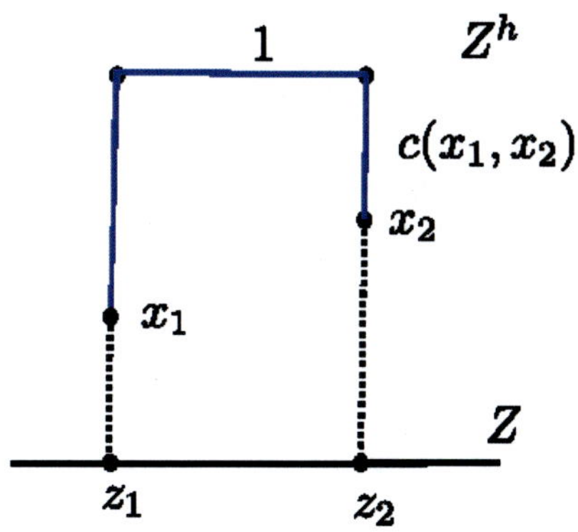

Figure 1.1. Combing of hyperbolic cone

Proposition 1.68. *The metric space Z^h is δ-hyperbolic for some uniform constant δ.*

PROOF. We will describe a proper slim combing on Z^h. Each combing path will be a concatenation of at most two vertical paths and at most one horizontal path. Consider two points $x_1 = (z_1, y_1), x_2 = (z_2, y_2)$ in Z^h. If $z_1 = z_2$ then the map $c(x_1, x_2)$ connecting x_1 to x_2 will be the unique vertical interval connecting these points. Suppose that $y_1 \le y_2$. Find the smallest $y \ge y_2$ such that

$$y^{-1} d_Z(z_1, z_2) \le 1/2.$$

Set $x'_i := (z_i, y)$, $i = 1, 2$. Then $c(x_1, x_2)$ is the concatenation

$$[x_1 x'_1] \star [x'_1 x'_2] \star [x'_2 x_2],$$

where the first and the last segments are vertical intervals between x_i, x'_i, $i = 1, 2$, and the middle segment $[x'_1 x'_2]$ is any path in $Z \times \{t\}$ connecting x'_1 to x'_2 and having length ≤ 1.

We now verify the proper slim combing properties of the paths c, as required by Corollary 1.64.

Uniform properness. First we make a few observations.

(1) Consider the coordinate projection $\pi_1 : Z^h \to Z$, i.e., $\pi_1(z, y) = z$ for all $(z, y) \in Z^h$. Then the composition $\log \circ \pi_2$ is 1-Lipschitz. Consequently for all $w_1, w_2 \in Z$ and $t \in [1, \infty)$ $d_{Z^h}((w_1, t), (w_2, t)) \le R$ implies $d_Z(w_1, w_2) \le e^R$.
(2) Any subpath of $c(x_1, x_2)$ is again a path in this path family. Thus it is enough to bound the length of $c(x_1, x_2)$ in terms of $d_{Z^h}(x_1, x_2)$.
(3) Suppose $D = d_{Z^h}(x_1, x_2)$ and $\epsilon > 0$ and α is a path in Z^h connecting x_1, x_2 which is a concatenation of finitely many vertical and horizontal paths such that $l(\alpha) \le D + \epsilon$. Then the sum of lengths of the vertical segments in α is at most $D + \epsilon$. Since this is true for all $\epsilon > 0$ it follows that $d_{Z^h}((z_1, y_1), (z_1, y_2)) \le D$. Let $x_3 = (z_1, y_2)$.

Thus $d_{Z^h}(x_3, x_2) \le 2D$. Hence by observation (1) we obtain equation (1.2):

$$d_{Z^h}(x_1, x_2) \le D \Rightarrow d_Z(z_1, z_2) \le e^{2D}. \tag{1.2}$$

Since y is determined by $d_Z(z_1, z_2)$ the proof follows.

Slim triangle condition. Consider three points $x_i = (z_i, y_i) \in Z^h$, $i = 1, 2, 3$. We let y_{ij} denote the maximum of $\pi_2 \circ c(x_i, x_i)$, $i \ne j$. After relabeling the points, we can assume that

$$y_{12} \le y_{23} \le y_{31}.$$

Define $x'_1 := (z_1, y_{12})$ and $x'_2 := (z_2, y_{12})$.

Replacing the points x_1, x_2, respectively, with $x'_1 = (z_1, y_{12})$ and $x'_2 = (z_2, y_{12})$ we see that the $c(x_1, x_2), c(x_2, x_3), c(x_3, x_1)$ is δ-slim if and only if the triangle formed by the paths $c(x'_1, x'_2), c(x'_2, x_3), c(x_3, x'_1)$ is δ-slim. Thus, without loss of generality, $x'_1 = x_1, x'_2 = x_2$ and the path $c(x_1, x_2)$ is horizontal, of length ≤ 1. We claim that in this situation, the paths $c(x_1, x_3), c(x_2, x_3)$ are uniformly Hausdorff-close, which, of course, will imply the uniform slimness. Denoting

$$x''_i := (z_i, y_{23}), i = 1, 2, 3,$$

we see that the parts $\mathrm{Hd}(c(x_1, x''_1), c(x_2, x''_2)) \le 1$, while the parts between x''_3 and x_3 of $c(x_1, x_3), c(x_2, x_3)$ are equal. Thus, we need to bound the Hausdorff distance between $c(x''_1, x''_3), c(x''_2, x''_3)$.

We have:

$$1 \ge y_{12}^{-1} d(z_1, z_2) \ge y_{23}^{-1} d(z_1, z_2).$$

Applying the triangle inequality in (Z, d_Z), we obtain

$$2 = 1 + 1 \ge y_{23}^{-1}(d(z_1, z_2) + d(z_2, z_3)) \ge y_{23}^{-1} d(z_3, z_1).$$

It follows that $y_{23} \le y_{31} \le 2y_{23}$. Hence,

$$d(x_1'', (z_1, y_{31})) \le \log(2).$$

It follows that the length of $c(x_1'', x_3'')$ is $\le 1 + 2\log(2)$, while the length of $c(x_2'', x_3'')$ is ≤ 1. Clearly, these path are at the Hausdorff-distance $\le 2 + 2\log(2)$. □

REMARK 1.69. By Corollary 1.64, the slim combing paths $c(x_1, x_2)$ in Z^h defined in this proof are k-quasigeodesics in Z^h with the quasigeodesic constant k independent of Z.

LEMMA 1.70. *Suppose that $z \in Z$ is within distance $\le C$ from a segment $[z_1z_2]_{Z^h}$ with the endpoints z_1, z_2 in Z. Then $d_Z(z, \{z_1, z_2\}) \le C'$, where C' depends only on C.*

PROOF. The combing path $c(z_1, z_2)$ is uniformly Hausdorff-close to $[z_1z_2]_{Z^h}$. Hence, it suffices to prove the lemma with $[z_1z_2]_{Z^h}$ replaced by the combing path $c(z_1, z_2)$. Then the inequality $d_{Z^h}(z, c(z_1, z_2)) \le C$ implies that

$$d_{Z^h}(z, \{z_1, z_2\}) \le 2C + 2.$$ □

We note that in the course of the proof of Proposition 1.68 we have shown the following.

PROPOSITION 1.71. *The inclusion map $Z \to Z^h$ is η-uniformly proper, where $\eta(t) = a\exp(at)$ and a is a universal constant.*

PROOF. The proposition follows from the proof of equation (1.2). □

1.12. Geometry of hyperbolic triangles

Informally speaking, triangles in hyperbolic spaces resemble triangles in trees, i.e., tripods. The *comparison map* to trees makes this precise and allows one to reduce proofs of various statements about hyperbolic triangles to statements about tripods in trees. Below, by a *tree* we mean a regular simplicial tree T of valence ≥ 3, equipped with the standard graph-metric. However, any real tree (a 0-hyperbolic geodesic metric space) not isometric to an interval, will work just as well. For any three numbers a_1, a_2, a_3 satisfying the triangle inequalities $a_i \le a_j + a_k$, $\{i, j, k\} = \{1, 2, 3\}$, there exists a triangle $\Delta \subset T$ with the side-lengths a_1, a_2, a_3. Accordingly, for each triangle $\Delta = \Delta x_1x_2x_3$ in a metric space X, we define its *comparison triangle*

$$\bar{\Delta} = \Delta\bar{x}_1\bar{x}_2\bar{x}_3$$

in T, as a triangle in T such that

$$d(x_i, x_j) = d_T(\bar{x}_i, \bar{x}_j), \quad 1 \le i, j \le 3.$$

For each point p in the side x_ix_j of Δ, define its *comparison point* $\bar{p} \in \bar{x}_i\bar{x}_j \subset \bar{\Delta}$ by the condition

$$d(p, x_i) = d_T(\bar{p}, \bar{x}_i).$$

Thus, we get the *comparison map* $\theta : \Delta \to \bar{\Delta}$, $\theta(p) = \bar{p}$, which restricts to an isometry on each side of Δ. The internal points of Δ are the points of the θ-pre-image of the center of $\bar{\Delta}$. A triangle Δ is δ-thin if and only if the diameters of all fibers of θ are $\le \delta$. For each pair of points $p, q \in \Delta$ in a δ-thin triangle $\Delta \subset X$, triangle inequalities imply:

$$d(p, q) - 2\delta \le d_T(\bar{p}, \bar{q}) \le d(p, q) + \delta.$$

Thus, the map θ is a $(1, 2\delta)$-quasi-isometry $\Delta \to \bar{\Delta}$ for δ-thin triangles $\Delta \subset X$.

DEFINITION 1.72. Let (X, d) be a metric space. A point $p \in X$ is said to be a *C-center* of a geodesic triangle Δ if p lies within distance C from all three sides of the triangle. A 0-center is simply called a center of Δ. We will use this definition when (X, d) is a tree, in which case the 0-center is obviously unique.

However, we have the following.

LEMMA 1.73. *Suppose (X, d) is a metric space and Δ is a geodesic triangle with vertices x, y, z. If Δ has a 0-center then it is unique.*

PROOF. Suppose p is a 0-center of Δ. It follows that $d(x, p)+d(p, y) = d(x, y), d(y, p)+d(p, z) = d(y, z)$ and $d(z, p) + d(p, x) = d(z, x)$. It follows that $d(z, p) = (x.y)_z$. Since p is on the geodesic $[y, z]$, $d(z, p) = (x.y)_z$ uniquely determines p. □

Note that if a geodesic metric space X is δ-hyperbolic, every geodesic triangle Δ in X has a δ-center, e.g., a point on one side of Δ, within distance δ from the two other sides. The internal points of a δ-thin triangle are mapped via the comparison map θ to the center of the comparison triangle in the tree and each internal points is a δ-center of Δ.

DEFINITION 1.74. *A C-tripod* in X is the union $T_p(xyz)$ of three geodesic segments $px \cup py \cup pz$, where p is a C-center of a geodesic triangle Δxyz. The points x, y, z are called the *extremities* of the tripod and the segments px, py, pz are the *legs* of the tripod. If X is a tree and $C = 0$ then by a *tripod* in X we mean a 0-tripod and the center of a tripod means its 0-center.

As noted above, if a triangle $\Delta = \Delta xyz$ is δ-slim, then there exists a point $p \in xy$ such that the union $px \cup py \cup pz$ is a δ-tripod.

Lemma 1.75 follows immediately from the definitions.

LEMMA 1.75. *If X is δ-hyperbolic in the sense of Rips, then each C-tripod $T_p(xyz) \subset X$ is $(C + \delta)$-Hausdorff close to the triangle Δxyz.*

LEMMA 1.76. *If X is δ-hyperbolic in the sense of Rips, then for each C-tripod $T_p(xyz)$, the point p is a $C + 2\delta$-center of every triangle $\Delta x'y'z'$, where $x' \in px, y' \in py, z' \in pz$.*

PROOF. Consider a point $u \in xy$ within distance C from p. By the 2δ-slimness of the quadrilateral $xx'y'y$, either there exists a point $u' \in x'y'$ at distance $\leq 2\delta$ from u, or there is a point $v \in xx' \cup yy'$ at distance $\leq \delta$ from u. In the latter case, for some $u' \in \{x', y'\}$, we get $d(p, u') \leq C + \delta$. Thus, in each case, there is $u' \in x'y'$ within distance $C + 2\delta$ from p. □

LEMMA 1.77. *Suppose that X is a δ-hyperbolic geodesic metric space. If p, q are C-centers of the same geodesic triangle, then $d(p, q) \leq D_{1.77}(\delta, C)$.*

PROOF. We will use the comparison map $\theta : \Delta \to \bar{\Delta}$. Since X is δ-hyperbolic, the triangle Δ will be 6δ-thin (Lemma 1.45) and, hence, $\theta : x \mapsto \bar{x}$ satisfies the inequalities

$$d(a, b) - 12\delta \leq d_T(\bar{a}, \bar{b}) \leq d(a, b) + 6\delta,$$

$a, b \in \Delta$. We now prove the lemma. Let p_1, p_2, p_3 be the points on the sides of Δ within distance C from p. Then $d(p_i, p_j) \leq 2C$, $1 \leq i < j \leq 3$. We will estimate the distances from the points p_i to the internal points c_i of Δ, where we label the points so that p_i, c_i lie on the same side of Δ. Let $\bar{c} \in \bar{\Delta}$ denote the center of $\bar{\Delta}$, $\bar{c} = \theta(c_i)$, $i = 1, 2, 3$. Then

$$d_T(\theta(p_i), \theta(p_j)) \leq 2C + 6\delta.$$

It is easy to see that all three points $\theta(p_i)$ cannot lie on the same leg of the tripod $\bar{\Delta}$, unless one of them equals to the center of $\bar{\Delta}$. Thus, there exists i such that $d_T(\theta(p_i), \bar{c}) \le \frac{1}{2}(2C + 6\delta) = C + 3\delta$. Since θ is an isometry on each side of Δ, we then obtain

$$d(p_i, c_i) \le C + 3\delta$$

and, hence, $d(p, c_i) \le 2C + 3\delta$ for one of the internal points c_i of Δ. Since the internal points are distance $\le 6\delta$ apart, we conclude that for any two C-centers p, q of Δ

$$d(p, q) \le D_{1.77}(\delta, C) := 4C + 12\delta. \qquad \square$$

LEMMA 1.78. *Suppose that X is δ-hyperbolic. For a geodesic triangle Δxyz suppose that $x' \in xz, y' \in yz$ are equidistant from z and satisfy the inequality $d(x', y') > 2\delta$. Then the path $xx' \star x'y' \star y'y$ is $C_{1.78}(\delta)$-quasigeodesic in X.*

PROOF. First we show that $x' \in N_\delta(xy)$. By δ-hyperbolicity of the Δxyz, $x' \in N_\delta(xy \cup yz)$. If possible suppose $d(w, x') \le \delta$ for some $w \in yz$. Then $|d(z, w) - d(x', z)| \le d(w, x') \le \delta$. This implies $d(y', w) \le \delta$, since $d(z, x') = d(z, y')$. It follows that $d(x', y') \le d(w, x') + d(w, y') \le 2\delta$, a contradiction. Hence, $x' \in N_\delta(xy)$. Let $x'' \in xy$ be any point with $d(x', x'') \le \delta$.

Next we claim that $y' \in N_{5\delta}(x''y)$. Since the quadrilateral $\square x'x''yz$ is 2δ-slim,

$$y' \in N_{2\delta}(x'x'' \cup x''y \cup x'z).$$

If $y' \in N_{2\delta}(x'x'')$ then clearly $d(x'', y') \le 3\delta$. As in the previous paragraph, if $y' \in N_{2\delta}(x'z)$ then $d(x', y') \le 4\delta$ and thus $d(y', x'') \le 5\delta$. If neither of these happen then $y' \in N_{2\delta}(x''y)$. This proves our claim. Let $y'' \in x''y$ be any point such that $d(y', y'') \le 5\delta$.

Let $\alpha : [0, l] \to X$ be the arc-length parametrization of $xx' * x'y' * y'y$. Suppose $s, t \in [0, l]$ and $s < t$. Let $x_1 = \alpha(s), y_1 = \alpha(t)$. We already have $d(x_1, y_1) \le t - s$.

If both x_1, y_1 are on the same segment then there is nothing to prove. So assume otherwise. Suppose $x_1 \in xx'$ and $y_1 \in x'y'$. Since the quadrilateral $\square x'y'y''x''$ is 2δ-slim, there is a point $y_1' \in x''y''$ such that $d(y_1, y_1') \le 7\delta$. Also by the δ-slimness of $\Delta xx'x''$, there is a point $x_1' \in xx''$ such that $d(x_1, x_1') \le 2\delta$. It follows that

$$\begin{aligned}
d(x_1, y_1) &\ge d(x_1', y_1') - d(x_1, x_1') - d(y_1, y_1') \\
&\ge d(x_1', x'') + d(x'', y_1') - 9\delta \\
&\ge d(x_1, x') - d(x, x_1') - d(x', x'') + d(x', y_1) - d(x', x'') - d(y_1, y_1') - 9\delta \\
&\ge d(x_1, x') + d(x', y_1) - 20\delta \\
&= t - s - 20\delta.
\end{aligned}$$

In the same way, if $x_1 \in x'y', y_1 \in y'y$ we have $d(x_1, y_1) \ge t - s - 20\delta$. Next suppose $x_1 \in xx'$ and $y_1 \in y'y$. Then there are points $x_1' \in xx''$ and $y_1' \in y''y$ such that $d(x_1, x_1') \le 2\delta$ and $d(y_1, y_1') \le 6\delta$ by the δ-slimness of the triangles $\Delta xx''x'$ and $\Delta y'y''y$ respectively. It follows that

$$\begin{aligned}
d(x_1, y_1) &\ge d(x_1', y_1') - d(x_1, x_1') - d(y_1, y_1') \\
&\ge d(x_1', x'') + d(x'', y'') + d(y'', y_1') - 8\delta \\
&\ge d(x_1, x') - d(x_1, x_1') - d(x', x'') + d(x', y') - d(x', x'') \\
&\quad - d(y', y'') + d(y', y_1) - d(y', y'') - d(y_1, y_1') - 8\delta \\
&\ge d(x_1, x') + d(x', y') + d(y', y_1) - 28\delta \\
&= t - s - 28\delta.
\end{aligned}$$

Hence, the arc-length parametrization of $xx' \star x'y' \star y'y$ is a $(1, 28\delta)$-quasigeodesic. In particular, we can take $C_{1.78}(\delta) = 1 + 28\delta$. □

As an application of the lemma we will prove that two geodesics xz, zy either are uniformly close to each other on "long subintervals" $x'z, y'z$, or their concatenation $xz \star zy$ is a uniform quasigeodesic in X:

LEMMA 1.79. *Suppose that $R \le \min(d(x,z), d(y,z))$ is such that points $x' \in xz, y' \in yz$ at the distance R from z satisfy $d(x', y') > 2\delta$. Then the concatenation $xz \star zy$ is an $L_{1.79}(R,\delta)$-quasigeodesic in X.*

PROOF. According to the previous lemma, the concatenation $xx' \star x'y' \star y'y$ is a $C_{1.78}(\delta)$-quasigeodesic in X. We regard $xz \star zy$, $xx' \star x'y' \star y'y$ as paths c, c' on intervals $[0,T]$, $[0,T']$, parameterized by arc-length and connecting x to y. We note that $T' \le T \le T' + 2R$ and for each $t \in [0, T']$

$$d(c(t), c'(t)) \le 2R.$$

Since c' is a $C_{1.78}(\delta)$-quasigeodesic, Lemma follows. □

Lemma 1.80 is a consequence of Lemma 1.25, but we will give a direct proof.

LEMMA 1.80. *Suppose that Δxyz is a geodesic triangle in a δ-hyperbolic space X and $d(y,z) \le C$. Then there is a monotonic map $f : xz \to xy$ such that $d(f, \mathrm{id}) \le 2(\delta + C)$.*

PROOF. We define f as follows.

(1) Points $p \in xz$ such that $d(x,p) \le \min(d(x,z), d(x,y))$, will be sent to $\bar{p} \in xy$ such that $d(x,\bar{p}) = d(x,p)$.
(2) Points p such that $d(x,p) > d(x,y)$ will be all sent to y.

Thus, the map $f : xz \to xy$ is monotonic and continuous. We next estimate the distances $d(p, f(p))$. Since X is δ-hyperbolic, there exists a point $p' \in xz \cup yz$ within distance δ from p.

CASE 1. $p' \in xy$ and $d(x,p) < \min(d(x,y), d(x,z))$. Then by the triangle inequalities,

$$d(x,p) - \delta \le d(x,p') \le d(x,p) + \delta.$$

In particular, $d(p', \bar{p}) \le \delta$ and, thus, $d(p, f(p)) \le 2\delta$.

CASE 2. $d(x,p) \ge \min(d(x,z), d(x,y))$, which implies that $d(p,z) \le C$, $d(p,y) \le 2C$. Thus, $d(p, f(p)) \le 2C$.

CASE 3. $p' \in yz$ and $d(x,p) < \min(d(x,y), d(x,z))$. Since $p' \in yz$ and $d(y,z) \le C$, we get $d(p,y) \le \delta + C$. By the same argument as in Case 1,

$$d(x,\bar{p}) - C - \delta \le d(x,y) \le d(x,\bar{p}) + \delta + C,$$

with the point y playing the role of p' in Case 1. Thus, $d(y,\bar{p}) \le \delta + C$, which implies

$$d(p, f(p)) \le 2(\delta + C).$$ □

We will apply this construction in the following situation. Let $\Delta = \Delta x_1x_2x_3$ be a geodesic triangle in X, z be its δ-center, $y \in x_1x_2$ a point within distance δ from z. We then use the maps $f_1 : x_1z \to x_1y$, $f_2 : x_2z \to x_2y$ as in the lemma. These maps combine to define a map $f : x_1z \cup zx_2 \to x_1x_2$. We parameterize concatenation $\alpha = x_1z \star zx_2$ by its arc-length. Then the map f is monotonic (with respect to the parameterization), fixes the endpoints of α and satisfies

$$d(p, f(p)) \le 4\delta, p \in \alpha.$$

We, thus, obtain Corollary 1.81.

COROLLARY 1.81. *Suppose that z is a δ-center of a geodesic triangle $\Delta = \Delta x_1x_2x_3$ in a δ-hyperbolic space X. Then there exists a monotonic map*

$$f : x_1z \star zx_2 \to x_1x_2$$

such that $d(f, \mathrm{id}) \le 4\delta$.

1.13. Ideal boundaries

In the book we will be mostly working with Gromov's notion of ideal boundary (Gromov-boundary) $\partial_\infty Z$ of a space Z which is hyperbolic in the sense of Gromov. The elements of $\partial_\infty Z$ are equivalence classes $[z_n]$ of *Gromov-sequences* (z_n) in Z (see, e.g., [**DK18, V05**]):

A sequence (z_n) in Z is called a *Gromov-sequence* if

$$\lim_{m,n\to\infty} (z_m.z_n)_z = \infty$$

for some $z \in Z$.

REMARK 1.82. It is easy to see that for any metric space X and $u, v, w, x \in X$ we have $|(u.v)_w - (u.v)_x| \le d(w, x)$. Thus in the above definition if $z' \in Z$ were another point then $\lim_{m,n\to\infty}(z_m.z_n)_z = \infty$ iff $\lim_{m,n\to\infty}(z_m.z_n)_{z'} = \infty$. Hence, whether a sequence (z_n) is a Gromov sequence or not does not depend on the base point z. The same remark applies to the next definition too.

Two Gromov-sequences $(w_m), (z_n)$ are equivalent if

$$\lim_{m,n\to\infty} (w_m.z_n)_z = \infty.$$

One extends the definition of the Gromov-product to the elements ξ, ζ of the Gromov-boundary (equivalence classes of Gromov-sequences) ξ, ζ by

$$(\xi.\zeta)_z := \inf \liminf_{m,n\to\infty} (w_m.z_n)_z,$$

where the infimum is taken over all sequences $(w_n), (z_n)$ representing ξ, η. (Here we follow the definition in [**CDP90, ABC$^+$91, V05**], which differs (by $\le 2\delta$) from the one in [**BH99**] and [**Gd90**], where the supremum is taken instead of the infimum.)

Similarly, one defines

$$(w.\zeta)_z := \inf \liminf_{n\to\infty} (w.z_n)_z,$$

for $w \in Z, \zeta \in \partial_\infty Z$. One topologizes the space $\bar{Z} := Z \cup \partial_\infty Z$ so that a neighborhood basis of $z \in \partial_\infty Z$ in $\bar{Z}$ is given by the subsets (with fixed $p \in Z$)

$$U_{z,\epsilon} = \{w \in Z \cup \partial_\infty Z : (z.w)_p > \epsilon\}.$$

In particular, a sequence (z_n) in Z converges to $\zeta \in \partial_\infty Z$ if and only if (z_n) is a Gromov-sequence representing ζ.

DEFINITION 1.83. Suppose that Z is a Gromov-hyperbolic metric space. We will use the notation $\partial_\infty Z$ for the Gromov-boundary of Z, equipped with the above topology. For a subset $Y \subset Z$, we will use the notation

$$\partial_\infty(Y, Z)$$

for the accumulation set of Y in $\partial_\infty Z$, the *relative ideal boundary* of Y in Z.

We also define $\partial_\infty^{(2)} Z \subset (\partial_\infty Z \times \partial_\infty Z)/\tau$, the set of unordered pairs of *distinct* elements of $\partial_\infty Z$: The involution τ swaps the two factors of $\partial_\infty Z \times \partial_\infty Z$.

Another common definition of the *visual boundary* of a hyperbolic geodesic metric space Z uses equivalence classes of geodesic rays in Z: Two rays are equivalent if they are at finite Hausdorff distance from each other. The two definitions agree if Z is a proper metric space; see, e.g., [**BH99**, III.H.3]. For nonproper spaces one can also use quasigeodesic rays; see, e.g., [**BS00**, Section 5], [**V05**].

For our purpose, it suffices to observe that if $\gamma : \mathbb{R}_+ \to Z$ is a quasigeodesic ray in a geodesic hyperbolic space Z, then for each sequence (t_n) in $\mathbb{R}_+$ diverging to ∞, the sequence $(\gamma(t_n))$ is a Gromov-sequence in Z; and any two such sequences are equivalent. This is a simple consequence of stability of quasigeodesics (Lemma 1.54). Thus, γ defines a point $\xi \in \partial_\infty Z$. We will say that γ is *asymptotic* to ξ and that γ *joins* $p = \gamma(0)$ and ξ. We will use the notation $\gamma = p\xi$ if γ is a geodesic ray joining p and ξ. If γ is a bi-infinite quasigeodesic, then it defines two quasigeodesic rays $\gamma_\pm$ (the restrictions of γ to $\mathbb{R}_+$ and to $\mathbb{R}_-$) and these are asymptotic to points $\xi_\pm$, also denoted $\gamma(\pm\infty)$. A hyperbolic space Z is said to be a *visibility space* if any two distinct ideal boundary points are connected by a bi-infinite geodesic. For instance, each proper geodesic hyperbolic space is a visibility space. Even if Z is a nonproper geodesic metric space, each point in X can be joined to each point in $\partial_\infty Z$ by a quasigeodesic ray and any two distinct points in $\partial_\infty Z$ are connected by a bi-infinite quasigeodesic; see, e.g., [**BS00**, Section 5].

A *generalized geodesic triangle* in a δ-hyperbolic geodesic metric space Z is defined by taking a triple of points $z_1, z_2, z_3 \in \bar{Z}$ (such that no two ideal boundary points in this triple are equal) and connecting them by geodesics in Z, the *sides* of the triangle $\Delta z_1 z_2 z_3$. An ideal triangle is a generalized triangle with all three vertices in $\partial_\infty Z$.

Lemma 1.84 is an application of the slim triangle property in Z, cf. [**DK18**, Section 11.11].

LEMMA 1.84. *[Slim generalized triangle property] Suppose that Z is a δ-hyperbolic geodesic metric space. Then:*

(1) *Every generalized geodesic triangle Δ in Z with two nonideal vertices is 2δ-slim: Each side of Δ is contained in the 2δ-neighborhood of the union of the two other sides.*
(2) *Every generalized geodesic triangle Δ in Z with one nonideal vertex is 3δ-slim.*
(3) *Every ideal triangle in Z is 4δ-slim.*

PROOF.

(1) Let $\Delta = \Delta xy\zeta$, where $x, y \in Z, \zeta \in \partial_\infty Z$. Take diverging sequences $p_n \in x\zeta$, $q_n \in y\zeta$, where $x\zeta, y\zeta$ are the infinite sides of Δ, such that $d(p_n, q_n) \le C$, where C is a constant. Since the quadrilateral xp_nq_ny is 2δ-slim (Lemma 1.48), it follows that for each point $z \in y\zeta$, if n is sufficiently large, then $z \in yq_n$ and, furthermore, $d(z, xp_n \cup xy) \le 2\delta$. (The point z can be 2δ-close to the side p_nq_n only for finitely many values of n.) In particular, z lies in the 2δ-neighborhood of $xy \cup x\zeta$. The same argument proves that each point $z \in xy$ also lies in the 2δ-neighborhood of $xy \cup x\zeta$.

(2) Suppose that $\Delta = \Delta x\eta\zeta$, where η, ζ are in $\partial_\infty Z$ and $x \in Z$. As before, consider diverging sequences $p_n \in x\eta$, $q_n \in x\zeta$. Find points $p'_n \in \eta\zeta$, $q'_n \in \eta\zeta$ within distance C from p_n, q_n respectively. Since the pentagon with the vertices x, p_n, p'_n, q'_nq_n is 3δ-slim, it follows that each point $z \in x\zeta$ lies in the 3δ-neighborhood of the union $\eta\zeta \cup x\eta$.

(3) The proof for ideal triangles is similar (we use 4δ-slimness of hexagons in X) and is left to the reader. □

LEMMA 1.85. *Suppose that Z is a δ-hyperbolic geodesic metric space, $\gamma_i : \mathbb{R}_+ \to Z, i = 1, 2$, are geodesic rays within finite Hausdorff distance from each other. Then there exist*

T_1, T_2 (depending on γ_1, γ_2) such that

$$\mathrm{Hd}(\gamma_1([T_1, \infty)), \gamma_2([T_2, \infty))) \le 4\delta$$

and, moreover, $d(\gamma_1(T_1), \gamma_2(T_2)) \le 2\delta$.

PROOF. Since the rays are at finite Hausdorff distance, $\gamma_1(\infty) = \gamma_2(\infty) = \xi$ for some $\xi \in \partial_\infty Z$. Set $p := \gamma(0)$, $D := d(p_1, p_2)$. Consider the generalized geodesic triangle Δ in Z with the vertices $p_1 = \gamma_1(0)$, $p_2 = \gamma_2(0)$ and the third vertex at infinity, ξ. Then by the slim triangle property, every point $x_i \in \gamma_i(\mathbb{R}_+)$ is within distance 2δ from $\gamma_{3-i}(\mathbb{R}_+) \cup p_1p_2, i = 1, 2$.

Take $T_1 := D + 2\delta$. The triangle inequality implies that $x_1 := \gamma_1(T_1)$ is within distance 2δ from some point $x_2 = \gamma_2(T_2)$. Since the generalized triangle with the vertices x_1, x_2, ξ is 2δ-slim, it follows that every point of $\gamma_1([T_1, \infty))$ is within distance 4δ from a point of $\gamma_2([T_2, \infty))$ and vice versa. □

COROLLARY 1.86. *Suppose that c_1, c_2 are k-quasigeodesic rays in Z such that $c_1(\infty) = c_2(\infty)$. Then there exist T_1, T_2 such that*

$$\mathrm{Hd}(\gamma_1([T_1, \infty)), \gamma_2([T_2, \infty))) \le 4\delta + 2D_{1.54}(\delta, k).$$

Moreover, $d(\gamma_1(T_1), \gamma_2(T_2)) \le 2\delta + 2D_{1.54}(\delta, k)$.

A similar result holds for bi-infinite quasigeodesics:

LEMMA 1.87. *Suppose that X is a δ-hyperbolic space, α, β are bi-infinite L-quasigeodesics in X such that*

$$\xi_\pm = \alpha(\pm\infty) = \beta(\pm\infty) \in \partial_\infty X.$$

Then $\mathrm{Hd}(\alpha, \beta) \le D_{1.87}(L, \delta)$.

PROOF. Using Corollary 1.86, we find subrays $\alpha_\pm$ in α and $\beta_\pm$ in β which are asymptotic to $\xi_\pm$ respectively, such that the respective initial points $x_\pm$ (of $\alpha_\pm$) and $y_\pm$ (of $\beta_\pm$) satisfy

$$d(x_\pm, y_\pm) \le r := 2\delta + 2D_{1.54}(\delta, k).$$

Removing the above rays from the quasigeodesics α, β, we are left with two finite quasi-geodesic subsegments $\alpha_0 \subset \alpha$ (between the points $x_\pm$) and $\beta_0 \subset \beta$ (between the points $y_\pm$). Now, applying Lemma 1.55 to α_0, β_0, we get:

$$\mathrm{Hd}(\alpha_0, \beta_0) \le D_{1.55}(\delta, L, r).$$

Hence, taking $D_{1.87}(L, \delta) = \max(r, D_{1.55}(\delta, L, r))$, concludes the proof of the lemma. □

Each qi embedding $f : X \to Y$ of geodesic hyperbolic spaces has an extension to a map of Gromov-boundaries, a topological embedding $\partial_\infty f : \partial_\infty X \to \partial_\infty Y$. The combined map

$$f \cup \partial_\infty f : X \cup \partial_\infty X \to Y \cup \partial_\infty Y$$

is continuous at each point of $\partial_\infty X$; see, e.g., [**DK18**, Exercise 11.109]. In Chapter 8 we will discuss the existence of such an extension in the case of more general coarse Lipschitz maps of hyperbolic spaces.

1.14. Quasiconvex subsets

DEFINITION 1.88 (Quasiconvex subset). Let X be a geodesic metric space, $Y \subseteq X$ and let $\lambda \geq 0$. We say that Y is λ-quasiconvex in X if every geodesic with endpoints in Y is contained in the λ-neighborhood of Y. A subset $Y \subset X$ is said to be quasiconvex if it is λ-quasiconvex for some $\lambda \geq 0$.

This definition generalizes in the setting of path-metric spaces where $Y \subset X$ is λ-quasiconvex if there is a function $\eta : \mathbb{R}_+ \to \mathbb{R}_+$ converging to 0 as $\epsilon \to 0+$, such that each ϵ-geodesic $\gamma \subset X$ with endpoints in Y, satisfies

$$\gamma \subset N_{\lambda+\eta(\epsilon)}(Y).$$

Convex subsets of, say, Hadamard spaces, have several characteristic properties such as:

- Intersections of convex subsets are again convex.
- The nearest-point projection to a nonempty closed convex subset is well-defined and 1-Lipschitz.

Below we discuss analogues of these properties for quasiconvex subsets of geodesic hyperbolic spaces.

The following two lemmata are straightforward and we omit the proofs:

LEMMA 1.89. *Every geodesic triangle in a δ-hyperbolic geodesic metric space is δ-quasiconvex.*

LEMMA 1.90.

(1) *Suppose that A is a λ-quasiconvex subset of a δ-hyperbolic space X. Then the R-neighborhood $N_R(A)$ is $(R + 2\delta + \lambda)$-quasiconvex in X.*
(2) *Suppose that A is a λ-quasiconvex subset of a δ-hyperbolic space X and $B \subset X$ is such that* $\mathrm{Hd}(A, B) \leq R$. *Then B is $(2R + 2\delta + \lambda)$-quasiconvex in X.*

We next discuss the relation between the quasiconvexity and qi embeddings.

LEMMA 1.91. *If A is a λ-quasiconvex subset of a geodesic hyperbolic metric space (X, d), then:*

(1) *The metric space $(N_\lambda(A), d)$ is a quasi-path metric space.*
(2) *When we equip $N_\lambda(A)$ with the path-metric d_p induced from X, the inclusion map $N_\lambda(A) \to X$ is a $(1, 4\lambda)$-qi embedding.*

PROOF. If $\lambda = 0$, then (A, d) is actually a geodesic metric space isometrically embedded in (X, d) and there is nothing to prove. Thus, we assume that $\lambda > 0$.

(1) Consider points $x, y \in N_\lambda(A)$ and $a \in A, b \in A$ be points within distance λ from x, y respectively. Let $\gamma = ab$ be a geodesic in X contained in $N_\lambda(A)$. Then divide γ into subsegments $x_i x_{i+1}$, $1 \leq i \leq n-1$ of equal length $\leq \lambda$. Then the finite sequence

$$x = x_0, a = x_1, x_2, \dots, x_n = b, x_{n+1} = y$$

is a λ-path connecting x to y and

$$\sum_{i=0}^{n} d(x_i, x_{i+1}) \leq d(x, y) + 4\lambda.$$

(2) By the same argument, for any two points $x, y \in N_\lambda(A)$,

$$d_p(x, y) \leq d(a, b) + 2\lambda \leq d(x, y) + 4\lambda.$$

Therefore, the inclusion map $(N_\lambda(A), d_p) \to (X, d)$ is a $(1, 4\lambda)$-qi embedding. □

LEMMA 1.92. *Suppose that Y, X are geodesic metric spaces, X is δ-hyperbolic and $f : Y \to X$ is a K-qi embedding. Then $f(Y)$ is $\lambda_{1.92}(\delta, K) = D_{1.54}(\delta, K)$-quasiconvex in X.*

PROOF. For a geodesic $\gamma = y_1y_2 \subset Y$, $f(\gamma)$ is a K-quasigeodesic in X connecting $x_i = f(y_i), i = 1, 2$. Therefore, by Lemma 1.54,

$$\mathrm{Hd}(f(\gamma), x_1x_2) \leq \lambda = D_{1.54}(\delta, K) = 92K^2(K + 3\delta),$$

which implies that $x_1x_2 \subset N_\lambda(f(Y))$, i.e., $f(Y)$ is λ-quasiconvex. □

Lemma 1.93 is a converse to Lemma 1.92:

LEMMA 1.93. *Suppose that (X, d) is hyperbolic, Y is a geodesic metric space and $f : Y \to X$ is a uniformly proper map with quasiconvex image. Then f is a qi embedding.*

PROOF. By Lemma 1.91, if $A := f(Y)$ is λ-quasiconvex, then $(N_\lambda(A), d_p)$ is qi embedded in X. Since f is uniformly proper, Lemma 1.18 then implies that $f : Y \to (N_\lambda(A), d_p)$ is a quasi-isometry. Therefore, $f : Y \to X$ is a qi embedding as a composition of two qi embeddings. □

1.15. Quasiconvex hulls

A common source of quasiconvex subsets in hyperbolic spaces is given by *quasiconvex hulls* which we discuss in this section.

DEFINITION 1.94. For a subset $U \neq \emptyset$ of a geodesic metric space X we define the *quasiconvex hull* $\mathrm{Hull}(U)$ as the union of all geodesics in X connecting all pairs of points in U. The *ϵ-quasiconvex hull* $\mathrm{Hull}_\epsilon(U)$ of U in X is defined as the closed ϵ-neighborhood of the quasiconvex hull $\mathrm{Hull}(U)$. Thus, $cl(\mathrm{Hull}(U)) = \mathrm{Hull}_0(U)$.

LEMMA 1.95. *Suppose that X is a δ-hyperbolic geodesic metric space. Then for every $Y \subset X$ we have the following:*

(1) *$\mathrm{Hull}(Y)$ is a 2δ-quasiconvex in X.*
(2) *$\mathrm{Hull}_\epsilon(Y)$ is 4δ-quasiconvex in X for all $\epsilon \geq 0$.*

PROOF.

(1) Suppose $x_1, x_2 \in \mathrm{Hull}(Y)$. Then there are $a_1, b_1, a_2, b_2 \in Y$ such that $x_1 \in a_1b_1, x_2 \in a_2, b_2$. Since the quadrilateral $a_1b_1b_2a_2$ is 2δ-slim (1) follows.

(2) If $z_1, z_2 \in \mathrm{Hull}_\epsilon(Y)$ then there are $y_1, y_2 \in \mathrm{Hull}(Y)$ such that $d(z_i, y_i) \leq \epsilon$, $i = 1, 2$. Using the 2δ-slimness of the quadrilateral $y_1z_1z_2y_2$ we have $z_1z_2 \subset N_{2\delta}(y_1z_1 \cup y_1y_2 \cup y_2z_2)$. However, $y_iz_i \subset \mathrm{Hull}_\epsilon(Y)$, $i = 1, 2$ and $y_1y_2 \subset N_{2\delta}(\mathrm{Hull}(Y))$ by the first part of this lemma. Hence, $z_1z_2 \subset N_{2\delta}(\mathrm{Hull}_\epsilon(Y))$. □

By the construction, if U is λ-quasiconvex, then

$$\mathrm{Hull}_\epsilon(U) \subset N_{\epsilon+\lambda}(U). \tag{1.3}$$

LEMMA 1.96. *If X is a δ-hyperbolic geodesic metric space, then for any nonempty subset U of X, the inclusion map $Y = \mathrm{Hull}_\delta(U) \to X$ is a $(1, \epsilon)$-quasi-isometric embedding with $\epsilon = 6\delta$ and Y equipped with the path-metric induced from X.*

PROOF. Suppose first that $y_1, y_2 \in \mathrm{Hull}(U)$ belong to geodesics a_1b_1, a_2b_2 with $a_i, b_i \in U, i = 1, 2$. The proof is divided into two cases.

CASE 1. Suppose $a_1 = a_2$ or $b_1 = b_2$, i.e., the quadrilateral $a_1b_1b_2a_2$ is degenerate. Without loss of generality, assume that $a_1 = a_2$. Consider the δ-slim triangle $\Delta a_1b_1b_2$. If $y_1 \in N_\delta(a_1b_2)$ or $y_2 \in N_\delta(a_1b_1)$ then $d_Y(y_1, y_2) \le d_X(y_1, y_2) + 2\delta$. Otherwise, $y_1, y_2 \in N_\delta(b_1b_2)$. In this case, clearly $d_Y(y_1, y_2) \le d_X(y_1, y_2) + 4\delta$.

CASE 2. $a_1 \neq a_2$ and $b_1 \neq b_2$. We consider the triangle $\Delta a_1b_1b_2$. There are two possibilities. (i) If there is $z_1 \in a_1b_2$ with $d(y_1, z_1) \le \delta$ then z_1, y_2 are on two sides of the triangle $\Delta a_1a_2b_2$. Then by Case 1 we have $d_Y(z_1, y_2) \le d_X(z_1, y_2) + 4\delta$ whence $d_Y(y_1, y_2) \le d_X(y_1, y_2) + 6\delta$. (ii) Otherwise there is a point $z_2 \in b_1b_2$ with $d(y_1, z_2) \le \delta$. Then z_2, y_2 are on two sides of the triangle $\Delta a_2b_1b_2$ and as in (i) we have $d_Y(y_1, y_2) \le d_X(y_1, y_2) + 6\delta$. □

COROLLARY 1.97. *Suppose X is a δ-hyperbolic geodesic metric space, $U \subset X$ is a λ-quasiconvex subset and $R \ge \lambda + \delta$. Then the inclusion map $N_R(U) \to X$ is a $(1, 6\delta + R)$-quasi-isometric embedding, where $N_R(U)$ equipped with the path-metric induced from X.*

LEMMA 1.98. *Let X be δ-hyperbolic geodesic metric space; assume that $U, V \subset X$ are λ-quasiconvex subsets with nonempty intersection. Then:*

(1) *For every $R \ge \lambda$ the subset*

$$Y = N_R(U \cup V) \subset X$$

is 2δ-quasiconvex.

(2) *Furthermore, for each $R \ge \lambda + 3\delta$, the inclusion map $Y \to X$ is a $(1, 6\delta + R)$-quasi-isometric embedding where we equip Y with the path-metric induced from X.*

PROOF.

(1) Suppose $x \in N_R(U), y \in N_R(V)$. Then there exist $a \in U, b \in V$ with $d(x, a) \le R$, $d(b, y) \le R$. Let $c \in U \cap V$. By δ-slimness of triangles $xy \subset N_\delta(xc \cup yc)$. Similarly, $xc \subset N_\delta(xa \cup ac)$ and $yc \subset N_\delta(yb \cup bc)$. Thus $xy \subset N_{2\delta}(xa \cup ac \cup yb \cup bc)$. However we note that $xa \cup ac \cup yb \cup bc \subset Y$. Thus $xy \subset N_{2\delta}(Y)$.

(2) Observe that

$$N_R(U \cup V) = N_{R-\lambda}(N_\lambda(U \cup V)).$$

Thus, by the first of this lemma in combination with Corollary 1.97 it follows that the inclusion map $Y \to X$ is a $(1, 6\delta + R - \lambda)$-qi embedding. In particular it is a $(1, 6\delta + R)$-qi embedding. □

1.16. Projections

We now discuss properties of quasiconvex subsets in hyperbolic metric spaces. For a (complete) metric space (X, d), a closed subset $Y \subset X$ and a point $x_0 \in X$, we denote by $\bar{x}_0$ a nearest point of Y from x_0, if it exists. When this is the case for every point $x \in X$ then there is a map $X \to Y, x \mapsto \bar{x}$ and such a map will be called a nearest point projection map.

REMARK 1.99.

(1) If X is a proper metric space and Y is a closed subspace or X is a metric graph where each edge of length 1 and Y is a subgraph then a nearest point projection map always $X \to Y$ exists. However, a nearest point projection $\bar{x} \in Y$ of x in the above situation is not unique in general; and it is easy to find such an example where X is a metric graph and Y is a subgraph.
(2) On the other hand in the absence of the properness a nearest point may not exist as the example below suggests.

(3) However, for us the assumptions that X is complete or proper or that Y is closed are all essentially irrelevant: If they fail, then instead of nearest-point projections $\bar{x} \in Y$ of $x \in X$, we can use ϵ-nearest point projections, i.e., points $\bar{x} \in Y$ such that for all $y \in Y$,

$$d(x, y) \geq d(x, \bar{x}) - \epsilon.$$

Since we are interested in coarse-geometric aspects of metric spaces, such "almost" nearest point projections will work just as well. In what follows, we, therefore, will use the same fudge as for path-metric spaces: We will talk about nearest point projections without imposing the assumption that a subset is closed and the ambient space is complete or proper.

EXAMPLE 1.100. Suppose X is the metric graph with vertex set $V = (\{0, 1\} \times \mathbb{N}) \sqcup \{p\}$. Suppose that there is an edge of length 1 joining the following pairs of points:

- (a, i) and $(a, i + 1)$ for each $a \in \{0, 1\}, i \in \mathbb{N}$.
- $(0, i)$ and $(1, i)$ for each $i \in \mathbb{N}$.
- $(1, i)$ and p for each $i \in \mathbb{N}$.

This defines X as a metric graph. Now, for Y we take the union of the subgraph of X spanned by the vertices $\{0\} \times \mathbb{N}$ and the ball of radius $1 - 1/i$ centered at $(0, i)$ for all $i \in \mathbb{N}$. Then clearly X is a complete geodesic space and Y is a closed subspace but a nearest point projection of p on Y does not exist.

Notation for nearest point projections. Suppose that Y is a closed subset in a complete metric space X. We shall denote by $P_{X,Y}$ or P_Y (when the choice of X is clear) or sometimes simply by P (when the choices of X and Y are clear) a nearest-point projection $X \to Y$. We will see below (Lemma 1.102) that in hyperbolic spaces, nearest-point projections to quasiconvex subsets are "coarsely well-defined," thus, justifying the notation.

LEMMA 1.101. *Suppose that X is a geodesic metric space, $Y \subset X$, $x \in X$ and $\hat{x} \in Y$ is such that each geodesic $xy, y \in Y$, passes within distance r from $\hat{x}$. Then $d(\bar{x}, \hat{x}) \leq 2r$, where $\bar{x} = P_Y(x)$.*

PROOF. We take $y = \bar{x}$ and let $z \in xy$ be a point within distance r form $\hat{x}$. Since $d(x, \hat{x}) \geq d(x, y)$, it follows that $r \geq d(z, \hat{x}) \geq d(z, y)$. Therefore, $d(\hat{x}, y) \leq 2r$. □

We now turn to quasiconvex subsets of hyperbolic spaces.

LEMMA 1.102 (See, e.g., [**DK18**, Lemma 11.53]). *For each $\lambda \geq 0$ there is a constant $L_{1.102} = L_{1.102}(\delta, \lambda)$ such that the following holds:*

If X is a δ-hyperbolic metric space and Y is a λ-quasiconvex subset, then the projection map $P = P_{X,Y} : X \to Y$ is coarsely $L_{1.102}(\delta, \lambda)$-Lipschitz, i.e., for all $x, y \in X$ we have $d_X(P(x), P(y)) \leq L_{1.102}(\delta, \lambda)(d_X(x, y) + 1)$. In particular, P is coarsely well-defined: For different choices $\bar{x}_1, \bar{x}_2$ of points in Y nearest to x we have $d_X(P(x), P(y)) \leq L_{1.102}(\delta, \lambda)$.

REMARK 1.103. We will frequently use this lemma when Y is a geodesic in X, in which case $\lambda = \delta$ and $d(P(x), P(y)) \leq d(x, y) + 12\delta$, i.e., $L_{1.102}(\delta, \delta) = 12\delta$, compare Lemma 1.123. In general:

$$L_{1.102}(\delta, \lambda) = \max(2, 2\lambda + 9\delta);$$

see [**DK18**, Lemma 11.53].

Lemma 1.104 is a converse to Lemma 1.102.

LEMMA 1.104. *Suppose that a metric space (X, d_X) is δ-hyperbolic, $Y \subset X$ is a rectifiably connected subset (equipped with the induced path-metric d_Y) such that there exists a k-coarse-Lipschitz retraction*

$$P : X \to Y.$$

Then Y is $\lambda = \lambda_{1.104}(k, \delta)$-quasiconvex in X.

PROOF. The inclusion map $i : (Y, d_Y) \to (X, d_X)$ is 1-Lipschitz. Combining this with the existence of the retraction P, we conclude that i is a k-quasi-isometric embedding. For each $\epsilon > 0$ any two points $y_1, y_2 \in Y$ can be connected by a path α whose length is $\leq d_Y(y_1, y_2) + \epsilon$, in particular, α is $(1, \epsilon)$-quasigeodesic. We will take $\epsilon = 1$. The composition $i \circ \alpha$ in X is $L = k(1 + \epsilon)$-quasigeodesic. By the Morse lemma (Lemma 1.54), the image of $i \circ \alpha$ is within distance $\lambda = D_{1.54}(\delta, L)$ from a geodesic $\gamma = y_1 y_2$ in X connecting y_1 to y_2. Hence, γ is contained in the λ-neighborhood of Y. □

Lemma 1.105 is a converse to Lemma 1.101 in the context of hyperbolic spaces; it will be used repeatedly in the book.

LEMMA 1.105. *Suppose that X is δ-hyperbolic geodesic metric space, $Y \subset X$ is a λ-quasiconvex subset of X. Let $\bar{x} \in Y$ be a nearest-point projection of $x \in X$ to Y and let $y \in Y$ be an arbitrary point. Then:*

(i) *$\bar{x}$ lies within distance $\lambda + 2\delta$ from (any) geodesic $xy \subset X$.*
(ii) $\mathrm{Hd}(xy, x\bar{x} \cup \bar{x}y) \leq \lambda + 3\delta$.
(iii) *The concatenation $x\bar{x} \star \bar{x}y$ is a $2(\lambda + 2\delta)$-quasigeodesic.*

PROOF.
(i) We consider a geodesic triangle $\Delta x\bar{x}y$. Since this triangle is δ-slim, there exist points $z \in xy, z' \in x\bar{x}, z'' \in y\bar{x}$ such that $d(z, z') \leq \delta, d(z', z'') \leq \delta$. Then $d(z'', Y) \leq \lambda$. Since $\bar{x}$ is a nearest point to x in Y, it follows that $d(z', \bar{x}) \leq \lambda + \delta$, hence, $d(\bar{x}, z) \leq \lambda + 2\delta$.
(ii) As in the proof above, let $z \in xy$ be such that $d(x, z) \leq \lambda + 2\delta$. Then by δ-hyperbolicity of X, it is clear that $\mathrm{Hd}(\bar{x}y, zy) \leq \lambda + 3\delta$ and $\mathrm{Hd}(x\bar{x}, xz) \leq \lambda + 3\delta$. From these it follows that $\mathrm{Hd}(xy, x\bar{x} \cup \bar{x}y) \leq \lambda + 3\delta$.
(iii) Taking $z \in xy$ as above, we see that

$$d(x, y) \leq d(x, \bar{x}) + d(\bar{x}, y) + 2(\lambda + 2\delta).$$

Hence, $x\bar{x} \star \bar{x}y$ is a $(1, 2(\lambda + 2\delta))$-quasigeodesic. □

Below are several corollaries of Lemma 1.105.

COROLLARY 1.106. *If Y is a geodesic in X, then $\bar{x}$ lies within distance 3δ from (any) geodesic $xy \subset X$. In particular, for each geodesic triangle $\Delta = \Delta xyz$ in X, the projection $P_{yz}(x)$ is a 3δ-center of Δ.*

COROLLARY 1.107 (Almost nearest-point projection). *Suppose that Y is a geodesic in X and $y \in Y$ satisfies*

$$d(x, y) \leq d(x, \bar{x}) + C.$$

Then $d(\bar{x}, y) \leq C + 2\lambda + 4\delta$.

PROOF. Let $z \in xy$ be a point within distance $\lambda + 2\delta$ from $\bar{x}$ and let D denote $d(y, z)$. Then

$$d(x, \bar{x}) + C - D \geq d(x, y) - D = d(x, z) \geq d(x, \bar{x}) - (\lambda + 2\delta),$$

which implies that $D \leq C + \lambda + 2\delta$. Hence, $d(\bar{x}, y) \leq d(\bar{x}, z) + d(y, z) \leq C + 2\lambda + 4\delta$. □

COROLLARY 1.108. *Let $U, V \subset X$ be a pair of closed λ-quasiconvex subsets such that* $\mathrm{Hd}(U, V) \le D$. *Then $d(P_U, P_V) \le D_{1.108}(\delta, \lambda, D)$.*

PROOF. Let $u \in U, v \in V$ denote $P_{X,U}(x)$ and $P_{X,V}(x)$ respectively. Since $\mathrm{Hd}(U, V) \le D$, there exist points $u' \in V, v' \in U$ such that

$$d(u, u') \le D, \quad d(v, v') \le D.$$

Hence,

$$d(x, u') - D \le d(x, u) \le d(x, v') \le d(x, v) + D$$

By Corollary 1.107, $d(v, u') \le 2D + 2\lambda + 4\delta$. Hence, $d(u, v) \le D_{1.108}(\delta, \lambda, D) := 3D + 2\lambda + 4\delta$. □

COROLLARY 1.109. *Suppose that $Y \subset X$ is a λ-quasiconvex subset of a δ-hyperbolic space X and $x, \hat{x}$ are points in X, Y respectively connected by a geodesic $c = x\hat{x}$ such that:*
There exists a function $R \mapsto \hat{R}$ satisfying $y \in Y, d(y, c) \le R \Rightarrow d(y, \hat{x}) \le \hat{R}$.
Then $d(\hat{x}, P_Y(x)) \le \widehat{\lambda + 2\delta}$.

PROOF. Let $\bar{x} = P_Y(x)$. Then, by Lemma 1.105(1), $d(\bar{x}, c) \le \lambda + 2\delta =: R$. It then follows that $d(\bar{x}, \hat{x}) \le \hat{R} = \widehat{\lambda + 2\delta}$. □

COROLLARY 1.110. *Suppose that $V \subset U \subset X$ are two λ-quasiconvex subsets of X. Then*

$$d(P_{X,V}, P_{U,V} \circ P_{X,U}) \le C_{1.110}(\delta, \lambda).$$

PROOF. The proof is by repeated use of Lemma 1.105. For $x \in X$ set $x_1 := P_U(x), x_2 := P_V(x), x_3 := P_V(x_1)$. Consider the triangle $\Delta x x_1 x_2$. By Lemma 1.105(i), there is a point $x_1' \in x x_2$ such that $d(x_1, x_1') \le \lambda + 2\delta$ because $x_2 \in U$ and x_1 is a nearest point projection of x to U. Now we note that x_2 is a nearest point projection of x_1' on V. Hence there is a point $x_2' \in x_1' x_3$ such that $d(x_2, x_2') \le \lambda + 2\delta$. Since $d(x_1, x_1') \le \lambda + 2\delta$ and the $\Delta x_1 x_1' x_3$ is δ-slim, there is a point $x_3' \in x_1 x_3$ such that $d(x_2', x_3') \le \lambda + 3\delta$. Hence,

$$d(x_2, x_3') \le d(x_2, x_2') + d(x_2', x_3') \le 2\lambda + 5\delta.$$

Finally, we note that x_3 is a nearest point projection of x_2' on V too. Hence, $d(x_3', x_3) \le d(x_3', x_2) \le 2\lambda + 5\delta$. Thus

$$d(x_2, x_3) \le C_{1.110}(\delta, \lambda) := d(x_2, x_3') + d(x_3', x_3) \le 4\lambda + 10\delta. \qquad \square$$

The following variation on this corollary will be used in the proof of Corollary 1.136 and Theorem 3.3.

LEMMA 1.111. *Suppose that X is δ-hyperbolic, $U, V \subset X$ are, respectively, λ and μ-quasiconvex subsets. Set $\epsilon := \mu + 2\delta$. Then for every $r \ge r_{1.111}(\lambda, \mu, \delta) = \epsilon + \delta + \lambda$, if $W = N_r(U) \cap V$ is nonempty, then the distance between the restrictions to V of the projections $P_{X,W}, P_{X,U}$, is at most $\epsilon + \delta + r$.*

PROOF. Take $x \in V$ and consider points $\bar{x} = P_{X,W}(x), \hat{x} := P_{X,U}(x)$. Since $\bar{x} \in N_r(U)$, there exists a point $x' \in U$ such that $d(x', \bar{x}) \le r$. By Lemma 1.105, there exists a point $y \in xx' \cap B(\hat{x}, \epsilon)$. Since the triangle $\Delta x x' \bar{x}$ is δ-slim, the point y is within distance δ either from a point $z \in \bar{x}x'$ or from a point $z \in \bar{x}x$. In the former case,

$$d(\bar{x}, \hat{x}) \le \epsilon + \delta + r,$$

as claimed. Suppose, therefore, that $z \in \bar{x}x$. Since V is λ-quasiconvex, there exists $w \in V$ within distance λ from z. Thus,

$$d(\hat{x}, w) \le \epsilon + \delta + \lambda.$$

The assumption that $r \geq \epsilon + \delta + \lambda$ implies that $w \in N_r(U) \cap V = W$. In particular,

$$d(x, w) \geq d(x, \bar{x}).$$

Setting $t = d(x, z)$, $s = d(z, \bar{x})$, we obtain $d(x, \bar{x}) = t + s$ and, hence,

$$t + s = d(x, \bar{x}) \leq d(x, w) \leq t + \lambda.$$

It follows that $s \leq \lambda$ and, therefore, $d(\hat{x}, \bar{x}) \leq \epsilon + \delta + s \leq \epsilon + \delta + \lambda \leq r$. Thus, in both cases, $d(\bar{x}, \hat{x}) \leq \epsilon + \delta + r$. □

Lemma 1.112 is another variation on Corollary 1.110 and Lemma 1.105(iii).

LEMMA 1.112. *Let X be a δ-hyperbolic space, and let $Y_1 \supset Y_2 \supset \cdots \supset Y_n$ be a chain of λ-quasiconvex subsets of X. For a point $x \in X$ define inductively points $y_1 = P_{X,Y_1}(x), \ldots, y_n = P_{Y_{n-1},Y_n}(y_{n-1})$. Then:*

(1) *The concatenation*

$$xy_1 \star \ldots . \star y_{n-1}y_n$$

is an $L_{1.112}(\delta, \lambda, n)$-quasigeodesic in X.

(2) $d_X(P_{X,Y_n}(x), x_n) \leq D_{1.112}(\delta, \lambda, n)$.

PROOF. The proof is by induction on n. If $n = 1$, there is nothing to prove. Suppose that the claim holds for $n - 1$, we will prove it for n.

By Lemma 1.105(iii), the concatenation $xy_1 \star y_1y_n$ is a $2(\lambda + 2\delta)$-quasigeodesic in X. By the induction hypothesis, the concatenation $y_1y_2 \star \cdots \star y_{n-1}y_n$ is a $L(\delta, \lambda, n - 1)$-quasigeodesic. Now, Part 1 follows from the stability of quasigeodesics in hyperbolic spaces.

Part (2) of the lemma follows from the inductive application of Corollary 1.110. □

We conclude the section with two technical lemmata that will be used in Section 3.3.1. The lemmata generalize the obvious fact that if $Z \subset Y \subset X$ are inclusions of subsets in a metric space X such that Z is convex in X, then Z is also convex in Y, while if Z is convex in Y and Y is convex in X then Z is convex in X.

LEMMA 1.113. *Suppose that $Z \subset Y \subset X$ are inclusions of metric spaces such that Y is a δ-hyperbolic geodesic metric space, X is a geodesic metric space, $Z \subset X$ is λ-quasiconvex, and the inclusion map $Y \to X$ is an L-quasi-isometric embedding, where $\lambda \geq \frac{3}{2}\delta$. Then $Z \subset Y$ is λ'-quasiconvex with $\lambda' = 1500(L\lambda)^3$.*

PROOF. Pick any pair of points $p, q \in Z$, let pq be a geodesic in X contained in $N_\lambda(Z)$, let $\gamma : I = [a, b] \to X$ be its arc-length parameterization. Pick a maximal finite sequence of 1-separated points $x_i = \gamma(t_i)$ and project these points to a sequence $z_i = P_Z(x_i) \in Z$. Then the maps $[t_i, t_{i+1}) \to \{z_i\}$ define a $k = 2L\lambda$-quasigeodesic $\beta : I \to Z \subset Y$. Let β^* denote a geodesic in Y connecting p to q. By Theorem 1.54,

$$\text{Hd}(\beta, \beta^*) \leq 92k^2(k + 3\delta) \leq 92 \cdot 16(L\lambda)^3 \leq 1500(L\lambda)^3.$$ □

REMARK 1.114. One can give a faster proof of the lemma using the restriction to Y of the coarse Lipschitz retraction $P_{X,Z}$ (and quoting Lemma 1.104), but we prefer to get an explicit estimate.

LEMMA 1.115. *Suppose that $Z \subset Y \subset X$ are inclusions of metric spaces such that X, Y are geodesic metric spaces (with the path-metric on Y is induced by that of X), X, Y are δ_1 and δ_2-hyperbolic respectively, Y is λ_1-quasiconvex in X, $Z \subset Y$ is λ_2-quasiconvex, then Z is $\lambda_{1.115}(\lambda_1, \lambda_2, \delta_1, \delta_2)$-quasiconvex in X.*

PROOF. The projections $P_1 = P_{X,Y} : X \to Y, P_2 = P_{Y,Z} : Y \to Z$ are $L_{1.102}(\delta_i, \lambda_i)$-coarse Lipschitz retractions, $i = 1, 2$. Hence, their composition $P_2 \circ P_1 : X \to Z$ is an k-coarse Lipschitz retraction with

$$k = L_{1.102}(\delta_1, \lambda_1) \cdot L_{1.102}(\delta_2, \lambda_2).$$

Therefore, by Lemma 1.104, $Z \subset X$ is λ-quasiconvex for

$$\lambda = \lambda_{1.104}(k, \delta_1).$$

□

1.17. Images and pre-images of quasiconvex subsets under projections

In this section we discuss the extent to which images and pre-images of projections of quasiconvex subsets to quasiconvex subsets are again quasiconvex.

LEMMA 1.116. *Let $\alpha = xy \subset X$ be a geodesic in a δ-hyperbolic geodesic metric space, $Y \subset X$ a λ-quasiconvex subset and $\bar{x} = P_Y(x), \bar{y} = P_Y(y)$. Then $\bar{\alpha} = P_Y(\alpha)$ is $D_{1.116}(\delta, \lambda)$-Hausdorff close to the geodesic $\bar{x}\bar{y}$.*

PROOF.

(i) Take $z \in \alpha$. Then z lies within distance 2δ from a point $w \in x\bar{x} \cup \bar{x}\bar{y} \cup \bar{y}y$. If $w \in \bar{x}\bar{y}$ then $d(z, Y) = d(z, \bar{z})$ ($\bar{z} = P_Y(z)$) satisfies $d(z, Y) \le \lambda + 2\delta$. In particular, $d(\bar{z}, w) \le 4\delta + \lambda$. Suppose that $w \in x\bar{x}$. Then, $d(w, Y) = d(w, \bar{x})$ and, without loss of generality, $P_Y(w) = \bar{x}$. Lemma 1.102 implies that $d(P_Y(z), \bar{x}) = d(P_Y(z), P_Y(w)) \le (2\delta + 1)L_{1.102}(\delta, \lambda)$. The case $w \in y\bar{y}$ is handled by relabelling x and y. To conclude:

$$\bar{\alpha} \subset N_D(\bar{x}\bar{y}), \quad D = \max((2\delta + 1)L_{1.102}(\delta, \lambda), 4\delta + \lambda).$$

(ii) Consider a point $z \in \bar{x}\bar{y}$. The point z is within distance 2δ from some $w \in xy \cup x\bar{x} \cup y\bar{y}$. If $w \in xy$ then $d(w, P_Y(w)) \le 2\delta + \lambda$ and, hence, $d(z, P_Y(w)) \le 4\delta + \lambda$. Suppose that $w \in x\bar{x}$. First note that $d(w, Y) \le d(w, z) + d(z, \bar{z}) \le 2\delta + \lambda$. Then $d(z, \bar{x}) \le d(z, w) + d(w, \bar{x}) = d(z, w) + d(w, Y) \le 4\delta + \lambda$. Similarly in case $w \in y\bar{y}$ we have $d(z, \bar{y}) \le 4\delta + \lambda$. Hence, in either case, we have $d(z, \bar{\alpha}) \le 4\delta + \lambda \le D$.

To conclude, for

$$D = D_{1.116}(\delta, \lambda) := \max((2\delta + 1)L_{1.102}(\delta, \lambda), 4\delta + \lambda),$$

the geodesic $\bar{x}\bar{y}$ and the set $\bar{\alpha}$ are D-Hausdorff close. □

COROLLARY 1.117. *If $d(\bar{x}, \bar{y}) \le D$ then* $\operatorname{diam}(\bar{\alpha}) \le 2D_{1.116}(\delta, \lambda) + D$.

REMARK 1.118. If $\delta \ge 1$, then

$$D_{1.116}(\delta, \lambda) \le 3\delta(2\lambda + 9\delta).$$

COROLLARY 1.119. *Let X be a δ-hyperbolic space, $Y \subset X$ a λ-quasiconvex subset and $\gamma \subset X$ be a K-quasigeodesic connecting points x and y and $P = P_{X,Y}$. Then $P(\gamma)$ is $C_{1.119}(\delta, \lambda, K)$-Hausdorff close to the geodesic segment $P(x)P(y)$.*

PROOF. This corollary follows from the Morse lemma, Lemmata 1.102 and 1.120.

As another application of the lemma, we obtain Lemma 1.120.

LEMMA 1.120. *Suppose that Y, Z are λ_Y, λ_Z-quasiconvex subsets respectively in a δ-hyperbolic space X. Then the projection $P_Y(Z)$ is $D_{1.120}(\delta, \lambda_Y, \lambda_Z)$-quasiconvex in X.*

PROOF. Take points $y_i = P_Y(z_i), z_i \in Z, y_i \in Y, i = 1, 2$. By Lemma 1.116,

$$\mathrm{Hd}(y_1y_2, P_Y(z_1z_2)) \le D_{1.116}(\delta, \lambda_Y).$$

Thus, for every $y \in y_1y_2$ there exists $x \in z_1z_2$ such that $d(y, P_Y(x)) \le D_{1.116}(\delta, \lambda_Y)$. Since Z is λ_Z-quasiconvex in X, there exists $z \in Z$ such that $d(x, z) \le \lambda_Z$. By Lemma 1.102,

$$d(P_Y(x), P_Y(z)) \le (\lambda_Z + 1)L_{1.102}(\delta, \lambda_Y).$$

Putting together these inequalities, we obtain:

$$\begin{aligned} d(y, P_Y(Z)) \le d(y, P_Y(z)) &\le D_{1.120}(\delta, \lambda_Y, \lambda_Z) \\ &:= (\lambda_Z + 1)L_{1.102}(\delta, \lambda_Y) + D_{1.116}(\delta, \lambda_Y). \end{aligned}$$

□

REMARK 1.121. If $\delta \ge 1, \lambda \ge 1$ then

$$D_{1.120}(\delta, \lambda, \lambda) \le (2\lambda + 3\delta)(2\lambda + 9\delta).$$

Thus, projections (to uniformly quasiconvex subsets) in hyperbolic spaces send uniformly quasiconvex subsets to uniformly quasiconvex subsets. Lemma 1.122, which we add only for the completeness of the picture and which is not used elsewhere otherwise, establishes a similar statement for pre-images. We need to warn the reader that pre-images of quasiconvex subsets under projections need not be quasiconvex, the true statement is more subtle.

LEMMA 1.122. *Suppose that X is geodesic, δ-hyperbolic, Y, Z are λ_Y, λ_Z-quasiconvex subsets in X respectively, such that $Z \subset Y$. Then*

$$\mathrm{Hull}(P_Y^{-1}(Z)) \subset P_Y^{-1}(N_D(Z))$$

for some $D = D_{1.122}(\delta, \lambda_Y, \lambda_Z)$.

PROOF. Take two points $x, y \in P_Y^{-1}(Z)$ and a geodesic $\alpha = xy \subset X$ connecting these points. Then, for $\bar{x} = P_Y(x), \bar{y} = P_Y(y)$, by Lemma 1.116, $P_Y(\alpha)$ is $D_{1.116}(\delta, \lambda_Y)$-Hausdorff close to the segment $\beta = \bar{x}\bar{y}$. Since $\bar{x}, \bar{y}$ are in Z and Z is λ_Z-quasiconvex in $X, \beta \subset N_{\lambda_Z}(Z)$. Thus,

$$\alpha \subset P_Y^{-1}(N_D(Z)), \quad D = D_{1.122}(\delta, \lambda_Y, \lambda_Z) = D_{1.116}(\delta, \lambda_Y) + \lambda_Z.$$

□

For a pair of λ-quasiconvex subsets U, V in a hyperbolic space X there is a basic dichotomy: Either $P_U(V)$ has uniformly bounded diameter (in terms of λ and δ) or it is uniformly close to a quasiconvex subset of V. We will discuss this and related issues in more detail in Sections 1.19 and 1.21; for now, we prove this statement in the context of projections of geodesics to geodesics:

LEMMA 1.123. *For any $\delta \ge 0$ and $\lambda \ge 0$ there is are constants $D = D_{1.123}(\delta, \lambda)$ $R = R_{1.123}(\delta, \lambda)$ such that the following holds:*

(1) *Suppose X is a δ-hyperbolic geodesic metric space and $Y \subset X$ is λ-quasiconvex. Let $x, y \in X$ and let $\bar{x}, \bar{y} \in Y$ be respectively their nearest-point projections to Y. If $d(\bar{x}, \bar{y}) \ge D$ then $\bar{x}\bar{y} \subset N_R(xy)$. One can take $D = 2\lambda + 7\delta$ and $R = \lambda + 5\delta$.*
(2) *When Y is a geodesic, $\lambda = \delta$ and we can take: $D_{1.123}(\delta, \delta) = 8\delta$, $R_{1.123}(\delta, \delta) = 6\delta$.*

PROOF. We prove Part (1) and leave computations in Part (2) (as a special case) to the reader. By Lemma 1.105(i) there is a point $z \in x\bar{y}$ such that $d(\bar{x}, z) \le \lambda + 2\delta$. Now we consider the geodesic triangle $\Delta xy\bar{y}$. Then $z \in N_\delta(xy \cup y\bar{y})$. Suppose $z \in N_\delta(y\bar{y})$ and let $w \in y\bar{y}$ such that $d(z, w) \le \delta$. Then $d(w, \bar{x}) \le \lambda + 3\delta$. Since $\bar{y}$ is a nearest point of Y

from w it follows that $d(w,\bar{y}) \leq \lambda + 3\delta$. Thus it follows that $d(\bar{x},\bar{y}) \leq 2\lambda + 6\delta$. Hence, if $d(\bar{x},\bar{y}) \geq \delta + 2\lambda + 6\delta$ then $\bar{x},\bar{y} \in N_{\lambda+3\delta}(xy)$. Since geodesic quadrilaterals in X are 2δ-slim, in that case it follows that $\bar{x}\bar{y} \subset N_{\lambda+5\delta}(xy)$. Hence, we may take $D = 2\lambda + 7\delta$ and $R = \lambda + 5\delta$. □

1.18. Modified projection

In Section 5.1 we will need a minor modification of the projection $P_{X,Y}$ to quasiconvex subsets $Y \subset X$ in the setting when $Y = T = T_p(xyz)$ is a C-tripod and also in the setting of quasigeodesic tripods.

DEFINITION 1.124. Suppose that T is a tripod as above in a δ-hyperbolic geodesic space X. For a subset $U \subset X$, we define its *modified projection* $\bar{P}_{X,T}(U) = \bar{P}_T(U) \subset T$ as

$$\bar{P}_T(U) = \mathrm{Hull}_0(P_T(U)), \tag{1.4}$$

where the (closed) hull is taken with respect to the intrinsic path-metric of T.

If the nearest-point projection were continuous and U were compact, then, of course, $\bar{P}_T(U) = P_T(U)$.

LEMMA 1.125. *Suppose that U is a λ-quasiconvex subset of a δ-hyperbolic space X, $T = T_p(xyz) \subset X$ is a C-tripod. Then*

$$\mathrm{Hd}_X(P_T(U), \bar{P}_T(U)) \leq C_{1.125}(\delta,\lambda,C) = D_{1.120}(\delta,\delta,\lambda) + C + 3\delta.$$

PROOF. First of all, each tripod $T \subset X$ is δ-quasiconvex. Therefore, according to Lemma 1.120, the projection $P_T(U)$ is a $\lambda' = D_{1.120}(\delta,\delta,\lambda)$-quasiconvex subset of X. Hence, for points $x' \in xp, y' \in yp$ which belong to $P_T(U)$, the segment $x'y' \subset X$ is contained in the λ'-neighborhood of $P_T(U)$. Since p is a $C + 2\delta$-center of the tripod $T_p(x'y'z)$ (Lemma 1.76), the geodesic segment $\gamma \subset T$ connecting x' to y' is $C + 3\delta$-Hausdorff close to $x'y'$. Thus, each point $u \in \gamma$ is within distance $C + 3\delta$ from some $u' \in x'y'$ and, by the λ''-quasiconvexity of $P_T(U)$, there exists $v \in P_T(U)$ such that $d(v,u') \leq \lambda'$. By the triangle inequality, $d(u,v) \leq \lambda' + C + 3\delta$. Thus, $\bar{P}_T(U) \subset N_{\lambda'+C+3\delta}(P_T(U))$. Since $P_T(U) \subset \bar{P}_T(U)$, Lemma 1.128 follows. □

REMARK 1.126. Assuming that $\delta \geq 1$, we can take:

$$\begin{aligned} C_{1.125}(\delta,\lambda,C) &= C + 3\delta + (L_{1.102}(\delta,\delta) + 1)\lambda + D_{1.116}(\delta,\delta) \\ &\leq C + 3\delta + (12\delta + 1)\lambda + 3\delta(2\lambda + 9\delta) \\ &\leq C + 30\delta^2 + 19\delta\lambda. \end{aligned}$$

REMARK 1.127. As a special case, we will use the modified projection when the tripod T is a single geodesic segment and U is also a geodesic segment. The estimate on the Hausdorff-distance in this situation is better; we leave it to the reader to verify (analogously to the proof of Lemma 1.123) that

$$C_{1.125}(\delta,\delta,0) = 4\delta.$$

This estimate will be used in Section 5.2.

We will need a generalization of $\bar{P}$ and the lemma in the following setting: We let Y be a union of three rectifiable arcs $\alpha \cup \beta \cup \gamma$ in X, connecting points x, y, z in X to a certain point $p \in X$ and parameterized by their arc-length. Thus, Y, equipped with its intrinsic path-metric d_Y, is an abstract tripod. We assume that the inclusion map $(Y, d_Y) \to X$ is a K-qi embedding. Note that if α, β, γ are all geodesics in X and Y is a C-tripod T, then the inclusion map $(T, d_T) \to X$ is a 4δ-qi embedding. As before, for $U \subset X$ we define $\bar{P}_Y(U)$ as the closed convex hull of $P_Y(U)$ with respect to the metric d_Y. The next lemma follows from the fact that Y is uniformly Hausdorff-close to the geodesic tripod $T = T_p(xyz)$; we leave a proof of the reader:

LEMMA 1.128. *Suppose that U is a λ-quasiconvex subset of a δ-hyperbolic space X. Then*

$$\mathrm{Hd}_X(P_T(U), \bar{P}_T(U)) \le C_{1.128}(\delta, \lambda, K).$$

1.19. Projections and coarse intersections

A basic fact of convex geometry is that intersections of convex subsets are again convex. In the context of quasiconvex subsets U, V of a hyperbolic space, one needs to modify the notion of intersection. The most esthetically pleasing way to do so is to intersect R-neighborhoods of U and V. However, most useful for us will be *asymmetric* coarse intersections $N_R(U) \cap V$. In this section we discuss these in conjunction with the projections $P_U(V)$ and $P_V(U)$.

LEMMA 1.129 (Coarse intersections of quasiconvex subsets are quasiconvex). *Suppose that $Y_i \subset X$ are λ_i-quasiconvex subsets in a δ-hyperbolic space X, $i = 1, 2$. Then for every $\epsilon \ge \lambda_1 + \lambda_2 + 2\delta$, the intersection*

$$Y = N_\epsilon(Y_1) \cap Y_2$$

is $\lambda_{1.129}(\epsilon, \delta)$-quasiconvex, with $\lambda_{1.129} = \epsilon + 2\delta$.

PROOF. Take two points $x, y \in Y$ and let $x_1, y_1 \in Y_1$, be points within distance ϵ from x, y respectively. In view of the 2δ-slimness of quadrilaterals in X, for each $z \in xy$ either $d(z, x_1y_1) \le 2\delta$ or $d(z, x) \le 2\delta + \epsilon$ or $d(z, y) \le 2\delta + \epsilon$. In the last two cases, $z \in N_{2\delta+\epsilon}(Y)$. Suppose, therefore, that $d(z, x_1y_1) \le 2\delta$. By the λ_i-quasiconvexity of Y_i, there exist points $z_i \in Y_i$, such that $d(z, z_1) \le \lambda_1 + 2\delta$ and $d(z, z_2) \le \lambda_2$. In particular, $d(z_1, z_2) \le \lambda_1 + \lambda_2 + 2\delta \le \epsilon$, i.e., $z_2 \in Y$. Since

$$d(z, z_2) \le \lambda_2 \le \epsilon + 2\delta,$$

we conclude that Y is $(\epsilon + 2\delta)$-quasiconvex. □

Lemma 1.130 will be used in the proof of Theorem 3.3:

LEMMA 1.130. *Define the function*

$$R' := R_{1.130}(R, \lambda, \delta) = 2\lambda + 3\delta + R.$$

Let U_1, U_2 be λ-quasiconvex subsets of a δ-hyperbolic space X such that $d(U_1, U_2) \le R$. Then

$$P_{U_2}(U_1) \subset N_{R'}(U_1) \cap U_2$$

and

$$\mathrm{Hd}(P_{U_1}(U_2), P_{U_2}(U_1)) \le R'.$$

PROOF. The key is to show that for every $a \in U_1$ its nearest-point projection $b = P_{U_2}(a)$ lies in the R'-neighborhood of U_1.

Suppose $a_1 \in U_1, b_1 \in U_2$ are such that $d(a_1, b_1) \le R$. By Lemma 1.105, there exists a point $c \in ab_1$ within distance $\lambda + 2\delta$ from b. Since $d(a_1, b_1) \le R$, the δ-slimness of the triangle Δaa_1b_1 implies existence of a point $c_1 \in aa_1$ within distance $R + \delta$ from c. In view of the λ-quasiconvexity of U_1, c_1 belongs to the λ-neighborhood of U_1. Thus,

$$b \in N_{2\lambda+3\delta+R}(U_1) = N_{R'}(U_1).$$

Therefore, the distance from b to $P_{U_1}(b)$ is at most R', verifying the inclusion

$$P_{U_2}(U_1) \subset N_{R'}(P_{U_1}(U_2)).$$

The reverse inclusion is proven by switching the roles of U_1 and U_2. □

Continuing with the notation of Lemma 1.130, we get Corollary 1.131.

COROLLARY 1.131. *If $d(U_1, U_2) \le R$, then*

$$\mathrm{Hd}(P_{U_1}(U_2), N_{R'}(U_2) \cap U_1) \le 2R' \text{ and } \mathrm{Hd}(N_{R'}(U_1) \cap U_2, N_{R'}(U_2) \cap U_1) \le R'.$$

PROOF.

(1) According to Lemma 1.130, $P_{U_1}(U_2) \subset N_{R'}(U_2) \cap U_1$. Conversely, given $x \in N_{R'}(U_2) \cap U_1$, there exists $y \in U_2$ with $d(x, y) \le R'$. Hence, $d(y, P_{U_1}(U_2)(y)) \le R$, implying

$$d(x, P_{U_1}(U_2)(y)) \le 2R'.$$

(2) The second claim is clear and holds for arbitrary $R' \ge 0$ and arbitrary subsets of arbitrary metric spaces. □

Thus, we proved that if $d(U_1, U_2) \le R$, then all four subsets

$$P_{U_1}(U_2), P_{U_2}(U_1), N_{R'}(U_1) \cap U_2, N_{R'}(U_2) \cap U_1$$

are within Hausdorff distance $2R'$ from each other.

1.20. Quasiconvex subgroups and actions

In this section we discuss quasiconvexity in the context of subgroups of hyperbolic groups and, more generally, group actions.

DEFINITION 1.132. A subgroup H of a hyperbolic group G is said to be *quasiconvex* if it is a quasiconvex subset of a Cayley graph of G for a finite generating set. More generally, a (metrically) proper isometric action of a discrete group H on a geodesic hyperbolic metric space X is *quasiconvex* if one (equivalently, every) H-orbit in X is a quasiconvex subset in X.

LEMMA 1.133. *If the action of H on X is* quasiconvex *then H is finitely generated, the orbit map $H \to H \cdot x \subset X$ is a qi embedding and H is a hyperbolic group.*

PROOF.

(1) Quasiconvexity of $Hx \subset X$ implies that Hx is coarsely connected. Hence, by the Milnor–Schwarz lemma, H is finitely generated. We, thus, equip H with a word metric corresponding to a finite generating set.
(2) Metric properness of the action implies that the orbit map $o_x : H \to Hx \subset X$ is uniformly proper. Since the image of this map is a quasiconvex subset of X, the orbit map is a qi embedding (see Lemma 1.93).
(3) Since X is assumed to be hyperbolic, in view of Lemma 1.56, the existence of a qi embedding o_x implies hyperbolicity of H. □

We now discuss the notion of coarse intersection in relation to quasiconvex subgroups and actions.

For general quasiconvex subsets U, V of hyperbolic spaces X, coarse intersections $N_R(U) \cap V$ might not be Hausdorff-close to the actual intersections $U \cap V$: For instance, $U \cap V$ might be empty while for some $R > 0$ the intersection $N_R(U) \cap V$ might be unbounded. As a specific example, consider $X = \mathbb{R}$ (which is 0-hyperbolic) and 1-quasiconvex subsets

U, V consisting of odd/even integers respectively. Then $N_1(U) \cap V = V$, while $U \cap V = \emptyset$. Nevertheless, in the group-theoretic setting we have

LEMMA 1.134. *Let G be a hyperbolic group, U, V be quasiconvex subgroups in G with $W := U \cap V$. Then for every $r > 0$ there exists $R = R_{1.134}(G, r)$ such that*

$$W \subset W_r := V \cap N_r(U) \subset N_R(W).$$

PROOF. The proof is quite standard, cf. [**Gro93**, pp. 164–165] or [**GMRS98**, Lemma 2.6]. Suppose that $u \in U, v \in V$ satisfy $d_G(u, v) \le r$, i.e., $u^{-1}v \in B_G(1, r)$. Since the ball $B_G(1, r)$ is finite, there exists a finite set of pairs $(u_i, v_i), i = 1, \dots, n$ such that for any pair $u \in U, v \in V$ within distance r from each other, the product $u^{-1}v$ equals $u_i^{-1}v_i$ for some $i \in \{1, \dots, n\}$. We have

$$u^{-1}v = u_i^{-1}v_i \Rightarrow uu_i^{-1} = vv_i^{-1} = w \in W = U \cap V.$$

Hence, $u = wu_i, v = wv_i$ and, therefore, for

$$R := \max\{|v_i| : i = 1, \dots, n\}$$

we have $d_G(v, W) \le R$. □

As an immediate consequence we obtain the standard result on intersections of quasiconvex subgroups of hyperbolic groups (see, e.g., [**Sho91**] or [**BH99**, Proposition 4.13]):

COROLLARY 1.135. *If G is a hyperbolic group and U, V are quasiconvex subgroups of G, then $U \cap V$ is also a quasiconvex subgroup of G, H and V.*

PROOF. This is a combination of Lemmata 1.134, 1.129, and 1.90(2). □

COROLLARY 1.136. *In the setting of Lemma 1.134, the distance between the restrictions to V of the projections $P_{X,W}, P_{X,U}$, is at most $C_{1.136}(G, \delta, \lambda)$, where δ is the hyperbolicity constant of G and λ is the maximum of the quasiconvexity constants of U, V in G.*

PROOF. We take $r := r_{1.111}(\lambda, \lambda, \delta)$. According to Lemma 1.111, the restrictions to V of the projections

$$P_{G,W_r}, P_{G,U}$$

are within distance $\mu + 3\delta + r$. By Lemma 1.134, the subsets $W, W_r \subset X$ are $R = R_{1.134}(G, r)$-Hausdorff-close. Therefore, by Corollary 1.108, the distance between the projections $P_{G,W_r}, P_{G,W}$ is $\le D_{1.108}(\delta, \lambda, R)$. Thus, we can take

$$C_{1.136}(G, \delta, \lambda) := D_{1.108}(\delta, \lambda, R) + \mu + 3\delta + r.$$ □

Essentially the same proofs as above work in the more general setting, when we have a quasiconvex metrically proper action of a hyperbolic group G on a δ-hyperbolic geodesic metric space X, and $Y \subset X$ is a quasiconvex subset with locally finite G-orbit (see Definition 1.6).

PROPOSITION 1.137. *Let $H < G$ denote the stabilizer of Y in G, $x \in Y$. Then:*

(a) *There exists a function $R = R_{1.137}(x, r)$ such that*

$$Hx \subset Gx \cap N_r(Y) \subset N_R(Hx)$$

(b) *Hx is a quasiconvex μ-subset of X, $\mu = \mu_{1.137}(x, \delta, \lambda)$, where λ is the maximum of quasiconvexity constants (in X) of Gx and Y.*

(c) *H is a quasiconvex subgroup of G.*

(d) *The restrictions of $P_{X,Y}$ and $P_{X,Hx}$ to the orbit $Gx \subset X$ are within distance $C = C_{1.137}(x, \delta, \lambda)$.*

PROOF.

(a) Suppose that $d_X(gx, Y) \le r$ for some $g \in G$; equivalently, $g^{-1}Y \cap B(x, r) \neq \emptyset$. Due to local finiteness of the G-orbit GY, there exists $S \subset G$- a finite subset depending only on x and r, such that $g^{-1}Y = g_iY$ for some $g_i \in S$. Thus $gg_iY = Y$ and so $gg_i \in H$. It follows that $d(gx, Hx) \le d(gx, gg_ix) = d(x, g_ix)$. Therefore, if we let $R = R(x, r)$ be the maximum of the distances $d(x, g_i(x))$ taken over all $g_i \in S$. Then $d(gx, Hx) \le R$. This proves (a).

(b) We take $r := 2\lambda + 2\delta$. By Lemma 1.129, the coarse intersection $Gx \cap N_r(Y)$ is $\lambda_{1.129}(r, \delta)$-quasiconvex in X. On the other hand, by Part (a),

$$\mathrm{Hd}(Hx, Gx \cap N_r(Y)) \le R.$$

Therefore, by Lemma 1.90(2), the subset Hx is μ-quasiconvex in X with

$$\mu = 2R + 2\delta + \lambda_{1.129}(r, \delta).$$

(c) Since the actions of G and H on X are quasiconvex, the orbit maps $o_x : G \to X$, $o_x : H \to X$ are qi embeddings (see Lemma 1.133). From this, we conclude that H is qi embedded and, hence, is quasiconvex in G.

(d) The proof of this part is identical to that of Corollary 1.136 and we omit it. □

We assume now that X is hyperbolic and that for each point $x \in X$ and each ideal boundary point $\xi \in \partial_\infty X$, there exists a geodesic ray $x\xi$ connecting x to ξ (e.g., X is a proper geodesic metric space).

DEFINITION 1.138. Suppose that G acts isometrically and properly on X. A point $\xi \in \partial_\infty X$ is called a *limit point* of this action (or, simply, a limit point of G) if there exists a sequence $g_i \in G$ such that for some (equivalently, every) $x \in X$, the sequence $(g_i(x))$ converges to ξ. A limit point ξ is called *conical* if the sequence (g_i) can be chosen so that for some (equivalently, all) $x \in X, y \in X$, there exists a constant R such that $d(g_iy, x\xi) \le R$ for all i.

The proof of the following result (a *Beardon–Maskit criterion* for quasiconvexity) can be found for instance in Swenson's paper [**Swe01**] (cf. also [**Bow95, Bow99, Tuk98**]); we will only need the easier direction (every limit point of a quasiconvex action is conical):

THEOREM 1.139. *Suppose that X is a proper geodesic hyperbolic metric space. Then a proper isometric action of a discrete group $G \curvearrowright X$ is quasiconvex if and only if every limit point of G is conical.*

1.21. Cobounded pairs of subsets

Recall that in Definition 1.26 we defined Lipschitz-cobounded pairs of subsets of general metric spaces. Informally in a metric space X two subspaces Y, Z are Lipschitz cobounded if there are coarse Lipschitz retractions $r_Y : X \to Y$ and $r_Z : X \to Z$ such that the diameters of $r_Y(Z)$ and $r_Z(Y)$ are small. Below, we establish a characterization of cobounded pairs in hyperbolic spaces.

PROPOSITION 1.140 (Characterizations of cobounded pairs). *The following are equivalent for λ-quasiconvex subsets $Y, Z \subset X$ in a δ-hyperbolic geodesic metric space X:*

(1) *Y, Z are C_1-Lipschitz cobounded.*
(2) *For every R there exists $D = D(R)$ such that if*

$$a_i \in Y, b_i \in Z, i = 1, 2,$$

satisfy $d(a_i, b_i) \le R$, $i = 1, 2$, then $d(a_1, a_2) \le D, d(b_1, b_2) \le D$.

(3) *The diameters of nearest-point projections* $P_{X,Y}(Z)$, $P_{X,Z}(Y)$ *are* $\leq C_2$.

Moreover, once a constant C_i *(or a function* $D(R)$*) in one of the items is chosen, this, together with* δ *and* λ*, determines the constant/function in the other two items.*

PROOF. The implication (1)⇒(2) is proven in Lemma 1.27 for arbitrary subsets of arbitrary metric spaces, with

$$D = 2C_1(R + 1) + C_1$$

For the implication (2)⇒(3), consider points $a_i \in Y, b_i \in Z$ such that $a_i \in P_{X,A}(b_i), i = 1, 2$. By Lemma 1.123, if $d(a_1, a_2) \geq D_{1.123}(\delta, \lambda)$ then there exists $R = R_{1.123}(\delta, \lambda)$ such that

$$a_1a_2 \subset N_R(b_1b_2).$$

In that case, there are points $b_i' \in Z$ within distance $\leq R + \lambda$ from a_i, $i = 1, 2$. Then, by (2),

$$d(a_1, a_2) \leq D(R + \lambda).$$

Hence, we can take $C_2 = \max\{D_{1.123}(\delta, \lambda), D(R + \lambda)\}$.

For the implication (3)⇒(1), we can take the retractions

$$r_A := P_{X,A}, \quad r_B := P_{X,B}.$$

□

In view of this proposition, for quasiconvex subsets of hyperbolic spaces we will adopt the following terminology.

DEFINITION 1.141. A pair of subsets $Y, Z \subset X$ in a hyperbolic space X is C-cobounded if the diameters of the projections $P_{X,Y}(Z), P_{X,Z}(Y)$ are $\leq C$.

LEMMA 1.142. *Given* $\delta \geq 0$ *and* $\lambda \geq 0$ *there are constants* $R = R_{1.142}(\delta, \lambda)$ *and* $D = D_{1.142}(\delta, \lambda)$ *such that the following holds.*

Suppose X *is a* δ*-hyperbolic metric space and* $Y, Z \subset X$ *are two* λ*-quasiconvex and* R*-separated subsets. Then* Y, Z *are* D*-cobounded. In fact, one can take* $D = 2\lambda + 7\delta$ *and* $R = 2\lambda + 5\delta$.

PROOF. We will show that the choice of $D = D_{1.123}(\delta, \lambda) = 2\lambda + 7\delta$ and

$$R = \lambda + R_{1.123}(\delta, \lambda) = 2\lambda + 5\delta$$

works. Let $R_1 = R_{1.123}(\delta, \lambda)$, so that $R = \lambda + R_1$. Suppose the diameter of $P_{X,Z}(Y)$ is greater than or equal to D. Let $x, y \in Y$ be such that $d(P_{X,Z}(x), P_{X,Z}(y)) \geq D$. Then by Lemma 1.123 $P_{X,Z}(x) \in N_{R_1}(xy)$. But Y is λ-quasiconvex and $x, y \in Y$. It follows that $P_{X,Z}(x) \in N_R(Y)$. Thus if Y, Z are R-separated then the diameter of $P_{X,Y}(Z)$ and $P_{X,Z}(Y)$ are both less than D. □

A consequence of this lemma allows one to simplify the verification that two subsets are cobounded; namely, it suffices to check only that one projection is bounded.

COROLLARY 1.143. *Suppose that* $U, V \subset X$ *are* λ*-quasiconvex subsets in a* δ*-hyperbolic space.*

(1) *If* $\text{diam}(P_U(V)) \leq D$*, then* $\text{diam}(P_V(U)) \leq C = C_{1.143}(\lambda, \delta, D)$*, where* $D \leq C$*. In particular, the pair* (U, V) *is* C*-cobounded.*
(2) *If the pair* $U, V \subset X$ *is not* $D_{1.142}(\delta, \lambda)$*-cobounded then*

$$\text{Hd}(P_U(V), P_V(U)) \leq D_{1.143}(\delta, \lambda) = R_{1.143}(\delta, \lambda) = 2\lambda + 3\delta + R_{1.142}(\delta, \lambda).$$

PROOF. There are two cases to consider:

(i) If $d(U, V) \geq R = R_{1.142}(\delta, \lambda)$, then the pair (U, V) is $D_{1.142}(\delta, \lambda)$-cobounded by Lemma 1.142.

(ii) Suppose that $d(U, V) \le R = R_{1.142}(\delta, \lambda)$. Then by Lemma 1.130,

$$\mathrm{Hd}(P_U(V), P_V(U)) \le R' = 2\lambda + 3\delta + R.$$

Since $\mathrm{diam}(P_U(V)) \le D$, it follows that

$$\mathrm{diam}(P_V(U)) \le D + R'.$$

Taking $C := \max(D + R', D_{1.142}(\delta, \lambda))$, concludes the proof of a. □

REMARK 1.144. Note that $C = \max(D + 4\lambda + 8\delta, D_{1.142}(\delta, \lambda)) = D + 4\lambda + 8\delta$.

By the argument in Part (a1), since the pair $U, V \subset X$ is not $D_{1.142}(\delta, \lambda)$-cobounded, $d(U, V) < R = R_{1.142}(\delta, \lambda)$. Thus, as in Part (a2),

$$\mathrm{Hd}(P_U(V), P_V(U)) \le R' = 2\lambda + 3\delta + R = 2\lambda + 3\delta + R_{1.142}(\delta, \lambda).$$

REMARK 1.145. If U_1, U_2 are geodesics in X, $\lambda = \delta$ and, by Lemma 1.123, one can take $D_{1.142}(\delta, \delta) = 8\delta$ and $R_{1.143}(\delta, \lambda) = 12\delta$.

Another application of Lemma 1.142 is Corollary 1.146:

COROLLARY 1.146. *Suppose that λ-quasiconvex subsets $U_1, U_2 \subset X$ are not $D = D_{1.142}(\delta, \lambda)$-cobounded. Then*

$$P_{U_2}(U_1) \subset N_{4\lambda+8\delta}(U_1) \cap U_2.$$

PROOF. By Lemma 1.142, since U_1, U_2 are not D-cobounded, then

$$d(U_1, U_2) \le R = R_{1.142}(\delta, \lambda) = 2\lambda + 5\delta.$$

According to Lemma 1.130,

$$P_{U_2}(U_1) \subset N_{R'}(U_1) \cap U_2,$$

where $R' = 2\lambda + 3\delta + R = 4\lambda + 8\delta$. □

LEMMA 1.147. *Given $\delta \ge 0$, $\lambda \ge 0$ and $C \ge 0$, there exists a constant*

$$D = D_{1.147}(\delta, \lambda, C)$$

such that the following holds.

Suppose X is a δ-hyperbolic metric space and $U, V \subset X$ are two nonempty λ-quasiconvex and C-cobounded subsets. Then there are points $x_0 \in U_0 = P_U(V) \subset U$, $y_0 \in V_0 = P_V(U) \subset V$, such that $x_0y_0 \subset N_D(xy)$, for all $x \in U$ and $y \in V$.

PROOF. Since the pair (U, V) is C-cobounded,

$$\mathrm{diam}(V_0) \le C, \quad \mathrm{diam}(U_0) \le C.$$

Choose any pair of points $x_0 \in U_0$, $y_0 \in V_0$. Take $x \in U, y \in V$ and consider the points $\bar{x} = P_V(x) \in V_0, \bar{y} = P_U(y) \in U_0$. By Lemma 1.105, the points $\bar{x}, \bar{y}$ are within distance $\lambda + 2\delta$ from xy. Therefore,

$$\max(d(x_0, xy), d(y_0, xy)) \le \lambda + 2\delta + C$$

and, hence, we can take $D = \lambda + 4\delta + C$. □

COROLLARY 1.148. *Given $\delta \ge 0$ and $\lambda \ge 0$, there are constants $R = R_{1.148}(\delta, \lambda)$ and $D = D_{1.148}(\delta, \lambda)$ such that the following holds.*

Suppose X is a δ-hyperbolic metric space and $U, V \subset X$ are two λ-quasiconvex and R-separated subsets. Then there are points $x_0 \in U$, $y_0 \in V$ such that $x_0y_0 \subset N_D(xy)$, for all $x \in U$ and $y \in V$.

Proof. By Lemma 1.142, there exists $R = R_{1.142}$ such that the pair (U, V) is $C = D_{1.142}$-cobounded whenever U, V are R-separated. Now, the claim follows from Lemma 1.147. □

CHAPTER 2

Graphs of groups and trees of metric spaces

2.1. Generalities

We presume that the reader is familiar with the Bass–Serre theory. However, we briefly recall some of the concepts that we shall need. For details we refer the reader to Serre's book [**Ser03**, Section 5.3]. Recall that graphs were defined in Section 1.1, see Definition 1.1. In particular, recall that for an edge e of a graph, $o(e), t(e)$ refer to the origin and the terminus of e, and $\bar{e}$ refers to the same edge with opposite orientation. Following Serre, we shall denote by $|e|$ the edge e without orientation, so that $|e| = |\bar{e}|$. Formally one could let $|e| = \{e, \bar{e}\}$. Let $|E(\Gamma)|$ denote the set of all unoriented edges of a graph Γ.

Definition 2.1 (Graph of groups). A **graph of groups** $(\mathcal{G}, \Gamma)$ consists of the following data:

(1) A connected graph Γ.
(2) An assignment to each vertex $v \in V(\Gamma)$ (and edge $e \in E(\Gamma)$) of a group G_v (respectively G_e) together with injective homomorphisms $\phi_{e,o(e)} : G_e \to G_{o(e)}$ and $\phi_{e,t(e)} : G_e \to G_{t(e)}$ for all $e \in E(\Gamma)$, such that the following conditions hold.
   (i) $G_e = G_{\bar{e}}$,
   (ii) $\phi_{e,o(e)} = \phi_{\bar{e},t(\bar{e})}$ and $\phi_{e,t(e)} = \phi_{\bar{e},o(\bar{e})}$.

We shall refer to the maps $\phi_{e,v}$ as the *canonical maps* of the graph of groups. We shall refer to the groups G_v and G_e, $v \in V(\Gamma)$ and $e \in E(\Gamma)$ as *vertex groups* and *edge groups* respectively. For topological motivations of graph of groups and the following definition of the fundamental group of a graph of groups one is referred to [**SW79**] or [**Hat02**]. In the terminology of [**BH99**], a graph of groups is a covariant functor from the graph Γ (regarded as a small category with set of objects $E \sqcup V$ and the set of morphisms consisting of the maps o and t) to the category of groups, sending morphisms o, t to group-monomorphisms. Functorially, in the case when Γ is a tree, one can define the group G, the *fundamental group* of $(\mathcal{G}, \Gamma)$, or the *pushout* of the diagram $\mathcal{G}$, by a universal property. Namely, there exist monomorphisms $G_e \to G, G_v \to G$ such that the diagram

$$\begin{array}{ccc} G_e & \longrightarrow & G_v \\ & \searrow & \downarrow \\ & & G \end{array}$$

commutes, and, whenever we have a group H and a compatible collection of homomorphisms $G_e \to H, G_v \to H$ forming commutative diagrams

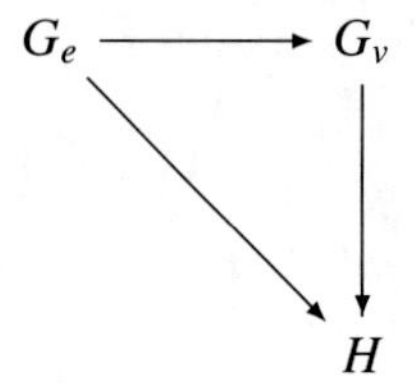

there is a unique homomorphism $G \to H$ forming commutative diagrams.

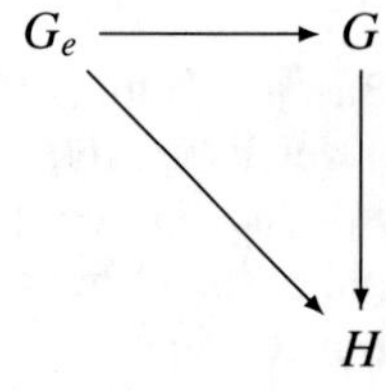

and

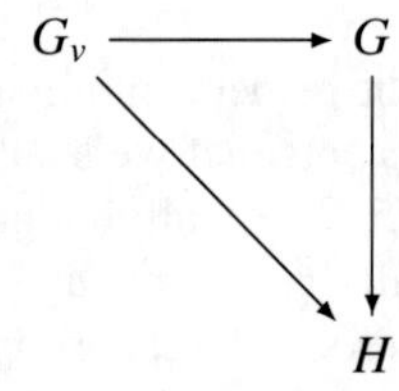

The general definition, i.e., when the graph is not a tree, is more complicated.

DEFINITION 2.2 (Fundamental group of a graph of groups). Suppose $(\mathcal{G}, \Gamma)$ is a graph of groups and let $S \subset \Gamma$ be a maximal (spanning) subtree. Then the fundamental group $G = \pi_1(\mathcal{G}, \Gamma, S)$ of $(\mathcal{G}, \Gamma)$ is defined in terms of generators and relators as follows:

The generators of G are the elements of the disjoint union of the generating sets of the vertex groups G_v, $v \in V(\Gamma)$ and the set $E(\Gamma)$ of *oriented* edges of Γ.

The relators are of four types:

(1) Those coming from the vertex groups;
(2) $\bar{e} = e^{-1}$ for all edge e;
(3) $e = 1$ whenever $|e|$ is an unoriented edge of S;
(4) $e\phi_{e,t(e)}(a)e^{-1} = \phi_{e,o(e)}(a)$ for all oriented edges e and $a \in G_e$.

The group G does not depend on the choice of S and it will be denoted $G = \pi_1(\mathcal{G})$ in what follows. We will also frequently suppress the letter Γ in the notation of a graph of groups.

DEFINITION 2.3 (Bass–Serre tree of a graph of groups). Suppose $(\mathcal{G}, \Gamma)$ is a graph of groups and let S be a maximal tree in Γ as in the above definition. Let $G = \pi_1(\mathcal{G}, \Gamma, S)$ be the fundamental group of this graph of groups. The *Bass–Serre tree*, denoted T, is the tree with the vertex set

$$\coprod_{v \in V(\Gamma)} G/G_v$$

and the edge set

$$\coprod_{e \in E(\Gamma)} G/G_e^e,$$

where $G_e^e = \phi_{e,t(e)}(G_e) < G_{t(e)}$. The origin/terminus maps are given by

$$t(gG_e^e) = gG_{t(e)}, \ o(gG_e^e) = gG_{o(e)}.$$

Note that whenever $|e|$ is a unoriented edge of S, then we have $e = 1$ in G. The group G acts on T via left multiplication.

Conversely, given an action *without inversions*[1] of a group G on a tree T, there exists a graph of groups $\mathcal{G}$ with $\pi_1(\mathcal{G}) \cong G$ such that T is equivariantly isomorphic to the Bass–Serre tree of $\mathcal{G}$, see [**Ser03**].

Since our main motivation comes from geometric group theory and, hence, finitely generated groups, we observe that for $G = \pi_1(\mathcal{G}, \Gamma, S)$ to be finitely generated, it suffices (not necessary!) to assume that each vertex group G_v is finitely generated and the graph Γ is finite. On the other hand, the edge groups G_e need not be finitely generated. Natural examples of the latter situation are given by amalgams

$$G = G_v \star_{G_e} G_w,$$

where G_e is an infinite rank free subgroup in two finitely-presented groups G_v, G_w: Such groups G are finitely generated but not finitely presentable. In the context of combination theorems for hyperbolic groups, one assumes that the graph Γ is finite, each vertex/edge group is hyperbolic and the monomorphisms ϕ are qi embeddings, i.e., have quasiconvex images.

Returning to the general setting with finitely generated vertex groups and finite graph Γ, we note that while it is meaningless to assume that the canonical maps ϕ are uniformly proper (as edge-groups do not have canonical qi classes of metrics), nevertheless, if we equip G_e with the pull-back word metric from $G_{o(e)}$, while $G_{t(e)}$ has a word metric coming from a finite generating set, then the monomorphism $G_e \to G_{t(e)}$ is uniformly proper. Since the graph Γ is finite, we conclude that each edge group has a left-invariant proper metric, such that the homomorphisms $\phi_{e,o(e)}$ and $\phi_{e,t(e)}$ are (η, L)-proper for some uniform function η and a constant L.

A *morphism* of graphs of groups, $\Psi : \mathcal{G} \to \mathcal{G}'$, consists of a morphism of the underlying graphs $\psi : \Gamma \to \Gamma'$ together with a collection of group homomorphisms

$$\Psi_v : G_v \to G_{\psi(v)}, v \in V(\Gamma), \quad \Psi_e : G_e \to G_{\psi(e)}, e \in E(\Gamma)$$

such that the following diagrams are commutative for $v = o(e)$ and $w = t(e)$ and their respective images $v' = \psi(v), w' = \psi(w), e' = \psi(e)$.

$$\begin{array}{ccc} G_e & \longrightarrow & G_{e'} \\ \downarrow{\scriptstyle \phi_{e,v}} & & \downarrow{\scriptstyle \phi_{e',v'}} \\ G_v & \longrightarrow & G_{v'} \end{array}, \qquad \begin{array}{ccc} G_e & \longrightarrow & G_{e'} \\ \downarrow{\scriptstyle \phi_{e,w}} & & \downarrow{\scriptstyle \phi_{e',w'}} \\ G_w & \longrightarrow & G_{w'} \end{array}$$

Given a graph of groups $(\mathcal{G}', \Gamma')$ and a graph-morphism $\psi : \Gamma \to \Gamma'$ from a connected graph Γ, there is a canonical *pull-back* graph of groups $(\mathcal{G}, \Gamma)$ and a morphism of graphs of groups $\Psi : \mathcal{G} \to \mathcal{G}'$, such that the underlying morphism of graphs $\Gamma \to \Gamma'$ equals ψ. In the special case when Γ is a connected subgraph of Γ', the graph of groups $(\mathcal{G}, \Gamma)$ is called the *restriction* of $\mathcal{G}'$ to Γ (see [**Bas93**, 2.15]). In this case, the Bass–Serre tree T of $(\mathcal{G}, \Gamma)$ admits a $G = \pi_1(\mathcal{G})$-equivariant embedding in the Bass–Serre tree T' of $(\mathcal{G}', \Gamma')$ and G equals the stabilizer of $T \subset T'$ in G'. We refer the reader to [**Bas93**] for further discussion of morphisms of graphs of groups.

In the book, on several occasions we will use the following definition from the theory of group actions on trees.

[1]Which means that if $g \in G$ preserves an edge $[v, w]$ of T, then it also fixes both v and w.

DEFINITION 2.4. An action of a group G on a tree T is said to be *k-acylindrical* if whenever a nontrivial[2] element $g \in G$ fixes element-wise an interval $J \subset T$, then J has length $\leq k$.

This terminology originates in Sela's paper [**Sel97**]. The definition of acylindrical actions on trees was later coarsified and generalized by Bowditch in [**Bow08**]; we will not use his generalization.

2.2. Trees of spaces

Each graph of groups yields a "tree of metric spaces" over its Bass–Serre tree; this was first formalized and used by Bestvina and Feighn in [**BF92**]. Below is our version of their definition.

We start with the simpler concept of a *tree of topological spaces*. One can regard a (simplicial) tree T (or a general graph) as a small category with object sets equal to $V(T) \sqcup E(T)$ and morphisms given by origin/terminus arrows. Then a tree of topological spaces over a tree T is a functor $\mathfrak{X}$ from T to the category of topological spaces. More explicitly:

DEFINITION 2.5. A tree of topological spaces over a tree T is a collection $\mathfrak{X}$ of nonempty topological spaces (vertex and edge spaces) $X_v, v \in V(T), X_t, t \in |E(T)|$, together with the following gluing data: Let $E^+(T) \subset E(T)$ be such that for any $e \in E(T)$ exactly one of e or $\bar{e}$ is in $E^+(T)$. Then for each oriented edge $e = [v.w]$ there are continuous *incidence maps* $f_{tv} : X_t \to X_v$ and $f_{tw} : X_t \to X_w$, where $t = |e|$. The *total space* X of $\mathfrak{X}$ is the mapping cylinder of the collection of these incidence maps, i.e., the quotient of the disjoint union

$$\coprod_{v \in V(T)} X_v \sqcup \coprod_{t \in |E(T)|} X_t \times [0,1]$$

by the equivalence relation

$$(x,0) \sim f_{tv}(x), (x,1) \sim f_{tw}(x), e = [v,w] \in E^+(T), t = |e|.$$

We note that the topology and the metric, as will be relevant later, of X is independent of the choice of $E^+(T)$. We will use trees of topological spaces in Section 8.9. For most of the book, we will work with trees of metric spaces defined below.

CONVENTION 2.6. In what follows for any oriented edge e we shall abuse notation and use X_e to denote $X_{|e|}$ and follow the convention that $f_{eo(e)} = f_{|e|o(e)}, f_{et(e)} = f_{|e|t(e)}$. Note that from this convention it follows that $f_{\bar{e}o(\bar{e})} = f_{et(e)}$ and $f_{\bar{e}t(\bar{e})} = f_{eo(e)}$. Moreover, a choice of $E^+(T)$ will be made implicitly whenever the total space is constructed.

Again, regarding a tree T as a small category, to *some degree*, a tree of metric spaces $\mathfrak{X}$ over a tree T is a functor from T to the coarse category C, see Remark 1.17. The actual definition is somewhat more restrictive:

DEFINITION 2.7 (Abstract tree of spaces). An *abstract tree of (metric) spaces* $\mathfrak{X}$ over a simplicial tree T, is a collection of nonempty metric spaces (vertex and edge-spaces) $X_v, v \in V(T), X_e, e \in E(T)$, together a collection of ψ-uniformly proper coarse L-Lipschitz *incidence maps* $f_{ev} : X_e \to X_v$ defined for each oriented edge $e = [v,w]$. The constant L and the function ψ are the *parameters* of the abstract tree of spaces $\mathfrak{X}$. The tree T is the *base* of $\mathfrak{X}$.

---

[2] In the literature, acylindricity is sometimes defined by requiring only that G-stabilizers of intervals of length $\geq k$ are finite, rather than trivial, subgroups.

We remark that it follows from the Convention 2.6 that for the edge $\bar{e} = [w, v]$ we have $f_{\bar{e}v} = f_{ev}$ and $f_{\bar{e}w} = f_{ew}$. Throughout the book, we will be assuming that all vertex-spaces X_v are path-metric spaces.

In view of the approximation lemmata (Lemma 1.36 and Lemma 1.37), one can replace general path-metric spaces X_v and incidence maps f_{ev} by (connected) metric graphs (equipped with graph-metrics) and simplicial incidence maps. Below we define the *total space* X of a tree of spaces and a projection $\pi : X \to T$. Thus, we will frequently refer to trees of spaces as $\mathfrak{X} = (\pi : X \to T)$, since the map π records the most important information about $\mathfrak{X}$.

In important class of trees of spaces consists of *metric bundles*. We refer to [**MS09**] for the general definition; for the purpose of this book Definition 2.8 will suffice.

DEFINITION 2.8. An abstract tree of spaces $\mathfrak{X} = (\pi : X \to T)$ is a *metric bundle* if the incidence maps f_{ev} are uniform quasi-isometries, i.e., there exists $\epsilon \geq 0$ such that for each edge $e = [v, w] \in E(T)$, the image $f_{ev}(X_e)$ (and, hence, $f_{ew}(X_w)$, by reversing the orientation on e) is ϵ-dense in X_v.

We note that in Definition 2.8, equivalence of f_{ev} being uniform quasi-isometries with their being uniformly coarsely dense follows from Lemma 1.19. However, while the main motivation for trees of spaces comes from graphs of groups, the main group-theoretic examples of metric bundles over trees are short exact sequences

$$1 \to K \to G \to H \to 1,$$

where K is a finitely generated group and H is a free group of finite rank.

DEFINITION 2.9. The *total space*, or the *push-out*, of a tree of spaces $\mathfrak{X}$ is a metric space X admitting a collection of L'-coarse Lipschitz maps $X_e \to X$, $e \in E(T)$, $X_v \to X, v \in V(T)$, and satisfying the following universal property: For every metric space Y and a compatible collection of L_1-coarse Lipchitz maps $X_e \to Y, X_v \to Y$, there exists a unique, up to a uniformly bounded error, L_2-coarse Lipschitz map $X \to Y$ forming diagrams which commute up to a uniform error C.

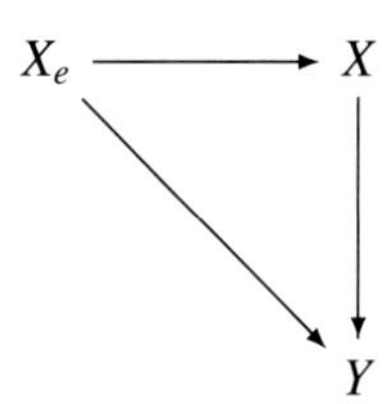

and

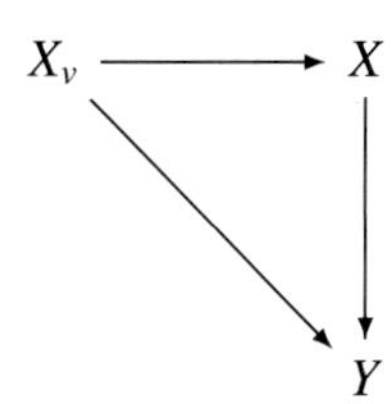

Here L_2 and C depend on L_1.

This definition implies uniqueness (up to a quasi-isometry) of the total space X. We will prove the existence of X below (Theorem 2.15).

DEFINITION 2.10. An abstract tree of spaces is said to be *retractible* (or *retractive*), if there exists a collection of (uniformly) L-coarse Lipschitz maps (retractions) $f_{ve} : X_v \to X_e$ defined for oriented edges $e = [v, w]$, which are uniformly coarse left-inverses to the incidence maps f_{ev}, i.e.,

$$\text{dist}(f_{ve} \circ f_{ev}, \text{id}_{X_e}) \leq \epsilon,$$

for some uniform constants $L \geq 1, \epsilon \in [0, \infty)$.

Under the retractibility assumption, the incidence maps are not only uniformly proper but are also uniformly quasi-isometric embeddings. While the definition is general, in this book, vertex and edge-spaces *mostly* will be uniformly hyperbolic, images of edge-spaces in vertex-spaces will be uniformly quasiconvex and the retractions f_{ve} will be given by nearest-point projections $P_{X_v,X_e} : X_v \to X_{ev}$.

Morphisms. Let $\mathfrak{X}, \mathfrak{X}'$ be abstract trees of spaces over trees T, T' respectively with the respective vertex/edge spaces $X_v, X'_{v'}, X_e, X'_{e'}$. A *morphism* of abstract trees of spaces $\mathfrak{X} \to \mathfrak{X}'$ is a graph-morphism $T \to T'$, $v \mapsto v', e \mapsto e'$, together with a collection of uniformly coarse Lipschitz maps, between respective vertex and edge-spaces

$$h_v : X_v \to X'_{v'}, \quad h_e : X_e \to X'_{e'}$$

such that the diagram (where the horizontal arrows are the incidence maps)

$$\begin{array}{ccc} X_e & \longrightarrow & X_v \\ \downarrow{\scriptstyle h_e} & & \downarrow{\scriptstyle h_v} \\ X'_{e'} & \longrightarrow & X'_{v'} \end{array}$$

commutes up to uniformly bounded errors. An *isomorphism* of abstract trees of spaces is an invertible morphism, equivalently, it is an isomorphism of trees $T \to T'$ and a collection of uniform quasi-isometries $X_v \to X'_{v'}, X_e \to X'_{e'}$.

REMARK 2.11. In this book we will be only considering *monic* morphisms of trees of spaces, i.e., ones for which the graph-morphism $T \to T'$ is injective and the maps $X_v \to X_{v'}, X_e \to X'_{e'}$ are ζ-proper for some uniform function ζ.

EXAMPLE 2.12. The most common examples of morphisms of trees of spaces used in this book are *subtrees of spaces*. Namely, suppose $S \subset T$ is a subtree, $\mathfrak{X}$ is a trees of spaces over T. Then the *pull-back* of $\mathfrak{X}$ over S is a tree of spaces $\mathfrak{Y}$ such that $Y_v = X_v, Y_e = X_e$, $v \in V(S), e \in E(S)$. The collection of identity maps $Y_v \to X_v, Y_e \to X_e$ defines a morphism of trees of spaces $\mathfrak{Y} \to \mathfrak{X}$. We will use the notation X_S for the total space of the tree of spaces $\mathfrak{Y}$. In the case when S is an interval (resp. tripod) in T, we will refer to X_S as an *interval-space* (resp., *tripod-space*).

While the above definition is the main definition used in this book, we now connect the notion of an abstract trees of spaces to the notion of a tree of spaces as defined by Mitra in [**Mit98**]. According to Mitra's definition, a tree of metric spaces is a path-metric space equipped with a certain auxiliary data, such as a map to a simplicial tree and a collection of maps to X from certain spaces. Below, we use the ℓ_1-metric on $X_e \times [v, w]$. Recall that for edges e of T, $\dot{e}$ denotes the edge minus its endpoints; below we will use the notation $m(e)$ for the midpoint of e.

DEFINITION 2.13 (Tree of spaces). A *tree of metric spaces*, denoted by $\mathfrak{X}$, is a path-metric space (X, d) equipped with a 1-Lipschitz surjective map $\pi : X \to T$ onto a simplicial tree T, such that there is a constant $L > 0$ and a function $\eta : [0, \infty) \to [0, \infty)$ satisfying $\eta(t) \geq t$ for all $t \in [0, \infty)$, with the following properties:

(1) For each $v \in V(T)$, the corresponding *vertex-space* $X_v := \pi^{-1}(v) \subset X$ is rectifiably connected.

(2) For every edge $e \in E(T)$, the *edge-space* $X_e := \pi^{-1}(m(e))$ is rectifiably connected. Every oriented edge $e = [v, w]$ comes equipped with an L-Lipschitz, η-proper map
$$f_e : X_e \times [v, w] \to X,$$
such that $f_e(X_e \times (v, w)) = \pi^{-1}((v, w))$ and that the diagram below commutes,

$$\begin{array}{ccc} X_e \times [v,w] & \xrightarrow[f_e]{} & \pi^{-1}([v,w]) \\ {\scriptstyle pr_2}\downarrow & \swarrow {\scriptstyle \pi} & \\ [v,w] & & \end{array}$$

where pr_2 denotes the second coordinate projection.

(3) We will use the notation f_{ev} for the composition
$$X_e \to X_e \times \{v\} \xrightarrow{f_e} X_v$$
and $X_{ev} := f_{ev}(X_e)$. We also assume that for an oriented edge $e = [v, w]$ we have $f_{ev} = f_{\bar{e}v}$ and $f_{ew} = f_{\bar{e}w}$.

By abusing the notation, we will denote a tree of spaces by $\pi : X \to T$.

REMARK 2.14.

(1) The Lipschitz condition is absent in Mitra's paper [**Mit98**], but it holds in all natural examples. On the other hand, Mitra assumes that each restriction $f_e|_{X_e \times \dot{e}}$ is an isometry onto $\pi^{-1}(\dot{e})$, equipped with its path-metric induced from X. We find this assumption unnecessarily restrictive.

(2) Mitra also assumes that inclusion maps $X_v \to X$ to be ζ-proper for some function ζ. We will see below that this is a consequence of uniform properness of the maps $X_e \to X_v$.

Observe that each tree of spaces $\mathfrak{X}$ yields naturally an abstract tree of spaces $\mathfrak{X}^{ab}$, the *abstraction* of $\mathfrak{X}$, with the incidence maps f_{ev}. Theorem 2.15 is a converse to this *abstraction* procedure.

THEOREM 2.15 (An existence theorem for trees of spaces). *For each abstract tree of spaces $\mathfrak{X}$ over a tree T, there exists a (unique up to an isomorphism) tree of spaces $(\pi : X \to T)$, called a* concretization *of $\mathfrak{Y}$, such that $\mathfrak{X}$ is isomorphic to the abstraction $\mathfrak{X}^{ab}$ of $(\pi : X \to T)$. The total space X of $\mathfrak{X}$ satisfies the universal property in the Definition* 2.9.

PROOF. Our proof mimics the definition of the underlying topological space (equipped with the weak topology) of a cell complex, where the latter is defined via an inductively defined collection of attaching maps. We let X denote the topological space obtained by attaching the products $X_e \times [0, 1]$ to the disjoint union
$$\mathcal{X} = \coprod_{v \in V(T)} X_v$$
via the attaching maps $f_{ev} : X_e \times \{v\} \to X_v, f_{ew} : X_e \times \{w\} \to X_w$, $e = [v, w] \in E(T)$. In other words, X is the mapping cylinder
$$Cyl(f : \mathcal{X}_E \to \mathcal{X}_V)$$

of the map

$$f : \mathcal{X}_E := \coprod_{e\in E(T)} X_e \to \mathcal{X}_V := \coprod_{v\in V(T)} X_v,$$

given by the collection of incidence maps f_{ev}. For each edge e of T we will identify $X_e \times \dot{e}$ with its image in X.

We define *admissible* paths in X (see Section 1.4) to be the continuous maps $c : [a,b] \to X$ which are concatenations of *vertical* paths, which are rectifiable (with respect to the metrics on vertex-spaces) paths contained in the vertex-spaces of X and *horizontal* paths, which are rectifiable paths contained in the intervals of the form $x \times [0,1]$, $x \in X_e, e \in E(T)$.[3] For every admissible path c, we let $length(c)$ be the sum of measures of lengths of its vertical and horizontal components. We leave it to the reader to verify that this defines a *length-structure* on X and, hence, a path-metric d. *We retopologize X using this path-metric.* By the construction, each inclusion map $X_e \times \dot{e} \to X$ is an isometry to its image, each vertex-space is rectifiably–connected in X, each inclusion map $X_v \to X$ is 1-Lipschitz and the projection map

$$\pi : X \to T$$

is 1-Lipschitz as well. The verification that the space X satisfies the universal property is rather straightforward. Given a collection of compatible coarse L-Lipschitz maps $h_v : X_v \to Y, h_e : X_e \to Y$ to a metric space Y, we define a map $h : X \to Y$ by sending each open interval $\{x\} \times (0,1) \subset X_e \times (0,1)$ to the point $h_e(x)$. The uniqueness of h (up to a bounded error) follows from the fact that the union

$$\coprod_{v\in V(T)} X_v \sqcup \coprod_{e\in E(T)} X_e$$

forms a 1/2-net in X. We will leave it to the reader to check that $\mathfrak{X}$ is isomorphic to the abstraction of $(\pi : X \to T)$. □

REMARK 2.16.

(1) A definition similar to our abstract tree of spaces and a construction analogous to the one in the proof of Theorem 2.15 appear in the work of Cashen and Martin [**CM17**, 2.4]. However, they work in the category of proper metric spaces and the metric spaces they produce do not satisfy all the properties in Definition 2.13 and, hence, we cannot directly use their work.

(2) Throughout the book, we will work with geometric realizations of trees of spaces constructed in the proof of Theorem 2.15. In particular, the inclusion maps

$$X_v \to X$$

are 1-Lipschitz. For every edge $e = [u,v] \in E(T)$ we will be frequently using the path-metric spaces

$$X_{uv} = X_{[\![u,v]\!]} = \pi^{-1}([\![u,v]\!]).$$

The inclusion maps $X_u \to X_{uv} \leftarrow X_v$ are also 1-Lipschitz.

(3) One drawback of our construction is that even if vertex and edge-spaces are complete and geodesic, the tree of spaces we construct in the proof of Theorem 2.15 is a only a path-metric space and, a priori, is not a geodesic metric space and need not be complete. There are two ways to rectify this issue which we describe below.

[3]Our usage of the vertical/horizontal terminology is opposite to the one in [**Gau16**].

(a) In the book, when we say "a geodesic" we really mean a path which is ϵ-short for a suitably chosen sufficiently small $\epsilon > 0$. Similarly, when dealing with nearest-point projections, we frequently project to nonclosed subsets. Then a nearest-point projection of $x \in X$ to $Y \subset X$ is a point $\bar{x} \in Y$ such that for a suitable chosen, sufficiently small $\epsilon > 0$,

$$d(x, \bar{x}) \leq d(x, Y) + \epsilon.$$

(b) For a reader uncomfortable with such a fudge, we describe an alternative approach to rectifying the issue with geodesics and nearest-point projections.

First of all, as we noted earlier (Lemmata 1.36, 1.37) without loss of generality, we may assume that all vertex-spaces and edge-spaces X_v, X_e are connected graphs equipped with standard graph-metrics. We will replace each X_e with its vertex set. Then the space X defined in the proof of Theorem 2.15 is a connected graph and the path-metric on this graph defined in the proof is the standard graph-metric. The drawback of this approach is the need to keep track of combinatorial issues which, are, ultimately, irrelevant.

From now on, we will work with abstract trees of spaces $\mathfrak{X}$ and their concretizations $\pi : X \to T$. The metric space X is the *total space* of $\mathfrak{X}$. There is nothing particularly canonical about our choice of X in this construction, it is just something we find convenient to work with. The reader could alternatively work for instance with, say, the ℓ_1-metric coming from the products $X_e \times [v, w]$ in the mapping cylinder X. In fact, most of our arguments deal with vertex-spaces and pull-backs X_{vw}: We will be using the fact that the natural inclusion maps $X_v \to X_{vw} \leftarrow X_w$ are 1-Lipschitz and either uniformly proper or, for trees of hyperbolic spaces, uniform qi embeddings.

Example 2.17. One motivation for our construction of X comes from Cayley graphs of fundamental groups G of graphs of groups. We assume that $(\mathcal{G}, Y)$ is a finite graph of finitely generated groups, $S \subset Y$ is a spanning tree, and $G = \pi_1(\mathcal{G}, Y, S)$ is the fundamental group. We will identify S with a subtree in the Bass–Serre tree T of $(\mathcal{G}, Y)$. Then form a graph Γ using the generators of G as described in Definition 2.2, except:

(a) We fix an orientation of the edges of Y and use only one generator for each edge (not two).

(b) We use the given generating sets of the vertex-groups G_v instead of the entire G_v.

Thus, in the graph Γ there are vertical edges (corresponding to translates $\Gamma_v, v \in V(T)$, of Cayley graphs of vertex groups) and horizontal edges (corresponding to the generators coming from the edges of Y). The vertex-spaces X_v are, then the graphs Γ_v. The edge-spaces are the translates of the edge-groups, gG_e, $g \in G$, $e \in E(Y)$. The incidence maps f_{ev}, f_{ew} for the oriented edges $e = [v, w]$ in S come from the monomorphisms $\phi_{e,o(e)}$ and $\phi_{e,t(e)}$. For general edges e of T (which are translates of the edges $e' \in S$), the incidence maps are obtained by composing with the action of G by left multiplication. Thus, we obtain a tree of spaces $\mathfrak{X}$ over T with vertex-spaces isometric to Cayley graphs of the vertex-groups G_v, $v \in V(Y)$, and edge-spaces isometric to edge-groups $G_e, e \in E(Y)$, with metrics obtained via pull-backs of word-metrics on the incident vertex-groups G_v, $v = t(e)$. Note that in the Cayley graph of G there are no edges corresponding to generators of the edge-groups. This is consistent to our use of only horizontal paths over the edges of T in the construction of the total space X in the proof of Theorem 2.15. We leave it to the reader to check that the Cayley graph Γ as above is G-equivariantly isometric to the total space X of the tree of spaces $\mathfrak{X}$ defined in the proof of Theorem 2.15.

PROPOSITION 2.18. *There exists a continuous function $\eta_{2.18}$ depending on the parameters of an abstract tree of spaces $\mathfrak{X}$, such that for every subtree $S \subset T$, the inclusion map*

$$X_S \to X$$

is an $\eta_{2.18}$-uniformly proper embedding.

PROOF. The key case to understand is when T has a single edge $e = [u, v]$ and $S = \{u\}$. We let Y denote the total space of the corresponding tree of spaces. It suffices to estimate (from below, in terms of $d_{X_u}(x, x')$) lengths of paths c in Y connecting $x = x_1, x' = x_n \in X_u$, such that c is a concatenation of the form

$$c(x_1, y_1) \star c(y_1, z_1) \star c(z_1, z_2) \star c(z_2, y_2) \star c(y_2, x_2) \star c(x_2, x_3) \star \cdots \star c(y_n, z_n),$$

where $x_i = f_{eu}(y_i), z_i = f_{ev}(y_i)$ and paths $c(x_i, y_i), c(y_i, z_i)$ are horizontal, while the paths $c(x_j, x_{j+1}), c(z_k, z_{k+1})$ are vertical geodesics in the vertex-spaces X_u, X_v. The lengths of this path is

$$\text{length}(c) = \sum_{i=\text{even}} d_{X_u}(x_i, x_{i+1}) + n + \sum_{j=\text{odd}} d_{X_v}(z_j, z_{j+1}).$$

Assume that $\text{length}(c) \leq D$. Then $n \leq D$ and $d_{X_v}(z_j, z_{j+1}) \leq D$ for each odd index j. We have (for j odd):

$$L^{-1} d_{X_u}(x_j, x_{j+1}) \leq d_{X_e}(y_j, y_{j+1}) \leq \psi(d_{X_v}(z_j, z_{j+1}))$$

and, hence,

$$d_{X_u}(x_j, x_{j+1}) \leq L\psi(d_{X_u}(z_j, z_{j+1})) \leq L\psi(D).$$

Thus, the concatenation c_u of vertical geodesics $[x_i x_{i+1}]_{X_u}$ connecting x to x' has total length $\text{length}(c_u)$ satisfying

$$\begin{aligned}
d_{X_u}(x, x') \leq \text{length}(c_u) &= \sum_{j=\text{odd}} d_{X_u}(x_j, x_{j+1}) + \sum_{i=\text{even}} d_{X_u}(x_i, x_{i+1}) \\
&\leq L \sum_{j=\text{odd}} \psi(D) + \sum_{i=\text{even}} d_{X_u}(x_i, x_{i+1}) \\
&\leq LD\psi(D) + D.
\end{aligned}$$

It follows that $d_{X_u}(x, x') \leq LD\psi(D)+D$ and, hence, the inclusion map $X_u \to X_{uv}$ is η-proper, for $\eta(D) := D(L\psi(D) + 1)$.

REMARK 2.19. Assuming that the map $f_{ev} : X_e \to X_v$ is an L-qi embedding (which eventually will be our assumption for trees of hyperbolic spaces), we obtain a better estimate:

$$\begin{aligned}
d_{X_u}(x, x') \leq \text{length}(c_u) &= \sum_{j=\text{odd}} d_{X_u}(x_j, x_{j+1}) + \sum_{i=\text{even}} d_{X_u}(x_i, x_{i+1}) \\
&\leq \sum_{i=\text{even}} d_{X_u}(x_i, x_{i+1}) + L^2 \sum_{j=\text{odd}} d_{X_v}(z_j, z_{j+1}) + L^3 n \\
&\leq L^3 \,\text{length}(c).
\end{aligned}$$

Thus, we conclude that each inclusion map $X_u \to X_{uv}$ in this case is an $(L^3, 0)$-qi embedding.

We now deal with the general case. Consider an admissible path $\beta : [0,1] \to X$ connecting $x, y \in X_S$. The projection $\pi \circ \beta$ is a path p in T connecting $\pi(x)$ to $\pi(y)$ whose length is $\le$ length(β). Without loss of generality, we may assume that $\pi(x), \pi(y)$ are vertices in S and p is a simplicial path in T. We now construct inductively a sequence of paths

$$\beta_0 = \beta, \beta_1, \ldots, \beta_n$$

in X with simplicial projections to T, all connecting x to y, such that:

(1) β_n is a path in X_S.
(2) The length of $\pi \circ \beta_{i+1}$ is at most length$(\pi \circ \beta_i) - 2$.
(3)
$$\text{length}(\beta_{i+1}) \le \eta(\text{length}(\beta_i)),$$
where $\eta(D) := D(L\psi(D) + 1)$ as above.

Assume that β_i is defined. If this path is contained in X_S, then $n = i$ we are done. Otherwise, there exists an edge $e = [v, w]$ in the tree $\pi\beta_i([0,1])$ such that β contains a subpath β' connecting points $x', y' \in X_v$ and contained in the subspace X_{vw}. We then replace β' with a geodesic in X_v connecting the endpoints x', y' of β'. By the above estimate in X_{vw},

$$d_{X_v}(x', y') \le \eta(\text{length}(\beta'))$$

and, hence, the new path β_{i+1} satisfies the required conditions.

Clearly, $n \le$ length(β), hence,

$$\text{length}(\beta_n) \le \eta^{(n)}(\text{length}(\beta)),$$

where $\eta^{(n)}$ is the n-fold iteration of the function η. Therefore, for $\eta_{2.18} = \eta^{(n)}$,

$$n = \lceil d_X(x, y) \rceil$$

we obtain

$$d_{X_S}(x, y) \le \eta_{2.18}(d_X(x, y)). \qquad \square$$

Applying the arguments of the proof of the proposition with the linear estimate in Remark 2.19 we obtain Corollary 2.20.

COROLLARY 2.20. *If each incidence map f_{ev} is an L-qi embedding, then each X_S is at most exponentially distorted in X, i.e., is η-uniformly properly embedded in X with $\eta(t) = \exp(at)$ for some $a \ge 1$ depending only on L.*

We omit the proof of this corollary since it is straightforward and the result is not used elsewhere.

DEFINITION 2.21. Let $\mathfrak{X} = (\pi : X \to T)$ be a tree of spaces.

(1) By a *K-qi section* over a subtree $S \subset T$ (or a *K-qi lift* of S) we mean a map $\sigma : S \to X$ such that for each vertex $v \in S$, $\sigma(v) \in X_v$, for any pair of adjacent vertices $u, v \in S$, we have $d_{X_{uv}}(\sigma(u), \sigma(v)) \le K$ and the restriction of σ to the interval uv is a parameterization of a geodesic $\sigma(u)\sigma(v)$ in X_{uv}.
(2) K-qi lifts of geodesic segments of T will be referred to as *K-qi leaves* in X and denoted by γ or γ_x or γ_{xy}, provided they start at x and end at y. We will refer to such γ as *horizontal* paths in X.
(3) A *vertical path* in X is a path contained in one of the vertex-spaces.

(4) If Y is a subset of X and $r \geq 0$ then the *fiberwise r-neighborhood* of Y in X (denoted $N_r^{fib}(Y)$) is the union

$$\bigcup_{v \in V(T)} N_r(Y \cap X_v),$$

where the latter neighborhood is taken with respect to the (intrinsic) metric of X_v.

Let $\mathfrak{X}$ be an abstract tree of spaces. A *subtree of spaces* in $\mathfrak{X} = (\pi : X \to T)$ is a tree of spaces $\mathfrak{X}' = (\pi' : X' \to T')$ whose base tree is a subtree $T' \subset T$, and vertex/edge spaces X'_v, X'_e are rectifiably connected uniformly properly embedded subsets of X_v, X_e respectively, so that the incidence maps of $\mathfrak{X}'$ are uniformly close to restrictions of incidence maps of $\mathfrak{X}$.

2.3. Coarse retractions

In this section we prove a general existence theorem of coarse Lipschitz left-inverses (*retractions*) for morphisms of trees of spaces.

Let T' be a subtree of T and let $\mathfrak{X} = (\pi : X \to T)$, $\mathfrak{X}' = (\pi' : X' \to T')$ be trees of spaces. We say that a morphism $h : X' \to X$ of these trees of spaces is a *relative K-qi embedding* for some constant $K \geq 1$, if for each $v \in V(T'), e \in E(T')$ incident on v, the maps $h_v : X'_v \to X_v, h_e : X'_e \to X_e$ are K-qi embeddings. Similarly, one can define a *relatively retractible* morphism of trees of spaces to be a morphism h such that for some constant $L \geq 0$, for each $v \in V(T'), e \in E(T')$ incident on v, the maps $h_v : X'_v \to X_v, h_e : X'_e \to X_e$ admit L-coarse Lipschitz left-inverses $h'_v : X_v \to X'_v$, $h'_e : X_e \to X'_e$. If $\mathfrak{X}, \mathfrak{X}'$ are trees of δ-hyperbolic spaces then the two notions are equivalent and, moreover, the subspaces $h_v(X'_v) \subset X_v, h_e(X'_e) \subset X_e$ are λ-quasiconvex for $\lambda = \lambda_{1.104}(L, \delta)$. Our goal is to prove that, under some conditions, a relatively retractive morphism is *absolutely retractive*, i.e., admits a coarse left-inverse $h' : X \to X'$. (Recall that the morphism h' is a collection of maps $h'_v : X_v \to X'_v, h'_e : X_e \to X'_e$ satisfying certain compatibility properties.) This result is motivated by Mitra's construction of a coarse retraction in [**Mit98**, Theorem 3.8]. For relatively retractible morphisms of trees, by abusing the notation, we will identify the vertex/edge spaces X'_v, X'_e of $\mathfrak{X}'$ with their images $h_v(X'_v) \subset X_v$ and $h_e(X'_e) \subset X_e$ respectively.

The following theorem is inspired by Mitra's coarse retraction theorem in [**Mit98**, Theorem 3.8] and its proof closely follows Mitra's argument. The next theorem and the corollary following it give sufficient conditions for the image of a retractive morphism of tree of spaces to be a coarse Lipschitz retract.

THEOREM 2.22 (Existence of a retraction). *Suppose that for some constants C, D, a relatively retractive morphism of trees of spaces $h : \mathfrak{X}' \to \mathfrak{X}$ satisfies the following conditions:*

(i) *For every boundary edge e of T', $e = [v, w], v \in V(T'), w \in V(T) - V(T')$,*

$$\operatorname{diam}_{X'_v}(h'_v \circ f_{ev}(X_e)) \leq D.$$

(ii) *For every edge $[v, w] = e \in E(T')$*

$$\operatorname{dist}_{X'_v}(h'_v \circ f_{ev}, f'_{ev} \circ h'_e) \leq C.$$

Then the map $h : X' \to X$ admits a coarse $L_{2.22}$-Lipschitz retraction $h' : X \to X'$ whose restriction to X_v equals h'_v for each $v \in V(T')$. Here $L_{2.22}$ depends only on C, D, coarse Lipschitz constants of the maps h'_v, h'_e, and the parameters of trees of spaces $\mathfrak{X}, \mathfrak{X}'$.

PROOF. We let L denote the maximum of Lipschitz constants of the projections $\pi : X \to T, \pi' : X' \to T'$. For each $v \in V(T')$ then we let $h'(x) := h'_v(x)$. Let $p : T \to T'$ denote the nearest-point projection.

Suppose $x \in X_w$, $w \in V(T) \setminus V(T')$; then $v = p(w) \in T'$ is the vertex nearest to w. Let $e \in E(T)$ be the edge incident to v and contained in the geodesic wv. Thus, e is a boundary edge of the subtree $T' \subset T$. By the assumption (i), the projection $h'_v(X_{ev}) \subset X'_v$ has the diameter $\leq D$. We let $h'(x)$ be any point x' of this projection (we will use the same point x' for all vertices w in each component of $T - T'$).

In order to verify that h' is (uniformly) coarse Lipschitz it suffices to find a uniform upper bound on distances $d(h'(x), h'(y))$ for points

$$x, y \in \mathcal{X} = \coprod_{v \in V(T)} X_v$$

which are within distance 1 from each other. If x, y belong to the same vertex-space X_v, then $d(h'(x), h'(y)) \leq L$, the upper bound for coarse Lipschitz constants of the maps $h'_v : X_v \to X'_v$. Suppose that x, y belong to X_v, X_w, $v, w \in V(T')$ are vertices spanning an edge $e \in E(T')$. Then, necessarily,

$$x \in X_{ev}, y \in X_{ew}, x = f_{ev}(z), y = f_{ew}(z)$$

for some $z \in X_e$. The condition (ii) then implies the estimates

$$d(h'_v(x), f'_{ev} \circ h'_e(z)) \leq C, \quad d(h'_w(y), f'_{ew} \circ h'_e(z)) \leq C,$$

hence $d(h'(x), h'(y)) \leq 2C$.

If $v, w \in V(T) - V(T')$ then the inequality $d_T(v, w) \leq 1$ implies that $p(v) = p(w) = u \in V(T')$ and there is a common boundary edge e of T' contained in the geodesics $uv, uw \subset T$. In particular, both $h'(x), h'(y)$ belong to the subset

$$h'_u \circ f_{eu}(X_e) \subset X'_u,$$

and, hence, $d(h'(x), h'(y)) \leq D$ by the condition (i).

Lastly, consider the case when $x \in X_v, y \in X_w$, where $v \in V(T'), w \in V(T) - V(T')$ and v, w span a boundary edge e of T'. Since $d(x, y) \leq 1$, it follows that $x \in X_{ev}, y \in X_{ew}$. Then $p(w) = v$ and, by the definition of h', $h'(x), h'(y) \in h'_v(X'_{ev})$ and, therefore,

$$d_{X'_v}(h'(x), h'(y)) \leq D. \qquad \square$$

An easy corollary of Theorem 2.22 is Corollary 2.23.

COROLLARY 2.23. *Suppose that $\mathfrak{X} = (\pi : X \to T)$ is a retractive tree of spaces. For every edge $e = [u, v] \in E(T)$ there exists an $r = r_{2.23}$-coarse retraction $X_{uv} \to X_u$, where r depends only on the parameters of $\mathfrak{X}$ and its retractivity constant.*

PROOF. We have a retractive tree of spaces $\mathfrak{Y} = (\pi : X_{uv} \to [\![u, v]\!])$. In $\mathfrak{Y}$ we have a subtree of spaces $\pi' : \mathfrak{Y}' = (Y' \to [\![u, v]\!])$, whose vertex-spaces are $Y'_u = Y_u = X_u, Y'_e = Y_e = X_e, Y'_v = X_e, f'_{ev} = \mathrm{id}, f'_{eu} = f_{eu}$, and the morphism $h : \mathfrak{Y}' \to \mathfrak{Y}$ is defined by using $\mathrm{id} : Y'_u \to Y_u$ and $f_{ev} : Y'_v \to Y_v$. Since $\mathfrak{X}$ is retractive, the morphism h is relatively retractive. Hence, by Theorem 2.22, the identity map $X_u \to X_u$ and the retraction $X_v \to X_e$ define a coarse Lipschitz retraction $X_{uv} \to Y'$. Since Y' is Hausdorff-close to X_u, we obtain a coarse Lipschitz retraction $Y' = Y_{uv} \to X_u$. The reader will verify that the coarse Lipschitz bound for this retraction depends only on the parameters of $\mathfrak{X}$ and its retractivity constant. $\square$

Another useful application of Theorem 2.22 is in the setting of trees of hyperbolic spaces (which we will discuss in more detail in the next section):

COROLLARY 2.24. *Suppose that the trees of spaces $\mathfrak{X}$, $\mathfrak{X}'$ and a morphism $\mathfrak{X}' \to \mathfrak{X}$ which is a fiberwise L-qi embedding, have the following properties.*

(1) *For some δ, all vertex and edge-spaces X_v, X_e are δ-hyperbolic. (Accordingly, the images vertex and edge-spaces $h_v(X'_v) \subset X_v$, $h'_e(X'_e) \subset X_e$ are λ-quasiconvex subsets in X_v, X_e respectively, where $\lambda = \lambda_{1.56}(\delta, L)$.)*
(2) *The retractions h'_v, h'_e are "nearest-point projections" in the sense that*

$$h'_v = P_{X_v, X'_v} \circ h_v, h'_e = P_{X_e, X'_e} \circ h_e.$$

(3) *There is constant K such that for every edge $e = [v, w] \in T'$ the Hausdorff distances $\mathrm{Hd}_{X'}(X'_v, X'_e)$ and $\mathrm{Hd}_{X'}(X'_w, X'_e)$ are $\leq K$.*
(4) *$T' = T$. Then the fiberwise nearest point projections h'_v, h'_e extend to an $L_{2.24}(\delta, L, K)$-coarse retraction $h' : X \to X'$, where (without loss of generality)*

$$L_{2.24}(\delta, L, K) \geq \max(L, K).$$

PROOF. For vertices v in T incident to an edge e, the images $h_v(X'_v)$, $f_{ev} \circ h_e(X'_e)$ are uniformly Hausdorff-close to each other. Therefore, the nearest-point projections (in X_v) to these uniformly quasiconvex subsets are also uniformly close to each other (see Corollary 1.108). Now, the claim follows from Theorem 2.22. □

2.4. Trees of hyperbolic spaces

We now introduce *hyperbolicity conditions* for trees of spaces.

DEFINITION 2.25. A tree of spaces $\mathfrak{X}$ satisfies Axiom **H** if there are constants δ_0 and L_0 such that:

(1) Each vertex/edge space X_v, X_e of $\mathfrak{X}$ is a δ_0-hyperbolic geodesic metric space.
(2) Each incidence map $f_{ev} : X_e \to X_v$ is an L_0-qi embedding.

We will refer to such $\mathfrak{X}$ as a *tree of hyperbolic spaces.*

A finite graph of finitely generated groups $\mathcal{G}$ satisfies Axiom **H** if the the corresponding tree of spaces does. In other words, all vertex and edge-groups have to be hyperbolic and edge-groups are quasiconvex in the incident vertex-groups.

A word of caution: Our terminology does not mean that a tree of hyperbolic spaces $\mathfrak{X} = (\pi : X \to T)$ has δ-hyperbolic total space X. Simple examples are given by Euclidean plane and Cayley complexes of Baumslag–Solitar groups. One needs to add a suitable *flaring condition* on $\mathfrak{X}$ to ensure hyperbolicity of X, as discussed in Section 2.5. Note also that our terminology requires not only uniform hyperbolicity of vertex and edge-spaces but also uniform qi embedding condition for the incidence maps.

DEFINITION 2.26. We will refer to δ_0 and L_0 as the *primary parameters* of a tree of hyperbolic spaces $\mathfrak{X}$.

In general, throughout the book, we will suppress the dependence of various constants and functions on the parameters of $\mathfrak{X}$.

DEFINITION 2.27. Suppose that $\mathfrak{X}$ is a tree of hyperbolic spaces. Let $A, B \subset X$ with $\pi(A) \subset \pi(B)$. If $X_v \cap A$ and $X_v \cap B$ are uniformly quasiconvex in X_v for all $v \in \pi(A)$, we define the nearest projections in X_v of $A \cap X_v$ to $B \cap X_v$. This gives us a map $A \to B$. We refer to this map as the *fiberwise projection* of A to B.

It is immediate that for every tree of hyperbolic spaces, for every edge $e = [v, w] \in E(T)$, the subset $X_{ev} \subset X_v$ is $\lambda_0 = \lambda_{1.92}(\delta_0, L_0)$-quasiconvex. In particular, every tree of hyperbolic spaces is retractive (see Definition 2.10) with retractions

$$f_{ve} : X_v \to X_e, e = [v, w],$$

given by the nearest-point projections $P = P_{X_v, X_{ev}}$ to the quasiconvex subsets

$$X_{ev} = f_{ev}(X_e) \subset X_v.$$

More precisely: $f_{ve}(x)$ is defined to be an arbitrary point in $f_{ev}^{-1}(P(x))$.

As an application of Remark 2.19 or, alternatively, of Corollary 2.23, we obtain Lemma 2.28.

Lemma 2.28. *Suppose that $\mathfrak{X}$ is a tree of hyperbolic spaces with the primary parameters δ and L. Then for every edge $e = [u, v] \in E(T)$, the inclusion maps $X_u \to X_{uv}, X_v \to X_{uv}$ are $L_0' = L_{2.28}(\delta, L)$-qi embeddings where L_0' is the maximum of 2 and of the coarse Lipschitz constant for a retraction $X_{uv} \to X_v$ (see Corollary 2.23).*

Remark 2.29. In this lemma we ensured that $L_0' \geq 2$. This, somewhat artificial, convention will be used in the proof of Lemma 2.38.

Suppose that $\mathfrak{X}' = (\pi : X' \to T)$ is a tree of hyperbolic spaces, $G < \mathrm{Isom}(X')$ is a subgroup acting by automorphisms of $\mathfrak{X}'$, such that the quotient graph T/G is finite and for every vertex $v \in V(T)$ (resp. edge $e \in E(T)$) the action of the corresponding stabilizer $G_v < G$ (resp. $G_e < G$) on X_v' (resp. X_e') is quasiconvex (see Definition 1.132). Thus, the group G also has structure of a graph of finitely generated groups $\mathcal{G}$ (with the underlying graph T/G); in particular, G is finitely generated. (Note that we are not assuming hyperbolicity of the space X.)

Since G acts via automorphisms of $\mathfrak{X}'$, for each edge $e = [v, w] \in E(T)$ the subspace $X_{ev}' \subset X_v'$ is G_e-invariant. We will also assume that for each $v \in V(T)$ the G_v-orbit of X_{ev}' is *locally finite* in X_v' (see Definition 1.6). Note that the local finiteness assumption is automatic for instance if there exists a larger discrete group G_v' (containing G_v) acting on X_v' geometrically (Lemma 1.7).

Proposition 2.30. *Under the above assumptions, there exists a coarse Lipschitz retraction $X' \to Gx$ for each G-orbit in X'. In particular, each orbit map $G \to Gx \subset X'$ is a qi embedding.*

Proof. The proof is similar to that of Corollary 2.24. Let $\mathfrak{X} = (\pi : X \to T)$ denote the tree of hyperbolic spaces corresponding to the graph of groups $\mathcal{G}$. The isometric action of G via automorphisms of $\mathfrak{X}'$ defines a morphism of trees of spaces $\mathfrak{X} \to \mathfrak{X}'$. This morphism is relatively retractive in view of the quasiconvexity assumption for the actions $G_v \curvearrowright X_v', G_e \curvearrowright X_e'$. In view of Proposition 1.137(4), the local finiteness assumption implies that for $y \in X_v'$, the restriction to X_e' of the nearest-point projection $P_{X_v', G_v y}$ is within uniformly bounded distance from the projection $P_{X_{ev}', G_e y}$. Thus, Theorem 2.22 applies and the coarse Lipschitz retractions $X_v' \to X_v, X_e' \to X_e$ together give rise to a coarse Lipschitz retraction $X' \to X$. Since G acts cocompactly on X, we, thus, obtain a coarse Lipschitz retraction $X' \to Gx$. ☐

Corollary 2.31. *Suppose that $\mathcal{G}'$ is a finite, connected graph of hyperbolic groups satisfying Axiom **H**, with $\pi_1(\mathcal{G}') = G'$ and let T be the Bass–Serre tree of $\mathcal{G}'$. Let $G < G'$ be a subgroup such that:*

(1) *For every vertex v (resp. edge e) of T, the G-stabilizer $G_v < G$ of v (resp. the G-stabilizer $G_e < G$ of e) is a quasiconvex subgroup of the G'-stabilizer $G'_v < G'$ of v (resp. of the G'-stabilizer $G'_e < G'$ of e).*
(2) *The quotient-graph T/G is finite.*

There exists a coarse Lipschitz retraction $G' \to G$. In particular, the subgroup G is qi embedded in G'.

EXAMPLE 2.32. Let $H = \pi_1(S_1)\star\pi_1(S_2)$, where S_1, S_2 are closed connected hyperbolic surfaces, and let $\phi_i : \pi_1(S_i) \to \pi_1(S_i), i = 1, 2$, be automorphisms. Then ϕ_1, ϕ_2 define an automorphism $\phi : H \to H$ and we obtain subgroups $G_i = \phi_1(S_i) \rtimes_{\phi_i} \mathbb{Z}$ in $G' = H \rtimes_\phi \mathbb{Z}$. The subgroups $G_i < G'$ clearly satisfy the assumptions of the corollary (where T is the line) which implies that they are coarse Lipschitz retracts of G'. Note, furthermore, that if ϕ_1, ϕ_2 are induced by pseudo-Anosov homeomorphisms of the surfaces S_1, S_2, then the group G' is isomorphic to the amalgam of hyperbolic groups $G_1 \star_{\mathbb{Z}} G_2$, where $\mathbb{Z}$ is a malnormal subgroup of both G_1, G_2. Hence, the group G' is hyperbolic (see Corollary 2.57) and the subgroups G_1, G_2 are quasiconvex in G'.

Below, H is a quasiconvex subgroup of G'_v for some vertex v in T.

LEMMA 2.33. *For each edge e and vertex w in T', the H-stabilizer of e (resp. w) is a quasiconvex subgroup of both H and G'_e (resp. G'_w).*

PROOF. Consider first an edge $e = [v, w]$. Then $H_e = H \cap G'_e$ is the intersection of two quasiconvex subgroups of G'_v, hence, is quasiconvex in G'_v, G'_e and H (see Corollary 1.135). The general case follows from induction on the length of the edge-path connecting e (resp., w) to v. □

LEMMA 2.34. *Suppose, in addition, that the H-stabilizers of edges incident to v are all finite. Then for each $R \geq 0$, $x \in X'_v$ and all the edges e incident to v, the coarse intersections intersections $Hx \cap N_R(X'_{ev})$ are uniformly bounded, with bound independent of e. In other words, the pairs Hx, X'_e are uniformly cobounded.*

PROOF. By properness of the action, there exists an edge e incident to v such that the diameter of the intersection $Hx \cap N_R(X'_{ev})$ is maximal. The subset X'_{ev} is Hausdorff-close to the orbit $G'_e x$. According to Proposition 1.137, the coarse intersection $Hx \cap N_R(X'_{ev})$ is Hausdorff-close to the orbit $H_e x$, where $H_e = H \cap G'_e$. Since, by the hypothesis of the lemma, the subgroup H_e is finite, the coarse intersection $Hx \cap N_R(X'_{ev})$ is bounded. □

COROLLARY 2.35. *If the pair Hx, X'_e is not cobounded then the intersection $H \cap G'_e$ is infinite.*

We will prove in Corollary 2.69 that there exists a function $\delta(n)$ (depending also on the constants δ_0 and L_0) such that for each interval J of length n, the pull-back space X'_J (with its intrinsic path-metric) is $\delta(n)$-hyperbolic. Thus, we can talk about cobounded pairs of subspaces in vertex-spaces X'_v, X'_w, $v, w \in V(J)$.

Continuing with the notation of Lemma 2.34, and applying Corollary 2.35 inductively (with Lemma 2.33), we obtain Lemma 2.36

LEMMA 2.36. *Suppose that for some vertex $w \in T, x \in X'_v$, the subsets Hx, X'_w are not cobounded in X'_J, $J = [\![v, w]\!]$. Then the H-stabilizer of the segment J is an infinite subgroup of H.*

Below are a few more easy consequences of Axiom **H** for trees of spaces.

LEMMA 2.37. *Assume that $\mathfrak{X}$ is a tree of hyperbolic spaces. Then for every edge $e = [v_1, v_2]$ of T, if $\alpha_i = [x_i y_i]_{X_{v_i}} \subset X_{v_i}$ are vertical geodesics such that*

$$d_{X_{v_1 v_2}}(x_1, x_2) \le C, \quad d_{X_{v_1 v_2}}(y_1, y_2) \le C,$$

then the Hausdorff distance between these vertical geodesics in $X_{v_1 v_2}$ is at most $C_1 = C_{2.37}(C)$.

PROOF. Geodesics x_1x_2, y_1y_2 in X_J, $J = [\![v_1, v_2]\!]$, have to cross X_e (separating X_{vw}) at some points $x, y \in X_e$. Lemma 1.55 applied to the geodesic α_i and the L_0-quasigeodesic $\alpha_i' = f_{ev_i}([xy]_{X_e})$ implies that

$$\mathrm{Hd}_{X_{v_i}}(\alpha_i, f_{ev_i}(\alpha)) \le D = D_{1.55}(\delta_0, L_0, C).$$

Since

$$\mathrm{Hd}_{X_{v_1 v_2}}(\alpha, f_{ev_i}(\alpha)) \le 1,$$

we conclude:

$$\mathrm{Hd}_{X_{v_1 v_2}}(\alpha_1, \alpha_2) \le 2(1 + D).$$

□

LEMMA 2.38. *Let $I = [\![v, w]\!] \subset T$ be a subinterval, we denote its consecutive vertices $v_0 = v, v_1, \ldots, v_n = w$. Let γ_0, γ_1 be K-qi sections over I. Then the function*

$$\ell(i) := d_{X_i}(\gamma_{0,k}(v_i), \gamma_{1,k}(v_i)), i \in [0, n] \cap \mathbb{Z},$$

satisfies

$$\ell(n) \le a^n \ell(0) + \frac{a^n - 1}{a - 1} b < a^n(\ell(0) + b),$$

where $a = L_0'$, $b = 2L_0'K + L_0'$.

PROOF. Consider an edge $e = [v_i, v_{i+1}] \subset I$. The points

$$\gamma_0(v_{i+1}), \gamma_1(v_{i+1}) \in X_{v_i}$$

are connected by a path of length $\le 2K + \ell(i)$ in $X_{[\![v_i, v_{i+1}]\!]}$, obtained by concatenating a vertical geodesic $[\gamma_0(v_i)\gamma_1(v_i)]_{X_{v_i}}$ with two geodesics of length $\le K$. Since $X_{v_{i+1}}$ is L_0'-qi embedded in $X_{[\![v_i, v_{i+1}]\!]}$, we have

$$\ell(i + 1) \le L_0'(2K + \ell(i)) + L_0'.$$

Then

$$\ell(n) \le a^n \ell(0) + (a^{n-1} + \ldots + 1)b = a^n \ell(0) + \frac{a^n - 1}{a - 1} b < a^n(\ell(0) + b).$$

□

COROLLARY 2.39. *If $\ell(0) \ge a(M + b)$, then for all*

$$n \in [0, N], N = \left\lfloor \log_a \left(\frac{\ell(0)}{M + b} \right) \right\rfloor,$$

we have

$$\ell(n) > M.$$

PROOF. We first reverse the role of $\ell(0)$ and $\ell(n)$ and obtain from the lemma that

$$\ell(n) < a^{-n} \ell(0) - b, n \in \mathbb{N}.$$

The inequality $\ell(n) > M$ then follows from

$$n \le N \le \log_a \left(\frac{\ell(0)}{M + b} \right).$$

The assumption that $\ell(0) \ge a(M + b)$ ensures that $N \ge 1$. □

Another corollary (or, rather, a special case of the lemma) is Corollary 2.40.

COROLLARY 2.40. *For every edge $e = [u, v]$ in T, any pair points $x, y \in X_u$, and a pair of K-qi sections γ_0, γ_1 over the interval uv, we have*

$$d_{X_v}(\gamma_0(v), \gamma_1(v)) \le D_{2.40}(K, d_{X_u}(x, y)) = L_0'(2K + d_{X_u}(x, y)).$$

The following construction of *metric interval bundles* determined by two qi sections of $\mathfrak{X}$ will be used often in what follows. Let $S \subset T$ be a subtree and $\Pi = (\gamma_0, \gamma_1)$ is a pair of K-qi sections of $\mathfrak{X}$ over S. We form a tree of spaces $\mathfrak{Y}_\Pi = \mathfrak{Y} = (Y \to S)$ as follows:

The vertex-spaces Y_v of Y are geodesic segments in X_v joining $\gamma_0(v), \gamma_1(v)$, $v \in V(S)$. The edge-spaces Y_e, $e = [v, w]$, of $\mathfrak{Y}$ are geodesic segments in X_e with endpoints within distance K from the respective endpoints of Y_v. The incidence maps of $\mathfrak{Y}$ are obtained by composing the incidence maps of f_{ev}, $v \in V(J)$, composed with the nearest-point projections to Y_v (taken in X_v).

LEMMA 2.41.

(1) *$\mathfrak{Y}$ is a metric bundle over S.*
(2) *For every point $x \in Y_u$, $u \in V(S)$, there exists a $K_{2.41}(K)$-qi section γ of $\mathfrak{Y}$ over the tree S, passing through x, where the qi constant is defined with respect to the metrics X_{vw}, $[v, w] \in E(S)$.*

PROOF. We orient edges $e = [v, w]$ of S away from u and define $\gamma(w)$ inductively, assuming that $\gamma(v)$ is defined. Since the quasigeodesic quadrilateral in X_{vw} formed by the paths Y_v, Y_w and geodesics $[\gamma_0(v)\gamma_1(v)]_{X_{vw}}$, $[\gamma_0(w)\gamma_1(w)]_{X_{vw}}$ (of lengths $\le K$) are uniformly slim (with respect to the metric of X_{vw}), the point $\gamma(v) \in Y_v$ is $K_{2.41}(K)$-close to a point $\gamma(v) \in Y_v$. This proves both claims of the lemma. □

2.5. Flaring

Geodesics (and, hence, uniform quasigeodesics) in hyperbolic spaces diverge (exponentially fast). Since k-qi leaves in hyperbolic trees of spaces $\pi : X \to T$ are uniform quasigeodesics, they should also diverge if X is hyperbolic. In this section we discuss several divergence conditions, called *flaring conditions*[4], one can impose on qi-leaves in trees of spaces. These conditions involve pairs $\Pi = (\gamma_0, \gamma_1)$ of k-sections γ_0, γ_1 over a common geodesic segment $J = [\![t_{-n}, t_n]\!] \subset T$ of length $2n$ and prescribe the nature of growth of the vertical distances

$$d_{X_{v_i}}(\gamma_0(v_i), \gamma_1(v_i))$$

for $i > 0$ or $i < 0$. The *girth* Π_0 of the pair (γ_0, γ_1) is the vertical distance

$$\Pi_0 = d_{X_0}(\gamma_0(0), \gamma_1(0)).$$

REMARK 2.42. Π_0 need not be equal to

$$\min_{v \in V(J)} d_{X_v}(\gamma_0(v), \gamma_1(v)).$$

We will frequently use the notation $\Pi_{\max}$ for the *maximal separation of the ends* of the pair $\Pi = (\gamma_0, \gamma_1)$,

$$\Pi_{\max} := \max\left(d_{X_{t_{-n}}}(\gamma_0(t_{-n}), \gamma_1(t_{-n})), d_{X_{t_n}}(\gamma_0(t_n), \gamma_1(t_n))\right),$$

describing the rate of growth of the above vertical distances (in one of the directions).

[4]Flaring conditions do not require Axiom **H**.

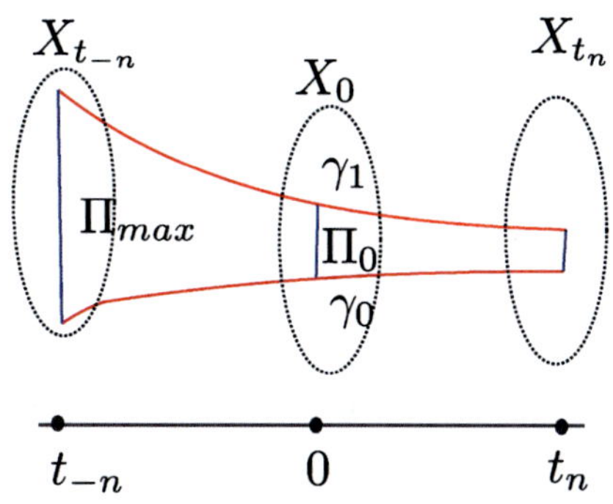

FIGURE 2.1. Flaring

2.5.1. Proper and uniform flaring conditions. The *uniformly proper flaring* condition requires qi sections over the same geodesic in T to diverge at some uniform rate in at least one direction. More precisely:

DEFINITION 2.43 (Uniformly proper flaring). A tree of spaces $\mathfrak{X} = (\pi : X \to T)$ is said to satisfy the *uniformly proper κ-flaring condition* if there exists $m_\kappa \geq 0$ and a positive proper function $\phi_\kappa : \mathbb{N} \to \mathbb{R}_+$ such that for every pair Π of κ-qi sections γ_0, γ_1 of girth $> m_\kappa$, over an interval of length $2n$ in T, we have

$$\Pi_{\max} \geq \phi_\kappa(n).$$

In other words, κ-sections have to diverge uniformly fast but the rate of divergence is allowed to be, say, sublinear (unlike the the Bestvina–Feighn flaring condition where one has an exponential rate of flaring).

It is clear from the definition that if $\mathfrak{X}$ satisfies the uniformly proper K-flaring condition, then it also satisfies the uniformly proper κ-flaring condition for all $\kappa \in [1, K]$: We simply take $\phi_\kappa := \phi_K$ and $m_\kappa := m_K$. Note also that it would be too much to ask for

$$\Pi_{\min} = \min\left(d_{X_{-n}}(\gamma_0(-n), \gamma_1(-n)), d_{X_n}(\gamma_0(n), \gamma_1(n))\right) \geq \phi_\kappa(n),$$

for all $n \in \mathbb{N}$ and all pairs of κ-qi sections $\Pi = (\gamma_0, \gamma_1)$ for some (uniform) proper function ϕ_κ.

DEFINITION 2.44. We will say that a pair $\Pi = (\gamma_0, \gamma_1)$ of sections over an interval $[\![-N, N]\!]$ in T is *flaring in the positive/negative direction* if, respectively,

$$d_{X_n}(\gamma_0(n), \gamma_1(n)) \geq \phi_\kappa(n),$$

or

$$d_{X_{-n}}(\gamma_0(-n), \gamma_1(-n)) \geq \phi_\kappa(n)$$

for all $n \in \mathbb{N} \cap [1, N]$.

We will see in Lemma 2.59 and Corollary 2.61, that uniformly proper flaring (for all $\kappa \geq 1$) implies uniformly proper flaring in positive or negative direction (after a possible change of the function ϕ_κ).

Below is a minor variation on the definition of uniformly proper flaring. For fixed κ and Π_0 define the function

$$\phi_{\kappa,\Pi_0}(n) := \inf\{\Pi_{\max}(n)\}$$

where the infimum is taken over all κ-qi sections (γ_0, γ_1) over length $2n$ intervals in T such that $\Pi_0(\gamma_0, \gamma_1) = \Pi_0$.

Definition 2.45 (Proper κ-flaring). A tree of spaces $\mathfrak{X} = (\pi : X \to T)$ is said to satisfy the *proper κ-flaring condition* if there exists m_κ such that for each $\Pi_0 \geq m_\kappa$ each function $\phi_{\kappa,\Pi_0}(n)$ is proper.

Clearly, uniformly proper flaring implies proper flaring. The next lemma (which will not be used elsewhere) shows that the converse is also true, provided that we are willing to change the qi constant:

Lemma 2.46. *There exists a function $K = K_{2.46} = K_{2.46}(\kappa)$ such that if $\mathfrak{X}$ satisfies the proper K-flaring condition, then it also satisfies the uniformly proper κ-flaring condition.*

Proof. Let $\Pi = (\gamma_0, \gamma_1)$ be a pair of κ-qi sections over an interval $J \subset T$ of length $2n$, such that $\Pi_0 \geq m_\kappa$. Consider a point $x \in [\gamma_0(0)\gamma_1(0)]_{X_0}$ at the distance m_k from $\gamma_0(0)$. By Lemma 2.41, there exists a $K = K_{2.41}(\kappa)$-section γ over the interval J such that $\gamma(0) = x$. In view of the proper flaring assumption (up to interchanging the endpoints of J), the pair of sections (γ_0, γ) satisfies

$$d_{X_n}(\gamma_0(n), \gamma(n)) \geq \phi_{\kappa,m_K}(n).$$

It follows that

$$\Pi_{\max}(\gamma_0, \gamma_1) \geq d_{X_n}(\gamma_0(n), \gamma(n)) \geq \phi_{\kappa,m_K}(n). \qquad \square$$

In the book we will be mostly using an alternative form of the proper flaring condition established in the next proposition. For the ease of the notation, in this section we will identify a geodesic, say $[\![v, w]\!]$, of length ℓ in T, where $v, w \in V(T)$ with an interval $[a, b] \subset \mathbb{R}$ of length ℓ, where $a, b \in \mathbb{Z}$ through an implicit isometry $[a, b] \to [\![v, w]\!]$; this means, in particular, that integers correspond to the vertices in $[\![v, w]\!]$.

Proposition 2.47. *The following are equivalent:*

(1) *A tree of spaces $\mathfrak{X} = (\pi : X \to T)$ satisfies the uniformly proper κ-flaring condition.*
(2) *There exist M_κ such that for all $D \geq 0$, there is $\tau = \tau_{2.47}(\kappa, D)$ satisfying the following:*

For every pair Π of κ-qi sections γ_0, γ_1 over a geodesic interval $[-m, n] \subset T$ ($m, n \in \mathbb{N}$), if

$$d_{X_i}(\gamma_1(i), \gamma_2(i)) > M_\kappa, \forall i \in [-m + 1, n - 1],$$

and

$$\Pi_{\max} = \max\left(d_{X_{-m}}(\gamma_0(-m), \gamma_1(-m)), d_{X_n}(\gamma_0(n), \gamma_1(n))\right) \leq D,$$

then

$$n + m \leq \tau.$$

Proof. First of all, we leave it to the reader to check that if (2) holds for all $m = n$ then it holds for all $m, n \in \mathbb{N}$. Therefore, in what follows, in (2) we will be always assuming that $n = m$.

(i) Assume that the uniformly proper flaring condition holds. Take $M_\kappa := m_\kappa$ and consider a pair $\Pi = (\gamma_0, \gamma_1)$ of κ-qi sections over an interval $[-n, n] \subset T$ as in part (2). In particular, girth of (γ_0, γ_1) is $> M_\kappa$. By the uniformly proper flaring condition, we have

$$D \geq \Pi_{\max} \geq \phi_\kappa(n).$$

Since $\phi_\kappa(t)$ is proper, the pre-image $\phi_\kappa^{-1}([0, D])$ is contained in an interval $[0, t_{\kappa,D}]$. Then we take

$$\tau(\kappa, D) := 2t_{\kappa,D}.$$

(ii) Conversely, suppose that (2) holds but the uniform proper flaring fails. Then there exist a constant $D > 0$ and a sequence Π^m of pairs of κ-qi sections γ_0^m, γ_1^m over some intervals $[\![s_m, t_m]\!] \subset T$ of length $2n_m$ with the midpoint vertex r_m such that $\Pi_0^m \to \infty, n_m \to \infty$, but

$$\Pi_{\max}^m = \max\left(d_{X_{t_m}}(\gamma_0^m(t_m), \gamma_1^m(t_m)), d_{X_{s_m}}(\gamma_0^m(s_m), \gamma_1^m(s_m))\right) \le D.$$

We will isometrically parameterize the geodesic $[s_m, t_m]$ by the interval $[-n_m, n_m] \subset \mathbb{Z}$ so that r_m corresponds to 0. Set $\tau := \tau(\kappa, M)$ where

$$M = \max(D, M_\kappa).$$

Define the function

$$\ell_m(i) := d_{X_i}(\gamma_0^m(i), \gamma_1^m(i)), i \in [-n_m, n_m]; \ell_m(0) = \Pi_0^m.$$

Then for sufficiently large m we have

$$\Pi_0^m = \ell_m(0) > a(M + b); a = L_0', b = 2L_0'\kappa,$$

and, hence, according to Corollary 2.39, for all $n \in [-N_m + 1, N_m - 1]$, we have

$$\ell_m(n) > M.$$

Here

$$N_m = \left\lfloor \log_a\left(\frac{\Pi_0^m}{M + b}\right)\right\rfloor.$$

Observe that the right-hand side diverges to infinity as $m \to \infty$. Therefore, for sufficiently large m, $N_m > \tau/2$. Thus, we obtain a contradiction with (2). □

While the (uniformly) proper flaring condition is quite natural, it is the condition (2) in Proposition 2.47 that we will use throughout the book.

Definition 2.48. We will say that a tree of spaces $\mathfrak{X}$ satisfies the *uniform κ-flaring* condition with the parameter M_κ if the condition (2) in the proposition holds.

Convention 2.49. In what follows, unless indicated otherwise, κ-flaring always means uniform κ-flaring.

Lemma 2.50. *Suppose that $\mathfrak{X} = (\pi : X \to T)$ is a tree of hyperbolic spaces with δ-hyperbolic total space X. Then $\mathfrak{X}$ satisfies the uniform κ-flaring condition for all $\kappa \ge 1$.*

Proof. As noted earlier, it suffices to consider the case $n = m$. Since γ_0, γ_1 are κ-quasigeodesics in X, they are within Hausdorff distance $D_{1.54}(\delta, \kappa)$ from geodesics γ_i^* in X connecting the endpoints of γ_0, γ_1 respectively. Take $x_0 = \gamma_0(0)$ and $x_0^* \in \gamma_0^*$ a point within distance $D_{1.54}(\delta, \kappa)$ from x_0. The projections to T of the geodesics $[\gamma_0(-n)\gamma_1(-n)]_X$, $[\gamma_0(n)\gamma_1(n)]_X$ each have length $\le D$. Thus,

$$d(x_0^*, [\gamma_0(\pm n)\gamma_1(\pm n)]_X) \ge D.$$

Suppose for a moment that $n - D > 2\delta$. By the slim quadrilateral property, there is a point $x_1^* \in \gamma_1^*$ within distance 2δ from x_0^*. (A priori, this could have been a point on one of two other sides of the geodesic quadrilateral with the vertices $\gamma_i(\pm n), i = 0, 1$, but this possibility is ruled out by our assumption that $n - D > 2\delta$.) Thus, we find a point $x_1 \in \gamma_1 \cap X_v$, within distance

$$D_0 = 2(\delta + D_{1.54}(\delta, \kappa)) + \kappa$$

from x_0. While v need not be equal to the vertex $0 \in [\![-n, n]\!] \subset T$, we still have

$$d_T(0, v) \le D_0.$$

In particular,

$$d_{X_0}(\gamma_0(0), \gamma_1(0)) = \Pi_0 \geq d_X(\gamma_0(0), \gamma_1(0)) \leq D_1 = D_0(\kappa + 1).$$

We, therefore, set

$$M_\kappa = D_1$$

and $\tau(\kappa, D) = \delta + \frac{1}{2}D$. Since in the uniform κ-flaring property, it is assumed, in particular, that

$$d_{X_0}(\gamma_0(0), \gamma_1(0)) > M_\kappa,$$

we obtain a contradiction with the above estimates, unless the inequality $n - D \geq 2\delta$ is violated, i.e., unless $n \leq D + 2\delta$, equivalently, the length of the interval $[\![-n, n]\!]$ is at most τ, as required. □

The uniform flaring condition has an immediate consequence that we will use on few occasions.

LEMMA 2.51 (Three flows lemma). *Suppose that $\pi : X \to T$ satisfies the uniform K-flaring condition. Suppose that $\gamma_1, \gamma_2, \gamma_3$ are K-qi sections of $\pi : X \to T$ over an interval $[\![s, t]\!]$ such that for all $r \in]\!]s, t[\![$,*

$$d_{X_r}(\gamma_1(r), \gamma_3(r)) > M_K$$

while

$$\max_{i,j} d_{X_s}(\gamma_i(s), \gamma_j(s)) \leq C, \quad \max_{i,j} d_{X_t}(\gamma_i(t), \gamma_j(t)) \leq C.$$

Then the length of the interval $[\![s, t]\!]$ is uniformly bounded, i.e is $\leq \tau_{2.51}(K, C)$.

DEFINITION 2.52.

(1) Given $K \geq 1$ and $C \geq 0$, by a (K, C)-*bigon* we will mean a pair of K-qi sections $\Pi = (\gamma_1, \gamma_2)$ such that $\Pi_{\max} \leq C$. Such a bigon said to be D-*thin* for some $D \geq 0$ if

$$\forall t \in V(I),\ d_{X_t}(\gamma_1(t), \gamma_2(t)) \leq D.$$

Here, as before,

$$\Pi_{max} = \max\left(d_{X_v}(\gamma_1(v), \gamma_2(v)), d_{X_w}(\gamma_1(w), \gamma_2(w))\right).$$

(2) Given $K \geq 1$, $C \geq 0$, we say that a tree of spaces $\mathfrak{X}$ *satisfies the* (K, C)-*thin bigon property* if there is a constant $R = R(K, C)$ such that any (K, C)-bigon is R-thin.

The next corollary will be used quite often in our book:

COROLLARY 2.53. *There is a function $R(K, C) = R_{2.53}(K, C)$ defined for $K \geq 1, C \geq 0$ such that the following holds:*

A tree of spaces $\mathfrak{X} = (X \to T)$ satisfies the uniform K-flaring condition if and only if for every $C \geq 0$, each (K, C)-bigon in X is $R(K, C)$-thin, or, equivalently, X satisfies the (K, C)-thin bigon property.

PROOF.

(1) Assume that $\mathfrak{X}$ satisfies the uniform K-flaring condition. Consider a pair of K-qi sections over an interval $I \subset T$. If for every vertex $t \in I$, $d_{X_t}(\gamma_1(t), \gamma_2(t)) \le M_K$, then we are done. Otherwise, let $I' = [\![v', w']\!] \subset [\![v, w]\!]$ be a maximal subinterval such that for all vertices $t \in I'$ we have

$$d_{X_t}(\gamma_1(t), \gamma_2(t)) > M_K.$$

Then there are edges $[v'', v'], [w', w'']$ in I (not contained in I') such that

$$d_{X_s}(\gamma_1(s), \gamma_2(s)) \le C' := \max(M_K, C, 3\delta_0), s \in \{v'', w''\}.$$

By Lemma 2.51 applied to K-qi sections $\gamma_1, \gamma_2 = \gamma_3$, restricted to $I'' := [\![v'', w'']\!]$, we obtain:

$$d_T(v'', w'') \le \tau := \tau_{2.51}(K, C').$$

By Lemma 2.38, we get that for all $t \in V(I')$,

$$d_{X_t}(\gamma_1(t), \gamma_2(t)) \le R_{2.53}(K, C) := a^\tau \left(C' + \frac{b}{a-1}\right),$$

with $a = L_0', b = 2L_0'K$. (Recall that $L_0' \ge 2$.)

(2) We argue as in the proof of Proposition 2.47. Suppose that the uniformly proper κ-flaring fails. Then there exist a constant $D > 0$ and a sequence Π^m of pairs of κ-qi sections γ_0^m, γ_1^m over some intervals $J_m = [\![s_m, t_m]\!] \subset T$ of length $2n_m$ with the midpoint vertex r_m such that $\Pi_0^m \to \infty, n_m \to \infty$, but

$$\Pi^m_{\max} = \max\left(d_{X_{t_m}}(\gamma_{0,m}(t_m), \gamma_{1,m}(t_m)), d_{X_{s_m}}(\gamma_{0,m}(s_m), \gamma_{1,m}(s_m))\right) \le D.$$

Setting $C := D$, the hypothesis in Part 2 of the corollary means that

$$d_{X_t}(\gamma_1(t), \gamma_2(t)) \le R(K, C)$$

for all vertices $t \in J_m$. This contradicts $\Pi_0^m \to \infty$. □

2.5.2. Acylindrical trees of spaces. An easy, and frequently occurring, *sufficient* condition for uniform κ-flaring is *acylindricity*:

DEFINITION 2.54. Fix constants $\kappa \ge 1$ and $\tau \ge 1$. A tree of spaces $(\pi : X \to T)$ is (M, κ, τ)-*acylindrical* if for every pair of κ-sections γ_0, γ_1 over an interval $J \subset T$ of length $\ge \tau$, we have

$$d_{X_v}(\gamma_0(t), \gamma_1(t)) \le M, \forall t \in V(J).$$

We give a few geometric examples of acylindrical trees of spaces in Section 2.6.2. In order to see that acylindrical trees of spaces satisfy uniform flaring, we take $M_\kappa := M$ and $\tau(\kappa, D) := \tau + 2$. Then, regardless of D, if $\Pi = (\gamma_0, \gamma_1)$ is a pair of κ-qi sections over an interval $J = [\![u, v]\!] \subset T$ and

$$d_{X_i}(\gamma_1(i), \gamma_2(i)) > M, i \in V(]\!]u, v[\![),$$

then the length of $]\!]u, v[\![$ is $< \tau$ and, hence, J has length $< \tau(\kappa, D) = \tau + 2$.

The terminology *acylindrical* has its origin in 3-dimensional topology: A compact oriented 3-dimensional manifold with incompressible boundary M is called (homotopically) *acylindrical* if every map of an annulus $(A, \partial A) \to (M, \partial M)$ is homotopic (rel. ∂A) to a map $A \to \partial M$. Algebraically speaking, this condition means that if two elements of $\pi_1(\partial M, m)$ are conjugate in $\pi_1(M, m)$, then they are conjugate in $\pi_1(\partial M, m)$. If one glues two connected acylindrical 3-manifolds M_1, M_2 along their boundary surfaces to form a 3-manifold M, then every subgroup of $\pi_1(M)$ isomorphic to $\mathbb{Z}^2$ is contained (up to conjugation) in $\pi_1(M_1)$

or in $\pi_1(M_2)$. Algebraically speaking, topological acylindricity corresponds to acylindricity in the sense of group actions on trees (Definition 2.4) as follows. The decomposition $M = M_1 \cup M_2$ yields graph-of-groups decomposition of the fundamental $G = \pi_1(M)$. Let $G \times T \to T$ denote the action of G on the Bass–Serre tree T corresponding to this decomposition of G. Then the action of G on T is 1-acylindrical if and only if both manifolds M_1, M_2 are acylindrical. The proof of this is straightforward and is similar to the proof of Corollary 2.57. However, since the proof is irrelevant for the rest of the book, we omit it. Suppose again that G is the fundamental group of a finite graph of finitely generated groups $(\mathcal{G}, Y)$; let $G \times T \to T$ be the corresponding G-action on the Bass–Serre tree and $\mathfrak{X} = (X \to T)$ the tree of spaces with X equal to the Cayley graph of G as discussed in Example 2.17. We will see in Proposition 2.55 that in this setting the tree of spaces $\mathfrak{X}$ is (κ, τ)-acylindrical provided that the action of G on T is k-acylindrical for suitable values of κ, τ and k.

2.5.3. Group-theoretic examples. The following proposition was proved by Ilya Kapovich [**Kap01**]; below, we give a different proof.

PROPOSITION 2.55. *Suppose $(\mathcal{G}, Y)$ is a finite graph of hyperbolic groups satisfying Axiom **H** and $G := \pi_1(\mathcal{G})$. If the G-action on the Bass–Serre tree T of $\mathcal{G}$ is R-acylindrical in the sense of Sela* [**Sel97**], *then for all $\kappa \geq 1$ there is a constant M_κ such that the induced tree of metric spaces $\mathfrak{X} = (\pi : X \to T)$ is (M_κ, κ, R)-acylindrical. In particular, in view Theorem 2.62, G is hyperbolic.*

PROOF. The first part of the proof follows by the arguments in [**Sar18**, Section 3]. We will need some properties of the tree of spaces $\mathfrak{X}$ listed below.

(1) The vertex-spaces of $\mathfrak{X}$ are metric graphs which are isometric copies of various cosets of G_y in G, where $y \in V(Y)$. The map $\pi : X \to T$ is G-equivariant. The G-action on X is proper and cocompact, and the stabilizer of each $v \in V(T)$ acts on $V(X_v)$ transitively.
(2) Suppose that Γ is a Cayley graph of G with respect to a finite generating set. Let $f : G \to X$ be an orbit map. We know that for each $y \in V(Y)$ and $g \in G$, gG_y is a vertex of T. We have $\mathrm{Hd}(X_{gy}, f(gG_y)) \leq D$, where D is a constant independent of $g \in G, y \in V(Y)$.

Suppose that the claim of the proposition fails for some κ. Then there is a sequence of pairs of κ-qi sections $\gamma_{0,n}, \gamma_{1,n}$ over geodesic intervals

$$\beta_n : [0, R+1] \to T$$

of length $R + 1$, such that for some integer $t \in [0, R+1]$ we have

$$d_{X_{\beta_n(t)}}(\gamma_{0,n}(t), \gamma_{1,n}(t)) \geq n, \quad \forall n \in \mathbb{N}.$$

Note that for all integers $s \in [0, R+1]$

$$d_X(\gamma_{0,n}(s), \gamma_{1,n}(s)) \geq d_X(\gamma_{0,n}(t), \gamma_{1,n}(t)) - d_X(\gamma_{0,n}(s), \gamma_{0,n}(t)) - d_X(\gamma_{1,n}(s), \gamma_{1,n}(t)).$$

Since $\gamma_{0,n}, \gamma_{1,n}$ are κ-qi sections, we have

$$\begin{aligned} d_X(\gamma_{0,n}(s), \gamma_{1,n}(s)) &\geq d_X(\gamma_{0,n}(t), \gamma_{1,n}(t)) - 2\kappa|s-t| - 2\kappa \\ &\geq d_X(\gamma_{0,n}(t), \gamma_{1,n}(t)) - 2(R+1)\kappa - 2\kappa. \end{aligned}$$

Since vertex-spaces of $\mathfrak{X}$ are uniformly properly embedded in the ambient space X we see that $d_X(\gamma_{0,n}(t), \gamma_{1,n}(t)) \to \infty$ as $n \to \infty$. Thus, $d_X(\gamma_{0,n}(t), \gamma_{1,n}(t)) \to \infty$, which in turn implies that $d_{X_{\beta_n(s)}}(\gamma_{0,n}(s), \gamma_{1,n}(s)) \to \infty$ for all $s \in [0, R+1]$. Thus, passing to subsequence, if necessary, we may assume that $d_{X_{\beta_n(s)}}(\gamma_{0,n}(s), \gamma_{1,n}(s)) \geq n$ for all $n \in \mathbb{N}$ and

$s \in [0, R+1]$. Also, since the group G acts on T cocompactly, we can assume, by passing to subsequence if necessary, that $\beta_n(0)$ is a fixed vertex v and $\gamma_{0,n}(0)$ is a fixed point $x \in X_v$. Since X is quasi-isometric to G, by passing to a further subsequence, if necessary, we may assume that β_n is a fixed geodesic vw in T, where $d_T(v,w) = R+1$. We note that since $d_X(\gamma_{0,n}(v), \gamma_{1,n}(w)) \leq \kappa + \kappa R$, by Lemma 2.37 we have

$$\mathrm{Hd}([\gamma_{0,n}(v)\gamma_{1,n}(v)]_{X_v}, [\gamma_{0,n}(w)\gamma_{1,n}(w)]_{X_v}) \leq C_{2.37}(\kappa + (R+1)\kappa).$$

Now, by (2) above we have a constant D_1 and $y, y' \in V(Y)$, $g, g' \in G$ such that

(i) $v = gG_y$, $w = g'G_{y'}$ and

(ii) the diameter of $N_{D_1}(gG_y) \cap g'G_{y'}$ is infinite in Γ.

The rest of the argument is borrowed from [**Mit04**, Theorem 4.6]. Let $\{h_n\} \subset gG_y$ and $\{h'_n\} \subset g'G_{y'}$ be sequences of distinct elements such that $d_\Gamma(h_n, h'_n) \leq D_1$ for all $n \in \mathbb{N}$. Hence, $d_\Gamma(1, h_n^{-1}h'_n) \leq D_1$. But there are only finitely many elements of G inside $B(1; D_1)$. Hence, passing to a subsequence, we may assume that the sequence $\{h_n^{-1}h'_n\}$ is constant. Let $x = h_n^{-1}h'_n$. Consider the equations $x = h_m^{-1}h'_m = h_n^{-1}h'_n$; whence $h_m x = h'_m, h_n x = h'_n$. Thus, we have

$$x^{-1}h_m^{-1}h_n x = h'^{-1}_m h'_n \Rightarrow h_m^{-1}h_n = xh'^{-1}_m h'_n x^{-1}$$
$$\Rightarrow h_m(h_m^{-1}h_n)h_m^{-1} = (h_m x)h'^{-1}_m h'_n (h_m x)^{-1} = h'_m(h'^{-1}_m h'_n)h'^{-1}_m.$$

Clearly, $h_m^{-1}h_n \in G_y$ and, hence, $h_m(h_m^{-1}h_n)h_m^{-1} \in h_m G_y h_m^{-1} = gG_y g^{-1}$, since $h_m \in gG_y$. Similarly, $h'_m(h'^{-1}_m h'_n)h'^{-1}_m \in g'G_{y'}g'^{-1}$. This implies that

$$h_m(h_m^{-1}h_n)h_m^{-1} = h'_m(h'^{-1}_m h'_n)h'^{-1}_m \in gG_y g^{-1} \cap g'G_{y'}g'^{-1}.$$

However, gG_yg^{-1} is the stabilizer of the vertex $v = gG_y$ and $g'G_{y'}g'^{-1}$ is the stabilizer of $w = g'G_{y'}$. Since $\{h_n\}$ and $\{h'_n\}$ are sequences of distinct elements in gG_y and $g'G_{y'}$ respectively, the intersection $G_v \cap G_w$ is infinite. Since $d_T(v,w) = R+1$ this contradicts the R-acylindricity of the G-action. □

REMARK 2.56.

(1) The proof of Proposition 2.55 also works even if we assume that $G_v \cap G_w$ finite whenever $d_T(v,w) \geq k+1$.

(2) In fact, to conclude hyperbolicity of G in the proposition, one does not need the full power of the Combination Theorem, Theorem 2.62, one can derive the result from the cobounded quasiconvex chain-amalgamation, Theorem 2.65.

For the next corollary, we recall that a subgroup H in a group G is *weakly malnormal* if for every $g \in G \setminus H$ the intersection

$$gHg^{-1} \cap H$$

is finite.

COROLLARY 2.57. ([**KM98**, Theorem 2]) *If G_1, G_2 are hyperbolic groups and H is a common quasiconvex subgroup which is weakly malnormal either in G_1 or in G_2, then $G = G_1 *_H G_2$ is hyperbolic.*

PROOF. We claim that the action of G on the Bass–Serre tree for the given amalgam decomposition is 3-acylindrical. Without loss of generality, let us assume that $H < G_1$ is weakly malnormal. If T is the Bass–Serre tree and $v, w \in V(T)$ with $d_T(v,w) \geq 4$, then there are is a sequence of consecutive vertices on vw of the form $v_1 = xG_2, v_2 = yG_1, v_3 = zG_2$, where $x, y, z \in G$. Then $G_{v_1} \cap G_{v_3}$ is equal to the intersection of the stabilizers of the two edges:

(i) one edge connecting v_1, v_2, and
(ii) the one edge connecting v_2, v_3.

However, these are two distinct conjugates of H in the stabilizer of $v_2 = yG_1$, i.e., they are of the form $ygHg^{-1}y^1, yg'Hg'^{-1}y^{-1}$ in yG_1y^{-1} where $g, g' \in G_1$. Since

$$ygHg^{-1}y^1 \cap yg'Hg'^{-1}y^{-1} = y(gHg^{-1} \cap g'Hg'^{-1})y^{-1}$$

(weak) malnormality of H in G_1 proves our claim. Then the hyperbolicity follows from Proposition 2.55. □

An example analogous to the situation of the corollary above in the context of a tree of spaces is discussed in Section 2.6.2.

2.5.4. Exponential flaring (Bestvina–Feighn flaring condition).

Definition 2.58 (Exponential flaring condition). We say that a tree $\mathfrak{X}$ of metric spaces $\pi : X \to T$ satisfies the *Bestvina–Feighn κ-flaring condition* or the *exponential κ-flaring condition*, if there exist $\lambda_\kappa > 1, M_\kappa > 0$ and $n_\kappa \in \mathbb{N}$ such that the following holds.

For every pair $\Pi = (\gamma_0, \gamma_1)$ of κ-qi sections of $\mathfrak{X}$ over a length $2n_\kappa$ geodesic interval $[\![-n_\kappa, n_\kappa]\!] \subset T$, if the girth Π_0 of the pair (γ_0, γ_1) is $\geq M_\kappa$, then

$$\lambda_\kappa \cdot \Pi_0 \leq \Pi_{\max}.$$

A form of this flaring condition first appeared in the paper **[BF92]** of Bestvina and Feighn. Actually, the original Bestvina–Feighn flaring condition was a bit different from the exponential flaring condition above as they required not just two qi sections but a 1-parameter family of κ-qi sections interpolating these two, i.e., a κ-hallway; see Definition 3.15. The existence of such a family (with a different but uniform qi constant κ') follows from **[MS09]**. It will be also proven in Lemma 3.17(b).

We will see below that the exponential flaring implies proper flaring with an exponential function ϕ_κ and that if X is hyperbolic, then $\mathfrak{X}$ satisfies the exponential κ-flaring condition for all $\kappa \geq 1$. Note that while in their first paper **[BF92]** Bestvina and Feighn imposed the exponential flaring condition for all $\kappa \geq 1$, in the addendum **[BF96]** to their paper, the flaring condition was required only for some value of κ, cf. the statement of our main result, Theorem 2.62.

Lemma 2.59. *Bestvina–Feighn κ-flaring implies exponential proper κ-flaring. Moreover, the proper flaring condition holds either in the negative or in the positive direction (see Definition* 2.44*).*

Proof. We fix κ and set $n := n_\kappa, \lambda := \lambda_\kappa$. Suppose that $\Pi = (\gamma_0, \gamma_1)$ is a pair of κ-qi sections over a geodesic interval I of length $N = sn$ and of girth $\Pi_0 \geq M_\kappa$. For concreteness, we assume that

$$d_{X_n}(\gamma_0(n), \gamma_1(n)) \geq \lambda d_{X_0}(\gamma_0(0), \gamma_1(0)) = \Pi_0.$$

Then, applying the flaring inequality to the subinterval in I of length $2n$ centered at n, we obtain

$$\max(d_{X_{2n}}(\gamma_0(2n), \gamma_1(2n)), \Pi_0) \geq \lambda d_{X_n}(\gamma_0(n), \gamma_1(n)).$$

Since $\lambda > 1$, the maximum in this inequality is attained by $d_{X_{2n}}(\gamma_0(2n), \gamma_1(2n))$ and, thus,

$$d_{X_{2n}}(\gamma_0(2n), \gamma_1(2n)) \geq \lambda d_{X_n}(\gamma_0(n), \gamma_1(n)).$$

Applying this argument inductively, we obtain:

$$\lambda_\kappa^s \cdot \Pi_0 \leq d_{X_{sn}}(\gamma_0(sn), \gamma_1(sn)) \leq \Pi_{\max}(sn).$$

By reducing λ to $\mu > 1$ if necessary and using Lemma 2.38, we also get

$$\Pi_{\max}(m) \geq d_{X_m}(\gamma_0(m), \gamma_1(m)) \geq \mu^m \Pi_0, \forall m \geq n.$$

Since the function $m \mapsto \mu^m, m \in \mathbb{N}$, is proper, the exponential proper κ-flaring condition for $\mathfrak{X}$ follows. □

PROPOSITION 2.60. *If $\mathfrak{X}$ satisfies the proper κ-flaring condition for all $\kappa \geq 1$, then $\mathfrak{X}$ also satisfies satisfies an exponential κ-flaring condition for all $\kappa \geq 1$. In particular, if X is hyperbolic, then $\mathfrak{X}$ satisfies satisfies an exponential κ-flaring condition for all $\kappa \geq 1$.*

PROOF. Since X is hyperbolic, the tree of spaces $\mathfrak{X} = (X \to T)$ satisfies both the proper κ-flaring and the property obtained in Corollary 2.53 for all $\kappa \geq 1$. We will use both of these in the proof. The proof is inspired by, but is conceptually simpler than [**MS09**, Proposition 5.8]. For each $K \geq 1$, we inductively define $K_0 := K$ and $K_i := \max\{K_{i-1}, C_{2.37}(K_{i-1})\}$, $i \geq 1$. Given $\kappa \geq 1$ we set

$$L := \eta_{2.18}(2\kappa_3), \quad \epsilon = 3\eta_{2.18}(2\kappa_3), \quad R := \max\{1, m_{\kappa_3}, L(5\epsilon + 4L)\},$$

and $D := \max\{R, R_{2.53}(\kappa_3, R)\}$. Let $n = n_\kappa$ be any integer such that $\phi_{\kappa_3}(n) \geq 12D$; set $\lambda_\kappa := 2$ and $M_\kappa := D + 1$.

If $\Pi = (\gamma_0, \gamma_1)$ is a pair of κ-qi sections over an interval $J = [-n, n] \subset T$, $\Pi_0 \geq M_\kappa$, then we form an interval bundle $\mathfrak{Y} = (Y \to J)$ (see the construction preceding Lemma 2.41).

After that, the idea is to first decompose this interval bundle into a finite number of subbundles by constructing qi sections in Y (cf. [**MS09**, Proposition 3.14], also Proposition 4.14), where the subbundles intersect along the qi sections. We then use uniformly proper flaring to prove that the qi sections bounding each of the subbundles flare in at least one direction. Finally, as in the last step of the proof of [**MS09**, Proposition 5.8], we verify that at least half of these will flare in the same direction to finish the proof.

STEP 1 (Construction of qi sections in Y). We note that through any point of the metric bundle formed by two K_{i-1}-qi sections over J, there is a K_i-qi section, $i \geq 1$ (see Lemma 2.41). Let $\alpha_i = Y \cap X_i$, $i \in V(J)$. For two consecutive integers i, j we have a map $h_{ij} : \alpha_i \to \alpha_j$ such that for all $x \in \alpha_i$, $d_{X_{ij}}(h_{ij}(x), x) \leq \kappa_3$. This map is clearly $\eta_{2.18}(2\kappa_3)$-coarsely Lipschitz, with a similarly defined $\eta_{2.18}(2\kappa_3)$-coarse inverse $h_{ji} : \alpha_j \to \alpha_i$, which is also an $\eta_{2.18}(2\kappa_3)$-coarsely Lipschitz map. Hence, by Lemma 1.15, the maps h_{ij}, h_{ji} are both $(\eta_{2.18}(2\kappa_3), 3\eta_{2.18}(2\kappa_3))$-quasi-isometries. By Lemma 1.24, if $x, y, z \in \alpha_i$ and y is between x, z with $d_{X_i}(x, y) \geq L(5\epsilon + 4L)$, and $d_{X_i}(y, z) \geq L(5\epsilon + 4L)$, then $h_{ij}(y)$ is between $h_{ij}(x)$ and $h_{ij}(z)$. In particular, this is true if $d_{X_i}(x, y) \geq R$ and $d_{X_i}(y, z) \geq R$ by the choice of R.

Suppose $l(\alpha_0) = l$ and let α_0 also denote the parametrization of this geodesic in X_0 so that $\alpha_0(0) = \gamma_0(0), \alpha_0(l) = \gamma_1(0)$. Next, we inductively construct a sequence of numbers $s_0 = 0, \cdots, s_n = l$ and a sequence of κ_1-qi sections $\gamma_0 = \beta_0, \beta_1, \cdots, \beta_n = \gamma_1$ in Y such that each β_{i+1} is contained in the metric bundle formed by β_i and β_n, $0 \leq i \leq n-2$ as follows. Suppose $s_0, \cdots, s_i$ are chosen and so are $\beta_0, \cdots, \beta_i$ and $s_i < l$. To construct s_{i+1} and β_{i+1} consider the subset $S \subset (s_i, l]$ consisting of s such that there is a κ_2-qi section β through s satisfying

$$\min_j d_{X_j}(\beta(j), \beta_i(j)) \leq R.$$

If $S = \emptyset$ then define $s_{i+1} = s_n = \gamma_1$. Assume now that S is nonempty. Suppose there is $s \in S$ and a κ_1-qi section β in Y through $\alpha_0(s)$ such that $\min_j d_{X_j}(\beta(j), \beta_i(j)) = R$. In this case, if $s \neq l$, then define $s_{i+1} = s, \beta_{i+1} = \beta$. Otherwise, if $s = l$, then we define $s_{i+1} = s_n = l$

and $\beta_n = \gamma_1$. Suppose there is no such $s \in S$. Then let $s_{i+1} = \min\{l, 1 + \sup S\}$. If $s_{i+1} \neq l$, then let β_{i+1} be any κ_1-qi section in Y passing through s_{i+1}. Otherwise, define $s_n = s_{i+1}$ and $\beta_{i+1} = \gamma_1$. We note that $s_{i+1} - s_i \geq R$ unless $s_{i+1} = s_n$.

STEP 2 (Verification of properties of qi sections). Let $\Pi^i = (\beta_i, \beta_{i+1})$ and let Y^i denote the interval-bundle over J formed by these qi sections. We claim that $Y^i \cap Y^j = \emptyset$, unless $|i - j| \leq 1$, and $Y^i \cap Y^{i+1} = \beta_{i+1}$ for all permissible i. Both claims follow from Lemma 1.24, cf. [**MS09**, Lemma 3.12].

STEP 3 (Flaring of $\Pi^i = (\beta_i, \beta_{i+1})$). We know that there is a κ_2-qi section $\bar{\beta}_i$ through either s_{i+1} or $s_{i+1} - 1$ inside the subbundle Y^i, such that

$$d_{X_j}(\beta_i(u_i), \bar{\beta}_i(u_i)) \leq R$$

for some $u_i \in V(J)$. Without loss of generality, we may assume that $u_i < 0$. Now we construct a new set of κ_3-qi sections inside the bundle formed by β_i and $\bar{\beta}_i$ as follows. Let $r = \lfloor (s_{i+1} - s_i - 1)/D \rfloor$. Let β'_j, $0 \leq j \leq r$ be arbitrary κ_3-qi sections in the bundle formed by $\beta_i, \bar{\beta}_i$ such that $\beta'_0 = \beta_i$, and for $j \neq 0$ β'_j passes through $\alpha_0(s_i + jD)$. It follows from Lemma 2.53 that for all $j, k \geq 0$ $d_{X_j}(\beta'_k(j), \beta'_{k+1}(j)) \geq R$ and, thus, as in Step 2, by Lemma 1.24 we see that

$$\beta_i(j) = \beta'_0(j), \cdots, \beta'_m(0), \beta_{i+1}(j)$$

is a monotonic sequence in the geodesic interval $[\beta_i(j)\beta_{i+1}(j)]_{X_j}$ for all $j \geq 0$. Thus,

$$\begin{aligned} d_{X_n}(\beta_i(n), \beta_{i+1}(n)) &\geq \sum_{j=0}^{r-1} d_{X_n}(\beta'_j(n), \beta'_{j+1}(n)) \\ &\geq \sum_{j=0}^{r-1} 12D \\ &= \sum_{j=0}^{r-1} 12 d_{X_0}(\beta'_j(0), \beta'_{j+1}(0)) \\ &= 12 d_{X_0}(\beta'_0(0), \beta'_r(0)) \end{aligned}$$

by the uniformly proper flaring and by the choice of n. However,

$$d_{X_0}(\beta_i(0), \beta_{i+1}(0)) = d_{X_0}(\beta'_0(0), \beta'_r(0)) + d_{X_0}(\beta'_r(0), \beta_{i+1}(0)),$$

where $d_{X_0}(\beta'_r(0), \beta_{i+1}(0) \leq D + 1$. It follows that

$$d_{X_n}(\beta_i(n), \beta_{i+1}(n)) \geq \frac{12D}{2D+1} d_{X_0}(\beta_i(0), \beta_{i+1}(0)) \geq 4 d_{X_0}(\beta_i(0), \beta_{i+1}(0)),$$

since $D \geq 1$.

STEP 4 (Exponential flaring of γ_0, γ_1). We know that each pair $\Pi^i = (\beta_i, \beta_{i+1})$ exponentially flares in at least one direction (say, in the positive direction) by Step 3. Then there is a subset of induces $I \subset \{1, 2, \dots\}$, such that

$$\sum_{i \in I} \Pi^i_0 \geq \frac{1}{2} \Pi_0.$$

It follows that Π flares exponentially in the positive direction with $\lambda_\kappa = 2$. □

COROLLARY 2.61. *If $\mathfrak{X}$ satisfies the (uniformly) proper κ-flaring condition for all $\kappa \geq 1$, then (again, for all $\kappa \geq 1$) it satisfies the (uniformly) proper flaring either in positive or in the negative direction.*

2.6. Hyperbolicity of trees of hyperbolic spaces

2.6.1. The combination theorem. We are now ready to state our version of the combination theorem of Bestvina and Feighn [**BF92**]:

THEOREM 2.62. *There exist $K_* = K_{2.62}(\delta_0, L_0)$ and $\delta_* = \delta_{2.62}(\delta_0, L_0)$, depending only on δ_0 and L_0, such that the following holds. Suppose $\mathfrak{X} = (\pi : X \to T)$ is a tree of hyperbolic spaces (with primary parameters δ_0, L_0) satisfying the uniform K_*-flaring condition. Then X is a δ_*-hyperbolic metric space.*

The constants K_* and δ_* are computable. In Remark 6.12 we will give a formula for K_*, which is inductive in nature, as it relies upon earlier computations of various constants and functions scattered throughout the book. (We will not attempt to write a formula for δ_*.) The reader unwilling to keep track of such computations, can simply assume that $\mathfrak{X}$ satisfies the uniform κ-flaring condition *for all* $\kappa \geq 1$.

COROLLARY 2.63. *There exist $K = K_{2.63}(\delta_0, L_0)$ and $\delta = \delta_{2.63}(\delta_0, L_0)$, depending only on δ_0 and L_0, such that the following holds. Suppose $\mathfrak{X} = (\pi : X \to T)$ is a tree of hyperbolic spaces (with primary parameters δ_0, L_0) satisfying one of the following conditions:*

(1) *The uniformly proper K-flaring condition (Definition* 2.43*).*
(2) *The proper K-flaring condition (Definition* 2.45*).*
(3) *The Bestvina–Feighn K-flaring condition (Definition* 2.58*).*
(4) *The $R(K, C)$-bigon-property for some C (Definition* 2.52*).*

Then X is a δ-hyperbolic metric space.

PROOF. This is a direct consequence of Theorem 2.62, Propositions 2.47, 2.46, 2.60 and Corollary 2.53. □

2.6.2. Cobounded quasiconvex chain-amalgamation. In the book we will be frequently using the following very special case of Theorem 2.62 which is much easier to prove; see, e.g., [**MS09**, Proposition 1.51]. This special case was motivated by a result of Hamenstadt, [**Ham05**, Lemma 3.5]. Although Hamenstadt used much stronger assumptions, it is clear that the proof of Hamenstadt goes through with the weaker hypothesis as well. We include a proof along the lines of Hamenstadt's arguments for the sake of completeness and also since we want a description of geodesics.

We assume that X is a geodesic metric space that can be represented as a union of a finite chain of rectifiably connected subsets equipped with induced path-metrics

$$X = Q_0 \cup Q_1 \cup \cdots \cup Q_n,$$

such that for some constants C and δ the following hold.

(1) Each Q_i is δ-hyperbolic geodesic metric space (with respect to the induced length metric from X).
(2) For each $i < n$ the intersection $Q_{i,i+1} = Q_i \cap Q_{i+1}$ is rectifiably connected and L-qi embedded in Q_i, Q_{i+1}. We assume these spaces also to be geodesic metric space (with respect to the induced length metric from X).
(3) Each $Q_{i,i+1}$ separates (in Y) Q_i from Q_{i+1} in the sense that every path c in X connecting Q_i to Q_{i+1} has to cross $Q_{i,i+1}$.
(4) Each pair of intersections $Q_{i-1,i}, Q_{i,i+1}$ is C-cobounded in Q_i.
(5) $d_{Q_i}(Q_{i-1,i}, Q_{i,i+1}) \geq 1$.

We will say that such X is a *cobounded quasiconvex chain-amalgam* of Q_i. If $n = 1$, we will refer to $X = Q_0 \cup Q_1$ simply as a *quasiconvex amalgam.*

REMARK 2.64.

(1) We could also work with a path-metric space X satisfying the above conditions, since we can always replace X with its graph approximation (see Lemma 1.33).
(2) Clearly, the collection Q_i in a cobounded quasiconvex chain-amalgam gives X structure of a tree of hyperbolic spaces with vertex-spaces Q_i and edge-spaces $Q_{i,i+1}$, such that the tree T is isometric to the interval J of length $n + 1$ in $\mathbb{R}$ with integer vertices. Conversely, consider a tree of hyperbolic spaces $\mathfrak{X}$ over an interval T such that for each vertex v with the incident edges $e_\pm$, the corresponding subsets $X_{e_\pm v}$ are C'-cobounded in X_v. Then $\mathfrak{X}$ yields a cobounded quasiconvex chain-amalgam with subsets $Q_i = Q_v$, $v = v_i$, equal to the unions

$$X_{e_-} \times \left[\frac{1}{2}, 1\right] \cup_{f_{e_-v}} X_v \cup_{f_{e_+v}} X_{e_+} \times \left[0, \frac{1}{2}\right],$$

$$Q_{i-1,i} = X_{e_-} \times \frac{1}{2};$$

see Section 1.6 for the definition of mapping cylinders.

However, for each i pick points

$$x_i^- \in N_{C'}(P_{Q_i,Q_{i-1,i}}(Q_{i,i+1})) \cap Q_{i-1,i},\ x_i^+ \in N_{C'}(P_{Q_i,Q_{i,i+1}}(Q_{i-1,i})) \cap Q_{i,i+1},$$

where the C'-neighborhoods are taken with respect to the metric of Q_i, Q_i. Since both projections of $Q_{i-1,i}$ to $Q_{i,i+1}$ and of $Q_{i,i+1}$ to $Q_{i-1,i}$ have diameters $\leq C$, we obtain

$$d_{Q_i}(Q_{i-1,i}, Q_{i,i+1}) \leq d_{Q_i}(x_i^-, x_i^+) \leq d_{Q_i}(Q_{i-1,i}, Q_{i,i+1}) + 2(C + C'), \tag{2.1}$$

i.e., the pair of points x_i^-, x_i^+ "almost" realizes the minimal distance in Q_i between the subsets $Q_{i-1,i}$, $Q_{i,i+1}$.

We will simultaneously prove hyperbolicity of X and describe uniform quasigeodesics connecting points in X. For this description, given points $x \in Q_{i-1}, x' \in Q_{k+1}$, it will be convenient to name their nearest-point projections (in Q_{i-1}, Q_{k+1}) to $Q_{i-1,i}, Q_{k,k+1}$ as $\bar{x}, \bar{x}'$, respectively. Suppose, furthermore, that

$$c(x_i^+, x_{i+1}^-),$$

are L'-quasigeodesic paths in $Q_{i,i+1}$ connecting x_i^+ to x_{i+1}^- and

$$c(x_i^-, x_i^+), c(x, \bar{x}), c(\bar{x}', x')$$

are L'-quasigeodesic paths in Q_i connecting x_i^- to x_i^+, etc. We let $c^*(\cdot, \cdot)$ denote the corresponding geodesics paths in $Q_{i,i+1}$, Q_i, connecting the respective points.

THEOREM 2.65. *Under the above assumptions, X is $\delta_{2.65}(\delta, L, L', C, C')$-hyperbolic. Moreover, the following paths $c(x, x')$ are $K = K_{2.65}(\delta, L, C, D, L')$-quasigeodesics in X connecting $x \in Q_{i-1}, x' \in Q_{k+1}$, $i \leq k$:*

(1) *If x, x' belong to $Q_i' = Q_i \setminus (Q_{i-1,i} \cup Q_{i,i+1})$ for some i, then we assume that $c(x, x')$ is an L'-quasigeodesic in Q_i connecting x to x'.*
(2) *Otherwise,*

$$c(x, x') = c(x, \bar{x}) \star c(\bar{x}, x_i^-) \star c(x_i^-, x_i^+) \star c(x_i^+, x_{i+1}^-) \star \cdots \star c(x_k^+, \bar{x}') \star c(\bar{x}', x').$$

PROOF. This theorem is proven by verifying the assumptions of Corollary 1.64, i.e., axioms of a slim combing.

(a1) We will need to estimate the length of $c(x, x')$ in terms of $d(x, x')$. First of all,

$$\text{length}(c(x, x')) \le L'(\text{length}(c^*(x, x')) + 1),$$

hence, it suffices to get an estimate for c^*.

Let γ be any geodesic in X connecting x to x'. In view of the separation property (3) in the theorem, for each $i \le j \le k$, γ will contain subpaths $\gamma_j \subset Q_j$ (necessarily a geodesic in Q_j) connecting a point $p_j^- \in Q_{j-1,j}$ to some $p_j^+ \in Q_{j,j+1}$.

Let P_-, P_+ denote the projections $Q_j \to Q_{j-1,j}$, $Q_j \to Q_{j,j+1}$ respectively. According to Lemma 1.105, γ_j contains points $y_j^\pm$ satisfying

$$d_{Q_j}(y_j^\pm, P_\pm(p_j^\mp)) \le 2\delta + \lambda,$$

hence,

$$d_{Q_j}(y_j^\pm, x_j^\pm) \le D := C + C' + 2\delta + \lambda,$$

where $\lambda = \lambda_{1.92}(\delta, L)$ is the quasiconvexity constant of $Q_{j,j\pm1}$ in Q_j. Thus,

$$\text{length}([p_j^- x_j^-]_{Q_j} \star [x_j^- x_j^+]_{Q_j} \star [x_j^+ p_j^+]_{Q_j}) \le \text{length}(\gamma_j) + 4D.$$

Since $Q_{j-1,j}, Q_{j,j+1}$ are L-qi embedded in Q_j we also obtain

$$\text{length}(c(p_j^-, p_j^+)) \le L \cdot \text{length}(\gamma_j) + 4D(L + 1). \tag{2.2}$$

Since

$$\text{length}(\gamma) \ge d(x, p_i^-) + \sum_{j=i}^{k} \text{length}(\gamma_j) + d(p_k^+, x'),$$

by combining the inequalities (2.2), we get:

$$\text{length}(c^*(x_i^-, x_k^+)) \le \text{length}(c(p_i^-, p_k^+)) \le L \sum_{j=i}^{k} \text{length}(\gamma_j) + 4D(L + 1)(k - i + 1).$$

To estimate the term $4D(L + 1)m$, $m = k - i + 1$, note that $d(x, y) \ge m$ (in view of the assumption 5 in the theorem). Thus,

$$\text{length}(c^*(x_i^-, x_k^+)) \le Ld(x, x') + 4D(L + 1)d(x, x') = (L + 4D(L + 1))d(x, x').$$

Lastly, we deal with $d(x, x_i^-)$ and $d(x', x_k^+)$. Recall that the metric space (Q_i, d_{Q_i}) is η-properly embedded in X (Proposition 2.18). We obtain:

$$\begin{aligned}
\text{length}\, c^*(x, x_i^-) &= d_{Q_i}(x, \bar{x}) + d_{Q_{i-1,i}}(\bar{x}, x_i^-)\\
&\le d_{Q_{i-1}}(x, \bar{x}) + L + Ld_{Q_{i-1}}(\bar{x}, x_{i+1}^-)\\
&\le L + 2D + Ld_{Q_{i-1}}(x, x_i^-)\\
&\le L + 2D + L\eta(d(x, x_i^-))\\
&\le L + 2D + L\eta(d(x, y_i^-) + D)\\
&\le L + 2D + L\eta(d(x, x') + D).
\end{aligned}$$

Similarly,

$$\text{length}\, c^*(x', x_k^+) \le L + 2D + L\eta(d(x, x') + D).$$

Combining the inequalities, we obtain:

$$\begin{aligned}
\text{length}(c^*(x, x')) &= \text{length}\, c^*(x, x_i^-) + \text{length}(c(x_i^-, x_k^+) + \text{length}\, c^*(x', x_k^+)\\
&\le \eta_{2.65}(d(x, x'))\\
&:= 2(L + 2D + L\eta(d(x, x') + D)) + (L + 4D(L + 1))d(x, x').
\end{aligned}$$

(a2) Consider a triple of points $x \in Q_i, y \in Q_j, z \in Q_k$, $i \le j \le k$, and the "triangle"

$$\Delta_c = c(x, y) \cup c(y, z) \cup c(z, x).$$

By the definition of the paths c in X, the paths $c(x, y), c(y, z)$ coincide away from Q_j, the same applies to the pair of paths $c(y, z), c(z, x)$. Therefore, it suffices to consider the case when $i = j = k$.

We will use the notation pq for geodesics in Q_i. Our goal is to verify that each of the paths $c(p, q)$ connecting points $p, q \in Q_i$ are uniformly Hausdorff-close to a geodesic pq: Once we are done with this, then uniform slimness of Δ_c will follow from δ-hyperbolicity of Q_i.

If both p, q are in Q_i' or in $Q_{i,i+1}$ or in $Q_{i-1,i}$, there is nothing to prove. Hence, up to permutation of the points p, q and reversing the order in the interval $[0, n]$, there are two cases to consider, depending on the position of the points p, q with respect to the subsets $Q_{i-1,i}, Q_{i,i+1}$:

CASE 1. Suppose that $p \notin Q_{i-1,i}$ and $q \in Q_{i,i+1}$. We will be using the notation $\bar{p} = P_{Q_i, Q_{i,i+1}}(p)$. Then

$$c(p, q) = c(p, \bar{p}) \cup c(\bar{p}, p)$$

Since $Q_{i,i+1}$ is L-qi embedded in Q_i, this path is $D_{1.54}(\delta, LL')$-Hausdorff close to the union $p\bar{p} \cup \bar{p}q$. According to Lemma 1.103,

$$\mathrm{Hd}_{Q_i}(p\bar{p} \cup \bar{p}q, pq) \le \lambda + 2\delta,$$

concluding the proof in this case.

CASE 2. Suppose that $p \in Q_{i-1,i}$ and $q \in Q_{i,i+1}$. In view of the assumption that $Q_{i-1,i}, Q_{i,i+1}$ are L-qi embedded in Q_i, we will work with Q_i-geodesics connecting pairs of points points in $Q_{i-1,i}$ and pairs of points in $Q_{i,i+1}$ instead of the c-paths in $Q_{i-1,i}$ and $Q_{i,i+1}$. Continuing with the notation of Case 1,

$$\mathrm{Hd}_{Q_i}(pq, p\bar{p} \cup \bar{p}q) \le \lambda + 3\delta$$

(see Lemma 1.105) and

$$d(\bar{p}, x_i^+) \le C.$$

Thus,

$$\mathrm{Hd}_{Q_i}(pq, px_i^+ \cup x_i^+ q) \le C + \lambda + 4\delta.$$

Similarly,

$$\mathrm{Hd}_{Q_i}(px_i^+, px_i^- \cup x_i^- x_i^+) \le C + \lambda + 4\delta.$$

Combining the inequalities, we obtain

$$\mathrm{Hd}_{Q_i}(pq, px_i^+ \cup x_i^- x_i^+ \cup x_i^+ q) \le 2(C + \lambda + 4\delta). \qquad \square$$

REMARK 2.66. The flexibility of working with concatenations of quasigeodesics points $x_i^\pm$ uniformly close to nearest-point projections will be critical in several places in the book, e.g., in Chapter 8.

COROLLARY 2.67. *Assuming that $X \to T$ is a tree of hyperbolic spaces satisfying the assumptions of Theorem 2.65, for every subinterval $S \subset T$, the inclusion map*

$$X_S \to X$$

is an $L_{2.67}(\delta, L, C)$-qi embedding.

PROOF. This is an immediate consequence of the of the fact that for any pair of points $x, y' \in S_X$, the path $c_{X_S}(x, x')$ equals the path $c_X(x, x')$ where the subscript denotes the space in which we define the combing. □

COROLLARY 2.68. *Suppose that $\mathfrak{X} = (\pi : X \to T)$ is a tree of spaces satisfying Axiom **H**. Then, for every edge $e = [v, w]$ of T the space X_{vw} equipped with its natural path-metric, is δ_0'-hyperbolic with the hyperbolicity constant δ_0' depending only on the primary parameters of the tree of hyperbolic spaces $\mathfrak{X}$.*

We now return to the general quasiconvex chain-amalgamation and relate this class of trees of spaces to acylindrical trees of spaces. Suppose that γ is a κ-qi section of the tree of spaces $\mathfrak{X} = (\pi : X \to J)$, defined on an interval $I \subset J$. Thus, for each integer i we have a point $x_i \in Q_i$ and $d_Y(x_i, x_{i+1}) \le K$ for some K depending on κ. If the length of I is ≥ 3, it follows that for each triple of indices $i-1, i, i+1$ the point x_i is within uniform distance $D = D(K)$ from both $Q_{i-1,i}$ and $Q_{i,i+1}$: There exist $y_i^- \in Q_{i-1,i}$, $y_i^+ \in Q_{i,i+1}$ such that

$$d(x_i, y_i^{\pm}) \le K.$$

Such a point x_i might not even exist, which would mean that each κ-qi section γ of $\mathfrak{X}$ is defined only on an interval of length ≤ 2, and that would definitely ensure acylindricity of $\mathfrak{X}$. In general, one can say that

$$d(P_{Q_i,Q_{i-1,i}}(y_i^+), x_i^-) \le 2K, \quad d(P_{Q_i,Q_{i,i+1}}(y_i^-), x_i^+) \le 2K,$$

and, hence,

$$d(x_i, x_i^{\pm}) \le 3K.$$

It follows that any two κ-qi sections γ_0, γ_1 defined on I satisfy

$$d(\gamma_0(i), \gamma_1(i)) \le 6K, i \in V(I),$$

thereby ensuring $(6K, \kappa, 3)$-acylindricity of $\mathfrak{X}$.

2.6.3. Hyperbolicity of finite trees of hyperbolic spaces. We will also need the following version of Theorem 2.65 in the situation when the coboundedness condition is dropped:

COROLLARY 2.69. *Suppose that T is a finite tree, $\mathfrak{X} = (\pi : X \to T)$ is a tree of hyperbolic spaces (satisfying Axiom **H**). Then X is δ-hyperbolic with*

$$\delta = \delta_{2.69}(\delta_0, L_0, |V(T)|),$$

i.e., δ depends only on the parameters of $\mathfrak{X}$ and the cardinality of $|V(T)|$.

PROOF. The proof is induction on $|V(T)|$. For $|V(T)| = 1$, there is nothing to prove. For $n = 2$, the corollary is nothing but Theorem 2.65 for the quasiconvex amalgam of pairs. We, thus, assume that the corollary holds for all trees S with $|V(S)| = n \ge 2$. Let $\mathfrak{X} = (\pi : X \to T)$ be a tree of hyperbolic spaces with $|V(T)| = n+1$. Pick a valence 1 vertex w of T and let $e = [v, w]$ be the incidence edge. Set $Y_v := X_{vw}$. We then form a new tree of spaces $\mathfrak{Y} = (Y \to S)$, where S is obtained from T by removing w and e, hence, S has one less vertex than T. For vertices of S which are distinct from v and edges which are not incident to v, we use the same incidence maps for $\mathfrak{Y}$ as we had for $\mathfrak{X}$. For edges $e_i = [v_i, v]$ incident to v we use incidence maps $Y_{e_i} = X_{e_i} \to Y_v = X_{vw}$ equal to the corestrictions of the incidence maps $X_{e_i} \to X_v$. The new tree of spaces still satisfies the assumptions of the corollary since $Y_v = X_{vw}$ is $\delta_1 = \delta(\delta_0, L_0, 2)$-hyperbolic and incidence maps

$$f_{e_i v} : X_{e_i} \to Y_v = X_{vw}$$

are $L_1 = L_0 \cdot L_0'$-qi embeddings, where $L_0' = L_{2.28}(\delta, L_0)$. Now, $\delta = \delta(\delta_0, L_0, n)$-hyperbolicity of X follows from the induction hypothesis. □

2.6.4. Secondary parameters of trees of hyperbolic spaces. In addition to the primary parameters of trees of hyperbolic spaces $\mathfrak{X} = (X \to T)$, we will be frequently using *secondary parameters*, which are functions of the primary parameters. Since these secondary parameters are used so often, we will give them special names. First of all, we recall some constants defined earlier, namely, λ_0, the quasiconvexity constant of the images X_{ev} of incidence maps $X_e \to X_v$ and $L_0' \geq 2$, the quasi-isometry constant for the inclusion maps $X_v \to X_{vw}$, where $e = [v, w]$ runs over all edges of T (Lemma 2.28). Also, δ_0' is the supremum of hyperbolicity constants of the spaces $X_{uv} = X_{[\![u,v]\!]}$ (Corollary 2.68). Let λ_0' denote an upper bound on the quasiconvexity constants for the images in X_{vw} of $4\delta_0$-quasiconvex subsets in X_v, X_e (in particular, each X_v, X_e is λ_0'-quasiconvex in X_{vw}). Explicitly, one can take

$$\lambda_0' = 92(L_0')^2(L_0' + 3\delta_0').$$

We will also use the notation L_1' for an upper bound of coarse Lipschitz constants of projections $P = P_{X_{uv}, X_v} : X_{uv} \to X_v$, $L_1' = (L_0' + 1) \cdot D_{1.102}(\delta_0', \lambda_0')$ (Lemma 1.102). Last, but not least, we define the constant

$$K_0 := (15(2\lambda_0' + 5\delta_0')L_0')^3. \tag{2.3}$$

The importance of this constant will become clear in Chapter 3 during the discussion of flows of quasiconvex subsets of vertex-spaces of $\mathfrak{X}$. This constant will be critical in computing the flaring constant K_* in Theorem 2.62.

2.7. Flaring for semidirect products of groups

The purpose of this section is to illustrate the concept of flaring in the case of semidirect products of groups, $G = H \rtimes \mathbb{Z}$.

Suppose H is a nonelementary finitely generated group (which we will eventually assume to be hyperbolic) with a finite generating set S and the corresponding word-metric d_H. Recall that the word-length of an element $h \in H$, denoted $|h|_H$ or $|h|_S$, when the generating set is to be stressed, is related to d_H by $|h|_H = d_H(1, h)$.

Let $f : H \to H$ be an automorphism and $G = H \rtimes_f \langle t\rangle$ the corresponding semidirect product. Let $S_G = S \cup \{t\}$ be a generating set of G, where t is the stable letter corresponding to the infinite cyclic factor in the semidirect product. Let X be the Cayley graph $\Gamma(G, S)$; define the linear tree $T = \Gamma(\mathbb{Z}, 1)$. Then we have a tree of metric spaces $\pi : X \to T$, where the vertex-spaces are various left cosets $X_i := t^i H, i \in \mathbb{Z}$, of H in G. (Strictly speaking, X is only quasi-isometric to the 2-dimensional complex which is the total space of the abstract tree of spaces whose vertex-spaces are isometric copies of the Cayley graph of H.) We shall denote the standard metric on X by d_G and the metrics on the left cosets $t^i H \subset G$ by d_{t^iH}; the latter are isometric to the word-metric on H corresponding to the generating set S.

Given $m \in \mathbb{Z}, n \in \mathbb{N}$, a κ-qi section over the interval $[\![m-n, m+n]\!]$ in T is a sequence[5] $\{h_i t^i\}$, $m \leq i \leq n$, such that for each $i \in [m-n, m+n] \cap \mathbb{Z}$, $d_{X_{i,i+1}}(h_i t^i, h_{i+1} t^{i+1}) \leq \kappa$, where we identify integers $i \in [m-n, m+n]$ with the corresponding vertices of T. This inequality is satisfied, in particular, when $d_{X_i}(h_i t^i, h_{i+1} t^i) \leq \kappa - 1$. Since the vertex-spaces X_i, X_{i+1} are qi

[5]For further computations, we find it notationally convenient to write elements of $t^i H$ as $h_i t^i, h_i \in H$, which is possible since H is normal in G.

embedded in $X_{i,i+1}$, after changing κ if necessary, we can (and will) identify κ-qi sections with sequences $\{h_i t^i\}$ satisfying the inequality

$$d_{X_i}(h_i t^i, h_{i+1} t^i) = d_H(1, t^{-i} h_i^{-1} h_{i+1} t^i) = d_H(1, f^{-i}(h_i^{-1} h_{i+1})) = |f^{-i}(h_i^{-1} h_{i+1})|_H \le \kappa,$$

equivalently,

$$d_H(f^{-i}(h_i), f^{-i}(h_{i+1})) \le \kappa. \tag{2.4}$$

Here is an explicit example.

EXAMPLE 2.70. Fix $h \in H$. Then $i \mapsto ht^i, i \in \mathbb{Z}$, is a 1-qi section over T.

Now, let us see what, respectively, exponential and proper flaring conditions in this context mean in group-theoretic terms. Suppose γ, γ' are two κ-qi sections over $[\![m - n, m + n]\!]$, where $m, n \in \mathbb{Z}$, given by maps $i \mapsto a_i t^i$ and $i \mapsto b_i t^i$. Then for each integer $i \in [m - n, m + n]$, the fiber-distance equals

$$d_{t^iH}(\gamma(i), \gamma'(i)) = d_{t^iH}(a_i t^i, b_i t^i) = d_H(t^{-i} a_i t^i, t^{-i} b_i t^i) = d_H(1, t^{-i} a_i^{-1} b_i t^i) = |f^{-i}(a_i^{-1} b_i)|_H.$$

If we denote the pair of sections (γ, γ') by Π, then

$$\Pi_{max} = \max\{|f^{-m+n}(a_{m-n}^{-1} b_{m-n})|_H, |f^{-m-n}(a_{m+n}^{-1} b_{m+n})|_H\}.$$

In the special case, when $m = 0, n = 1$,

$$\Pi_{max} = \max\{|f(a_{-1}^{-1} b_{-1})|_H, |f^{-1}(a_1^{-1} b_1)|_H\}. \tag{2.5}$$

EXAMPLE 2.71. If γ, γ' are given by the maps $i \mapsto t^i$ and $i \mapsto ht^i$ respectively, where $h \in H$, then

$$\Pi_{max} = \max\{|f^{-m+n}(h)|_H, |f^{-m-n}(h)|_H\}.$$

Since G acts on itself isometrically via the left multiplication, in order to formulate flaring conditions, without loss of generality, we may assume that the qi sections γ, γ' are defined over intervals of the form $[\![-n, n]\!]$ (i.e., $m = 0$) and $\gamma(0) = 1$.

One can also reformulate the above conditions and quantities using the notion of *pseudo-orbits* coming from the theory of dynamical systems.

DEFINITION 2.72. Let (Y, d) be a metric space and $\phi : (Y, d) \to (Y, d)$ be a homeomorphism. For a number K, a *K-pseudo-orbit* of ϕ in Y is a bi-infinite sequence $(y_i)_{i\in\mathbb{Z}}$ in Y such that for each i

$$d(y_{i+1}, \phi(y_i)) \le K.$$

For instance, if $K = 0$ then 0-pseudo-orbits are just orbits of ϕ (or, more precisely, the cyclic group generated by ϕ) in Y. The element y_i is called the i-th member of the pseudo-orbit $(y_i)_{i\in\mathbb{Z}}$.

A *partial* K-pseudo-orbit is the restriction of a K-pseudo-orbit sequence to an interval in $\mathbb{Z}$.

Given an automorphism f of the group H, set $\phi := f^{-1}$. We will use (H, d_H) as our metric space (Y, d). For a section γ, $\gamma(i) = h_i t^i$, we define $g_i := \phi^i(h_{i+1})$. In particular, $g_0 = h_1$. Then the inequality (2.4) is equivalent to

$$d_H(\phi(g_i), g_{i+1}) \le \kappa,$$

the κ-pseudo-orbit condition. In other words, instead of working with κ-sections, we can work with (partial) κ-pseudo-orbits. The case of a 1-qi section corresponds to the case when (g_i) is the (partial) ϕ-orbit of g_0. Given two sections γ, γ', we note that the corresponding partial pseudo-orbit sequences (g_i), (g_i'), satisfy

$$d_{t^iH}(\gamma(i), \gamma'(i)) = d_H(\phi(g_i), \phi(g_i')).$$

In particular, for fixed ϕ, a uniform bound on $d_{t^iH}(\gamma(i), \gamma'(i))$ is equivalent to a (possibly different) uniform bound on $d_H(\phi(g_i), \phi(g_i'))$.

We can now restate various flaring conditions:

(1) The linear (Bestvina–Feighn) κ-flaring condition is equivalent to:

There exist constants $M_\kappa \geq 0$, $\lambda_\kappa > 1$ and $n_\kappa \in \mathbb{N}$ such that for every pair of maps $i \mapsto a_i \in H$ and $i \mapsto b_i \in H$, $i \in [-n_\kappa, n_\kappa] \cap \mathbb{Z}$, satisfying:

(a) $a_0 = 1$, $|b_0|_H \geq M_\kappa$,

(b) $d_H(f^{-i}(a_i), f^{-i}(a_{i+1})) \leq \kappa$, $d_H(f^{-i}(b_i), f^{-i}(b_{i+1})) \leq \kappa$, $i \in [-n, n]$,

we have

$$\max\{d_H(f^n(a_{-n}), f^n(b_{-n})), d_H(f^{-n}(a_n), f^{-n}(b_n))\} \geq \lambda |b_0|_H.$$

(2) The proper κ-flaring condition is equivalent to:

There exists a constant $M_\kappa \geq 0$ and a proper function $\phi_\kappa : \mathbb{N} \to \mathbb{R}_+$ such that for every pair of maps $i \mapsto a_i \in H$ and $i \mapsto b_i \in H$, $i \in \mathbb{Z}$, satisfying:

(a) $a_0 = 1$, $|b_0|_H \geq M_\kappa$,

(b) $d_H(f^{-i}(a_i), f^{-i}(a_{i+1})) \leq \kappa$, $d_H(f^{-i}(b_i), f^{-i}(b_{i+1})) \leq \kappa$, $i \in [-n, n]$,

we have

$$\max\{d_H(f^n(a_{-n}), f^n(b_{-n})), d_H(f^{-n}(a_n), f^{-n}(b_n))\} \geq \phi_\kappa(n).$$

It is also useful to spell out the negation of the proper κ-flaring condition, which is most apparent as the negation of the bigon property from Corollary 2.53:

There exists elements $g, h \in H$ and pairs sequences of partial κ-pseudo-orbits $(g_{i,n})_{n\in\mathbb{N}}$, $(g'_{i,n})_{n\in\mathbb{N}}$ of f in H defined for $i \in [0, N_n]$, such that:

(a) For all n, $g_{0,n} = 1$, $g'_{N_n,n} = g$, $g'_{N_n} = hg_{N_n}$.

(b) $\lim_{n\to\infty} \max_{i\in[0,N_n]} d_H(g_{i,n}, g'_{i,n}) = \infty$.

Note that, by possibly increasing κ to $K := \kappa + C$, where $C = \max\{|g|, |h|\}$, and working with partial K-pseudo-orbits, we can even ensure that $g = 1, h = 1$ and, hence, $g'_{0,n} = 1$, $g_{N_n,n} = g'_{N_n,n}$.

We next relate flaring to *hyperbolicity properties* of the automorphism f.

DEFINITION 2.73. (Bestvina, Feighn, [**BF92**]) Suppose H is a finitely generated group and S is finite generating set for H. Suppose $f : H \to H$ is an automorphism. We say that f is *weakly hyperbolic* if there is $m \in \mathbb{N}$, $\lambda > 1$ and a finite subset $E \subset H$, such that for all $h \in H \setminus E$ we have

$$\lambda|h| \leq \max\{|f^m(h)|, |f^{-m}(h)|\}.$$

We say that E is the *exceptional* subset. An automorphism is called *hyperbolic* if the above inequality holds with $E = \emptyset$.

Some remarks are in order regarding this definition.

REMARK 2.74.

(1) The notion of *hyperbolicity* of an automorphism was introduced by Bestvina and Feighn in [**BF92**] (the exceptional subset E was absent). They also proved hyperbolicity of semidirect products $H \rtimes_f \langle t\rangle$ of hyperbolic automorphisms of hyperbolic groups, see Corollary in [**BF92**, section 5]. However, the original Bestvina–Feighn definition is too restrictive if the purpose is to conclude hyperbolicity of semidirect products. Lemma 2.77 below (which is already present in Gersten's paper [**Ger98**, Corollary 6.9]) shows that hyperbolicity of the semidirect product is equivalent to the weak hyperbolicity of the automorphism.

(2) If $f : H \to H$ is a weakly hyperbolic automorphism, then for any nontrivial finite group H_1, the automorphism $f' = f \times \mathrm{id}$ of $H' = H \times H_1$ is also weakly hyperbolic but fixes the subgroup H_1 element-wise and, hence, is not hyperbolic.
(3) In Corollary 2.79 we will prove that for automorphisms of torsion-free hyperbolic groups weak hyperbolicity is equivalent to hyperbolicity.
(4) The only hyperbolic groups which admit weakly hyperbolic automorphisms are the ones commensurable to free products of surface groups and free groups, as follows for instance from [**RS94**].
(5) The hyperbolicity inequality trivially holds for the trivial element $h = 1 \in H$. Suppose that the exceptional subset E is a ball $B(1, r) \subset H$ and that (with $\lambda > 1$) for $h \notin E$, $|f^m(h)| \geq \lambda|h|$. Then $f^m(h) \notin E$ and, thus, we can apply the same hyperbolicity inequality to $f^m(h)$. Clearly,

$$|f^{2m}(h)| = \max\{|f^{2m}(h)|, |f^{-m+m}(h)|\} \geq \lambda^2|h|.$$

Repeating this argument, we see that for each $i \geq 1$,

$$|f^{im}(h)| \geq \lambda^i|h|.$$

LEMMA 2.75. *Hyperbolicity and weak hyperbolicity of an automorphism $f : H \to H$ are independent of the finite generating set of H.*

PROOF. We will verify the claim for the weak hyperbolicity property since the proof for hyperbolicity is identical (with $E = \emptyset$).

Suppose f is weakly hyperbolic with respect to a finite generating set S, i.e., there exist $m \in \mathbb{N}$, $\lambda > 1$ and a finite subset $E \subset H$ such that

$$\lambda|h|_S \leq \max\{|f^m(h)|_S, |f^{-m}(h)|_S\}$$

for all $h \in H \setminus E$.

Suppose S' is another finite generating set of H. For any $h \in H$ let $|h|_S$ and $|h|_{S'}$ denote the word-lengths of h with respect to S and S' respectively. Then there is a constant $C > 0$ such that $\frac{1}{C}|h|_S \leq |h|_{S'} \leq C|h|_S$ for all $h \in H$. Also, we note that for all $r \in \mathbb{N}$ we have $\lambda^r|h|_S \leq \max\{|f^{mr}(h)|_S, |f^{-mr}(h)|_S\}$ for all $h \in G \setminus E_r$ where

$$E_r = \bigcup_{-(r-1)\leq i\leq r-1} f^{im}(E).$$

Hence,

$$\lambda^r|h|_{S'} \leq C\lambda^r|h|_S \leq C\max\{|f^{mr}(h)|_S, |f^{-mr}(h)|_S\} \leq C^2\max\{|f^{mr}(h)|_{S'}, |f^{-mr}(h)|_{S'}\}$$

for $h \in H \setminus E_r$. It follows that $C^{-2}\lambda^r|h|_{S'} \leq \max\{|f^{mr}(h)|_{S'}, |f^{-mr}(h)|_{S'}\}$ for $h \in H \setminus E_r$. Thus, if we choose $r = r_1$ large enough, we get $\lambda_1 = C^{-2}\lambda^{r_1} > 1$ and setting $m_1 = r_1 m$, we obtain

$$\lambda_1|h|_{S'} \leq \max\{|f^{m_1}(h)|_{S'}, |f^{-m_1}(h)|_{S'}\}$$

for all $h \in H \setminus E_{r_1}$. Whence f is weakly hyperbolic with respect to the generating set S'. □

EXAMPLE 2.76.

(1) Suppose $H = \pi_1(\Sigma)$ where Σ is a closed connected hyperbolic surface. Then an automorphism f of H is hyperbolic if and only if it is induced by a pseudo-Anosov automorphism of Σ, if and only if it has no nontrivial periodic conjugacy classes, if and only if the semidirect product $H \rtimes_f \mathbb{Z}$ is hyperbolic. (The equivalence of the last three properties is due to William Thurston; see, e.g., [**CB88, Ota01**]. The equivalence with hyperbolicity of f can be seeing either as

a consequence of the pseudo-Anosov property or of the combination of Lemma 2.77 and Corollary 2.79.)

(2) If $H = \mathbb{F}_n, n \geq 2$, then any automorphism $f \in Aut(H)$ with no periodic conjugacy classes is hyperbolic (in the sense of Bestvina–Feighn). See [**BFH97**, Theorem 5.1].

We refer the reader to [**Bri00, DT18, Mut21**] for other results in this direction.

LEMMA 2.77 ([**BF92, Ger98**]). *If f is weakly hyperbolic, then the tree of metric spaces $X = \Gamma(H \rtimes_f \mathbb{Z}, S_G) \to T = \Gamma(Z, 1)$— as constructed in the beginning of this subsection—satisfies the Bestvina–Feighn flaring condition. The converse is also true.*

PROOF.

(1) Suppose f is weakly hyperbolic. Let $R = \max\{d_H(1, h) : h \in E\}$ where $E \subset H$ is a finite exceptional subset as in Definition 2.73. Then for all $x \in H$ with $|x|_H > R$ we have

$$\lambda|x| \leq \max\{|f^m(x)|, |f^{-m}(x)|\}.$$

First of all, since the Bestvina–Feighn flaring condition is equivalent to hyperbolicity of G and the latter is equivalent to the hyperbolicity of the semidirect product $H \rtimes_{f^m} \mathbb{Z}$ (commensurable to the original group G), it suffices to consider the case when $m = 1$. We will also assume that in the definition of a weakly hyperbolic automorphism the maximum is attained by $\phi = f^{-1}$ rather than f (otherwise, we replace f with f^{-1}). Then the weak hyperbolicity inequality reads

$$\lambda|x| \leq |\phi(x)| \tag{2.6}$$

for all $x \in H \setminus E$.

Take $\kappa \geq 1$. As we noted earlier, it suffices to verify the Bestvina–Feighn flaring condition for pairs of κ-qi sections $\Pi = (\gamma, \gamma')$ defined over intervals of the form $[\![-m, m]\!]$, satisfying $\gamma(0) = 1, h := \gamma'(0)$, where, as above, $m = 1$, such that $\Pi_0 = |h| \geq M_\kappa$ for a suitable uniform constant M_κ. Pick any λ' in the open interval $(1, \lambda)$; for concreteness, we take $\lambda' = \frac{1}{2}(\lambda + 1)$. We claim that

$$\lambda'|h| = d_H(\gamma(0), \gamma'(0)) \leq d_{tH}(\gamma(m), \gamma'(m)).$$

Set $\gamma(1) = h_1 t, \gamma'(1) = h_1' t$. Then (as we noted earlier, after changing κ) the κ-qi section condition for γ, γ' over the interval $[-1, 0]$ is equivalent to the inequalities

$$d_H(\phi(h_1), 1) \leq \kappa, \quad d_H(\phi(h), \phi(h_1')) \leq \kappa. \tag{2.7}$$

We will estimate from below the distance (in H) between $\gamma(1), \gamma'(1)$; according to the computation in (2.5), we need to estimate from below the distance

$$d_H(\phi(h_1), \phi(h_1')).$$

By the triangle inequality, combined with the inequalities (2.6) and (2.7), we have

$$d_H(\phi(h_1), \phi(h_1')) \geq d_H(1, \phi(h)) - 2\kappa \geq \lambda|h| - 2\kappa.$$

Then the desired inequality $\lambda'|h| \leq d_{tH}(\gamma(1), \gamma'(1))$ is equivalent to

$$(\lambda - \lambda')|h| \geq 2\kappa$$

By out choice of λ', this amounts to

$$|h| \geq \frac{2\kappa}{\lambda - 1}.$$

Therefore, taking M_κ equal to the maximum of $\frac{2\kappa}{\lambda-1}$ and $1 + \mathrm{diam}_H(E)$, we ensure the Bestvina–Feighn flaring condition.

(2) For the converse one applies the flaring condition to $\kappa = 1$-qi sections. More precisely, we work with pairs of sections as in Example 2.71. We take the finite set in the definition of weak hyperbolicity to be $E = \{h \in H : |h|_H \leq M_1\}$. The rest is straightforward and hence left as an exercise for the reader. □

LEMMA 2.78. *Suppose that H is a hyperbolic group, $f : H \to H$ is weakly hyperbolic. Then the exceptional subset of E can be chosen to contain only finite order elements.*

PROOF. Let $E \subset H$ be an exceptional subset of f, i.e., there exist $m \in \mathbb{N}$, $\lambda > 1$ such that

$$\lambda|h| \leq \max\{|f^m(h)|, |f^{-m}(h)|\}$$

for all $h \in H \setminus E$. After replacing f with f^m, we can assume that $\lambda|h| \leq \max\{|f(h)|, |f^{-1}(h)|\}$ unless $h \in E$. After enlarging E is necessary, we can assume that it equals to the ball of certain radius r in H (centered at $1 \in H$). We define $E' \subset E$, the subset consisting of infinite order elements.

Suppose that $h \in E$ is such that for infinitely many values of $m \geq 1$,

$$|f^m(h)| \leq r.$$

We claim that h has finite order. Indeed, then there exist two numbers $n > m \geq 1$ such that

$$f^m(h) = f^n(h), f^{n-m}(h) = h.$$

It follows that in the group $G = H \rtimes_f \mathbb{Z}$ we have

$$t^{n-m}h = ht^{n-m}.$$

Since f is weakly hyperbolic, the semidirect product G is a hyperbolic group, see Lemma 2.77. Hence, the abelian subgroup $A < G$ generated by t^{n-m} and h is virtually cyclic, i.e., h has finite order.

Thus, as noted above, for each $h \in E'$ there exists a smallest natural number $n = n_h$ such that $|f^n(h)| > r$, which, in particular, implies that $f^n(h) \notin E'$. Thus,

$$\max\{|f \circ f^n(h)|, |f^{-1} \circ f^n(h)|\} \geq \lambda|f^n(h)| \geq \lambda|h|.$$

Since n_h was chosen to be smallest, it follows that the above inequality holds for $f \circ f^n(h) = f^{n+1}(h)$. By the same argument as in the proof of Lemma 2.59, we see that for each $m \geq \max\{n_h : h \in E'\}$, and $h \in E'$,

$$\lambda|h| \leq |f^m(h)|.$$

Since the hyperbolicity inequality holds for all $h \in H \setminus E$, we conclude that f satisfies the hyperbolicity condition except for the subset $E'' \subset E$ consisting of torsion elements. □

Lemma 2.78 immediately implies Corollary 2.79.

COROLLARY 2.79. *If H is torsion-free then each weakly hyperbolic automorphism of H is hyperbolic.*

CHAPTER 3

Carpets, ladders, flow-spaces, metric bundles, and their retractions

In this chapter we introduce and analyze four classes of subtrees of spaces in hyperbolic trees of spaces:

- Ladders
- Metric bundles
- Carpets
- Flow-spaces

These spaces play key role in proving hyperbolicity and describing geodesics in trees of spaces $(\pi : X \to T)$: Uniform quasigeodesics in X will be inductively described as concatenations of uniform quasigeodesics in carpets, ladders and flow-spaces. The main result of this and the next chapter is that all ladders, carpets and flow-spaces are hyperbolic and admit coarse Lipchitz retractions from X. We note that our definitions of ladders and flow-spaces are inspired by the *ladder* construction of Mitra, [**Mit98**], while the notion of metric bundles is adapted from [**MS09**]. We conclude the chapter with some examples illustrating the above concepts.

3.1. Semicontinuous families of spaces

All four classes of spaces discussed in this (and the next) chapter are special cases of *semicontinuous families* of spaces, which are certain subtrees of spaces $\mathfrak{Y} \subset \mathfrak{X}$. In what follows, given a subtree of spaces $\mathfrak{Y} = (\pi : Y \to S) \subset \mathfrak{X} = (\pi : X \to T)$, it will be notationally convenient to extend $\mathfrak{Y}$ to a tree of spaces (still denoted $\mathfrak{Y}$) over the entire tree T by declaring $Y_v = \emptyset, Y_e = \emptyset$ for each $v \in V(T) - V(S)$ and $e \in E(T) - E(S)$.

DEFINITION 3.1. Suppose that $\mathfrak{X} = (\pi : X \to T)$ is a tree of hyperbolic spaces. Fix constants $\lambda \in [0, \infty)$, $E, K \in [1, \infty), D \in [0, \infty]$. We will say that a subtree of spaces $\mathfrak{Y} = (\pi : Y \to S), S \subset T$, in $\mathfrak{X}$ forms a (K, D, E, λ)*-semicontinuous family*, relative to a vertex $u \in V(S)$, called the *center* of $\mathfrak{Y}$, if the following conditions hold:

(1) Each vertex/edge space $Y_v \subset X_v, Y_e \subset X_e, v \in V(S), e \in E(S)$, is λ-quasiconvex.
(2) Each $y \in \mathcal{Y}$ is connected to Y_u by a K-leaf γ_y in Y.
(3) For each edge $e = [v, w] \in E(T)$ we define the (possibly empty!) projection

$$Y_w^v := P_{X_{vw}, X_w}(Y_v) \subset X_w. \tag{3.1}$$

We require that whenever $e = [v, w] \in E(S)$ is oriented away from u,

$$\mathrm{Hd}_{X_{vw}}(Y_w^v, Y_w) \leq E \tag{3.2}$$

and

$$\mathrm{Hd}_{X_{vw}}(Y_w, Y_e) \leq K \tag{3.3}$$

(4) For every edge $e = [v, w]$ such that $v \in V(S), w \notin V(S)$, we require the pair of quasiconvex subsets (Y_v, X_w) in X_{vw} to be D-cobounded.

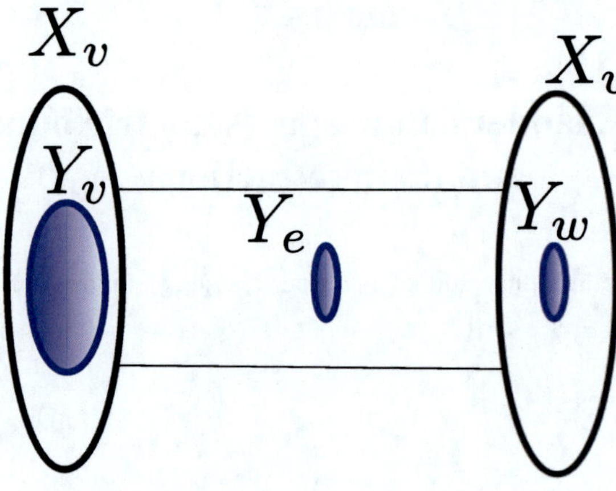

FIGURE 3.1. A semicontinuous family

REMARK 3.2.

(i) Condition 2 implies that $Y_w \subset N_K^e(Y_v)$, hence, $Y_w \subset N_{\eta_0(2K)}(Y_w^v)$ with respect to the metric of X_w. (Recall that η_0 is the distortion function of X_v in X, hence, in X_{uv}.)

(ii) Condition 2 in this definition ensures that Y_w cannot be "much larger" than Y_v, while Condition 3 ensures a certain lower bound on Y_w. Thus, as we move away from X_u, vertex-spaces of $\mathfrak{Y}$ can shrink substantially (even disappear) but they cannot substantially increase.

(iii) In most examples in our book, $\lambda = 4\delta_0$, hence, we will be suppressing the dependence on this parameter and record only the triple of numbers (K, D, E).

(iv) To ensure uniform coboundedness of the pairs (Y_v, X_w) in Axiom 4, it suffices to get a uniform upper bound C on the diameters of Y_w^v: It will then follow that the pair (Y_v, X_w) is D-cobounded for some $D = D(\delta_0', \lambda_0', C)$, see Corollary 1.143.

(v) We do not insist on the converse implication in Axiom 4: There will be important situations when we have to consider $\mathfrak{Y}$ with uniformly bounded fibers over nonboundary vertices of S.

(vi) Axioms 3 and 4 will be needed in order to have a uniform coarse Lipschitz retraction $X \to Y$, see Theorem 3.3.

(vii) The edge-spaces Y_e of are largely irrelevant for our discussion.

(viii) The projections P_{X_{vw}, X_w} restricted to X_v are at uniformly bounded distance (as measured in X_{vw}) from the projections $P_{X_v, X_{ev}}$. The same, of course, applies to restrictions of the projections P_{X_{vw}, X_v}. However, we decided to work with the projections P_{X_{vw}, X_w} as computations tend to be simpler in this setting.

THEOREM 3.3. *Suppose that $\mathfrak{Y}$ is a (K, D, E, λ)-semicontinuous family of spaces with $D < \infty$. Then there exists an $L_{3.3}(K, D, E, \lambda)$-coarse Lipschitz retraction $\rho_{\mathfrak{Y}} : X \to Y$.*

PROOF. We will verify that the subtree of spaces $\mathfrak{Y}$ is (uniformly) retractible; we use Theorem 2.22 as follows.

(i) We let $h'_v : X_v \to Y_v$ denote the restriction of the nearest-point projections P_{X_{vw},Y_v}. According to Theorem 2.22(i), we need to bound the diameter of the image of X_{ev} in Y_v under h'_v. However, X_{ev} is contained in the unit neighborhood of X_w (taken in X_{vw}); thus, we need to bound the diameter of the projection (in X_{vw}) of X_w to Y_v, which is the content of Axiom 4 of the definition of a semicontinuous family of spaces: This diameter is $\le D$.

(ii) Assuming that $e = [v, w]$ is an edge of S oriented away from u, we need to get a uniform bound

$$\mathrm{dist}_{X'_v}(h'_v \circ f_{ev}, f'_{ev} \circ h'_e) \le Const.$$

The subsets $Y_{ev}, Y_e \subset X_{vw}$ are within unit Hausdorff distance, while

$$\mathrm{Hd}_{X_{vw}}(Y_e, Y_w) \le E \text{ and } \mathrm{Hd}_{X_{vw}}(Y^v_w, Y_w) \le E.$$

Since $d_{X_{vw}}(Y_w, X_v) \le K$, by applying Lemma 1.130 to the subsets $U_1 = X_v$, $U_2 = Y_w$ in X_{vw}, we conclude that

$$\mathrm{Hd}_{X_{vw}}(P_{X_{vw},Y_v}(X_w), P_{X_{vw},X_w}(Y_v)) \le 2\lambda'_0 + 3\delta'_0 + K.$$

Taking into account that the projection P_{X_{vw},Y_v} is uniformly coarse Lipschitz, we conclude that Y_{ev} is uniformly close to the image of X_e under the nearest-point projection P_{X_{vw},Y_v} and, accordingly, the nearest-point projection $h'_e : X_e \to Y_e$ is uniformly close to the restriction of the nearest-point projection $X_{vw} \to P_{X_v,Y_v}(X_e)$ (see Lemma 1.111). Also taking into account that the map f'_{ev} moves points by distance $\le K$ in X_{vw}, we can replace $f'_{ev} \circ h'_e$ with the restriction of the projection $X_{vw} \to P_{X_v,Y_v}(X_e)$ to X_e. Similarly, the map f_{ev} moves points distance ≤ 1 in X_{vw} and, hence (in view of the uniform coarse Lipschitz property of h'_v), we can replace the composition $h'_v \circ f_{ev}$ with the restriction of the nearest-point projection P_{X_{vw},Y_v} to X_e. But now, the projections $X_{vw} \to P_{X_v,Y_v}(X_e)$ and P_{X_{vw},Y_v} are uniformly close to each other according to Corollary 1.110 applied to the λ'_0-quasiconvex subsets Y_v and X_e in the ambient hyperbolic space $Z = X_{vw}$. □

Corollary 3.4. *If $\mathfrak{Y} = (\pi : Y \to S)$ is a (K, D, E, λ)-semicontinuous family of spaces with $D < \infty$, then the inclusion map $Y \to X$ is an $L_{3.3}(K, D, E, \lambda)$-qi embedding.*

We next describe a class of semicontinuous subtrees of spaces, called *metric bundles*. The theory of metric bundles was developed in **[MS09]** in a more general setting when the base is allowed to be an arbitrary geodesic metric space but we will not need that in our book. The following definition of *metric bundles* is adapted from **[MS09]** in a form suitable for our purposes. It is easy to verify that the two definitions (ours and that of **[MS09]**) are equivalent when the base is a tree. The reader should also compare this definition with the notion of an abstract metric bundle given in Definition 2.8: Each metric bundle defined below is also an abstract metric bundle.

Definition 3.5 (Metric bundles). A subtree of spaces $\mathfrak{Y} = (\pi : Y \to S) \subset \mathfrak{X} = (\pi : X \to T)$ is called a *K-metric bundle* if:

(1) Every vertex/edge space of $\mathfrak{Y}$ is λ-quasiconvex in the respective vertex/edge space of $\mathfrak{X}$.
(2) For every vertex $u \in V(S)$ and edge $e = [v, w] \in E(S)$ (directed away from u), and $x \in Y_w, x \in Y_e$, there exist K-qi sections γ_x in $\mathfrak{Y}$ on $[\![w, u]\!]$, such that $\gamma_x(w) = x$.

It follows immediately that each K-metric bundle forms a (K, ∞, E, λ)-semicontinuous family of spaces in $\mathfrak{X}$ (relative to any vertex $u \in S$), with $E = \eta_0(2K)$. The reader uncomfortable with using $D = \infty$ here can simply restrict $\mathfrak{X}$ to S, then one can take $D = 0$.

While Theorem 3.3 does not directly apply to metric bundles $\mathfrak{Y} \subset X$ (unless $S = T$), we will see in Corollary 3.63 that under certain extra assumption weakening condition 4 in Definition 3.1, these too admit uniform coarse Lipschitz retraction from $\mathfrak{X}$.

3.2. Ladders

Ladders are certain semicontinuous subtrees of spaces $\mathfrak{L} = (\pi : L \to S) \subset \mathfrak{X}$ whose fibers are geodesic segments. However, in addition to the properties of a semicontinuous family of spaces, we will impose a certain extra structure on $\mathfrak{L}$.

Each ladder $\mathfrak{L} = (\pi : L \to S)$ comes equipped with certain parameters and two pieces of extra data: An orientation of the fibers (hence, ladders generalize oriented line bundles) and a canonical choice of a maximal K-qi section $\Sigma_x \subset L$ through each point $x \in L$. The choice of $\Sigma_\bullet$ can be regarded as a "connection" on $\mathfrak{L}$. Thus, ladders can be regarded as "oriented line semi-bundles equipped with connections."

We will be primarily interested in ladders such that $\mathcal{L}$ is contained in the $4\delta_0$-fiberwise neighborhood of a k-flow space $\mathcal{F}l_k(Q_u) \subset X$ (these will be defined in Section 3.3). For the ease of notation, we will be ignoring the flow-spaces for now; formally speaking, one can regard the flow-space $\mathfrak{Fl}_k(Q_u)$ as a tree of hyperbolic spaces satisfying a uniform flaring condition.

We now begin with an axiomatic definition of ladders. Let $\mathfrak{X} = (\pi : X \to T)$ be a tree of hyperbolic spaces satisfying Axiom **H**. Fix positive numbers K, D, E and a vertex $u \in T$; these are the *parameters* of a ladder $\mathfrak{L}$. A *ladder* with these parameters (a *(K, D, E)-ladder centered at u*) is a subtree of oriented geodesic intervals in $\mathfrak{X}$, $\mathfrak{L} = (\pi : L \to S)$, $S = \pi(L)$ which satisfies further axioms described below. Each fiber $L_v := L \cap X_v, v \in V(T), L_e = L \cap X_e, e \in E(T)$, of $\mathfrak{L}$ is an *oriented* geodesic segment denoted $[x_v y_v]_{X_v}$ or $[x_e y_e]_{X_e}$.

Furthermore, we fix once and for all a family $\Sigma_\bullet$ of *maximal partial K-qi sections* Σ_x of $\pi : L \to S$, whose domains T_x are subtrees in S containing the vertex u (and $\pi(x)$, of course). Maximality here is understood in the sense that if Σ'_x is another partial K-qi section containing Σ_x, then $\Sigma_x = \Sigma'_x$. The subscript x in Σ_x indicates that $x \in \Sigma_x$. We will assume that x belongs to a vertex-space of $\mathfrak{X}$.

Axiom L0. We will require the family of sections $\Sigma_\bullet$ to be *consistent* in the sense that whenever $v = \pi(y)$ is between u and $w = \pi(x)$, the sections Σ_y and Σ_x agree on the interval $[\![u, v]\!] \subset T$.

DEFINITION 3.6. Let $\mathfrak{L}$ be a ladder centered at the vertex u, $L_u = [x_u y_u]_{X_u}$. We will refer to the subsets $\Sigma_{x_u} = bot(\mathfrak{L}), \Sigma_{y_u} = top(\mathfrak{L})$ as, respectively, the *bottom* and the *top* of the ladder $\mathfrak{L}$.

Thus, $\Sigma_\bullet$ defines a family of maps

$$\Pi_{w,v} : L_w \to L_v,$$

for every vertex $v \in V(S)$ between u and $w \in V(S)$:

$$\Pi_{w,v}(x) = \Sigma_x \cap L_v, x \in \mathcal{L}_w.$$

(Note that this intersection is nonempty since $\pi(\Sigma_x)$ is a subtree containing both w and u, hence, also containing v.) These maps can be regarded analogues of *parallel transport maps* in the conventional theory of connections on bundles. Consistency of sections implies the following *semigroup property*:

$$\Pi_{w_1,w_3} = \Pi_{w_2,w_3} \circ \Pi_{w_1,w_2}$$

whenever w_2, w_3 belong to the interval $[\![u, w_1]\!]$ and appear there in the following order:

$$u \leq w_3 \leq w_2 \leq w_1.$$

As we will see below, axioms of a ladder require each map $\Pi_{w,v}$ to be either constant or an orientation-preserving topological embedding. The maps $\Pi_{w,v}$ need not be surjective; for every oriented edge $e = [v, w]$ in S (oriented away from u) we have an oriented subsegment

$$L'_v = [x'_v y'_v]_{X_v} := \Pi_{w,v}(L_w)$$

in L_v. Here $x'_v = \Pi_{w,v}(x_w), y'_v := \Pi_{w,v}(y_w)$. The orientation of the segment L'_v is then consistent with that of L_v (since $\Pi_{w,v}$ is orientation-preserving). Observe also that the mere existence of K-qi sections Σ_x implies *some semicontinuity* of the ladder $\mathfrak{L}$: For every edge $e = [v, w] \subset S$ (oriented away from u)

$$L_w \subset N^e_K(L'_v) \subset N^e_K(L_v), \tag{3.4}$$

where the K-neighborhood is taken in the subspace X_{vw} (which is what the superscript e indicates). However, L_w can be much smaller than L_v.

For $\mathfrak{L}$ (equipped with $\Sigma_\bullet$) to be a ladder, it has to satisfy three further axioms listed below. Note, however, that the assumption that the fibers L_v, L_e of $\mathfrak{L}$ are geodesic segments ensures Property 1 in Definition 3.1 with $\lambda = \delta_0$, while Axiom L1 implies Property 2 in that definition, thus making Axiom L3 somewhat redundant.

We now fix $K \in [1, \infty), E \geq 1$ and $D \in [0, \infty]$. While all other two parameters in the triple (K, D, E) are real numbers, as with general semicontinuous subtrees of spaces, it is convenient to allow for infinite values of the parameter D.

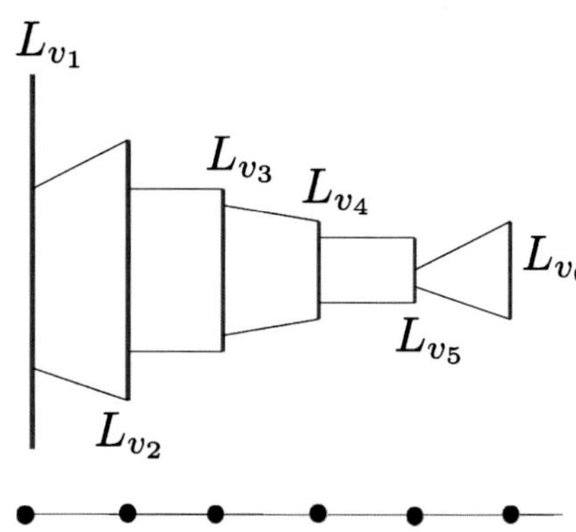

FIGURE 3.2. Ladder

Ladder Axioms.

- L1 Each $x \in L$ belongs to some Σ_x.
- L2 Each map $\Pi_{w,v}$ is either constant or is an orientation-preserving topological embedding.
- L3 $\mathfrak{L}$ is a (K, D, E, δ_0)-semicontinuous family of spaces.

REMARK 3.7. In general, $\pi(top(\mathfrak{L}))$ and $\pi(bot(\mathfrak{L}))$ are smaller than the base S. If

$$\pi(top(\mathfrak{L})) = \pi(bot(\mathfrak{L})) = S$$

then $\mathfrak{L}$ will be a K-metric bundle. This will happen in the important case of *carpets*.

DEFINITION 3.8. A (K, D, E)-*ladder* in $\mathfrak{X}$ is a subtree of spaces

$$\mathfrak{L} = (\pi : L \to S) \subset \mathfrak{X} = (\pi : X \to T)$$

whose vertex and edge-spaces are oriented geodesic segments, equipped with a family of K-qi sections $\Sigma_\bullet$ and satisfying Axioms L0—L3.

EXAMPLE 3.9. Let x be a point in X_u and let Σ_x be a K-qi section of $\pi : X \to T$ defined over a subtree $S \subset T$ such that $x \in \Sigma_x$. Then $L = \Sigma_x$ is the total space of a $(K, 0, \eta_0(2K))$-ladder.

DEFINITION 3.10. A *subladder* in $\mathfrak{L}$ is a ladder $\mathfrak{L}' = \mathfrak{L}(\alpha') \subset \mathfrak{L} = \mathfrak{L}(\alpha)$ with the same center u as $\mathfrak{L}$, such that the sections $\Sigma'_\bullet$ of $\mathfrak{L}'$ are restrictions of the sections $\Sigma_\bullet$ of $\mathfrak{L}$. In particular, top and the bottom of $\mathfrak{L}'$ are contained in sections of $\mathfrak{L}$ through the endpoints of α'.

In what follows, given a ladder $\mathfrak{L} = \mathfrak{L}_K(\alpha)$, $\alpha \subset X_u$, for each point $x \in \mathcal{L}$ let $\gamma_x \subset \Sigma_x \subset L$ denote the section over the interval $[\![u, \pi(x)]\!]$, connecting x to a point in α. Similarly, given two points $x, y \in \mathcal{L}$, if $\Sigma_x \cap \Sigma_y \neq \emptyset$ and the restriction of π to $\Sigma_x \cup \Sigma_y$ is 1-1, then there exists a unique K-leaf $\gamma_{x,y}$ in $\Sigma_x \cup \Sigma_y \subset \mathfrak{L}$ connecting x to y.

We omit a proof of the next lemma as it is straightforward.

LEMMA 3.11 (Bisecting a ladder). *Suppose $u \in V(T)$, $\alpha = [xy]_{X_u} \subset X_u$ and we are given a ladder $\mathfrak{L} = \mathfrak{L}_{K,D,E}(\alpha)$. Then for every point $z \in [xy]_{X_u}$ the K-qi section $\Sigma_z \subset \mathcal{L}$ decomposes $\mathfrak{L}$ into two (K, D, E)-subladders $\mathfrak{L}^+, \mathfrak{L}^-$ such that*

(1) $L^+_u = [zy]_{X_u} \subset \alpha$, $L^-_u = [xz]_{X_u} \subset \alpha$,

(2)

$$top(\mathfrak{L}^-) = \Sigma_z = bot(\mathfrak{L}^+).$$

Applying this lemma twice, we obtain Corollary 3.12.

COROLLARY 3.12 (Trisecting a ladder). *Suppose $u \in V(T)$, $\alpha = [xy]_{X_u} \subset X_u$ and we are given a ladder $\mathfrak{L} = \mathfrak{L}_{K,D,E}(\alpha)$. Then for every subsegment $\alpha' = [x'y']_{X_u} \subset \alpha$ there exists a subladder $\mathfrak{L}' = \mathfrak{L}_{K,D,E}(\alpha') \subset \mathfrak{L}$ bounded by the K-qi sections $\Sigma_{x'}, \Sigma_{y'} \subset \mathfrak{L}$ (its bottom and top respectively).*

Since a (K, D, E)-ladder $\mathfrak{L} = (\pi : L \to S)$ is a (K, D, E, δ_0)-semicontinuous subtree of spaces in $\mathfrak{X}$, as an application of the retraction Theorem 3.3 we obtain Corollary 3.13.

COROLLARY 3.13 (Retraction to ladders). *For every (K, D, E)-ladder $\mathfrak{L} = (\pi : L \to S)$ there exists a $L_{3.3}(K, D, E, \delta_0)$-coarse Lipschitz retraction $\rho_{\mathfrak{L}} : X \to L$.*

We next define *carpets* which are both ladders and metric bundles. While in Axiom L3 of a ladder we assume that fibers over *all* boundary vertices of S have uniformly bounded diameter when projected to adjacent vertex-spaces X_w, $w \notin V(S)$, in the definition of carpets (where the base S is an oriented interval $[\![u, w]\!]$) we will allow one of the boundary vertices of S (namely the vertex u) to violate this property (which is why $D = \infty$). However, instead, we will add a stronger requirement on the other boundary vertex w and a requirement on the top and the bottom.

DEFINITION 3.14. A $(K, \infty, \eta_0(2K))$-ladder $\mathfrak{A} = (\pi : A \to S) \subset \mathfrak{X}$ is called a (K, C)-*narrow carpet* or just a (K, C)-*carpet* if:

(1) S is an interval $[\![u, w]\!]$ and, furthermore, the top and the bottom of $\mathfrak{A}$ connect the two endpoints of A_u to that of A_w, i.e.,

$$\pi(top(\mathfrak{A})) = \pi(bot(\mathfrak{A})) = S.$$

In this case, we will say that $\mathfrak{A}$ is bounded by the K-qi sections $\gamma_1 = bot(\mathfrak{L}), \gamma_2 = top(\mathfrak{L})$ of the carpet. We will refer to $\beta = A_w$ as the *(narrow) end* of the carpet and will say that $\mathfrak{A}$ is *from* $\alpha = A_u$ *to* $\beta = A_w$.

(2) A_w has length $\leq C$.

We will use the notation $\mathfrak{A} = \mathfrak{A}_K(\alpha)$ for such carpets to indicate the two key parameters.

DEFINITION 3.15. A (K, ∞)-carpet is called a K-hallway.

REMARK 3.16.

(1) Every K-hallway is a K-metric bundle.
(2) The pair of sections γ_1, γ_2 determines a hallway $\mathfrak{A}$ "coarsely uniquely": The ambiguity in the definition comes from the choice of the vertical geodesics $A_t, t \in V(S)$, and, hence, is uniformly controlled. Therefore, in what follows, we will ignore this ambiguity.

The next lemma establishes existence of ladder and hallway structures on subsets in X which are unions of vertical geodesic segments.

LEMMA 3.17. *Suppose that $\mathfrak{X}$ is a tree of hyperbolic spaces. There exists a function $K' = K'_{3.17}(K)$ such that the following holds.*

(a0) *Suppose that $\mathcal{L} \subset X$ is a subset whose projection to T is the vertex-set of a subtree $S \subset T$ containing a vertex u satisfying:*
(a1) *Every fiber $L_v = \mathcal{L} \cap X_v, v \in V(S)$, is an oriented geodesic segment $[x_v y_v]_{X_v}$ in X_v.*
(a2) *$\mathcal{L}$ satisfies Property 4 of a semicontinuous family of spaces with the parameter D. Furthermore, in line with Property 3, for every edge $e = [v, w] \in E(S)$ which is oriented away from u, $\mathrm{Hd}_{X_{vw}}(L^v_w, L_w) \leq E$, where, as before,*

$$L^v_w = P_{X_{vw}, X_w}(L_v) \subset X_w.$$

(a3) *Points x_w, y_w are within distance K (in X_{vw}) from points $x'_v, y'_v \in L_v$ respectively, so that*

$$x_v \leq x'_v \leq y'_v \leq y_v$$

on the oriented segment L_v.

Then $\mathcal{L} \subset X$ is the union of vertex-spaces of a (K', D, E)-ladder $\mathfrak{L} \subset \mathfrak{X}$ centered at u.

(b0) *Suppose that $\mathcal{A} \subset X$ is a subset whose projection to T is the vertex-set of an interval $S = [\![u, w]\!] \subset T$ such that:*
(b1) *Every fiber $A_v, v \in V(S)$, of $\mathcal{A}$, is an oriented geodesic segment $[x_v y_v]_{X_v}$ in X_v.*
(b2) *For every edge $[v_1, v_2]$ in S, $d_{X_{v_1 v_2}}(x_{v_1}, x_{v_2}) \leq K$, $d_{X_{v_1 v_2}}(y_{v_1}, y_{v_2}) \leq K$.*

Then $\mathcal{A}$ is the union of vertex-spaces of a K'-hallway $\mathfrak{A} \subset X$.

PROOF. Our first goal is to define the function K'.

We let $r' := D_{1.55}(\delta'_0, L'_0, K)$ be given by Lemma 1.55. For $k = K_{1.14}(r', L'_0)$ given by Lemma 1.14, we let $\lambda' = k_{1.25}(k)$ be given by Lemma 1.25. Lastly, set

$$K' := r' + D_{1.25}(k).$$

We now prove the lemma.

(a) We define inductively the projections Π_{v_1,v_2} (where $e = [v_1, v_2]$ is an edge in S oriented away from u), as well as the edge-spaces L_e.

Suppose that for the subtree $B_n \subset S$, which is the closed n-ball centered at u, we defined (partial) K-qi sections Σ and maps Π satisfying all the requirements of a ladder with respect to the parameter K'.

We extend the definitions of these sections and maps to the vertices in the ball $B_{n+1} \subset S$ as follows. Let $e = [v_1, v_2]$ be an edge of S with $v_1 \in B_n, v_2 \notin B_n$. Let L'_{v_1} denote the oriented subsegment of L_{v_1} bounded by x'_{v_1}, y'_{v_1} respectively. Similarly, we define the edge-space L_e as the oriented geodesic segment in X_e spanned by the nearest-point projections of the endpoints x_{v_2}, y_{v_2} of L_{v_2}.

According to Lemma 1.55, we have

$$\mathrm{Hd}_{X_{v_1v_2}}(L'_{v_1}, L_{v_2}) \le r' = D_{1.55}(\delta'_0, L'_0, K) \le K',$$

$$\mathrm{Hd}_{X_{v_1v_2}}(L_e, L_{v_2}) \le r'.$$

These conditions ensure Property 3 of Definition 3.1, i.e., Axiom L3 of a ladder.

Using Lemma 1.14, we extend the map $x_{v_2} \mapsto x'_{v_1}, y_{v_2} \mapsto y'_{v_1}$ to a $k = K_{1.14}(r', L'_0)$-quasi-isometry of geodesic segments $q : L_{v_2} \to L'_{v_1}$, which moves each point distance $\le r'$ (with respect to the metric of $X_{v_1v_2}$). Applying Lemma 1.25, we can replace the quasi-isometry q by an increasing homeomorphism $\tilde{q}$ (or a constant function) within distance $D_{1.25}(k)$ from q, such that $\tilde{q}$ is a $k_{1.25}(k)$-quasi-isometry.

Since q was moving every point of L_{v_2} at most distance r', it follows that $\tilde{q}$ moves every point within distance $K' = r' + D_{1.25}(k)$ in $X_{v_1v_2}$. We set

$$\Pi_{v_2,v_1} := \tilde{q}.$$

Thus, we obtain maps $\Pi_{w,v} : L_{v_2} \to L_{v_1}$ for oriented edges $[v_1, v_2]$ of the tree $S = \pi(\mathcal{L})$, such that $d_{X_{v_1v_2}}(x, \Pi_{v_2,v_1}(x)) \le K', x \in L_{v_2}$. For vertices v_1, v_2 of S such that v_1 is between u and v_2 we define the map $\Pi_{v_2,v_1} : L_{v_2} \to L_{v_1}$ as the composition of maps defined for the sequence of edges connecting v_2 to v_1. If Π_{v_2,v_1} is injective, then for $z \in L'_{v_1}$ we define the section $\Sigma_z \cap L_{v_2}$ as

$$\Pi^{-1}_{v_2,v_1}(z).$$

If the map is not injective, i.e., constant, we choose an arbitrary point in L_{v_2} as $\Sigma_z \cap L_{v_2}$.

(b) The proof of this part is exactly the same as of Part a, except that we use $x'_{v_1} = x_{v_1}, y'_{v_1} = y_{v_1}$. □

3.3. Flow-spaces

3.3.1. *K*-flow spaces and Mitra's retraction. Suppose that $\mathfrak{X} = (\pi : X \to T)$ is a tree of hyperbolic spaces. We fix a vertex $u \in T$, the *center of the flow* and orient all edges $e = [v, w]$ of T so that v is closer to u than w. For each $4\delta_0$-quasiconvex subset $Q_u \subset X_u$ we will define the *K-flow-space*

$$\mathfrak{Fl}_K(Q_u) = (\pi : Fl_K(Q_u) \to S) \subset \mathfrak{X},$$

which, unlike ladders and carpets, depends only on Q_u and on K, and which will be a $(K, D, E, 4\delta_0)$-semicontinuous family of spaces (relative to the vertex u), with the parameter E depending only on K and $D = D_0$, where

$$D_0 = D_{1.142}(\delta'_0, \lambda'_0) \tag{3.5}$$

is independent of K.

However, for the construction to work, the parameter K has to be sufficiently large, specifically, $K \geq K_0$, where K_0 (which depends only on the parameters of the tree of spaces $\mathfrak{X}$) is given by the equation (2.3). As before, we will use $\mathcal{F}l_K(Q_u)$ to denote the union of vertex-spaces of $\mathfrak{F}l_K(Q_u)$. We first compute the auxiliary parameter E and a certain parameter R (depending on K) which will be used to define the K-flow.

Suppose that $\lambda \geq \frac{3}{2}\delta_0$. Recall (Lemma 1.113) that if the image of a subset Q of X_v is λ-quasiconvex in X_{vw} then Q is $\hat{\lambda}$-quasiconvex in X_v with

$$\hat{\lambda} = 1500(L_0'\lambda)^3. \tag{3.6}$$

Take

$$R \geq R_0 := \max(2(\lambda_0' + \delta_0'), R_{1.148}(\delta_0', \lambda_0')) = 2\lambda_0' + 5\delta_0'. \tag{3.7}$$

Set (cf. (3.6))

$$\lambda' := 1500(L_0'(R + 2\delta_0'))^3, \tag{3.8}$$

$$E := 2(2\lambda_0' + 3\delta_0' + R) + (\lambda' + \delta_0). \tag{3.9}$$

While our proofs will work whenever

$$K \geq R + \lambda' + \delta_0, \tag{3.10}$$

concretely, we will use

$$K = R^{\wedge} := (15L_0'R)^3, \quad \text{i.e.,} \quad R = K^{\vee} := \frac{1}{15L_0'}K^{1/3} \tag{3.11}$$

(The reader will verify that this K satisfies the inequality (3.10).) Thus, the inequality (3.7) translates into the inequality

$$K \geq K_0 = 15^3(2\lambda_0' + 5\delta_0')^3(L_0')^3. \tag{3.12}$$

Note also that (3.9) makes E a function of K

$$E = E_{(3.13)}(K), \tag{3.13}$$

and R also becomes a function of K.

We inductively define $4\delta_0$-quasiconvex subsets $Q_v \subset X_v, Q_e \subset X_e, v \in V(T), e \in E(T)$, and, at the same time, verify conditions of Definition 3.1 for the collection of subsets Q_v, Q_e, aiming eventually to prove Theorem 3.20. Assuming that for $v \in V(T)$ a $4\delta_0$-quasiconvex subset $Q_v \subset X_v$ is defined, for the oriented edge $e = [v, w]$ of T (oriented away from u) we set

$$Q_w^v := P_{X_{vw}, X_w}(Q_v), \quad Q_w' := N_R^e(Q_v) \cap X_w.$$

According to Corollary 1.131,

$$\mathrm{Hd}_{X_{vw}}(Q_w^v, Q_w') \leq 2(2\lambda_0' + 3\delta_0' + R).$$

Note that both X_w, Q_v are λ_0'-quasiconvex in X_{vw}.

Furthermore, by Lemma 1.129, since $R \geq R_0 \geq 2\lambda_0' + 2\delta_0'$, the intersection $Q_w' := N_R^e(Q_v) \cap X_w$ is $\lambda_{1.129} = R + 2\delta_0'$-quasiconvex in X_{vw}. Hence, Q_w' is $\lambda' = \widehat{R + 2\delta_0'}$-quasiconvex in X_w, where

$$\lambda' = 1500(L_0'(R + 2\delta_0'))^3;$$

see Lemma 1.113.

Therefore, by (1.3), the δ_0-hull, taken in X_w,

$$Q_w := \mathrm{Hull}_{\delta_0}(Q_w')$$

is $(\lambda' + \delta_0)$-Hausdorff close to Q'_w, thus,

$$\mathrm{Hd}_{X_{vw}}(Q^v_w, Q_w) \le E = 2(2\lambda'_0 + 3\delta'_0 + R) + (\lambda' + \delta_0),$$

verifying the condition (3.2) in Part 3 of a semicontinuous family of spaces (in the case when $Q'_w \neq \emptyset$, equivalently, $Q_w \neq \emptyset$).

We define the edge-space Q_e as the δ_0-hull (in X_e) of the projection

$$P_{X_{vw},X_e}(Q_v).$$

Thus,

$$\mathrm{Hd}_{X_{vw}}(Q^v_w, Q_e) \le \delta_0 + 1.$$

At the same time, since each point of Q'_w is within distance R from Q_v, each point of Q_w is within distance

$$R + \lambda' + \delta_0$$

from Q_v, where both distances are computed in X_{vw}. Since

$$K = (15L'_0R)^3 \ge R + \lambda' + \delta_0,$$

we conclude that each point of Q_w is within distance K from Q_v. From this, since Q_e was defined as the projection of Q_w to X_e, it also follows that $\mathrm{Hd}_{X_{vw}}(Q_w, Q_e) \le K$. Thus, we verified Part 3 of Definition 3.1 (for the edge e). Since the subsets Q_w, Q_e were defined as δ_0-hulls in δ_0-hyperbolic spaces, we conclude that $Q_e \subset X_e, Q_w \subset X_w$ are $4\delta_0$-quasiconvex, verifying Part 1 of Definition 3.1.

Lastly, we turn to Part 4 of Definition 3.1. As we noted earlier, $Q_w = \emptyset$ if and only if $Q'_w = N^e_R(Q_v) \cap X_w = \emptyset$. In other words, the λ'_0-quasiconvex subsets $Q_v, X_w \subset X_{vw}$ are R-separated. Since R was chosen to satisfy

$$R \ge R_0 = R_{1.142}(\delta'_0, \lambda'_0) = 2\lambda'_0 + 5\delta'_0,$$

Lemma 1.142 implies that subsets $Q_v, X_w \subset X_{vw}$ are $D = D_{1.142}(\delta'_0, \lambda'_0)$-cobounded. This verifies Part 4 of Definition 3.1.

Definition 3.18. We define the *K-flow space* $\mathfrak{Fl}_K(Q_u)$ of Q_u as the subtree of spaces in $\mathfrak{X}$ as follows. The nonempty subsets Q_v, Q_e defined by the inductive procedure above will be the vertex/edge spaces of $\mathfrak{Fl}_K(Q_u)$. The incidence maps g_{ev} of $\mathfrak{Fl}_K(Q_u)$ are the compositions of the incidence maps f_{ev} with fiberwise nearest-point projections in X_v to Q_v. The vertex and edge-spaces of $\mathfrak{Fl}_K(Q_u)$ are equipped with path-metrics induced from the ambient path-metrics on vertex and edge-spaces of $\mathfrak{X}$. We let $Fl_K(Q_u) \subset X$ denote the total space of $\mathfrak{Fl}_K(Q_u)$, set $S := \pi(Fl_K(Q_u))$; we will use the notation $\mathcal{F}l_K(Q_u)$ for the disjoint union

$$\coprod_{v \in V(S)} Q_v,$$

which is the union of vertex-spaces of $\mathfrak{Fl}_K(Q_u)$. We will equip $Fl_K(Q_u)$ with the standard path-metric of a tree of spaces.

Sometimes it will be convenient to restrict flow-spaces to subtrees $T' \subset T$. We will denote such "subflows" by

$$\mathfrak{Fl}^{T'}_K(Q_u).$$

Remark 3.19.

(1) The δ_0-neighborhoods in the definition of flow-spaces are taken in order to ensure that the each inclusion map $Q_w \to X_w$ is a $(1, C_{1.96}(\delta_0))$-quasi-isometric embedding, where Q_w is equipped with the path-metric induced from X_w, see Lemma 1.96.

(2) In general, it is not true that for Hausdorff-close subsets $A, B \subset X_u$, the K-flow spaces of A and B are Hausdorff-close to each other. However, if

$$B \subset N_r^{fib}(Q_u) \subset X_u$$

then (by the very definition of a flow-space)

$$Fl_K(B) \subset Fl_{K+r}(Q_u).$$

Similarly,

$$N_r^{fib}(Fl_K(Q_u)) \setminus N_r^{fib}(Q_u) \subset Fl_{K+r}(Q_u).$$

The discussion preceding the definition of flow-spaces proves:

THEOREM 3.20. *For every $K \geq K_0$, the flow-space $\mathfrak{Fl}_K(Q_u)$ is a (K, D, E)-semicontinuous family of spaces in $\mathfrak{X}$, where $D = D_0 = D_{1.142}(\delta_0', \lambda_0')$ and $E = E_{3.20}(K)$ is given by the equation* (3.9). *In particular, every $x \in Fl_K(Q_u)$ belongs to a K-leaf γ_x in $Fl_K(Q_u)$ connecting x to Q_u.*

Combining with with the existence of uniform coarse Lipschitz retractions to semicontinuous subtrees of spaces (Theorem 3.3), we conclude:

THEOREM 3.21 (Mitra's retraction). *Suppose that $\mathfrak{X}$ is a tree of hyperbolic spaces. Then for each $K \geq K_0$, there exists an $L_{3.21}(K)$-coarse Lipschitz retraction, called* Mitra's retraction, *$\rho = \rho_{Fl_K(Q_u)} : X \to Fl_K(Q_u)$, where*

$$L_{3.21}(K) = L_{3.3}(K, D_0, E_{3.20}(K), 4\delta_0').$$

Below we collect several consequences of Theorem 3.21.

COROLLARY 3.22. *The inclusion map $Fl_K(Q_u) \to X$ is an $L_{3.21}(K)$-qi embedding.*

COROLLARY 3.23 ([**Mit98**]). *If X were a hyperbolic metric space then for all $u \in V(T)$ and $4\delta_0$-quasiconvex subsets $Q_u \subset X_u$, the flow-spaces $Fl_K(Q_u)$ would be uniformly quasiconvex subsets in X.*

COROLLARY 3.24. *If $\pi(Fl_K(X_{u_1})) \cap \pi(Fl_K(X_{u_2})) = \emptyset$, then the flow-spaces $Fl_K(X_{u_1})$, $Fl_K(X_{u_2})$ are $L_{3.21}(K)$-Lipschitz-cobounded in X (cf. Definition* 1.26*).*

PROOF. We will be using Mitra's projections

$$\rho_i = \rho_{Fl_K(X_{u_i})}, i = 1, 2.$$

Since these projections are $L_{3.21}(K)$-coarsely Lipschitz, it suffices to show that diameters of $\rho_i(Fl_K(X_{u_{3-i}}))$, $i = 1, 2$, are uniformly bounded. Let $v_i \in \pi(Fl_K(X_{u_i}))$, $i = 1, 2$, denote the vertices realizing the minimal distance between these subtrees of T. Let $e_1 = [v_1, w_1]$, $e_2 = [v_2, w_2]$ be the edges incident to v_1, v_2 and contained in the interval $[\![v_1, v_2]\!]$ (it is possible that $w_1 = v_2$, $w_2 = v_1$). Then, by the definitions of Mitra's projection (see the proof of Theorem 2.22) and the K-flow, for $i = 1, 2$,

$$\rho_{3-i}(Fl_K(X_{u_i})) \subset \rho_{3-i}(X_{v_i}) = \{x_{3-i}\} \subset Fl_K(X_{u_{3-i}}) \cap X_{v_{3-i}},$$

i.e., $\rho_{3-i}(Fl_K(X_{u_i}))$ is the singleton $\{x_{3-i}\}$. Thus, the flow-spaces $Fl_K(X_{u_1})$, $Fl_K(X_{u_2})$ are $L_{3.21}(K)$-Lipschitz-cobounded in X, see Definition 1.26. □

EXAMPLE 3.25. One can realize the hyperbolic plane as the total space (up to a quasi-isometry) of a metric line bundle over a line. Namely, let $T = \mathbb{R}$, where the vertices are the integer points. We will identify T, via an isometry, with the y-axis in the upper half-plane model of the hyperbolic plane (of course, we parameterize the y-axis in $\mathbb{H}^2$ by the hyperbolic arc-length). The projection $\pi : \mathbb{H}^2 \to T$ is given by the y-coordinate of the points in $\mathbb{H}^2$. Examples of 1-qi sections of the bundle $\pi : \mathbb{H}^2 \to T$ are given by hyperbolic geodesics in $\mathbb{H}^2$ which are vertical rays in the half-plane model. For $K = 1$, the K-flow space of a singleton $Q_u = \{q\}$ is then such a hyperbolic geodesic through this point. Mitra's retraction to $L = Fl_K(Q_u)$ in this example is within finite distance from the horocyclic projection $\mathbb{H}^2 \to L$. While it is 1-Lipschitz, it is not close to the nearest-point projection $\mathbb{H}^2 \to L$. For a general $K \geq 1$, the K-flow of a singleton $Q_u = \{q\}$ is Hausdorff-close to a geodesic in $\mathbb{H}^2$, equivalently, to a K-qi section through q.

3.3.2. Basic properties of flow-spaces. Most of the time, besides Mitra's retractions and the fact that each flow-space forms a semicontinuous subtree of spaces in $\mathfrak{X}$, instead of the definition of flow-spaces we will use their properties summarized in the next proposition (we recall the equations (3.7) and (3.11) defining the constants K_0, R_0 and the function $R \mapsto R^\wedge = K$):

PROPOSITION 3.26. *Suppose that $Q_u \subset X_u$ is a $4\delta_0$-quasiconvex subset.*

(1) *Suppose that a vertex w lies between vertices u and v. Then for every $r \geq 0$, and all $K \geq K_0$,*

$$N_r^{fib}(Fl_K(X_u)) \cap N_r^{fib}(Fl_K(X_v)) \subset N_r^{fib}(Fl_K(X_w)), \tag{3.14}$$

and

$$N_r(Fl_K(X_u)) \cap N_r(Fl_K(X_v)) \subset N_r(Fl_K(X_w)). \tag{3.15}$$

(2) *Suppose that $R \geq R_0$ and let γ is a R-qi leaf in X emanating from some $\gamma(u) \in Q_u$. Then $\gamma \cap X$ is contained in $\mathcal{F}l_K(Q_u)$, where $K = R^\wedge$.*

(3) *For all $K \geq R_0$ and $Q_v := Fl_K(Q_u) \cap X_v$, we have*

$$Q_u \subset Fl_{K^\wedge}(Q_v).$$

(4) *For every boundary edge $e = [v, w]$ of $S = \pi(Fl_K(Q_u))$, the subsets Q_v, X_w are D_0-cobounded in X_{vw}, where D_0 is given by (3.5).*

PROOF.

(i) The first containment follows from the inductive nature of the definition of flows. We now prove the second inclusion. Let $x \in Fl_K(X_u), y \in Fl_K(X_v)$ and $z \in \bar{B}(x, r) \cap \bar{B}(y, r) \subset X$. Up to relabeling, there are two cases:

a. w lies in the geodesic segment $\pi(y)v \subset T$. Then $y \in Fl_K(X_w)$ (according to (3.14) with $r = 0$), which implies that $z \in N_r(Fl_K(X_w))$.

b. The vertex w is not in $u\pi(x) \cup v\pi(y)$, hence, w separates $\pi(x), \pi(y)$ in T. Therefore, after relabeling, w separates $\pi(x)$ from $\pi(z)$. In particular the geodesic $xz \subset X$ crosses X_w and, hence, $z \in N_r(X_w) \subset N_r(Fl_K(X_w))$.

(ii) The proof is by induction on the distance from vertices of $\pi(\gamma)$ to u. We will use the notation Let Q_v, Q'_w, etc. as in the definition of the flow-space $Fl_K(Q_u)$.

The base of induction (for the vertex u) is clear. Suppose that $e = [v, w]$ is an edge in $\pi(\gamma)$ oriented away from u such that $x = \gamma(v) \in Q_v, y = \gamma(w) \in X_w$. Arguing inductively, we assume that the claim holds for the point x, i.e.,

$$x \in Q_v = \mathcal{F}l_K(Q_u) \cap X_v.$$

Since γ is an R-qi leaf in X,

$$d_{X_{vw}}(x, y) \leq R,$$

i.e., by the definition of the function $R \mapsto R^{\wedge} = K$,

$$y \in N^e_R(Q_v) \cap X_w = Q'_w \subset Q_w.$$

Part (3) is an immediate consequence of Part (2). □

NOTATION 3.27. In what follows, we will refer to K-qi leaves in $Fl_K(Q_u)$ as *leaves* in $Fl_K(Q_u)$ and denote them by γ with some subscripts. We will use the notation γ_x, $\gamma_{x,y}$ for such leaves provided that γ has x as one of its endpoints (in the former case) or γ connects x and y.

LEMMA 3.28. *Given any point $x \in Fl_K(X_u)$, there is a maximal*[1] *K-qi section Σ_x in $Fl_K(X_u)$ over a subtree T_x of T such that $x \in \Sigma_x$ and $u \in T_x$.*

PROOF. Let $w = \pi(x)$. By Theorem 3.20, there exists a K-qi section $\Sigma_{[\![u,w]\!],x}$ over the geodesic $[\![u, w]\!] \subset T$ connecting u and w such that $x \in \Sigma_{[\![u,w]\!],x}$. We define a poset consisting of K-qi sections $\Sigma_{S,x}$ over subtrees $S \subset T$ and containing $\Sigma_{[\![u,w]\!],x}$. Define the partial order $\Sigma_{S,x} \le \Sigma_{S',x}$ if the qi section $\Sigma_{S',x}$ extends $\Sigma_{S,x}$. This poset is clearly nonempty. The existence of a maximal element in this poset follows from Zorn's lemma. □

In Chapter 8 we will also use the following property of flow-spaces:

LEMMA 3.29. *Suppose that $Q_w = Fl_K(X_u) \cap X_w \neq \emptyset$. Then $N_R(Fl_K(X_u)) \cap X_w$ is contained in*

$$N^{fib}_D(Q_w),$$

where $D = D_{3.29}(R, K)$.

PROOF. Let $x \in N_R(Fl_K(X_u)) \cap X_w$ and let $y \in Fl_K(X_u)$ such that $d(x, y) \le R$. Let γ be a geodesic in X joining x, y. Let $\rho : X \to Fl_K(X)$ denotes the Mitra's projection. We know that ρ is an $L_{3.21}(K)$-coarse Lipschitz retraction. We note that $\rho(x) \in Q_w$ and $\rho(y) = y$. Hence, we have $d(\rho(x), y) \le (R + 1)L_{3.21}(K)$. Therefore,

$$d(x, \rho(x)) \le d(x, y) + d(\rho(x), y) \le R + (R + 1)L_{3.21}(K).$$

Finally, since the vertex-spaces are uniformly properly embedded in X, we have a uniform bound on $d_{X_w}(x, \rho(x))$ which proves the lemma. □

3.3.3. Generalized flow-spaces. In this section we discuss several generalizations of flow-spaces and retractions to these, generalizing Mitra's projection. We recall (see Definition 3.18) that superscript in the form of a subtree $S \subset T$ in the notation of a flow-space, means that the flow-space is taken in the subtree of spaces $\mathfrak{X}_S \subset \mathfrak{X}$ over the subtree $S \subset T$.

DEFINITION 3.30. Assume that $K \ge K_0$.

(1) We define flow-spaces of quasiconvex subsets $Q_e \subset X_e$ of edge-spaces, $e = [v, w]$. Define subtrees T_v (resp, T_w) in T as the maximal subtree in T containing v (resp., w) and disjoint from w (resp., v). We define the flow-space $\mathfrak{Fl}_K(Q_e)$ of such Q_e so that its union of vertex-spaces is

$$\mathcal{F}l_K(Q_e) := \mathcal{F}l^{T_v}_K(\mathrm{Hull}_{\delta_0} f_{ev}(Q_e)) \cup \mathcal{F}l^{T_w}_K(\mathrm{Hull}_{\delta_0} f_{ew}(Q_e)).$$

[1]Here maximal means we can not find a K-qi section Σ'_x over a subtree $T'_x \subset T$ containing T_x such that $\Sigma_x \subsetneq \Sigma'_x$.

(2) Similarly, we define flow-spaces in X of $X_S \subset X$, where $S \subset T$ is a subtree. For each $w \in V(S)$ we define the maximal subtree $T_w \subset T$ whose intersection with S equals $\{w\}$. We then define the K-flow-space $\mathfrak{Fl}_K(X_S)$ in X so that $X_S \subset \mathfrak{Fl}_K(X_S)$ and the union of vertex-spaces of $\mathfrak{Fl}_K(X_S)$ equals

$$\mathcal{F}l_K(X_S) = \bigcup_{v \in V(S)} \mathcal{F}l_K^{T_v}(X_v). \tag{3.16}$$

COROLLARY 3.31. *For oriented edges $e = [u, v]$ of T the inclusion maps*

$$Fl_K(X_e) \to Fl_K(X_u)$$

are uniformly quasi-isometric embeddings.

PROOF. Apply Theorem 3.21 to the δ_0-quasiconvex hull of $Q_u := f_{eu}(X_e)$ and observe that $Fl_K(X_e) \subset Fl_K(Q_u) \subset Fl_K(X_u)$. □

PROPOSITION 3.32. *There exists $L = L_{3.32}(K)$ such that for every subtree $S \subset T$ and $K \geq K_0$, the flow-space $Fl_K(X_S)$ is L-qi embedded in X.*

PROOF. We continue with the notation introduced in Definition 3.30. For each $w \in V(S)$ we define Mitra's retraction $\rho_w = \rho_{X_w} : X_{T_w} \to Fl_K^{T_w}(X_w)$. Hence, the collection of maps ρ_w is uniformly coarsely Lipschitz. We then obtain a (uniformly) coarse Lipschitz retraction

$$X \to Fl_K(X_S)$$

whose restriction to $\pi^{-1}(S)$ is the identity and whose restriction to each X_{T_w} equals ρ_w. □

We now define flow-spaces of metric bundles. Let $S \subset T$ be a subtree and let

$$\mathfrak{Q} = (\pi : Q \to S)$$

be a metric K_1-bundle in X whose vertex-spaces $Q_v \subset X_v$ are $4\delta_0$-quasiconvex subsets. Let $Q = \bigcup_v Q_v$ denote the union of vertex-spaces of $\mathfrak{Q}$. For each vertex $v \in V(S)$, as above, we have the associated subtree $T_v \subset T$, which is the maximal subtree in T such that $T_v \cap S = \{v\}$. Accordingly, we have subtrees of spaces $X_{T_v} \subset X$.

For each vertex $v \in S$ we take the K_2-flow-space

$$\mathcal{F}_v := \mathcal{F}l_{K_2}(Q_v) \cap X_{T_v}$$

inside X_{T_v}. Lastly, set

$$\mathcal{F}l_{K_2}(Q_u) := \bigcup_{v \in V(S)} \mathcal{F}_v.$$

Then, as in the case when Q was the union of vertex-spaces X_v, $v \in V(S)$ (see in Definition 3.30(2)), $\mathcal{F}l_{K_2}(Q_u)$ is the union of vertex-spaces of a tree of spaces $\mathfrak{Fl}_{K_2}(\mathfrak{Q})$.

DEFINITION 3.33. We will refer to $\mathfrak{Fl}_k(\mathfrak{Q})$ (and its total space $Fl_k(\mathfrak{Q})$) as a *generalized k-flow-space*, or the *k-flow-space of a metric bundle*.

The following is an extension of Mitra's theorem to such generalized flow-spaces:

THEOREM 3.34. *For every $K_2 \geq K_0$, there exists an $L = L_{3.34}(K_1, K_2)$-coarse Lipschitz retraction $X \to Fl_{K_2}(\mathfrak{Q})$.*

PROOF. The proof is similar to that of Proposition 3.32. Over subtrees T_w, $w \in V(S)$, we use Mitra's retractions $X_{T_w} \to Fl_{K_2}(\mathfrak{Q}) \cap X_{T_w}$. The fact that these maps define a uniformly coarse Lipschitz retraction $X_S \to Q$ (over the three S) follows from the assumption that $\mathfrak{Q} = (\pi : Q \to S)$ is a K_1-bundle, cf. Corollary 2.24. □

This theorem will be used in Section 6.1.2.

3.3.4. Boundary flows. In this section we will define *ideal boundary flows*, which are ideal counterparts of flow-spaces. Ideal boundary flows will be used only in Chapter 7 (specifically, Section 7.3) and Chapter 8. The definition of boundary flows given below was latent in [**Sar18**, Definition 4.3].

Fix a vertex $u \in V(T)$ and consider a subset $Z_u \subset \partial_\infty X_u$. We will define the *flow-space* $\mathfrak{Fl}(Z_u)$ of Z_u in $\partial_\infty X$, analogously to the definition of flow-spaces $\mathfrak{Fl}_K(Q_u)$ of quasiconvex subsets $Q_u \subset X_u$ given in Section 3.3. Each $\mathfrak{Fl}(Z_u) = \mathfrak{Z}$ will be a tree of topological spaces over a subtree $S \subset T$, with vertex-spaces $Z_v \subset \partial_\infty X_v$ and edge-spaces $Z_e \subset \partial_\infty X_e$.

Our definition of subsets $Z_v \subset \partial_\infty X_v$, $Z_e \subset \partial_\infty X_e$, is inductive on $d_T(u, v)$, analogously to the definition of $\mathfrak{Fl}_K(Q_u) \subset X$. We assume that subsets Z_v, Z_e are defined for all vertices and edges of T contained in the ball (in T) of radius n centered at u. Consider an edge $e = [v, w]$ in T oriented away from u, such that $d_T(u, v) = n$. We have qi embeddings $f_{ev} : X_e \to X_v$ and $f_{ew} : X_e \to X_w$. They induce topological embeddings (*boundary maps*)

$$\partial_\infty f_{ev} : \partial_\infty X_e \to \partial_\infty X_v, \quad \partial_\infty f_{ew} : \partial_\infty X_e \to \partial_\infty X_w;$$

see Section 1.13. Define

$$Z_e := (\partial_\infty f_{ev})^{-1}(Z_v), \quad \text{and} \quad Z_w := \partial_\infty f_{ew}(Z_e).$$

The boundary maps $\partial_\infty f_{ev}, \partial_\infty f_{ew}$ provide incidence maps $Z_e \to Z_v, Z_e \to Z_w$. We continue inductively. Define S as the subtree in T spanned by the vertices $v \in T$ such that $Z_v \neq \emptyset$. Thus, we obtain a tree $\mathfrak{Z} = (Z \to S), Z = Fl(Z_u)$, of topological spaces with the vertex-spaces $Z_v, v \in V(S)$, the edge-spaces $Z_e, e \in E(S)$ and incidence maps $\partial_\infty f_{ev}$ as above, see Definition 2.5. We will use the notation $\mathcal{F}l(Z_u)$ for $\mathcal{Z}$, the union of vertex-spaces of $\mathfrak{Z}$. Note that since the maps $\partial_\infty f_{ev}$ are 1-1, the set $\mathcal{F}l(Z_u)$ breaks as a disjoint union of flow-spaces $\mathcal{F}l(\{z\})$ of singletons $\{z\} \subset Z_u$. For each $v \in V(T)$, the intersection

$$Fl_v(\{z\}) := \mathcal{F}l(\{z\}) \cap \partial_\infty X_v$$

is either empty or a singleton. Similarly, for each edge $e = [v, w] \in E(T)$, the intersection $Fl_e(\{z\}) := Fl(\{z\}) \cap Z_e$ is either empty or a singleton. In view of the inductive definition of $Fl(Z_u)$, if $Fl_w(\{z\}) \neq \emptyset$ for a vertex w, then for every vertex $v \in uw$, $Fl_v(\{z\}) = \{z'\}$ is also nonempty and we have

$$Fl_v(\{z'\}) = Fl_w(\{z\}).$$

Furthermore, for each edge e in the interval uv, $Fl_e(\{z\})$ is nonempty as well. For $z \in \partial_\infty X_u$ we define a tree $T_z \subset S$ such that $V(T_z)$ consists of vertices v for which $Fl_v(\{z\})$ is nonempty.

The next lemma is partially proven in [**Sar18**, Lemma 4.4]; we include a proof for the sake of completeness:

LEMMA 3.35. *Suppose that $\mathfrak{X} = (\pi : X \to T)$ is a tree of hyperbolic spaces. Consider a point $z \in \partial_\infty X_u$ and a geodesic ray $\alpha_u \subset X_u$ asymptotic to z. Then for a vertex $v \in V(T)$ the following are equivalent:*

(1) *$Fl_v(\{z\})$ is nonempty.*
(2) *There is a geodesic ray $\alpha_v \subset X_v$ such that* $\mathrm{Hd}(\alpha_u, \alpha_v) < \infty$. *In this case $Fl_v(\{z\}) = \{\alpha_v(\infty)\}$.*
(3) *α_u is Hausdorff-close to a subset of X_v.*

PROOF. It suffices to prove the equivalence in the case when u, v span an edge $e = [v, w]$ in T (the general case is proven by induction on $d_T(u, v)$). As we noted earlier, $Fl_v(\{z\}) =$

$\{z_v\} \neq \emptyset$ if and only if $Fl_e(\{z\}) = \{z_e\} \neq \emptyset$. The latter holds if and only if there is a geodesic ray α_e in X_e such that $\alpha_e(\infty) = z_e$ and

$$\partial_\infty f_{eu}(z_e) = z, \quad \partial_\infty f_{ev}(z_e) = z_v.$$

The paths $f_{eu}(\alpha_e)$, $f_{ev}(\alpha_e)$ are quasigeodesic rays in X_u and X_v respectively, which are asymptotic, respectively, to the points $\{z\}, \{z_v\}$. By the stability of quasigeodesics we have $\mathrm{Hd}(\alpha_u, f_{eu}(\alpha_e)) < \infty$ and there is a geodesic ray α_v in X_v Hausdorff-close to $f_{ev}(\alpha_e)$. This proves the equivalence of (1) and (2).

The implication (2) ⇒ (3) is clear. If (3) holds, then the image of α_u under the projection $P = P_{X_{uv}, X_v}$ is again a quasigeodesic in X_v (which is Hausdorff-close to α_u). By the stability of quasigeodesics, $P(\alpha_u)$ is Hausdorff-close to a geodesic α_v in X_v. □

COROLLARY 3.36. *The following is an equivalence relation on* $\coprod_{v \in V(T)} \partial_\infty X_v$*: Two points* $z_i \in \partial_\infty X_{u_i}$, $i = 1, 2$ *are related iff* $Fl_{u_2}(\{z_1\}) = \{z_2\}$.

We next relate boundary flow-spaces to the flow-spaces in $\mathfrak{X}$.

LEMMA 3.37. *There is a constant K depending on the parameters of* $\mathfrak{X}$ *such that the following holds:*

(1) *If* $\alpha_u \subset X_u$ *is a geodesic ray asymptotic to* $z \in \partial_\infty X_u$*, then*

$$Fl_v(\{z\}) = \partial_\infty(Fl_K(\alpha_u) \cap X_v, X_v)$$

for all $v \in V(T_z)$.

(2) $Fl_v(\partial_\infty X_u) = \partial_\infty(Fl_K(X_u) \cap X_v, X_v)$ *for all* $v \in V(T)$.

PROOF. The inclusion $\partial_\infty(Fl_K(\alpha_u) \cap X_v, X_v) \subset Fl_v(\{z\})$ is clear, we will prove the opposite inclusion. Suppose first that v_1, v_2 are vertices in the tree of T_z satisfying

$$d_T(u, v_2) = d_T(u, v_1) + 1,$$

and let $\alpha_i \subset X_{v_i}$, $i = 1, 2$, be geodesic rays such that $Fl_{v_i}(\{z\}) = \{\alpha_i(\infty)\}$, $i = 1, 2$. According to Lemma 3.35,

$$\mathrm{Hd}_{X_{v_1 v_2}}(\alpha_1, \alpha_2) < \infty.$$

Since $X_{v_1 v_2}$ is δ_0'-hyperbolic and α_1, α_2 are L_0'-quasigeodesics in $X_{v_1 v_2}$, it follows that there are subrays in $\alpha_1' \subset \alpha_1, \alpha_2' \subset \alpha_2$ which are R-Hausdorff close in $X_{v_1 v_2}$, where R depends only on the parameters of $\mathfrak{X}$ (see Corollary 1.86). Set $K := R^\wedge$. Then, if α_1 is contained in $Fl_K(\alpha_u)$, then so does α_2' (see Proposition 3.26(2)). Now, Part (1) of the lemma follows by induction on $d_T(u, v)$.

Part (2) follows immediately from Part (1). □

3.3.5. Geometry of the flow-incidence graph. In this section we will assume that $K \geq K_0$. We will analyze (to some degree) the intersection pattern of projections to T of the flow-spaces $Fl_K(X_u)$, $u \in V(T)$. Some of these results will be important (specifically, Subdivision Lemma and its corollary, Corollary 3.45) in Chapter 6.

DEFINITION 3.38. An interval $J = [\![u, v]\!] \subset T$ is *special* (more precisely, *K-special*) if one of its endpoints (say, u) has the property that $J \subset \pi(Fl_K(X_u))$. In this case, the vertex u is said to be *special* in J.

For instance, for every edge $e = [u, v] \in E(T)$, the interval $J = [\![u, v]\!]$ is special.

REMARK 3.39. The notion of a special interval can be refined and one can say that an *oriented* interval $J = [\![u, v]\!] \subset T$ is special if $J \subset \pi(Fl_K(X_u))$. We will not need this refinement.

The importance of special intervals comes from the following simple fact:

LEMMA 3.40. *Suppose that* $u, v \in V(T)$ *are such that* $\pi(Fl_K(X_u)) \cap \pi(Fl_K(X_v))$ *contains a vertex* t. *Then the center* w *of the triangle* Δtuv *subdivides the interval* $I = [\![u, v]\!]$ *in two special subintervals*

$$J = [\![u, w]\!], J' = [\![w, v]\!].$$

PROOF. Since w separates u and v from t, Proposition 3.26(1) implies that

$$J \subset \pi(Fl_K(X_u)), J' \subset \pi(Fl_K(X_v)).$$

□

Of prime importance for us will be whether intersections as in the lemma are empty or not, since empty intersection will imply that the pair of subsets $Fl_K(X_u)$, $Fl_K(X_v)$ of X is $L_{3.21}(K)$-Lipschitz cobounded, see Corollary 3.24. This observation motivates the following definition:

DEFINITION 3.41. For each K we define the *flow-incidence graph* $\Gamma = \Gamma_K$. Its vertex-set is $V(T)$. Two vertices u, v of Γ are connected by an edge $e \in E(\Gamma)$ if and only if $\pi(Fl_K(X_u)) \cap \pi(Fl_K(X_v)) \neq \emptyset$.

Lemma 3.40 implies Corollary 3.42.

COROLLARY 3.42. $d_\Gamma(u, v) \leq 1$ *if and only if the interval* $[\![u, v]\!]$ *is the union of two special subintervals.*

While the graph Γ is not necessarily a tree, we will see that it is a *quasi-tree*. Recall that a geodesic metric space Y is called a *quasi-tree* (see Manning's paper [**Man05**]) if it is quasi-isometric to a simplicial tree. According to [**Man05**], Y is a quasi-tree if and only if there exists a constant r such that for every geodesic segment $xy \subset Y$, the (closed) ball $B(m, r)$ centered at the midpoint m of xy separates x and y. (Here and below, separation is understood in the sense that every path connecting x and y has to pass through $B(z, r)$.) Alternatively, one characterizes quasi-trees by the existence of r such that for every geodesic segment $xy \subset Y$ and every $z \in xy$, the ball $B(z, r)$ separates x and y.

LEMMA 3.43. *Suppose that* p *is a point*[2] *in an interval* $J = [\![u, v]\!] \subset T$. *Then:*

(1) *The closed unit ball* $B_\Gamma(p, 1) \subset \Gamma$ *separates* u *and* v.
(2) $d_\Gamma(u, v) \geq d_\Gamma(u, p)$.

PROOF. Both parts of the lemma are proven by the same argument. Suppose that $w_0 = u, w_1, \ldots, w_n, w_{n+1} = v$ is a vertex-path in Γ connecting u to v. Since p separates u from v, there exists $i \leq n$ such that w_i, w_{i+1} are separated by p in T and, of course, $d_\Gamma(w_i, w_{i+1}) = 1$. Lemma 3.40 implies that

$$d_\Gamma(w_i, p) \leq 1, d_\Gamma(w_{i+1}, p) \leq 1.$$

This proves Part (1) of the lemma. To prove Part (2) we take the above vertex-path to be geodesic in Γ and observe that

$$d_\Gamma(p, u) \leq d_\Gamma(p, w_i) + d_\Gamma(w_i, u) \leq 1 + d_\Gamma(u, w_i) \leq d_\Gamma(u, v).$$

□

Lemma 3.43 thus implies that for every K and a tree of hyperbolic spaces spaces $\mathfrak{X}$, the graph $\Gamma = \Gamma_K$ is a quasi-tree with the constant $r = 1$.

We are now ready to prove the *horizontal subdivision lemma*, which will play an important role in Chapter 6, when we establish uniform hyperbolicity of K-flows of interval-spaces X_J in $\mathfrak{X}$.

[2] This is not necessarily a vertex.

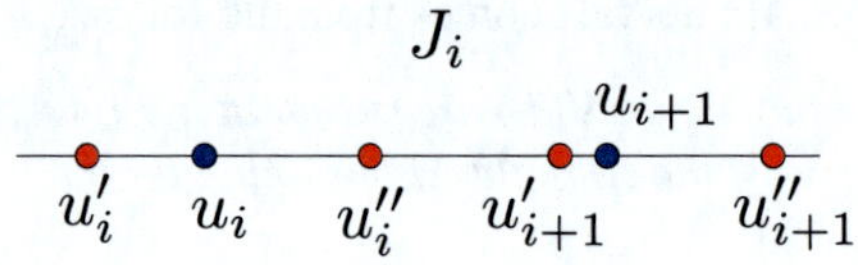

FIGURE 3.3. Horizontal subdivision

LEMMA 3.44 (Horizontal subdivision lemma). *For any pair of distinct vertices $u, v \in T$, the interval $J = [\![u, v]\!]$, can be subdivided into nondegenerate subintervals*

$$J = [\![u_0, u_1]\!] \cup \dots \cup [\![u_{n-1}, u_n]\!] \cup [\![u_n, u_{n+1}]\!], \quad u = u_0, v = u_{n+1},$$

and each $J_i = [\![u_i, u_{i+1}]\!]$ can be further subdivided into subintervals (some of which could be degenerate),

$$[\![u_i, u''_i]\!] \cup [\![u''_i, u'_{i+1}]\!] \cup [\![u'_{i+1}, u_{i+1}]\!],$$

so that the following hold for all $i \le n$:

(1)
$$\pi(Fl_K(X_{u_i})) \cap J_i = [\![u_i, u''_i]\!],$$
(i.e., the interval $[\![u_i, u''_i]\!]$ is special) and
$$u''_i \notin \pi(Fl_K(X_{u_{i+1}})),$$
unless $i = n$ in which case we could *have $u''_i \in \pi(Fl_K(X_{u_{i+1}}))$.*

(2) *The interval $[\![u''_i, u'_{i+1}]\!]$ is special, it is contained in $\pi(Fl_K(X_{u'_{i+1}}))$.*

(3) *The interval $[\![u''_i, u'_{i+1}]\!]$ is nondegenerate unless $i = n$.*

(4) *$d_T(u'_{i+1}, u_{i+1}) \le 1$, thus, each interval $[\![u'_{i+1}, u_{i+1}]\!]$ is special.*

PROOF. We find the subdivision vertices inductively. Set $u_0 := u$. Inductively, we assume that u_i is defined. If $u_i = v$, we set $n + 1 = i$ and terminate the induction. Suppose, therefore, that this is not the case. We then define u''_i, u'_{i+1} and u_{i+1}:

We choose a vertex $u''_i \in]\!]u_i, v]\!]$ to be the farthest from u_i such that

$$Fl_K(X_{u_i}) \cap X_{u''_i} \neq \emptyset.$$

Note that such a vertex always exists since for the edge $[u_i, v_i] \in E([\![u_i, v]\!])$ we have

$$Fl_K(X_{u_i}) \cap X_{v_i} \neq \emptyset.$$

If it so happens that $u''_i = v$, we set $n = i$, and $u'_{i+1} := u_{i+1} = v$; this will conclude the induction. Suppose that this is not the case.

Then consider the vertices $s \in]\!]u''_i, v]\!]$ such that

$$Fl_K(X_s) \cap X_{u''_i} = \emptyset.$$

If such a vertex does not exist, then we set $n = i$, $u'_{i+1} = u''_i$ and $u_{i+1} = v$, and again conclude the subdivision process. Assume that this is not the case. Then we define $u_{i+1} \in]\!]u''_i, v]\!]$ to be the closest vertex to u''_i such that $Fl_K(X_{u_{i+1}}) \cap X_{u''_i} = \emptyset$. We define u'_{i+1} in this case to be the vertex in $J_i = [\![u_i, u_{i+1}]\!]$ adjacent to u_{i+1}, i.e., $d_T(u'_{i+1}, u_{i+1}) = 1$. Then, by the definition of u_{i+1},

$$Fl_K(u'_{i+1}) \cap X_{u''_i} \neq \emptyset.$$

Hence, the vertices u''_i, u'_{i+1}, u_{i+1} satisfies requirements of the lemma and we continue inductively. □

Corollary 3.45.

(1) *Each interval J_i as above is the union of (at most) three special intervals and the sequence*
$$\ldots u_i, u'_{i+1}, u_{i+1}, \ldots$$
is a vertex-path in Γ.

(2) *For any two consecutive vertices u_i, u_{i+1}, $i \le n-1$,*
$$d_\Gamma(u_i, u_{i+1}) = 2,$$
while $1 \le d_\Gamma(u_n, u_{n+1}) \le 2$.

(3) *For each pair of indices i, j, if $0 \le i+1 < j \ne n$, then*
$$d_\Gamma(u_i, u_j) \ge 2.$$
In particular, if $|i-j| \ge 2$, the flow-spaces $Fl_K(X_{J_i}), Fl_K(X_{J_j})$ are $L_{3.21}(K)$-Lipschitz cobounded in X.

3.4. Retractions to bundles

The main goal of this section is to prove Theorem 3.49, which is an analogue of Theorem 3.21, constructing coarse Lipschitz retractions from flow-spaces to certain K'-metric bundles $\mathfrak{Y} = (\pi : Y \to S) \subset \mathfrak{X}$ with λ-quasiconvex fibers $Y_v \subset X_v$. In Theorem 3.49 we will impose a stronger assumption on $\mathfrak{X}$, namely the κ-uniform flaring condition for a certain constant $\kappa \ge K$ (see (3.19) for the definition of this constant, which depends on K, K', on a quasiconvexity constant λ of $Y_v \subset X_v$, and on a constant D which is an upper bound on the diameter of Y_w for some $w \in V(S)$), that was not needed in Theorem 3.21. While the κ-flaring condition implies k-flaring for all $k \in [1, \kappa]$, it will be notationally convenient to also have the constants M_k, $k \le \kappa$, at our disposal, hence, we will be *assuming* the uniform k-flaring condition for all $k \in [1, \kappa]$.

The retractions $\rho_{\mathfrak{Y}} : Fl_K(Q_u) \to Y$ will be defined on flow-spaces $Fl_K(Q_u)$ whose $4\delta_0$-fiberwise neighborhoods contain Y, but composing $\rho_{\mathfrak{Y}}$ with Mitra's retraction $\rho : X \to Fl_K(Q_u)$, we then obtain retractions defined on the entire X. In the special case, when $Fl_K(Q_u)$ is δ-hyperbolic, the retraction $\rho_{\mathfrak{Y}}$ will be uniformly close to the nearest-point projection $Fl_K(Q_u) \supset Y$ (see Proposition 3.62). We recall that $\mathcal{Y}$ denotes the union of all the vertex-spaces in Y.

Remark 3.46.

(1) The condition that $\rho_{\mathfrak{Y}}$ is a retraction should be understood coarsely since Y is not quite contained in $Fl_K(Q_u)$: We can only guarantee that $\rho_{\mathfrak{Y}}$ fixes all points in $\mathcal{Y} \cap \mathcal{F}l_K(Q_u)$; the rest of the points of $Fl_K(Q_u)$ lie in the $\max(K, 4\delta_0)$-neighborhood of Y and $\rho_{\mathfrak{Y}}$ can move them only by a uniformly bounded amount.

(2) In the case when $\mathfrak{Y}$ is a K'-carpet, which is of the main interest, $\lambda = \delta_0$.

Lemma 3.47. *Fix λ and K'. Suppose that a subtree of spaces $\mathfrak{Y} \subset \mathfrak{X}$ is a K'-metric bundle over a subtree $S = \pi(Y) \subset T$. Assume also that vertex-spaces $Y_v = Y \cap X_v, v \in V(S)$, are λ-quasiconvex in X_v. Then the fiberwise nearest-point projection $X_S \to Y$ is a $D_{3.47}(\lambda, K')$-coarse Lipschitz retraction. In particular, if γ is a C-qi section over some interval $J \subset S$, then the fiberwise projection $\bar\gamma$ of γ to Y is a $K_{3.47}(\lambda, K', C) = CD_{3.47}(\lambda, K')$-qi section over J, where*
$$K_{3.47}(\lambda, K', C) \ge \max(C, K').$$

Proof. The lemma is an immediate corollary of Corollary 2.24. □

NOTATION 3.48.

(1) For the rest of this subsection we will use the notation $\bar{K}$ for $K_{3.47}(\lambda, K', K)$. We also set $\bar{\bar{K}} := K_{3.47}(\delta_0, K'_{3.17}(K), \bar{K})$. Note that
$$\bar{\bar{K}} \geq \max(\bar{K}, K'_{3.17}(K)).$$
In these notation we suppress the dependence on λ and K': In the most useful for us case, when $\mathfrak{Y}$ is a K-carpet, we will have $\lambda = \delta_0$.

(2) Define

$$r := r_{(3.17)} = 3\delta_0 + \lambda + R_{2.53}(\bar{\bar{K}}, 1) + R_{2.53}(\bar{K}, M_{\bar{K}}). \tag{3.17}$$

$$k = K' + R_{2.53}(\bar{K}, r). \tag{3.18}$$

$$\kappa := \kappa_{(3.19)}(\lambda, K, K') := \max(k, \bar{\bar{K}}). \tag{3.19}$$

Observe that $\kappa \geq \bar{\bar{K}} \geq \bar{K} \geq K$. The proof of the following theorem will need uniform $\bar{\bar{K}}$-flaring (in numerous places) as well as the uniform $(K' + R_{2.53}(\bar{K}, r_i))$-flaring for some numbers r_1, r_2, r_3 (subcases (i), (ii) and (iii) respectively in the proof of Proposition 3.51); the constant r above is chosen to be the maximum of the numbers r_1, r_2, r_3. We recall that $\mathcal{X}$ is the union of all the vertex-spaces in X. Also, below we have $\mathcal{Y} = \mathcal{X} \cap Y$ and $\mathcal{Z} = \mathcal{X} \cap Z$.

THEOREM 3.49. *Fix constants K, K' and λ and assume that, $\mathfrak{Y} = (\pi : Y \to S) \subset \mathfrak{X}$ is a K'-metric bundle with λ-quasiconvex fibers. We assume, furthermore, that*

(1) *There exists a vertex $w \in S$ such that* $\operatorname{diam}_{X_w}(Y_w) \leq D'$.

(2) *The tree of spaces $\mathfrak{X}$ satisfies the uniform flaring condition for all parameters in the interval $[1, \kappa]$, in particular, for $\bar{K}$, K' and $\bar{\bar{K}}$; see* (3.19) *for the definition of κ which depends on K, K' and λ.*

(3) *We assume that $\mathfrak{Y}$ is either a (K', D')-carpet $\mathfrak{A}(\alpha')$ contained in a K-ladder*[3] *$\mathfrak{Z} = \mathfrak{L}_K(\alpha)$,*
$$\alpha' \subset \alpha \subset X_u, \text{length}(\alpha') \geq \text{length}(\alpha) - M_{\bar{K}},$$
or $\mathfrak{Y}$ is a general K'-metric bundle contained in the fiberwise $4\delta'_0$-neighborhood of a K-flow-space $\mathfrak{Z} = \mathfrak{Fl}_K(Q_u)$, such that $Y_u = Q_u$.

In both cases, we let Z denote the total space of $\mathfrak{Z}$ and $\mathcal{Z} := Z \cap \mathcal{X}$.
Then there exists a coarse $L_{3.49}(\lambda, K, K', D')$-Lipschitz retraction $\rho = \rho_Y : Z \to Y$.

PROOF. The proof of this theorem is quite long and technical; it occupies most of the rest of this section.

STEP 1 (Construction of the map $\rho : \mathcal{Z} \to \mathcal{Y}$). For each $x \in \mathcal{Z}$ with $v = \pi(x) \in V(T)$, we fix a K-qi section γ_x of $\pi : Z \to \pi(Z) \subset T$ over $[\![u, v]\!]$, connecting x to some point in Z_u once and for all.

Let $b = b_x$ be the nearest point projection of v to S in T. We define the following important points:

- We let $t = t_x \in [\![u, b]\!]$ be the vertex farthest from u such that there exists a point $\tilde{x} \in \gamma_x(t)$ for which
$$d_{X_t}(\tilde{x}, Y_t) \leq M_{\bar{K}}.$$
(Note that it is possible that $t = u$ and $\tilde{x} \in Z_u$.)
- Let $\bar{x} \in Y_t$ be a nearest-point projection to $\tilde{x}$ to Y_t in the vertex-space X_t.

[3] The other two parameters, D, E, of $\mathfrak{L}$ play no role in this theorem.

Thus,

$$d_{X_{t_x}}(\tilde{x}, \bar{x}) \le M_{\bar{K}} \tag{3.20}$$

and if $x \in Y_v$, then $t_x = v$ and $\bar{x} = \tilde{x} = x$.

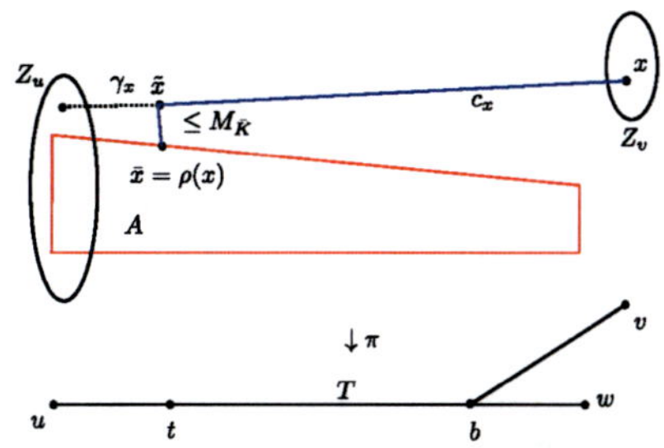

FIGURE 3.4. Projection to a bundle which is a carpet

DEFINITION 3.50.

(1) We define the retraction $\rho = \rho_{\mathcal{Y}} : \mathcal{Z} \to \mathcal{Y}$ by $\rho(x) := \bar{x}$. We extend this map to Z using the fact that $\mathcal{Z}$ is a K-net in Z (to define the extension we compose a nearest-point projection map $Z \to \mathcal{Z}$ with ρ). See Figure 3.4.

(2) We define a path $c_x = c_{x,Y}$ connecting $x \in \mathcal{Z}$ to $\rho(x) \in Y$ as the concatenation

$$\gamma_{x,\tilde{x}} \star [\tilde{x}\bar{x}]_{X_t},$$

where $t = t_x$, and

$$\gamma_{x,\tilde{x}} = \gamma_x|_{[\![v,t]\!]}$$

is the subpath of γ_x connecting x to $\tilde{x}$. We will refer to $\gamma_{x,\tilde{x}}$ as the *horizontal part* of c_x. The *vertical* part of c_x is the geodesic $[\tilde{x}\bar{x}]_{X_t}$; it is a path of uniformly bounded length (see (3.20)) connecting $\tilde{x}$ to $\bar{x}$ and contained in the $4\delta_0$-neighborhood of $Z_t = Z \cap X_t$.

STEP 2 (Verification of the properties of ρ). It suffices to verify the coarse Lipschitz property for the restriction of ρ to $\mathcal{Z}$. We note further that it is enough to get a uniform upper bound on $d(\rho(x), \rho(y))$ for two types of pairs (x, y):

(a) $x, y \in Z_v$, $d_{X_v}(x, y) \le 1$.

(b) The vertices $v_1 = \pi(x)$, $v_2 = \pi(y)$ span an edge in T and $d_{X_{v_1 v_2}}(x, y) \le K$.

These two cases are treated in Proposition 3.51 and Lemma 3.58 respectively. The former is the longest and hardest part of the proof.

For the following proposition we observe that, according to the 3rd assumption of Theorem 3.49, $\mathfrak{X}$ satisfies the uniform k-flaring condition for

$$k = K' + R_{2.53}(\bar{K}, r),$$

where

$$r = \max(M_{\bar{K}}, 6\delta_0 + \lambda + R_{2.53}(\bar{\bar{K}}, 1), 3\delta_0 + \lambda + R_{2.53}(\bar{\bar{K}}, 1) + R_{2.53}(\bar{K}, M_{\bar{K}})),$$

whence

$$r = 3\delta_0 + \lambda + R_{2.53}(\bar{\bar{K}}, 1) + R_{2.53}(\bar{K}, M_{\bar{K}}),$$

since $R_{2.53}(\bar{K}, M_{\bar{K}}) \ge \max(3\delta_0, M_{\bar{K}})$.

PROPOSITION 3.51. *Suppose that $x, y \in \mathcal{Z} \cap X_v$. If x, y are within distance 1 from each other in X_v, then $d_X(\rho(x), \rho(y)) \le C_{3.51}(\lambda, K, K', D')$. The bound is independent of the choice of the paths γ_x, γ_y as above.*

The proof of this proposition is long and is done through analyzing several cases and subcases. We use the notation preceding Definition 3.50 and note that in the setting of the proposition, $b_x = b_y$ (the projection in T of $v = \pi(x) = \pi(y)$ to S); we will denote the vertex $b_x = b_y$ simply by b.

Now we define certain auxiliary objects and make general remarks to be used in the proof, especially in Cases 2 and 3 below. We let $\bar{b}$ denote the nearest point projection (in T) of b to $[\![u, w]\!]$, i.e., $\bar{b}$ is the center of the triangle Δuwb.

REMARK 3.52.

(1) For each vertex $s \in T$, every geodesic $\alpha \subset X_s$ is δ_0-quasiconvex in X_s. Hence, the nearest-point projection (in X_s) to the geodesic α is coarsely $L_{1.102}(\delta_0, \delta_0)$-Lipschitz.

(2) We let $\bar{\gamma}_x$ and $\bar{\gamma}_y$ denote the fiberwise projections (to Y) of the restrictions to $[\![b, u]\!]$ of γ_x and γ_y respectively. These are $\bar{K}$-qi sections over $[\![b, u]\!]$. (See Lemma 3.47 and Notation 3.48.) Then, by the definition of t_x, t_y,

$$d_{X_s}(\gamma_x(s), \bar{\gamma}_x(s)) > M_{\bar{K}}, \forall s \in V(]\!]t_x, b]\!]), d_{X_{t_x}}(\gamma_x(t_x), \bar{\gamma}_x(t_x)) \le M_{\bar{K}},$$

$$d_{X_s}(\gamma_y(s), \bar{\gamma}_y(s)) > M_{\bar{K}}, \forall s \in V(]\!]t_y, b]\!]), d_{X_{t_y}}(\gamma_y(t_y), \bar{\gamma}_y(t_y)) \le M_{\bar{K}}.$$

(3) Since $\bar{K} \ge K$ and $\gamma_x(u) = \bar{\gamma}_x(u)$, by Corollary 2.53 we have

$$d_{X_s}(\gamma_x(s), \bar{\gamma}_x(s)) \le R_{2.53}(\bar{K}, M_{\bar{K}}),$$

for all $s \in V([\![t_x, u]\!])$ and, similarly,

$$d_{X_s}(\gamma_y(s), \bar{\gamma}_y(s)) \le R_{2.53}(\bar{K}, M_{\bar{K}}), \quad \forall s \in V([\![t_y, u]\!]).$$

(4) The carpet $\mathfrak{A}$ bounded by γ_x and γ_y (with the narrow end $[\gamma_x(v)\gamma_y(v)]_{X_v}$) is a $K'_{3.17}(K)$-carpet over $[\![u, v]\!]$; in particular, it is a $K'_{3.17}(K)$-metric bundle whose fibers are δ_0-quasiconvex in the corresponding vertex-spaces. We let $\bar{\bar{\gamma}}_x$ and $\bar{\bar{\gamma}}_y$ denote, respectively, the fiberwise projections of $\bar{\gamma}_x$ and $\bar{\gamma}_y$ to $\mathfrak{A}$. Thus, by Lemma 3.47, both $\bar{\bar{\gamma}}_x$ and $\bar{\bar{\gamma}}_y$ are $\bar{\bar{K}} = K_{3.47}(\delta_0, K'_{3.17}(K), \bar{K})$-qi sections over $[\![u, b]\!]$.

(5) Since $\mathfrak{A}$ is a $K'_{3.17}(K)$-carpet, we can join the points $\bar{\bar{\gamma}}_x(b)$ and $\bar{\bar{\gamma}}_y(b)$ to some points of $[xy]_{X_v}$ in $\mathfrak{A}$ via $K'_{3.17}(K)$-qi sections over $[\![v, b]\!]$. Since $\bar{\bar{K}} \ge K'_{3.17}(K)$, the concatenation of these qi sections with sections $\bar{\bar{\gamma}}_x$ and $\bar{\bar{\gamma}}_y$ (over $[\![u, b]\!]$) are both $\bar{\bar{K}}$-qi sections over $[\![u, v]\!]$, joining $\gamma_x(u)$ and $\gamma_y(u)$ to some points of $[xy]_{X_v}$. We retain the notation $\bar{\bar{\gamma}}_x$ and $\bar{\bar{\gamma}}_y$ for these concatenations.

(6) Notice that $\bar{\bar{\gamma}}_x(u) = \bar{\gamma}_x(u) = \gamma_x(u)$ and $\bar{\bar{\gamma}}_y(u) = \bar{\gamma}_y(u) = \gamma_y(u)$. At the same time, the other pair of endpoints (namely, points in $[xy]_{X_v}$) of γ_x and $\bar{\bar{\gamma}}_x$, respectively, of γ_y and $\bar{\bar{\gamma}}_y$, are within distance ≤ 1 in X_v. Therefore, by Corollary 2.53, we have

$$d_{X_s}(\gamma_x(s), \bar{\bar{\gamma}}_x(s)) \le R_{2.53}(\bar{\bar{K}}, 1), \forall s \in V([\![v, u]\!])$$

and

$$d_{X_s}(\gamma_y(s), \bar{\bar{\gamma}}_y(s)) \le R_{2.53}(\bar{\bar{K}}, 1), \forall s \in V([\![v, u]\!]).$$

Before proving the proposition we will need a technical lemma.

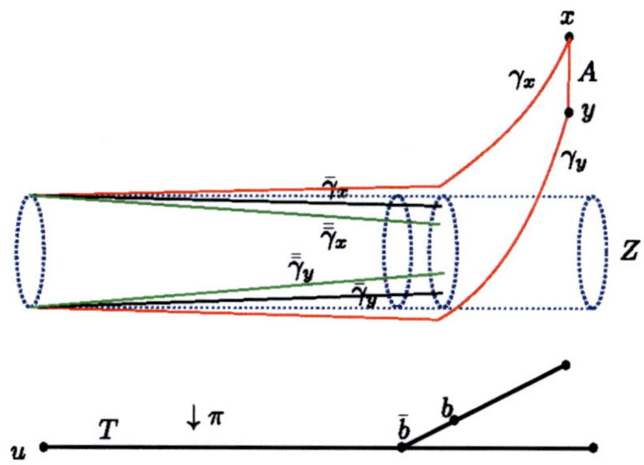

FIGURE 3.5. Projections $\bar{\gamma}_x, \bar{\gamma}_y, \bar{\bar{\gamma}}_x, \bar{\bar{\gamma}}_y$

LEMMA 3.53. *Suppose that r is such that $\mathfrak{X}$ satisfies the uniform k-flaring condition with*

$$k = k_{3.53} = K' + R_{2.53}(\bar{K}, r).$$

Then the following holds.

Suppose that there are vertices $v_1 \in [\![t_x, b]\!] \cap [\![\bar{b}, b]\!]$ and $v_2 \in [\![t_y, b]\!] \cap [\![\bar{b}, b]\!]$ such that

$$d_{X_{v_1}}(\gamma_x(v_1), \bar{\gamma}_x(v_1)) \le r \quad \textit{and} \quad d_{X_{v_2}}(\gamma_y(v_2), \bar{\gamma}_y(v_2)) \le r. \tag{3.21}$$

Then:

(i) $d_T(v_1, v_2) \le \tau_{3.53} = \tau_{3.53}(K, r, D')$.
(ii) $d(\gamma_x(v_1), \gamma_y(v_2)) \le R_{3.53} = R_{3.53}(K, r, D')$.
(iii)

$$d_{X_s}(\gamma_x(s), \gamma_y(s)) \le R_{2.53}(k, \max(1, D')), \forall s \in [\![v, \bar{b}]\!].$$

PROOF. Taking into account the fact that $K \le \bar{K} \le \kappa$, $\gamma_x(u) = \bar{\gamma}_x(u)$, $\gamma_y(u) = \bar{\gamma}_y(u)$, as well as the inequalities (3.21), we see that Corollary 2.53 applied to $\gamma_x|_{[\![u,v_1]\!]}$ and $\bar{\gamma}_y|_{[\![u,v_2]\!]}$ and the vertex $\bar{b} \in [\![u, v_1]\!] \cap [\![u, v_2]\!]$, implies that

$$d_{X_{\bar{b}}}(\gamma_x(\bar{b}), \bar{\gamma}_x(\bar{b})) \le R_{2.53}(\bar{K}, r)$$

and

$$d_{X_{\bar{b}}}(\gamma_y(\bar{b}), \bar{\gamma}_y(\bar{b})) \le R_{2.53}(\bar{K}, r).$$

We join $\bar{\gamma}_x(\bar{b})$ and $\bar{\gamma}_y(\bar{b})$ to Y_w by two K'-qi sections over $[\![\bar{b}, w]\!]$ contained in Y, $\gamma_{x,1}$ and $\gamma_{y,1}$ respectively.

We let $\bar{b}_1$ denote the vertex in $[\![z, w]\!]$ adjacent to $\bar{b}$, assuming that $\bar{b} \ne w$. We connect $\gamma_x(\bar{b})$ and $\gamma_{x,1}(\bar{b}_1)$ by a geodesic path $\gamma_{1,\bar{b},\bar{b}_1}$ in $X_{\bar{b}\bar{b}_1}$, similarly, we connect $\gamma_y(\bar{b})$ and $\gamma_{y,1}(\bar{b}_1)$ by a geodesic path $\gamma_{2,\bar{b},\bar{b}_1}$ in $X_{\bar{b}\bar{b}_1}$. Both paths have length $\le k = K' + R_{2.53}(\bar{K}, r)$.

Then we get that the concatenations

$$\gamma'_x := \gamma_x|_{[\![v,\bar{b}]\!]} \star \gamma_{1,\bar{b},\bar{b}_1} \star \gamma_{x,1}|_{[\![\bar{b}_1,w]\!]}$$

and

$$\gamma'_y := \gamma_y|_{[\![v,\bar{b}]\!]} \star \gamma_{2,\bar{b},\bar{b}_1} \star \gamma_{y,1}|_{[\![\bar{b}_1,w]\!]}.$$

These are k-qi sections over $[\![v,w]\!]$. (See the bold paths in Figure 3.6.) Their endpoints are at a distance at most $\max(1,D')$ of the respective vertex-spaces (since $d_{X_v}(x,y)\le 1$ and Y_w has diameter $\le D'$).

Hence (since $\mathfrak{X}$ is assumed to satisfy the uniform k-flaring condition), by Corollary 2.53,

$$d_{X_s}(\gamma'_x(s),\gamma'_y(s))\le R_{2.53}(k,\max(1,D')),\forall s\in [\![v,w]\!]$$

and, restricting to the subinterval $[\![v,\bar b]\!]$, we obtain

$$d_{X_s}(\gamma_x(s),\gamma_y(s))\le R_{2.53}(k,\max(1,D')),\forall s\in [\![v,z]\!]. \tag{3.22}$$

In the case $\bar b=w$, we will use the paths

$$\gamma'_x:=\gamma_x|_{[\![v,\bar b]\!]},\gamma'_y:=\gamma_y|_{[\![v,\bar b]\!]}$$

and obtain the same inequality (3.22).

Thus, we establish Part (iii) of the lemma.

Since the fiberwise projection to Y_s, $s\in [\![b,\bar b]\!]$ is $L_{1.102}(\delta_0,\lambda)$-Lipschitz, we get

$$d_{X_s}(\bar\gamma_x(s),\bar\gamma_y(s))\le L_{1.102}(\delta_0,\lambda)(R_{2.53}(k,\max(1,D'))+1),\forall s\in [\![b,z]\!]. \tag{3.23}$$

Without loss of generality (by switching the roles of x and y if necessary), we may assume that v_2 is a vertex in $[\![v_1,b]\!]$.

Combining the second inequality in (3.21) with the inequalities (3.22), (3.23) applied to $s=v_2$, by the triangle inequality in X_{v_2} we obtain:

(3.24)

$$d_{X_{v_2}}(\gamma_x(v_2),\bar\gamma_x(v_2))\le R_1$$
$$:=r+R_{2.53}(k,\max(1,D'))+L_{1.102}(\delta_0,\lambda)(R_{2.53}(k,\max(1,D'))+1).$$

Hence, by taking into account the fact that

$$d_{X_s}(\gamma_x(s),\bar\gamma_x(s))>M_{\bar K}$$

for all vertices s of $]\!]t_x,b]\!]$ (and the reverse inequality at t_x and the inequality (3.24)) and using the $\bar K$-uniform flaring property of the sections $\gamma_x,\bar\gamma_x$ over the interval $[\![t_x,v_2]\!]$, we obtain

$$d_T(v_1,v_2)\le d_T(t_x,v_1)\le \tau_{3.53}(K,r,D'):=\tau_{2.47}(\bar K,R_1).$$

This concludes the proof of Part (i).

(i)⇒(ii): $d(\gamma_x(v_1),\gamma_y(v_2))$ is bounded by the length of the concatenation of the paths $\gamma_x|_{[\![v_1,v_2]\!]}$ (whose length is estimated by (i)) and $[\gamma_x(v_2)\gamma_y(v_2)]_{X_{v_2}}$ (whose length is estimated by (3.22) since $v_2\in[\![b,z]\!]$) is which therefore at most

$$R_{3.53}(K,r,D'):=K\cdot\tau_{3.53}(K,r,D')+R_{2.53}(k,\max(1,D')).$$ □

We will be using this lemma in the proof of Proposition 3.51 with $r=r_1, r=r_2, r=r_3$ (defined below), where r_2 is the largest of the three parameters. The uniform flaring condition made in the statement of the proposition ensures that the uniform k-flaring condition in the lemma is satisfied for $r=r_2$ and, hence, for the two other (smaller) values of r.

PROOF OF PROPOSITION 3.51. There are several cases to consider depending on mutual position of various vertices in T.

CASE 1. Suppose $v\in V([\![u,w]\!])$, in which case $b=v$. Without loss of generality, we may assume that $t_y\in V([\![t_x,v]\!])$.

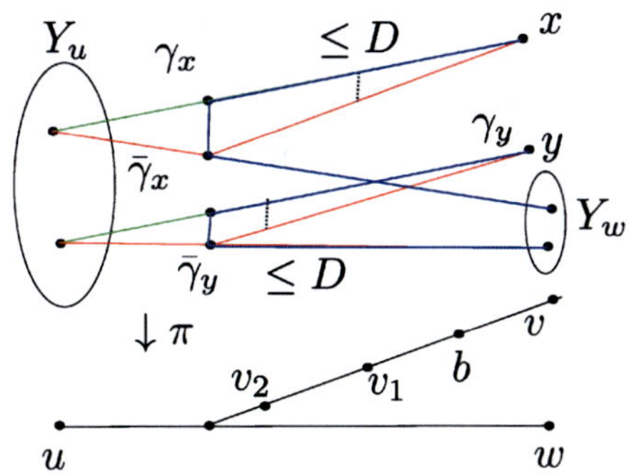

FIGURE 3.6. Illustration of the proof of Lemma 3.53

Consider the subset

$$W = \{s \in V([\![t_y, v]\!]) : d_{X_s}(\bar{\gamma}_x(s), \bar{\gamma}_y(s)) \geq D'_{1.123}(\delta_0, \delta_0)\}.$$

If $W \neq \emptyset$, let v_1 be the farthest vertex from t_y in W. If $W = \emptyset$ then define v_1 to be t_y. In other words, $v_1 = \sup W$ in the (oriented) interval $[\![t_y, v]\!]$.

CLAIM 3.54. All three distances

$$d_{X_{v_1}}(\gamma_x(v_1), \gamma_y(v_1)), d_{X_{v_1}}(\gamma_x(v_1), \bar{\gamma}_x(v_1)), d_{X_{v_1}}(\gamma_y(v_1), \bar{\gamma}_y(v_1))$$

are bounded above by

$$C_{3.54} = 2R_{2.53}(\bar{\bar{K}}, 1) + \max(M_{\bar{K}} + \delta_0(21 + 72\delta_0), 9\delta_0(1 + D_{2.40}(\bar{K}, 9\delta_0))).$$

See the bold curves in Figure 3.7.

The proof is divided into two subcases.

Subcase (i): Suppose that $W = \emptyset$, thus, $v_1 = t_y \in [\![u, v]\!]$. In this case

$$d_{X_{v_1}}(\bar{\gamma}_x(v_1), \bar{\gamma}_y(v_1)) < D_{1.123}(\delta_0, \delta_0).$$

By Remark 3.52(1),

$$d_{X_{v_1}}(\bar{\bar{\gamma}}_x(v_1)), \bar{\bar{\gamma}}_y(v_1)) \leq L_{1.102}(\delta_0, \delta_0)(1 + D_{1.123}(\delta_0, \delta_0)).$$

By combining this inequality with the two inequalities in Remark 3.52(6), we obtain (by the triangle inequality in X_{v_1})

$$\begin{aligned} d_{X_{v_1}}(\gamma_x(v_1), \gamma_y(v_1)) &\leq 2R_{2.53}(\bar{\bar{K}}, 1) + L_{1.102}(\delta_0, \delta_0)(1 + D_{1.123}(\delta_0, \delta_0)) \\ &= 2R_{2.53}(\bar{\bar{K}}, 1) + 12\delta_0(1 + 9\delta_0) \end{aligned} \tag{3.25}$$

c.f. Remark 1.103 and Lemma 1.123(2).

Since $v_1 = t_y$,

$$d_{X_{v_1}}(\gamma_y(v_1), \bar{\gamma}_y(v_1)) \leq M_{\bar{K}} \tag{3.26}$$

and it follows from the triangle inequality applied to the quadrilateral in X_{v_1} with the vertices

$$\gamma_x(v_1), \gamma_y(v_1), \bar{\gamma}_y(v_1), \bar{\gamma}_x(v_1),$$

that

$$\begin{aligned} d_{X_{v_1}}(\gamma_x(v_1), \bar\gamma_x(v_1)) &\le 2R_{2.53}(\bar{\bar{K}}, 1) + L_{1.102}(\delta_0, \delta_0)(1 + D_{1.123}(\delta_0, \delta_0)) \\ &\quad + M_{\bar K} + D_{1.123}(\delta_0, \delta_0) \\ &= 2R_{2.53}(\bar{\bar{K}}, 1) + M_{\bar K} + \delta_0(21 + 72\delta_0) \end{aligned}$$

(see Remark 1.103). By combining this inequality with (3.26) and (3.25), we obtain the upper bound

$$\begin{aligned} &\max\{d_{X_{v_1}}(\gamma_x(v_1), \gamma_y(v_1)), d_{X_{v_1}}(\gamma_x(v_1), \bar\gamma_x(v_1)), d_{X_{v_1}}(\gamma_y(v_1), \bar\gamma_y(v_1))\} \\ &\le 2R_{2.53}(\bar{\bar{K}}, 1) + M_{\bar K} + \delta_0(21 + 72\delta_0) \\ &\le C_{3.54}. \end{aligned}$$

This proves the inequality in the claim in the subcase (i).

Subcase (ii): Suppose $W \neq \emptyset$; hence,

$$d_{X_{v_1}}(\bar\gamma_x(v_1), \bar\gamma_y(v_1)) \ge D_{1.123}(\delta_0, \delta_0) = 9\delta_0.$$

By Lemma 1.123,

$$\max\left(d_{X_{v_1}}(\bar\gamma_x(v_1), \bar{\bar\gamma}_x(v_1)), d_{X_{v_1}}(\bar\gamma_y(v_1), \bar{\bar\gamma}_y(v_1))\right) \le R_{1.123}(\delta_0, \delta_0) = 6\delta_0.$$

Combining this inequality with Remark 3.52(6) and the triangle inequality, we obtain

$$\begin{aligned} d_{X_{v_1}}(\bar\gamma_x(v_1), \gamma_x(v_1)) &\le d_{X_{v_1}}(\bar\gamma_x(v_1), \bar{\bar\gamma}_x(v_1)) + d_{X_{v_1}}(\bar{\bar\gamma}_x(v_1), \gamma_x(v_1)) \\ &\le 6\delta_0 + R_{2.53}(\bar{\bar{K}}, 1) \end{aligned}$$

and, similarly,

$$d_{X_{v_1}}(\bar\gamma_y(v_1), \gamma_y(v_1)) \le 6\delta_0 + R_{2.53}(\bar{\bar{K}}, 1) \le C_{3.54}.$$

This establishes two out of three bounds in the claim.

Lastly, we get a bound on $d_{X_{v_1}}(\gamma_x(v_1), \gamma_y(v_1))$. If $v_1 = v$ then the bound definitely holds since $d_{X_v}(\gamma_x(v), \gamma_y(v)) = d_{X_v}(x, y) \le 1$ by the assumption of the proposition.

Otherwise, by the definition of v_1, if v_2 is the vertex in $[\![v_1, v]\!]$ adjacent to v_1, then

$$d_{X_{v_2}}(\bar\gamma_x(v_2), \bar\gamma_y(v_2)) < 9\delta_0 = D_{1.123}(\delta_0, \delta_0).$$

Hence, by Corollary 2.40, we have $d_{X_{v_1}}(\bar\gamma_x(v_1), \bar\gamma_y(v_1)) \le D_{2.40}(\bar K, 9\delta_0)$. By Remark 3.52(1)

$$d_{X_{v_1}}(\bar{\bar\gamma}_x(v_1)), \bar{\bar\gamma}_y(v_1)) \le 9\delta_0(1 + D_{2.40}(\bar K, 9\delta_0)).$$

Hence, using Remark 3.52(6), we obtain

$$d_{X_{v_1}}(\gamma_x(v_1), \gamma_y(v_1)) \le 2R_{2.53}(\bar{\bar{K}}, 1) + 9\delta_0(1 + D_{2.40}(\bar K, 9\delta_0)) \le C_{3.54}.$$

This completes the proof of the claim. □

We now prove the proposition in Case 1. By Corollary 2.53 applied to the $\bar K$-qi sections γ_x and $\bar\gamma_x$ over the interval $[\![t_x, v_1]\!]$, in view of the bound in Claim 3.54, since

$$M_{\bar K} \le C_{3.54},$$

we obtain:

$$d_T(t_x, v_1) \le R_{2.53}(\bar K, C_{3.54})$$

and

$$d_T(t_y, v_1) \le R_{2.53}(\bar K, C_{3.54}).$$

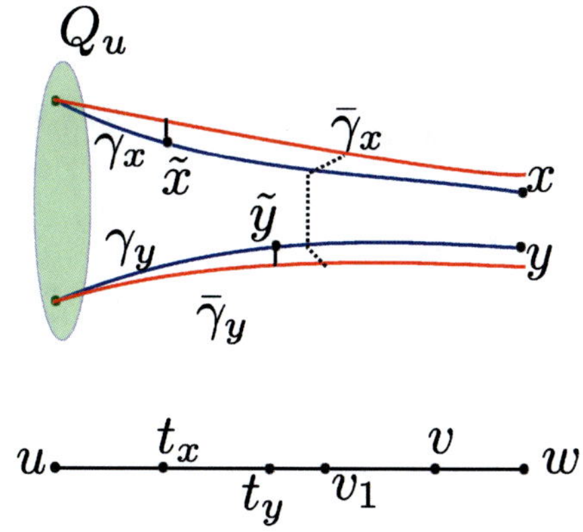

FIGURE 3.7. Case 1

By the triangle inequality, $d_X(\rho(x),\rho(y))$ is bounded by

$$d_{X_{t_x}}(\rho(x),\gamma_x(t_x)) + d_X(\gamma_x(t_x),\gamma_x(v_1)) + d_{X_{v_1}}(\gamma_x(v_1),\gamma_y(v_1)) \\ + d_X(\gamma_y(v_1),\gamma_y(t_y)) + d_{X_{t_y}}(\gamma_y(t_y),\rho(y)).$$

This, in turn, is bounded by

$$M_{\bar{K}} + Kd_T(t_x,v_1) + Kd_T(t_y,v_1) + d_{X_{v_1}}(\gamma_x(v_1),\gamma_y(v_1)) + Kd_T(t_x,v_1) + M_{\bar{K}} \\ \leq 2(M_{\bar{K}} + R_{2.53}(\bar{K},C_{3.54})) + C_{3.54}.$$

This concludes the proof of the proposition in Case 1.

CASE 2. Assume that $v \notin [\![u,w]\!]$. Let $\bar{b}$ denote the nearest point projection of b to $[\![u,w]\!]$ in T. There are three subcases to be dealt with.

Subcase (i): Suppose $t_x, t_y \in [\![b,\bar{b}]\!]$. Letting $v_1 = t_x, v_2 = t_y$ and applying Lemma 3.53(ii) with $r = r_1 = M_{\bar{K}}$ we obtain

$$d(\gamma_x(t_x),\gamma_y(t_y)) \leq R_{3.53}(K,M_{\bar{K}},D'),$$

and, hence,

$$d(\rho(x),\rho(y)) \leq 2M_{\bar{K}} + R_{3.53}(K,M_{\bar{K}},D').$$

Observe that we need uniform k-flaring for

$$k = k_{3.53} = K' + R_{2.53}(\bar{K},r_1), r_1 = M_{\bar{K}}.$$

Subcase (ii): Suppose $t_x \in [\![u,\bar{b}]\!]$ and $t_y \in [\![b,\bar{b}]\!]$, or vice versa.

Without loss of generality, we may assume that $t_x \in [\![u,\bar{b}]\!]$ and $t_y \in [\![b,\bar{b}]\!]$. We first get a uniform upper bound on $d_{X_{\bar{b}}}(\bar{\gamma}_x(\bar{b}),\bar{\bar{\gamma}}_x(\bar{b}))$.

By Lemma 1.105, since $Y_{\bar{b}}$ is λ-quasiconvex in $X_{\bar{b}}$, $\bar{\gamma}_x(\bar{b})$ lies in the $(\lambda+2\delta_0)$-neighborhood of $[\gamma_x(\bar{b})\bar{\gamma}_y(\bar{b})]_{X_{\bar{b}}}$ in $X_{\bar{b}}$.

Since

$$d_{X_{\bar{b}}}(\gamma_y(\bar{b}),\bar{\gamma}_y(\bar{b})) \leq R_{2.53}(\bar{K},M_{\bar{K}}),$$

by Remark 3.52(3), it follows from the δ_0-hyperbolicity of $X_{\bar{b}}$ that the Hausdorff distance between the geodesics $[\gamma_x(\bar{b})\gamma_y(\bar{b})]_{X_{\bar{b}}}$ and $[\gamma_x(\bar{b})\bar{\gamma}_y(\bar{b})]_{X_{\bar{b}}}$ in $X_{\bar{b}}$ is at most $\delta_0+R_{2.53}(\bar{K},M_{\bar{K}})$.

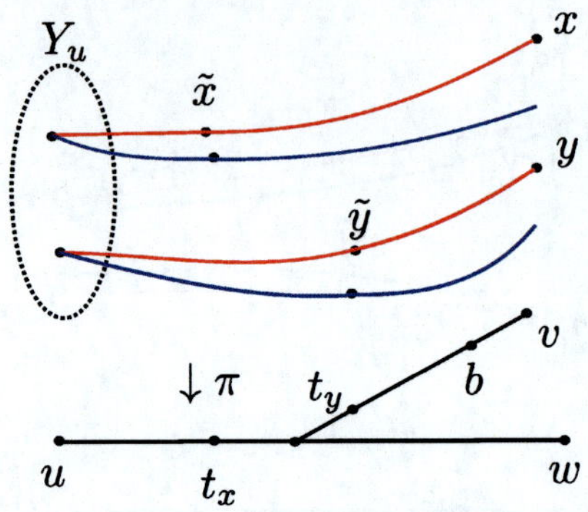

FIGURE 3.8. Case 3(ii)

Combining this with the earlier observation that

$$\bar{\gamma}_x(\bar{b}) \in N^{fib}_{\lambda+2\delta_0}([\gamma_x(\bar{b})\bar{\gamma}_y(\bar{b})]_{X_{\bar{b}}}),$$

we conclude that $\bar{\gamma}_x(\bar{b})$ belongs to the $(3\delta_0 + \lambda + R_{2.53}(\bar{K}, M_{\bar{K}}))$-neighborhood of the segment $[\gamma_x(\bar{b})\gamma_y(\bar{b})]_{X_{\bar{b}}}$ in $X_{\bar{b}}$. Therefore,

$$d_{X_{\bar{b}}}(\bar{\gamma}_x(\bar{b}), \bar{\bar{\gamma}}_x(\bar{b})) \leq 3\delta_0 + \lambda + R_{2.53}(\bar{K}, M_{\bar{K}}).$$

Since, by Remark 3.52(6),

$$d_{X_{\bar{b}}}(\gamma_x(\bar{b}), \bar{\bar{\gamma}}_x(\bar{b})) \leq R_{2.53}(\bar{\bar{K}}, 1),$$

by triangle inequality we obtain

$$d_{X_{\bar{b}}}(\gamma_x(\bar{b}), \bar{\gamma}_x(\bar{b})) \leq r_2 := R_{2.53}(\bar{\bar{K}}, 1) + 3\delta_0 + \lambda + R_{2.53}(\bar{K}, M_{\bar{K}}).$$

We will apply Lemma 3.53(ii) with $r = r_2$ and $v_1 = \bar{b}$, $v_2 = t_y$.

REMARK 3.55. Note that in order to apply Lemma 3.53(ii) we need the uniform k-flaring condition for

$$k = k_{3.53} = K' + R_{2.53}(\bar{K}, r_2), r_2 = 3\delta_0 + \lambda + R_{2.53}(\bar{\bar{K}}, 1) + R_{2.53}(\bar{K}, M_{\bar{K}}),$$

which is ensured by the choice of the constant r in (3.17), k in (3.18) and the uniform κ-flaring assumption in Theorem 3.49.

In view of the $\bar{K}$-uniform flaring condition, applied to $\gamma_x, \bar{\gamma}_x$, we have $d_T(t_x, \bar{b}) \leq \tau_{2.47}(\bar{K}, r_2)$. Hence,

$$d(\gamma_x(t_x), \gamma_x(\bar{b})) \leq K\tau_{2.47}(\bar{K}, r_2). \tag{3.27}$$

Since

$$d_{X_{t_y}}(\gamma_y(t_y), \bar{\gamma}_y(t_y)) \leq M_{\bar{K}} \leq r_2$$

and

$$d_{X_{\bar{b}}}(\gamma_x(\bar{b}), \bar{\gamma}_x(\bar{b})) \leq r_2,$$

we can apply Lemma 3.53(ii) with $v_1 = \bar{b}, v_2 = t_y$ and obtain:

$$d(\gamma_x(\bar{b}), \gamma_y(t_y)) \leq R_{3.53}(K, r_2, D').$$

Combining this estimate with the inequality (3.27) we get

$$d(\gamma_x(t_x), \gamma_y(t_y)) \leq K\tau_{2.47}(\bar{K}, r_2) + R_{3.53}(K, r_2, D')$$

This, in turn, implies that

$$d(\rho(x), \rho(y)) \le K\tau_{2.47}(\bar{K}, r) + R_{3.53}(K, r) + 2M_{\bar{K}}.$$

Subcase (iii): Suppose $t_x, t_y \in [\![u, \bar{b}]\!]$. Without loss of generality, after swapping x and y if necessary, we may assume that $t_y \in [\![t_x, \bar{b}]\!]$, see Figure 3.9. We will show that

$$d_{X_{\bar{b}}}(\gamma_x(\bar{b}), \gamma_y(\bar{b}))$$

is uniformly bounded in this subcase.

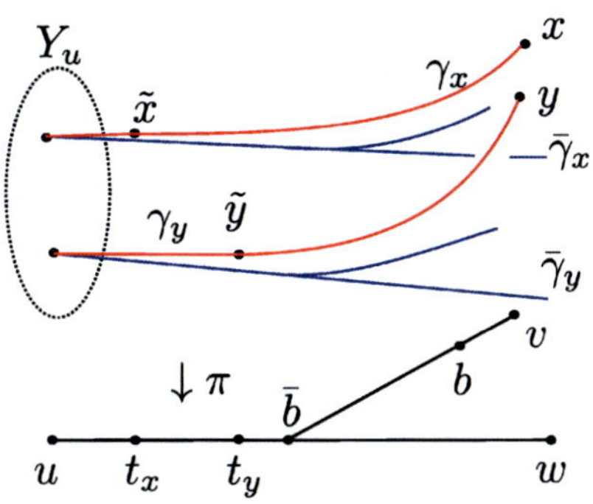

FIGURE 3.9. Case 3(iii)

Suppose first that

$$d_{X_{\bar{b}}}(\bar{\bar{\gamma}}_x(\bar{b}), \bar{\bar{\gamma}}_y(\bar{b})) \ge 9\delta_0 = D_{1.123}(\delta_0, \delta_0).$$

Then (by Lemma 1.123) both points $\bar{\bar{\gamma}}_x(\bar{b}), \bar{\bar{\gamma}}_y(\bar{b})$ belong to the $6\delta_0 = R_{1.123}(\delta_0, \delta_0)$-neighborhood of $[\bar{\gamma}_x(\bar{b})\bar{\gamma}_y(\bar{b})]_{X_{\bar{b}}}$ in $X_{\bar{b}}$.

However, by Remark 3.52(6),

$$d_{X_{\bar{b}}}(\bar{\bar{\gamma}}_x(\bar{b}), \gamma_x(\bar{b})) \le R_{2.53}(\bar{\bar{K}}, 1) \quad \text{and} \quad d_{X_{\bar{b}}}(\bar{\bar{\gamma}}_y(\bar{b}), \gamma_y(\bar{b})) \le R_{2.53}(\bar{\bar{K}}, 1).$$

Thus, since $Y_{\bar{b}}$ is λ-quasiconvex in $X_{\bar{b}}$ and both endpoints of $[\bar{\gamma}_x(\bar{b})\bar{\gamma}_y(\bar{b})]_{X_{\bar{b}}}$ are in $Y_{\bar{b}}$, the above inequalities imply:

$$d_{X_{\bar{b}}}(\gamma_x(\bar{b}), Y_{\bar{b}}) \le r_3 := \lambda + R_{2.53}(\bar{\bar{K}}, 1) + 6\delta_0,$$

and

$$d_{X_{\bar{b}}}(\gamma_y(\bar{b}), Y_{\bar{b}}) \le r_3.$$

Since $\bar{\gamma}_x(\bar{b})$ is a nearest-point projection (in $X_{\bar{b}}$) of $\gamma_x(\bar{b})$ to $Y_{\bar{b}}$, it follows that

$$d_{X_{\bar{b}}}(\gamma_x(\bar{b}), \bar{\gamma}_x(\bar{b})) \le r_3,$$

and, similarly,

$$d_{X_{\bar{b}}}(\gamma_y(\bar{b}), \bar{\gamma}_y(\bar{b})) \le r_3.$$

We are now again in position to apply Lemma 3.53(iii) with $r = r_3, v_1 = v_2 = \bar{b}$ and conclude that

$$d_{X_{\bar{b}}}(\gamma_x(\bar{b}), \gamma_y(\bar{b})) \le R_{2.53}(k, \max(1, D')),$$

where $k = K' + R_{2.53}(\bar{K}, r_3)$.

REMARK 3.56. Observe that in order to apply Lemma 3.53(iii), we need the uniform k-flaring condition for this k and

$$r_3 := \lambda + R_{2.53}(\bar{\bar{\bar{K}}}, 1) + 6\delta_0,$$

which is ensured by the choice of the parameter r in (3.17) and the uniform κ-flaring assumption in Theorem 3.49.

The uniform $\bar{K}$-flaring condition in $\mathfrak{X}$ applied to the pairs of $\bar{K}$-qi sections $(\gamma_x, \bar{\gamma}_x)$, $(\gamma_y, \bar{\gamma}_y)$, then implies the inequality

$$\max(d_T(t_x, \bar{b}), d_T(t_y, \bar{b})) \leq \tau_{2.47}(\bar{K}, r_3).$$

Hence,

$$\max(d(\rho(x), \gamma_x(\bar{b})), d(\rho(y), \gamma_y(\bar{b}))) \leq \bar{K}\tau_{2.47}(\bar{K}, r_3) + M_{\bar{K}}.$$

By the triangle inequality we get

$$d(\rho(x), \rho(y)) \leq 2(\bar{K}\tau_{2.47}(\bar{K}, r_3) + M_{\bar{K}}) + R_{2.53}(k, \max(1, D')).$$

This concludes the argument in subcase (iii) and, hence, the proof of Proposition 3.51.

The following is an immediate corollary of the proposition:

COROLLARY 3.57. *For each vertex $v \in [\![u, w]\!]$ the restriction of the retraction ρ to the subspace Z_v is $C_{3.51}(\lambda, K, K', D')$-coarse Lipschitz. This bound is independent of the choice of the paths γ_x, γ_y as above.*

LEMMA 3.58. *If $\pi(x) = v_1, \pi(y) = v_2$, $d_T(v_1, v_2) = 1$ and $d_{X_{v_1 v_2}}(x, y) \leq K$, then $d(\rho(x), \rho(y)) \leq C_{3.58}(\lambda, K, K', D')$.*

PROOF. Without loss of generality, we may assume that $d(u, v_2) = d(u, v_1) + 1$. Let $y_1 = \gamma_y(v_1)$. Since $d(x, y_1) \leq 2K$, we also have

$$d_{X_{v_1}}(x, y_1) \leq \eta_0(2K).$$

Applying Corollary 3.57 to the points $x, y_1 \in X_{v_1}$, we obtain:

$$d(\rho(x), \rho(y_1)) \leq C_{3.51}(\lambda, K, K', D')(\eta_0(2K) + 1).$$

We next note that, without loss of generality we may assume that γ_{y_1} is chosen to be the restriction of γ_y to the subinterval $[\![u, v_1]\!]$ since the Lipschitz bound in Corollary 3.57 holds regardless of the choice of the sections γ_y. Hence,

$$d(\rho(x), \rho(y)) \leq C_{3.51}(\lambda, K, K', D')(\eta_0(2K) + 1).$$ □

This completes the proof of Theorem 3.49.

The next corollary is immediate from Theorem 3.49.

COROLLARY 3.59. *The map ρ in Theorem 3.49 is "coarsely independent" of the choice of paths γ_x used in its construction. More precisely, if ρ, ρ' are two projections defined using different choices of paths γ_x, then*

$$d(\rho(x), \rho'(x)) \leq L_{3.49}(\lambda, K, K', D').$$

Recall that for $x \in \mathcal{Z}$ we defined a path c_x in Z connecting x to $\rho(x)$; see Definition 3.50.

COROLLARY 3.60. *Under the assumptions of Theorem 3.49, for any two points $x, y \in Z$ within distance C from each other, the Hausdorff distance between the paths c_x, c_y is $\leq D_{3.60}(\lambda, K, K', D', C)$.*

PROOF. As in the proof of Theorem 3.49, it suffices to verify the claim in two cases:

CASE 1. Suppose that $x, y \in Z_v$ and $d(x, y) \le C$. Without loss of generality we may assume that $t_y \in [\![t_x, v]\!]$. By Theorem 3.49 we have

$$d_T(\pi(t_x), \pi(t_y)) \le d(\rho(x), \rho(y)) \le L_{3.49}(\lambda, K, K', D') \cdot (C + 1).$$

Hence, the length of the portion of c_x between $\gamma_x(t_y)$ and $\rho(x)$ is at most

$$M_{\bar{K}} + K \cdot L_{3.49}(\lambda, K, K', D') \cdot (C + 1).$$

It follows that

$$d_X(\gamma_x(t_y), \gamma_y(t_y)) \le R_1 := L_{3.49}(\lambda, K, D') + 2M_{\bar{K}} + K \cdot L_{3.49}(\lambda, K, K', D') \cdot (C + 1).$$

Since X_{t_y} is η_0-uniformly properly embedded in X, we also obtain

$$d_{X_{t_y}}(\gamma_x(t_y), \gamma_y(t_y)) \le R_2 := \eta_0(R_1).$$

By Corollary 2.53, we obtain that for all vertices $s \in V([\![t_y, v]\!])$

$$d_{X_s}(\gamma_x(s), \gamma_y(s)) \le R_3 := R_{2.53}(K, \max(R_2, C)).$$

By combining this with the earlier estimate on the length of the portion of c_x between $\gamma_x(t_y)$ and $\rho(x)$, we obtain that the Hausdorff distance between c_x, c_y is at most

$$R_4 := M_{\bar{K}} + K \cdot L_{3.49}(\lambda, K, K', D') \cdot (C + 1) + M_{\bar{K}} + R_3.$$

This concludes the proof in Case 1. □

REMARK 3.61. This argument also proves that if in the definition of $\rho(x)$ and c_x we use different K-qi sections γ'_x, γ''_x, then the resulting paths c'_y, c''_y are within Hausdorff distance R_4 from each other.

CASE 2. Suppose that $\pi(x) = v_1, v_2 = \pi(y)$, $d_T(v_1, v_2) = 1$ and $d(x, y) \le K$. Without loss of generality we may assume that $d(u, v_2) = d(u, v_1) + 1$. Setting $y_1 := \gamma_y(v_1)$, according to Case 1 and the above remark, we obtain the bound R_4 on the Hausdorff distance between c_x and c_{y_1}. It follows that the Hausdorff distance between c_x and c_y is $\le K + R_4$. □

PROPOSITION 3.62. *Under the assumptions of Theorem* 3.49*, assuming, in addition, that Z is δ-hyperbolic, there exists $C = C_{3.62}(\delta, \lambda, K, K', D')$ such that $d(\rho_Y, P) \le C$, where $P = P_{Z,Y}$ is the nearest-point projection in Z.*

PROOF. By Theorem 3.49, there exists a coarse $L = L_{3.49}(\lambda, K, K', D')$-Lipschitz retraction $\rho : Z \to Y$. By Lemma 1.104, the subset Y is $\lambda' = \lambda_{1.104}(L, \delta)$-quasiconvex in Z.

By the construction of ρ, the path c_x connecting $x \in Q_v$ to $\bar{x} = \rho_Y(x)$ is the concatenation of the K-qi section $\gamma_{x,\bar{x}}$ and the vertical geodesic $[\tilde{x}\bar{x}]_{X_t}$ of length $\le M_{\bar{K}}$, $t = t_x$. Hence, considering the geodesic triangle $\Delta x\tilde{x}\bar{x}$ in Z and using the δ-hyperbolicity of Z, we conclude that

$$\mathrm{Hd}_Z(\gamma_{x,\bar{x}}, [x\bar{x}]_Z) \le \delta + M_{\bar{K}} + D_{1.54}(\delta, K).$$

Therefore, since Y is λ'-quasiconvex in Z, in order to prove the proposition, according to Corollary 1.109, it suffices to verify that there exists a function $R \mapsto R'$ (depending on δ, λ') such that for all $v \in V(\pi(Z))$, $x \in Q_v$, $y \in \mathcal{Y}$

$$d_Z(y, \gamma_{x,\bar{x}} \cap \mathcal{X}) \le R \Rightarrow d_Y(y, \bar{x}) \le R'.$$

Our goal then is to define such a function.

Take $y \in \mathcal{Y}$ such that $d_Z(y, \gamma_{x,\bar{x}}) \le R$. Pick a point

$$x' \in \gamma_{x,\bar{x}} \cap X_{v'}, v' \in [\![u, v]\!],$$

such that $d_Z(y, x') \le R$. In particular, $d_T(v', \pi(y)) \le R$.

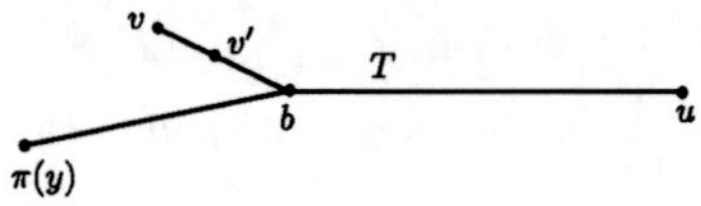

FIGURE 3.10. Tripod $\Delta v'\pi(y)u$ in T

Let b denote the center of the tripod $\Delta v'\pi(y)u \subset T$,

$$b \in [\![u, v']\!] \subset [\![u, v]\!],$$

see Figure 3.10. Then

$$d_T(v', b) + d_T(b, \pi(y)) = d_T(v', \pi(y)) \le R. \tag{3.28}$$

Set $x'' := \gamma_x(b)$ (this point is defined since $b \in [\![u, v]\!]$ and γ_x is a section over that interval) and let γ'' denote $\gamma_{x'',\tilde{x}}$, the restriction of γ_x to $[\![b, t]\!]$. (Note that the order in which the vertices t, b appear in the interval $[\![u, v']\!]$ is unclear.)

Let $\bar{\gamma}''$ denote the fiberwise projection of γ'' to $\mathcal{Y}_{tb}$; without loss of generality, $\bar{\gamma}''(b) = \bar{x}$. Then $\bar{\gamma}$ is a $\bar{K}$-qi leaf in Y connecting $\bar{x}$ to $\bar{x}'' = P_{X_b,Y_b}(x'')$.

Furthermore, let γ_y be a K'-qi section in Y over $[\![\pi(y), u]\!]$, connecting y to Y_u and let γ_y'' be its restriction to the interval $[\![\pi(y), b]\!]$ (recall that $b \in [\![\pi(y), u]\!]$). Set $y'' := \gamma_y(b) = \gamma_y''(b)$.

Now, consider the quadruple of points x'', x', y, y'': We have a K-qi section $\gamma_{x'',x'}$ connecting x'' to x', a geodesic $[x'y]_Z$ in Z connecting x' to y and the K'-qi section γ_y'' connecting y to y''. Therefore,

$$d_Z(y, y'') \le K' d_T(\pi(y), b)$$

and

$$d_Z(x'', y'') \le K d_T(b, v') + R + K' d_T(\pi(y), b).$$

Taking into account the inequality (3.28), we obtain

$$\begin{aligned} d_Z(x'', y'') &\le R + \max(K, K') d_T(\pi(y), v') \le \\ &R + R \max(K, K') = R(1 + \max(K, K')), \end{aligned} \tag{3.29}$$

$$d_Z(y, y'') \le K' d_T(\pi(y), v') \le K'R. \tag{3.30}$$

Since $x'', y'' \in X_b$, we also get

$$d_{X_b}(x'', y'') \le \eta_0 R(1 + \max(K, K'))). \tag{3.31}$$

Since $y'' \in Y_b$ and $\bar{x}''$ is the projection of x'' to Y_b, it follows that

$$d_{X_b}(x'', \bar{x}'') \le \eta_0(R(1 + \max(K, K'))). \tag{3.32}$$

The uniform $\bar{K}$-flaring condition applied to the restrictions of γ'' and $\bar{\gamma}$ to $[\![t, b]\!]$, and the inequalities

$$\begin{aligned} d_{X_b}(\gamma''(b), \bar{\gamma}''(b)) &\le \eta_0(R(1 + \max(K, K'))), \\ d_{X_t}(\gamma''(t), \bar{\gamma}''(t)) &= d_{X_t}(\tilde{x}, \bar{x}) \le M_{K'}, \end{aligned} \tag{3.33}$$

imply that

$$d_T(t, b) \le \tau_{2.47}(\bar{K}, \max(\eta_0 R(1 + \max(K, K'))), M_{K'}).$$

In particular,

$$d(\bar{x}, \bar{x}'') \leq \bar{K}\tau_{2.47}(\bar{K}, \max(\eta_0 R(1 + \max(K, K'))), M_{K'}). \tag{3.34}$$

Putting together the inequalities (3.34), (3.32), (3.31), (3.30), by the triangle inequality, we get:

$$\begin{aligned} d_Z(\bar{x}, y) &\leq d_Z(\bar{x}, \bar{x}'') + d_Z(\bar{x}'', x'') + d_Z(x'', y'') + d_Z(y'', y) \\ &\leq R' \\ &:= \bar{K}\tau_{2.47}(\bar{K}, \max(\eta_0 R(1 + \max(K, K'))), M_{K'}) \\ &\quad + \eta_0 R(1 + \max(K, K')) \\ &+ R(1 + \max(K, K')) + K'R. \end{aligned}$$

This proves the proposition. □

Combining Theorem 3.49 with the existence of uniformly coarse Lipschitz retractions $X \to Z$, where $Z = Fl_K(Q_u)$ or $Z = L_K(\alpha)$, (Theorem 3.21, Corollary 3.13), we obtain:

COROLLARY 3.63. *Suppose that* $Y \subset N^{fib}_{4\delta_0}(Fl_K(Q_u)) \subset X$ *or* $A_{K',D'}(\alpha') \subset L = L_{K,D,E}(\alpha)$ *satisfy the assumptions of Theorem* 3.49 *with parameters* λ, K, D, E, K' *and* D'. *Then there exists a coarse* $L_{3.63}(\lambda, K, K', D, D', E)$*-Lipschitz retraction* $\rho_Y : X \to Y$.

REMARK 3.64.

(1) It follows that if X were hyperbolic, then the total space A of each (K', D')-carpet $\mathfrak{A}_{K',D'}(\alpha')$ would be uniformly quasiconvex in X.
(2) Unlike the existence theorem for coarse Lipschitz retractions to semicontinuous subtrees of spaces, in order to get a uniform coarse Lipschitz retractions to carpets (and to bundles) we do not need lower bounds on K besides the obvious inequality $K \geq 1$.

3.5. Examples

In this section we give several examples of metric bundles, carpets, ladders, and flow-spaces. We will describe (up to a quasi-isometry) total spaces (which will be always hyperbolic) and vertex-spaces in these examples. We will leave it to the reader to verify that these examples indeed satisfy the claimed properties (this verification will be straightforward).

3.5.1. Metric bundles. The concept of a metric bundle is, probably, the easiest one to illustrate.

EXAMPLE 3.65. As we noted in Section 2.2, the main motivating example comes from short exact sequences

$$1 \to K \to G \to H \to 1,$$

where K is a finitely generated group (for concreteness, a free group) and H is a free group of finite rank, see [**Mos97**] for a proof of existence of hyperbolic groups G of this type. The tree T is the Cayley graph of the free group H. Since H is free, the sequence splits and, thus, we obtain a semidirect product decomposition $G = K \rtimes H$. Accordingly, we will identify H with a subgroup of G via splitting of the exact sequence. We equip G with a finite generating set which is the union of free generating sets of K and H. The total space of the bundle will be G (more precisely, a Cayley graph of G) and vertex-spaces will be the cosets $Kh \subset G$, $h \in H$, more precisely, isometric copies of the Cayley graph of K: Since K is a normal subgroup, each coset $Kh = hK$ is an isometric left-translate of K, which allows us to equip each Kh with a word metric isometric to the one of K (with respect to the free generating set of K). We suppress the description of the edge-spaces of our metric bundle since they are isometric copies of K and observe that for each free generator s of H we obtain a map of cosets $Kh_1 \to Kh_2 s$ given by the right multiplication; we use these maps in lieu of the incidence maps of the tree of spaces. See also the proof of Lemma 2.77 where we discuss in more detail the case when H is infinite cyclic.

3.5.2. The hyperbolic plane.

EXAMPLE 3.66. A geometric example is given by the hyperbolic plane X in the upper half-plane model, where the tree T is the y-axis, where the vertex-set is given by $\{2^n : n \in \mathbb{Z}\}$. The vertex-space of $X_v, v = 2^n$ is the horizontal line $\{(x, 2^n) : x \in \mathbb{R}\}$ (a hyperbolic horocycle) equipped with the induced path-metric (thus, each X_v is isometric to the real line). Of course, the vertex-spaces in this example are exponentially distorted in X.

EXAMPLE 3.67. A slight variation on the first example is obtained gluing (via an isometry) of two copies $X' = \mathbb{H}^2 \times \{p'\}$, $X'' = \mathbb{H}^2 \times \{p''\}$ of the hyperbolic plane $\mathbb{H}^2$ (fibered over the y-axis as before) over a closed horodisk. More precisely, let $B \subset \mathbb{H}^2$ denote the closed horodisk given by the inequality $y \geq 1$. We have natural isometric inclusion maps $B \to X', B \to X''$ defined via $z \mapsto (z, p'), z \mapsto (z, p'')$. We then let X denote the pushout of the diagram

$$X' \leftarrow B \to X''$$

with the natural path-metric. Accordingly, we amalgamate the base-trees (actually, lines, T', T'') of X', X'' over a common ray, which we denote R (the projection of the common horodisk). The result is a tree of spaces $X \to T$, where T is the union of three rays meeting at a common vertex u. This tree of spaces is a metric bundle over T. Topologically, of course, X is just the product $T \times \mathbb{R}$.

We return to the hyperbolic plane example, Example 3.66. Note that typical qi sections (flow-lines) of the hyperbolic plane (as a bundle over the y-axis) are given by vertical half-lines, which are, of course, hyperbolic geodesics in X. Given $x \in \mathbb{R}$, we have the section $y \mapsto \gamma_x(y) = (x, y)$. For $x_1, x_2 \in \mathbb{R}$, the sections converge as $y \to \infty$:

$$\lim_{y\to\infty} d_X(\gamma_{x_1}(y), \gamma_{x_2}(y)) = 0.$$

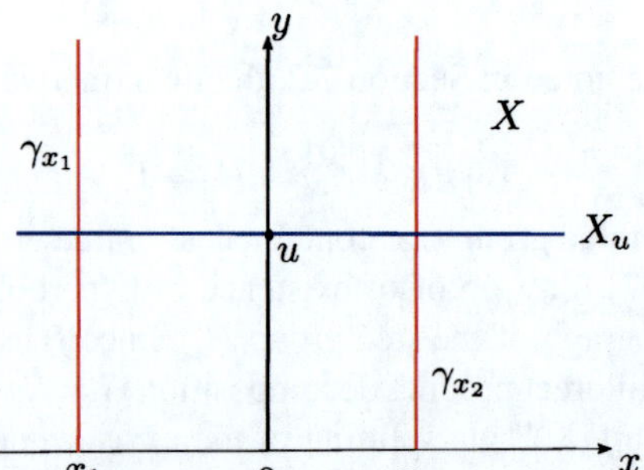

FIGURE 3.11. The hyperbolic plane as a metric bundle

3.5.3. The flipped hyperbolic plane. This is another variation on the hyperbolic plane example.

EXAMPLE 3.68. Consider the hyperbolic quarter-planes $X_{\pm} = \{(x, y) : y > 0, \pm x \geq 0\}$ with their hyperbolic metrics. Now, glue $X_{\pm}$ via the isometric map $y \mapsto y^{-1}$ of their boundary half-lines. Call the resulting metric space X'. It still has a natural projection π' to the y-axis T and the structure of a metric bundle. While, as a metric space, X' is still isometric to the hyperbolic plane X, the metric bundle structure is quite different. From the view point of the hyperbolic geometry, each vertex-space of $X' \to T$ is the union of two halves of hyperbolic horocycles and each X'_v has two limit points in $\partial_{\infty} X'$ which are the same as the limit points of the geodesic given by the y-axis. (See Figure 3.12 where we use the unit disk model of the hyperbolic plane and where the vertical line represents the y-axis.) Note that each vertex-space is still exponentially distorted in X' due to the exponential distortion of half-horocycles. Given two qi sections $\gamma_{x_1}, \gamma_{x_2}$ of $\pi' : X' \to T$ we observe the following convergence/divergence behavior:

$$\lim_{y \to \infty} d_X(\gamma_{x_1}(y), \gamma_{x_2}(y)) = 0, \quad \text{if} \quad x_1 x_2 > 0,$$

$$\lim_{y \to \pm\infty} d_X(\gamma_{x_1}(y), \gamma_{x_2}(y)) = \infty, \quad \text{if} \quad x_1 x_2 < 0.$$

3.5.4. Flow-spaces.

EXAMPLE 3.69. Let $X \to T$ be the metric bundle corresponding to a short exact sequence as in Example 3.65, where H is the infinite cyclic group (the latter assumption will simplify the description of flow-spaces). Then $G = K \rtimes_{\phi} \mathbb{Z}$, where ϕ is an automorphism of the free group K. Take a quasiconvex subset $Q \subset K$, say, a finitely generated subgroup $Q < K$. Let $u \in V(T)$ be the vertex corresponding to $0 \in H = \mathbb{Z}$. Then $Q_u = Q$. For $v_{\pm 1} \in (T)$ corresponding to $\pm 1 \in \mathbb{Z}$ take $Q_{v_{\pm 1}}$ equal to the translate (in G) of the convex hulls (in K) of $\phi^{\pm 1}(Q)$; these are the same as the quasiconvex hulls $\mathrm{Hull}(\phi^{\pm 1}(Q))$. For $v_{\pm 2}$ corresponding to $\pm 2 \in \mathbb{Z}$ take $Q_{v_{\pm 2}}$ equal to the translates of the convex hulls (in K) of, respectively, $\phi^{\pm 1}(\mathrm{Hull}(\phi^{\pm 1}(Q)))$. And so on.

EXAMPLE 3.70. To get examples of flow-spaces in the hyperbolic plane (regarded as a metric bundle over the y-axis), fix two real numbers $a < b$ and take the following subset

$$\{(x, y) : a \leq x \leq b, y > 0\}$$

of the upper half-plane. This will be a flow-space of the interval $\{(x, y) : a \leq x \leq b, y = 1\} \subset X_u \subset X$, where $u = 1$. This flow-space is bounded by two qi sections γ_a, γ_b which, of course, converge to each other as $y \to \infty$. The same flow-space example works for the *flipped hyperbolic plane* metric bundle $\mathfrak{X}' = (X' \to T)$ discussed above. The main difference is that the boundary lines γ_a, γ_b of the flow-space in $\mathfrak{X}'$ will diverge from each other in both directions.

3.5.5. Carpets.

EXAMPLE 3.71. Consider the hyperbolic plane regarded as a metric bundle over the y-axis. Take two qi sections $\gamma_{x_i}, i = 1, 2$ over an interval $[\![u, w]\!] \subset T$, $u < w$, where the sections are vertical line segments in the upper half-plane. The Euclidean rectangle defined by these line segments is a carpet $\mathfrak{A} = \mathfrak{A}(\alpha)$ with the narrow end β; see Figure 3.13. The same example works in the flipped hyperbolic plane as long as $x_1 x_2 > 0$.

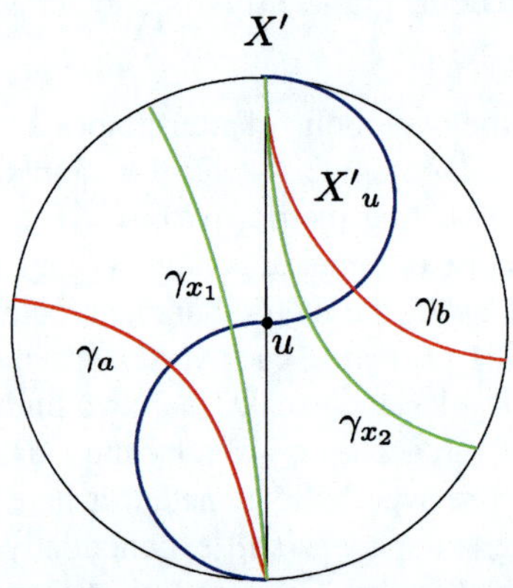

FIGURE 3.12. The flipped hyperbolic plane as a metric bundle

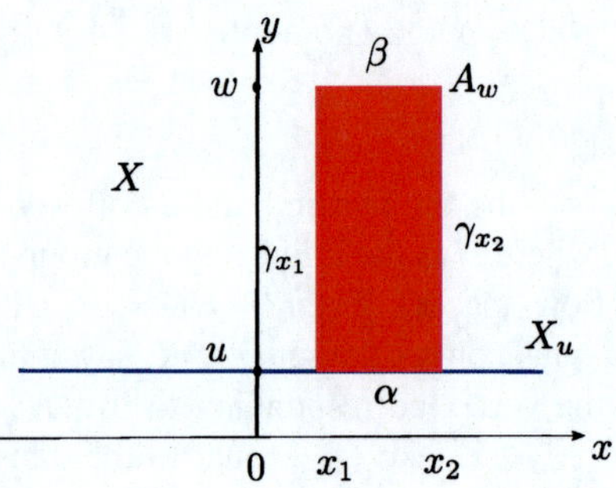

FIGURE 3.13. A carpet in the hyperbolic plane

EXAMPLE 3.72. To get a group-theoretic example, take a semidirect product as in a flow-space example, $G = K \rtimes_\phi H$, $H \cong \mathbb{Z}$. Start with a geodesic line segment $\alpha \subset K$, connecting two elements x_1, x_2 of the free group K such that $d_K(\phi^n(x_1), \phi^n(x_2)) > d_K(x_1, x_2) = C$ for all $n > 0$. Take vertices u, w in T corresponding to the integers $N > 0$ and 0 respectively. Take α to be the translate (in X_u) of the geodesic segment $[\phi^N(x_1)\phi^N(x_2)]_K$ and, respectively, β the geodesic segment $[x_1x_2]_K \subset X_u$. Furthermore, take qi sections $\gamma_{x_i}, i = 1, 2,$ of $\pi : G \to H$, where $\gamma_{x_i}(n) = x_in$ lies in the vertex-space X_{v_n} corresponding to the vertex $v_n = n \in [\![u, w]\!]$. (For the notation x_in, recall that we identify H with a subgroup of G.) Then the qi sections $\gamma_{x_i}, i = 1, 2$, define a carpet $\mathfrak{A} = \mathfrak{A}(\alpha)$ with the C-narrow end β. The vertex-spaces of the carpet are geodesic segments in the cosets nK ($n \in [0, N]$) connecting $n\phi^n(x_1), n\phi^n(x_2) \in nK$.

3.5.6. Ladders.

EXAMPLE 3.73. Start with the example (Example 3.67) of a metric bundle $\pi : X \to T$ which is the union of two hyperbolic planes X', X'' glued isometrically over a horodisk. The vertex-space X_u of X corresponding to the boundary of the common horodisk is a horocycle. Take a segment $\alpha = [x_1x_2] \subset X_u$ and consider (as in the example of a carpet, Example 3.71) two qi sections of $X' \to T', X'' \to T''$, which are geodesics $\gamma'_{x_i}, \gamma''_{x_i}$ in the hyperbolic planes X', X'', passing through the points x_1, x_2. Combining these sections we obtain sections γ_1, γ_2 of $X \to T$. These sections intersect each vertex-space X_v in two points $\gamma_i(v), i = 1, 2$. We then obtain a collection of intervals $[\gamma_1(v)\gamma_2(v)]_{X_v}$, $v \in V(T)$. Together, all these intervals define a ladder L in X.

Note that the sections γ_1, γ_2 in this example converge along the ray R (the projection to T of the common horodisk):

$$\lim_{t\to\infty} d_{X_t}(\gamma_1(t), \gamma_2(t)) = 0,$$

where t diverges to infinity in the ray R. Thus, we can obtain a carpeted subladder in L as follows. Pick a number C and assume that $d_{X_u}(x_1, x_2) > C$. Then, in view of convergence of the sections over the ray R, there exists a vertex w in R such that $d_{X_w}(\gamma_1(w), \gamma_2(w)) \le C$. The points $\gamma_1(w), \gamma_2(w) \in X_w$ bound an interval $\beta \subset X_w$. Now, remove from T the open subray of R emanating from w. The result is a subtree $S \subset T$; taking its pre-image Y in the space X we obtain a bundle $Y \to S$. Intersection of L with this bundle is a carpeted ladder L^Y in Y, it contains the carpet A bounded by the intervals $\alpha \subset X_u, \beta \subset X_w$, where β has length $\le C$. The top and the bottom boundary paths of A are the restrictions of the sections γ_1, γ_2 to the interval $[\![u, w]\!]$.

One gets a group-theoretic example of ladders by working with semidirect products $K \rtimes H$, where H is a free group of finite rank ≥ 2. Then one proceeds as in Example 3.72, except instead of acting repeatedly by the automorphism corresponding to just one generator of H, one acts on K by all the generators of H.

CHAPTER 4

Hyperbolicity of ladders

In this chapter we prove uniform hyperbolicity of (K, D, E)-ladders in X. The proof is divided in three main steps. We first prove (Section 4.1) hyperbolicity of carpets by exhibiting slim combings of carpets (combings satisfying the conditions of Corollary 1.66). We use these paths, in conjunction with the retractions to carpets (see Corollary 3.63) to construct combings of *carpeted ladders*, which are ladders $\mathfrak{L}(\alpha)$ containing carpets $\mathfrak{A}(\alpha')$ with $\alpha' \subset \alpha$ whose length almost equals that of α. This is done in Section 4.2. Lastly, in Section 4.3 we prove hyperbolicity of general ladders subdividing these (a "vertical subdivision") into carpeted ladders and then using quasiconvex amalgamation to prove hyperbolicity. Hyperbolicity of ladders is a key step for proving hyperbolicity of flow-spaces, which is done in the next chapter.

4.1. Hyperbolicity of carpets

The proof of the following simple proposition will serve as a model for more complex proofs of hyperbolicity of certain subspaces of X.

PROPOSITION 4.1. *For every $K \geq 1$, every (K, C)-narrow carpet $\mathfrak{A} = (\pi : A \to [\![u, w]\!])$ in $\mathfrak{X}$, equipped with its intrinsic metric, is $\delta_{4.1}(K, C)$-hyperbolic, provided that $\mathfrak{X}$ satisfies the uniform K-flaring condition.*

PROOF. Set $\mathcal{A} = A \cap (\cup_{v \in V(T)} X_v)$. Let $\beta = \mathcal{A} \cap X_w$ denote the C-narrow end of $\mathfrak{A}$. For each $x \in \mathcal{A}$ we have the K-qi section $\gamma_x \subset \Sigma_x$ of $\pi : A \to [\![u, w]\!]$ over $[\![\pi(x), w]\!]$, connecting x to β. Suppose $x, y \in \mathcal{A}$. Without loss of generality we can assume that $\pi(y) \in [\![w, \pi(x)]\!]$. Now, let $t_{xy} \in V([\![w, \pi(y)]\!])$ denote the supremum of the set

$$\{t \in V([\![w, \pi(y)]\!]) : d_{X_t}(\gamma_x(t), \gamma_y(t)) \leq M_K\},$$

unless the set is empty in which case we let $t_{xy} = w$. Set $t_0 := t_{xy}$. We then define a path $c(x, y)$ as the concatenation of the section γ_x restricted to $[\![\pi(x), t_0]\!]$ with the vertical segment $[\gamma_x(t_0)\gamma_y(t_0)]_{X_{t_0}}$, followed by the concatenation with the restriction of the section γ_y to the subinterval $[\![t_0, \pi(y)]\!]$. The assumption that $\mathcal{A}$ is (K, C)-narrow implies that the length of $[\gamma_x(t_0)\gamma_y(t_0)]_{X_{t_0}}$ is at most $\max(C, M_K)$.

We claim that this family of paths in A satisfies the conditions of Corollary 1.66 with constants depending only on K and C. The assumption that the paths $c(x, y)$ are uniformly coarse Lipschitz follows from the fact that each path $c(x, y)$ is a concatenation of K-qi sections and of vertical geodesics.

LEMMA 4.2. *The family of paths $c(x, y)$ is uniformly proper in A, with distortion function depending only on K and C.*

PROOF. Let $x, y \in \mathcal{A}$ be such that $d(x, y) \leq r$. Set $v_1 := \pi(x), v_2 := \pi(y)$. Then $d_T(v_1, v_2) \leq r$ as well. Without loss of generality, on the oriented interval $[\![w, u]\!]$ we have

$$w \leq t_{xy} \leq v_1 \leq v_2 \leq u.$$

Let $y_1 \in A_{v_1}$ be the intersection point with γ_y. Then the subpath γ_{yy_1} in γ_y between y, y_1 has length $\le Kr$ and $d(y_1, x) \le r_1 = r(K+1)$. Furthermore, $d_{X_{v_1}}(y_1, x) \le \eta_0(r_1)$. By the uniform K-flaring condition,

$$d_T(v_1, t_{xy}) \le \tau_{2.47}(K, \max(M_K, C, \eta_0(r_1))).$$

Therefore, the overall length of $c(x, y)$ is

$$\le \max(M_K, C) + 2K\tau_{2.47}(K, \max(M_K, C, \eta_0(r_1))) + Kr. \qquad \square$$

REMARK 4.3. This lemma is the only place where the constant C plays any role in the proof of the proposition.

We next verify the condition (a2) of Corollary 1.66 for the family of paths $c(x, y)$. The following remark and lemma are new.

REMARK 4.4. For the proof of slimness of triangles formed by the paths $c(x, y)$ it is helpful to prove the Hausdorff fellow traveling property, i.e., if x, x' are nearby and y, y' are nearby then the Hausdorff distance of $c(x, y)$ and $c(x', y')$ is uniformly small. However, to show this, it is enough to consider the case $y = y'$. Then again it is enough to consider the two cases $\pi(x) = \pi(x')$ and $d(\pi(x), \pi(x')) = 1$. This reduction arguments appear a number times later.

LEMMA 4.5. *Given $D \ge 0$ there is $D' \ge 0$ such that the following holds: If $d(y, z) \le D$ then $Hd(c(x, y), c(x, z)) \le D'$.*

PROOF. Clearly, it is enough to prove the lemma for the two cases (1) $\pi(y) = \pi(z)$ and (2) $d(\pi(y), \pi(z)) = 1$, $\pi(z) \in [\![\pi(y), w]\!]$. However, clearly (2) follows from (1) and the case (2') where $d(\pi(y), \pi(z)) = 1$, $\pi(z) \in [\![\pi(y), w]\!]$ and $z \in \gamma_y$. On the other hand, even when we are in case (1) or (2') there are two possibilities; either $\pi(x) \in [\![\pi(y), w]\!]$ or $\pi(y) \in [\![\pi(x), w]\!]$. We shall assume that $\pi(x) \in [\![\pi(y), w]\!]$ leaving the other one for the reader since that is analogous.

Case 1. In this case, we have $\pi(y) = \pi(z)$ and $\pi(y) \in [\![\pi(x), w]\!]$. By Corollary 2.53 it follows that portions of γ_y and γ_z over $[\![\pi(y), w]\!]$ are at a uniformly bounded Hausdorff distance. Since γ_x is chosen independent of y, z it is enough to show that $d(t_{xy}, t_{xz})$ is uniformly small. Without loss of generality, we may assume that $t_{xz} \in [\![t_{xy}, w]\!]$. Set $t_1 = t_{xy}, t_2 = t_{xz}$. The rest of the argument follows from the uniform flaring (see Proposition 2.47) applied to γ_x and γ_z over $[\![t_{xy}, t_{xz}]\!]$.

Case 2'. In this case, we have $d(\pi(y), \pi(z)) = 1$, $\pi(z) \in [\![\pi(y), w]\!]$ and $z \in \gamma_y$ and $\pi(y) \in [\![\pi(x), w]\!]$. In this case, we note that $\gamma_y = \gamma_z$. Then there are two possibilities: If $t_{xy} \in [\![\pi(z), w]\!]$ then $t_{xz} = t_{xy}$ are done. Otherwise, $t_{xy} = \pi(y)$ in which case the lemma is immediate. $\square$

Consider now three points $x, y, z \in \mathcal{A}$ and the triangle Δ formed by the paths $c(x, y)$, $c(y, z)$, $c(z, x)$ connecting them. After relabelling the points x, y, z we can assume that the vertices t_{xy}, t_{yz}, t_{zx} appear in the interval $[\![u, w]\!]$ in the following order:

$$u \le t_{xy} \le t_{yz} \le t_{zx} \le w.$$

CASE 1. Suppose first that $t := t_{xy} < w$. Then

$$d_{X_t}(\gamma_x(t), \gamma_y(t)) \le M_K.$$

Therefore, we replace x, y with $x' := \gamma_x(t), y' := \gamma_y(t)$ respectively; $d_{X_t}(x', y') \le M_K$, i.e., the length of the path $c(x', y')$ is $\le M_K$. Thus, $\delta_{4.1}(K)$-slimness of the triangle

$\Delta' = \Delta x'y'z$ formed by the paths $c(x', y'), c(y', z), c(z, x')$ follows from the uniform fellow-traveling property of the paths c in $\mathfrak{A}$. Since, without loss of generality we may assume that $c(y', z), c(z, x')$ are subpaths in $c(y, z), c(z, x)$, we conclude the $\delta_{4.1}(K)$-slimness of the original triangle Δ.

CASE 2. It remains to consider the case $t_{xy} = t_{yz} = t_{zx} = w$. Then, as in Case 1, we replace the points x, y, z with the points $x' := \gamma_x(w), y' := \gamma_y(w), z' = \gamma_z(w)$. The triangle Δ' formed by the paths $c(x', y'), c(y', z'), c(z', x')$ is contained in the geodesic segment β; hence Δ' is δ_0-slim. We conclude that Δ is $\delta_{4.1}(K)$-slim as well. □

REMARK 4.6. Lemma 4.2 establishes that the paths $c(x, y)$ are uniformly proper (with distortion depending only on K and C). Hence, by Corollary 1.66(b), the paths $c(x, y)$ are uniformly (in terms of K and C) close to geodesics in A.

4.2. Hyperbolicity of carpeted ladders

DEFINITION 4.7. Let $\bar{K}$ be defined as in Notation 3.48, with $K' = K, \lambda = \delta_0$,

$$\bar{K} := K_{3.47}(\delta_0, K, K). \tag{4.1}$$

Set

$$\kappa := \kappa_{4.7}(K) = \kappa_{(3.19)}(\delta_0, K, K). \tag{4.2}$$

A (K, D, E)-ladder $\mathfrak{L}(\alpha)$ containing a (K, C)-carpet $\mathfrak{A} = \mathfrak{A}_{K,C}(\alpha')$, $\alpha' \subset \alpha$, as a subladder[1], satisfying

$$\text{length}(\alpha') \geq \text{length}(\alpha) - M_{\bar{K}},$$

will be called *carpeted.*

In this section, we will prove that carpeted ladders are uniformly hyperbolic. Only the parameters K and C will play a role in the proof, the parameters D and E will be irrelevant, just as in the proof of the existence of a coarse Lipschitz retraction $L_K(\alpha) \to A_{K,C}(\alpha')$.

THEOREM 4.8 (Hyperbolicity of carpeted ladders). *Carpeted ladders in $\mathfrak{X}$ are hyperbolic. More precisely: Fix $K \geq 1$ and suppose that that $\mathfrak{X}$ satisfies the uniform κ-flaring condition, where $\kappa = \kappa(K)$ is as in Definition* 4.7, (4.2). *Let $\mathfrak{L} = \mathfrak{L}_K(\alpha)$ be a K-ladder containing a (K, C)-carpet $\mathfrak{A}(\alpha')$ as in Definition* 4.7. *Then $L_K(\alpha)$ (with its intrinsic metric) is $\delta_{4.8}(K, C)$-hyperbolic.*

PROOF. The proof of this theorem is divided in two steps: We first define a family of paths $c(x, y)$ in $L_K(\alpha)$ connecting points $x, y \in \mathcal{L}$, *using the family of uniform quasi-geodesics for the carpet $\mathfrak{A}$ as a blackbox*, and then check that these paths satisfy the conditions of Corollary 1.66. We let $[\![u, w]\!] \subset S \subset T$ denote the base-interval of the carpet $\mathfrak{A}$, where $S = \pi(L_K(\alpha))$ is the base of $\mathfrak{L} = (\pi : L_K(\alpha) \to S)$. Then u is the center of both $\mathfrak{A}$ and $\mathfrak{L}$, $\alpha' = A_u$ and $\beta = A_w$ is the narrow end of $\mathfrak{A}$.

For each $x \in \mathcal{L}$ we let $b_x \in V(T)$ denote the center of the triangle $\Delta uw\pi(x) \subset T$.

According to Theorem 3.49, there exists $\rho = \rho_{\mathfrak{L},\mathfrak{A}} : \mathcal{L} \to \mathcal{A}$, a coarsely k-Lipschitz retraction with

$$k = L_{3.49}(\delta_0, K, K, C).$$

We first review some facts about $\rho(x)$. The point $\rho(x) = \bar{x}$ belongs to the interval $A_s \subset \mathcal{A}$, for a certain vertex

$$s \in [\![u, \pi(x)]\!] \cap [\![u, w]\!],$$

[1]See Definition 3.10.

such that $\bar{x}$ is within vertical distance $M_{\bar{K}}$ (in the vertex-space X_s) from a point $\tilde{x} = \gamma_x(s) \in X_s$. Here $\gamma_x \subset \Sigma_x$ is the K-section in $\mathcal{L}$ over $[\![u, \pi(x)]\!]$ connecting x to α. (Such point $\bar{x}$ always exists in view of the assumption on the length of α'.)

The paths $\gamma'_x = \gamma_{x\tilde{x}} \star [\tilde{x}\bar{x}]_{X_s}$ defined in the proof of Theorem 3.49, connecting x to $\bar{x}$, satisfy the Hausdorff fellow-traveling property with respect to variations of x; see Corollary 3.60. Here $\gamma_{x\tilde{x}}$ is a subpath of γ_x connecting x to $\tilde{x}$.

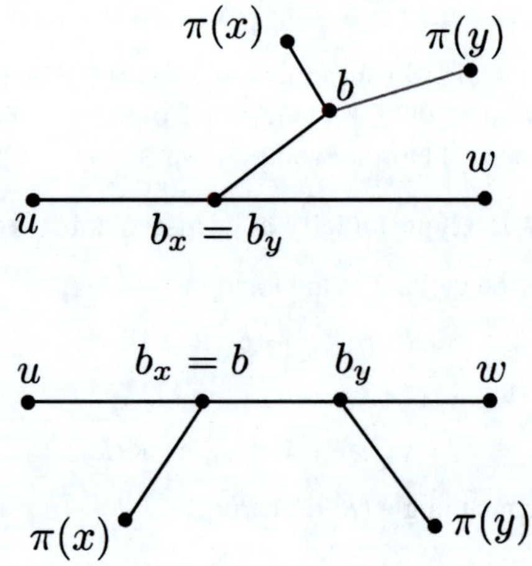

FIGURE 4.1. Up to relabeling x, y there are two possible configurations of the points $u, w, \pi(x), \pi(y)$.

STEP 1 (Definition of the paths $c_{x,y} = c(x, y)$). For $x, y \in \mathcal{L}$ we let $b = b_{xy}$ be the center of the triangle $\triangle u\pi(x)\pi(y) \subset T$. There are two cases to consider, depending on which we get two types of paths $c(x, y)$.

Paths of type 1: There exists $t \in V([\![\pi(x), \pi(\bar{x})]\!] \cap [\![\pi(y), \pi(\bar{y})]\!]) \subset V([\![u, b]\!])$ such that

$$d_{X_t}(\gamma_x(t), \gamma_y(t)) \leq M_{\bar{K}},$$

i.e., the paths γ_x, γ_y "come sufficiently close" in some common vertex-space.

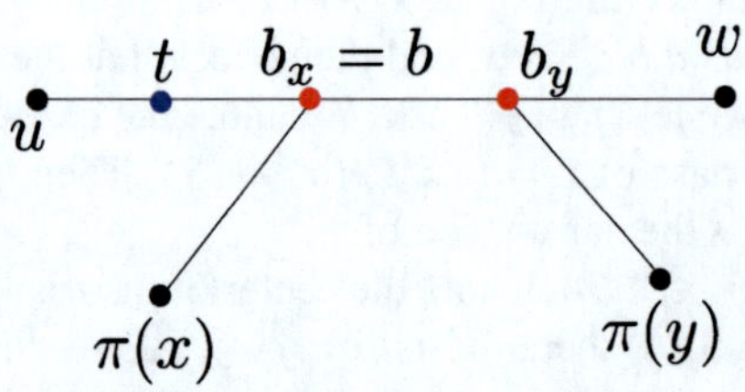

FIGURE 4.2. Paths of type 1: $t = t_{x,y}$

Let t_{xy} be the maximal vertex in $[\![u, b]\!]$ with this property. Then define $c(x, y)$ to be the concatenation of the portions of γ_x and (the reverse of) γ_y over $[\![t_{xy}, \pi(x)]\!]$ and $[\![t_{x,y}, \pi(y)]\!]$ respectively with the subsegment of $L_{t_{xy}}$ joining their endpoints.

Paths of type 2: Suppose t as in type 1 does not exist, i.e., the paths γ_x, γ_y "stay far apart" in every vertex-space they both enter.

Then define $c(x, y)$ to be the concatenation of γ'_x and the reverse of γ'_y with a geodesic $[\bar{x}\bar{y}]_A$ in $A = A_{(K,C)}(\alpha')$ connecting $\bar{x}$ to $\bar{y}$.

REMARK 4.9. More precisely, instead of geodesics $[\bar{x}\bar{y}]_A$ one should use uniform quasigeodesic paths in A defined in the previous section. Since the two families are uniformly close to each other, we will work with geodesics for the ease of the notation. We will do the same repeatedly later in the book.

We observe that each path of type 1 is a concatenation of (at most) three uniformly quasigeodesics paths (the middle one of which has) uniformly bounded length $\leq M_{\bar{K}}$, while each path of type 2 is a concatenation of (at most) five uniformly quasigeodesics paths (two of which have uniformly bounded lengths $\leq M_{\bar{K}}$).

Our next task is to establish a uniform Hausdorff fellow-traveling property of the family of paths $c(x, y)$ (Lemma 4.10). Even though this property is not required by Corollary 1.66, it will play key role in verifying the other conditions of the corollary.

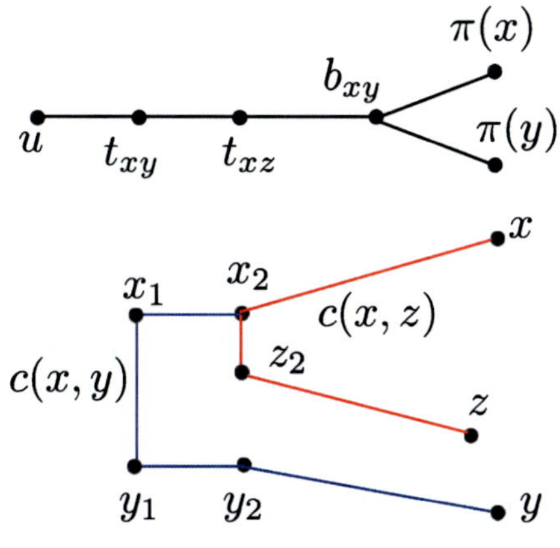

FIGURE 4.3. Case 1-1

LEMMA 4.10. *The paths $c(x, y)$ satisfy the Hausdorff fellow–traveling property, i.e., if y, z are uniformly close to each other (in the total space L of the ladder $\mathfrak{L}$), so are the images of the paths $c(x, y), c(x, z)$. More precisely, there is a function $D_{4.10}(C, K, r)$ such that if $d_L(y, z) \leq r$, then*

$$\mathrm{Hd}(c_{x,y}, c_{x,z}) \leq D_{4.10}(C, K, r),$$

PROOF. As in the proof of Theorem 3.49, (see also Remark 4.4) there are two cases to consider: $\pi(y) = \pi(z)$ (see part of the proof covered in Proposition 3.51) and $d(\pi(y), \pi(z)) = 1$ (see part of the proof covered in Lemma 3.58). The second case follows from the first one just as in the proof of Lemma 3.58, so we assume that $\pi(y) = \pi(z)$. There are three cases to check depending on the types of the paths $c(x, y), c(x, z)$.

CASE 1–1. Both paths $c(x, y), c(x, z)$ have type 1. The paths $c(x, y), c(x, z)$ agree over the interval $[\![t, \pi(x)]\!]$ where $t \in \{t_{xy}, t_{xz}\}$ is the vertex closer to b_{xy}; after swapping the roles of y and z we may assume that $t = t_{xz}$; see Figure 4.3.

Define the points

$$x_1 := \gamma_x(t_{xy}), y_1 := \gamma_y(t_{xy}), x_2 := \gamma_x(t_{xz}), y_2 = \gamma_y(t_{xz}), z_2 := \gamma_z(t_{xz}).$$

They satisfy the inequalities

$$d_{X_{t_{xy}}}(x_1, y_1) \le M_{\bar{K}}, \quad d_{X_{t_{xz}}}(x_2, z_2) \le M_{\bar{K}}.$$

Except for the point y_2, these are the points where the paths $c(x, y), c(x, z)$ switch from vertical to horizontal. The points x_1, y_1, y_2 lie in the image of $c(x, y)$, while the points x_2, z_2 lie in the image of $c(x, z)$.

We will show that the length of the interval $[\![t_{xy}, t_{xz}]\!]$ is uniformly bounded (in terms of r and the other parameters in the theorem). Since

$$d_{X_{t_{xy}}}(x_1, y_1) \le M_{\bar{K}},$$

it will follow, according to Lemma 2.38 that the part of $c(x, y)$ lying between x_2 and y_2 is uniformly Hausdorff-close to the vertical segment $[x_2 y_2]_{X_{t_{xz}}}$. In particular, the length of that segment will be uniformly bounded. However, the points x_2 and z_2 are within vertical distance $M_{\bar{K}}$ from each other. Hence, the vertical distance between z_2 and y_2 is also uniformly bounded. Since $d_L(y, z) \le r$, by Corollary 2.53, it will follow that the vertical distance between γ_y, γ_z over the interval $[\![t_{xz}, \pi(y)]\!]$ is also uniformly bounded, thereby, concluding the proof. Thus, it remains to bound the length of the interval $[\![t_{xy}, t_{xz}]\!]$.

By the definition of the projection $\rho : \mathcal{L} \to \mathcal{A}$,

$$\rho(x_1) = \rho(x_2) = \bar{x}, \rho(y_1) = \bar{y}, \rho(z_2) = \bar{z}.$$

Since ρ is k-coarse Lipschitz, we have

$$d_A(\bar{x}, \bar{y}) \le k(M_{\bar{K}} + 1), \quad d_A(\bar{x}, \bar{z}) \le k(M_{\bar{K}} + 1), \quad d_A(\bar{y}, \bar{z}) \le k(r + 1).$$

Lemma 2.51 now implies that

$$d_T(t_{xy}, t_{xz}) \le \tau_{2.51}(K, \max(k(M_{\bar{K}} + 1), k(r + 1))).$$

This concludes the proof in Case 1-1. □

Case 1–2. The path $c(x, y)$ is of type 1 while $c(x, z)$ is of type 2.

Since $c(x, y)$ has type 1, we define the vertex

$$t = t_{xy} \in [\![u, \pi(x)]\!] \cap [\![u, v]\!], v = \pi(y) = \pi(z).$$

At this vertex

$$d_{X_t}(\gamma_x(t), \gamma_y(t)) \le M_{\bar{K}}. \tag{4.3}$$

All three paths $\gamma_x, \gamma_y, \gamma_z$ enter the same vertex-space X_t, $t = t_{xy}$ at points x_1, y_1, z_1 respectively.

Since ρ is k-coarse Lipschitz, and $\bar{x} = \rho(x) = \rho(x_1), \bar{y} = \rho(y) = \rho(y_1)$, we have

$$d_A(\bar{x}, \bar{y}) \le k(M_{\bar{K}} + 1).$$

Similarly, since $d_L(y, z) \le r$, we obtain

$$d_A(\bar{y}, \bar{z}) \le k(r + 1).$$

Define $v_y := \pi(\rho(y)), v_z := \pi(\rho(z))$. Thus,

$$d_T(v_y, v_z) \le k(r + 1)$$

$$d(\gamma_y(v_y), \gamma_z(v_z)) \le k(r + 1) + M_{\bar{K}}.$$

Let $\bar{v} \in \{v_y, v_z\}$ denote the vertex further away from u. Thus,

$$d(\gamma_y(\bar{v}), \gamma_z(\bar{v})) \le k(r + 1) + 2M_{\bar{K}} + 2kK(r + 1).$$

By Corollary 2.53, the fiberwise distance between γ_y, γ_z over the interval $[\![\bar{v}, v]\!]$ is at most $\tau_{2.53}(K, \max(r, k(r+1) + 2M_{\bar{K}} + 2kK(r+1))) = \tau_{2.53}(K, k(r+1) + 2M_{\bar{K}} + 2kK(r+1))$.

In particular, this inequality holds at the vertex t since it belongs to the interval $[\![\bar{v}, v]\!]$:

$$d_{X_t}(\gamma_y(t), \gamma_z(t)) \leq \tau_{2.53}(K, k(r+1) + 2M_{\bar{K}} + 2kK(r+1)).$$

Hence, by (4.3),

$$d_{X_t}(\gamma_x(t), \gamma_z(t)) \leq M_{\bar{K}} + \tau_{2.53}(K, k(r+1) + 2M_{\bar{K}} + 2kK(r+1)).$$

Define a vertex v' by

$$[\![\pi(\bar{x}), t]\!] \cap [\![\pi(\bar{z}), t]\!] = [\![v', t]\!].$$

Since the path $c(x, z)$ has type 2, for all $s \in V([\![v', t[\![)$ we have the inequality

$$d_{X_s}(\gamma_x(s), \gamma_z(s)) > M_{\bar{K}}.$$

Therefore, as in Case 1-1, Lemma 2.51 implies a uniform upper bound in the lengths of the intervals $[\![\pi(\bar{x}), t]\!], [\![\pi(\bar{z}), t]\!]$. Thus, just as in Case 1-1, we obtain a uniform upper bound on the distances $d(\bar{x}, x_1), d(\bar{z}, z_1)$ and, hence, the paths $c(x, y), c(x, z)$ uniformly Hausdorff fellow-travel.

CASE 2–2. Both paths $c(x, y), c(x, z)$ have type 2. The points $\bar{y}, \bar{z} \in \mathcal{A}$ are within distance $\leq k(r+1)$ from each other and, hence, by Corollary 2.53, the paths γ'_y, γ'_z uniformly fellow–travel. The same holds for geodesics $[\bar{y}\bar{x}]_A, [\bar{z}\bar{x}]_A$ since A is $\delta_A = \delta_{4.1}(K, C)$-hyperbolic. Hence, the paths $c(x, y), c(x, z)$ uniformly fellow-travel as well. □

The proof of the next lemma is very similar to that of Lemma 4.10 and we omit it.

LEMMA 4.11. *There are functions $\xi_{4.11}(C, K, r, \tau)$ (which is linear in the variable τ) and $\nu_{4.11}(C, K, r)$ satisfying the following: Suppose that $c = c(x, y)$ has domain* $[0, D]$ *and $x', y' \in L$ are such that $d_L(x', c(s)) \leq r, d_L(x', c(t)) \leq r$ for some $s, t \in$* $[0, D]$. *Then: 1.* $\text{length}(c(x', y')) \leq \xi_{4.11}(C, K, r, \text{length}(c|_{[s,t]}))$. *2.* $\text{Hd}(c(x', y'), c([s, t])) \leq \nu_{4.11}(C, K, r)$.

As a consequence, we obtain:

LEMMA 4.12. *The paths $c(x, y)$ are uniformly proper. More precisely, there is a function $\zeta_{4.12}(r, K, C)$ such that for each path $c(x, y)$ defined above, for any two points x', y' on $c(x, y)$, if $d_L(x', y') \leq r$ then the length of the portion of $c(x, y)$ between x', y' is $\leq \zeta_{4.12}(r, K, C)$.*

PROOF. In view of Lemma 4.11, it suffices to consider the case $x = x', y = y'$. Consider a geodesic xy in L of length $d_L(x, y)$ and subdivide it into subsegments $[x_i x_{i+1}]$ of length ≤ 1, $i = 1, ..., N$. The number N of these subsegments is at most r. Each pair of points x_i, x_{i+1} project to the same vertex or edge of the tree T. By the the definition of the path $c(x_i, x_{i+1})$, it is has length bounded from above by a constant R depending only on the parameters of $\mathfrak{X}$. We will apply this to $i = 1$. Consider the paths $c_i := c(x_1, x_i)$, $i = 2, ..., N$. By Lemma 4.11,

$$\text{length}(c_{i+1}) \leq \xi_{4.11}(C, K, 1, \text{length}(c_i)), i = 1, ..., N.$$

Applying these inequalities inductively and taking into account that $\text{length}(c_1) \leq R$, we obtain the lemma. □

STEP 2. We shall now check that the paths c defined in Step 1 satisfy the conditions of Corollary 1.66 to conclude the proof of the theorem.

CONDITION (a1). This is an immediate consequence of Lemma 4.12.

CONDITION (a2). The verification of the condition of uniform slimness of the triangle Δ in X formed by the paths $c(x, y), c(y, z), c(z, x)$ is broken into several cases depending on the relative positions of the three points x, y, z and the types of the paths $c(x, y), c(y, z), c(z, x)$ (type 1 or type 2). The trick is to reduce the proof to a simpler case by replacing x, y, z with some other suitable points, analogously to the proof of Proposition 4.1 and then appeal to Proposition 4.8. For instance, suppose that there is a constant $r = r(K, C)$ such that a triangle Δ as above is r-thin, i.e., there exists a point $x \in X$ within distance r from all three sides of Δ. Then the Hausdorff-fellow-traveling condition (Lemma 4.10) will imply that Δ is $D_{4.10}(C, K, 2r)$-thin.

CASE 1. Suppose all three paths are of type 2. Then we replace x, y, z by their ρ-projections to A: The points $\bar{x}, \bar{y}, \bar{z}$ respectively. The subtriangle in $\Delta = c(x, y) \cup c(y, z) \cup c(z, x)$ which is the union $\bar{\Delta} = c(\bar{x}, \bar{y}) \cup c(\bar{y}, \bar{z}) \cup c(\bar{z}, \bar{x}) \subset A$ is $\delta_{\mathfrak{A}}$-hyperbolic, where $\delta_{\mathfrak{A}}$ (depending only on K and C) is a uniform bound on the hyperbolicity constant of the carpet $\mathfrak{A}$ (Proposition 4.1). Thus, Δ' is $r = \delta_{\mathfrak{A}}$-thin, therefore, as we noted above, Δ itself is $D_{4.10}(C, K, 2r)$-slim.

CASE 2. Suppose we have a triangle Δ formed by three paths exactly two of which are of type 2; say, $c(x, y), c(y, z)$ are of type 2 and $c(x, z)$ is of type 1. Since $c(x, z)$ has type 1, the vertex $t = t_{xz}$ satisfies the inequalities

$$d_T(\pi(x), \pi(\bar{x})) \geq d_T(\pi(x), t), \quad d_T(\pi(z), \pi(\bar{z})) \geq d_T(\pi(z), t).$$

Thus, we can replace the points x, z by $x' := \gamma_x(t)$ and $z' := \gamma_z(t)$ respectively (as they belong to $c(x, y)$ and $c(z, y)$ respectively). Now, by the definition of $t = t_{xz}$,

$$d(x', z') \leq d_{X_t}(x', z') \leq M_{\bar{K}}.$$

Hence, the subtriangle $\Delta' \subset \Delta$ formed by the paths $c(x', y), c'(y, z'), c(z', x')$ is $r = M_{\bar{K}}$-thin, which, in turn, implies that Δ is $D_{4.10}(C, K, 2r)$-slim.

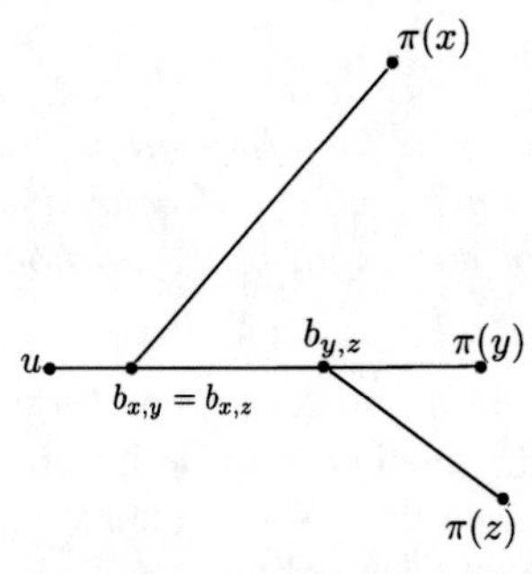

FIGURE 4.4. Three centers

CASE 3. Suppose we have a triangle Δ formed by three paths exactly one which, say, $c(x, y)$ is of type 2. After swapping x and y we can assume that on the interval $[\![u, \pi(z)]\!]$ the following vertices appear in this order:

$$u \leq \pi(\bar{z}) \leq t_{yz} \leq t_{xz} \leq \pi(z).$$

(Since $c(y, z), c(z, x)$ are of type 1, the vertex $\pi(\bar{z})$ is closer to u than both t_{yz}, t_{xz}.) As in Case 2, for $t = t_{x,z}$ we replace x, z by the points

$$x' := \gamma_x(t), \quad z' := \gamma_z(t)$$

respectively. This defines a subtriangle $\Delta' \subset \Delta$ formed by the paths $c(x', y)$, $c(y, z')$, $c(z', x')$. By the definition of type 1 paths, $d_{X_t}(x', z') \le M_{\bar{K}}$. Thus, again, the subtriangle $\Delta' \subset \Delta$ is $r = M_{\bar{K}}$-thin and, therefore, Δ is $D_{4.10}(C, K, 2r)$-slim.

CASE 4. Suppose all three paths $c(x, y), c(y, z), c(z, x)$ forming a triangle Δ are of type 1. This is the most interesting of the four cases.

Without loss of generality we may assume that in the tree T,

$$(\pi(x).\pi(y))_u \le (\pi(y).\pi(z))_u.$$

In particular,

$$b_{x,y} = b_{x,z};$$

see Figure 4.4.

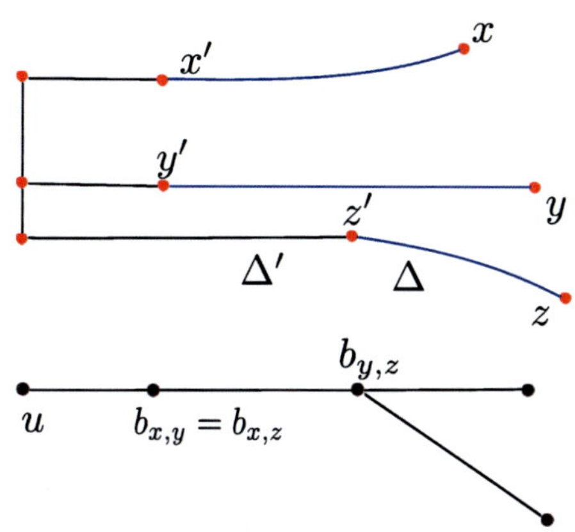

FIGURE 4.5. Triangles Δ, Δ'

Consider the points

$$x' \in c(x, z) \cap X_{b_{x,z}}, y' \in c(x, y) \cap X_{b_{x,y}}, z' \in c(x, z) \cap X_{b_{y,z}}$$

which are closest to, respectively, x, y, z along the above paths; see Figure 4.5. The point z' then equals the point of the intersection

$$c(y, z) \cap X_{b_{y,z}}$$

which is closest to z along the path $c(y, z)$.

Moreover, again by the definition of the paths c, the triangle $\Delta := c(x, y) \cup c(y, z) \cup c(z, x)$ is obtained from $\Delta' := c(x', y') \cup c(y', z') \cup c(z', x')$ by attaching (to its vertices) the following segments of the K-flow-lines:

$$\gamma_{x,x'} \subset \gamma_x, \gamma_{y,y'} \subset \gamma_y, \gamma_{z,z'} \subset \gamma_z.$$

Hence, the r-slimness of the triangle Δ' will imply r-slimness of the triangle Δ.

Thus, after replacing $x \to x', y \to y', z \to z'$, it suffices to consider the case when $v' = \pi(x) = \pi(y)$ and $v' \in [\![u, \pi(z)]\!]$. Below, *we will not be using the property that* $\pi(x) = \pi(y)$, only that all three projections $\pi(x), \pi(y), \pi(z)$ belong to a common oriented interval $J = [\![u, u']\!] \subset T$. Therefore,

$$\{t_{zy}, t_{yx}, t_{xz}\} \subset J.$$

After a permuting the points x, y, z, we can assume that

$$t_{zy} \le t_{yx} \le t_{xz}$$

on the oriented interval J. Therefore, all three paths γ_x, γ_y, γ_z contains the subpaths $\gamma_{xx''}$, $\gamma_{yy''}$, $\gamma_{zz''}$ with $x'', y'', z'' \in X_t$, $t = t_{xz}$ and these subpaths are in the respective sides $c(x, y), c(y, z), c(z, x)$ of Δ; see Figure 4.6.

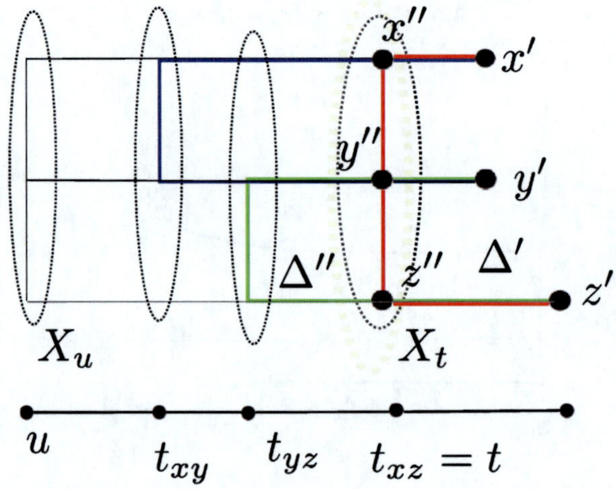

FIGURE 4.6. Triangles Δ', Δ''

Thus, we perform one more reduction, replacing the triangle Δ with the subtriangle Δ'' formed by the paths $c(x'', y''), c(y'', z''), c(z'', x'')$, such that $x'', y'', z'' \in X_t, t = t_{xz}$. By the definition of the vertex t_{xz},

$$d_{X_t}(x'', z'') \le M_{\bar{K}}.$$

Thus, the triangle Δ'' is $M_{\bar{K}}/2$-thin, which concludes the proof in Case 4. This also completes the proof of Theorem 4.8. □

REMARK 4.13. Since the paths $c(x, y)$ are uniformly proper (Corollary 4.12), by Part (b) of Corollary 1.66, up to a reparameterization, they are uniform quasigeodesics in $L_K(\alpha)$.

4.3. Hyperbolicity of general ladders

In this section we prove that all ladders contained in X are uniformly hyperbolic. The idea is to decompose the given ladder as a union of subladders, each of which is carpeted, and then make use of the quasiconvex chain-amalgamation, Theorem 2.65.

We define $\bar{K}$ and κ as in the previous section, Definition 4.7, (4.1) and (4.2). Recall that each K-ladder comes equipped with a family of K-sections $\Sigma_\bullet$. These are the sections which will be used in the next proposition.

PROPOSITION 4.14 (Vertical subdivision of general ladders). *Fix numbers K and C such that $M_{\bar{K}} \le C$, and suppose that $\mathfrak{X}$ satisfies the uniform κ-flaring condition. Consider a (K, D, E)-ladder*[2] *$\mathfrak{L} = \mathfrak{L}_K(\alpha)$ over a subtree $S \subset T$, where $\alpha = [x_u y_u]_{X_u}$.*

Then the geodesic segment α can be subdivided into subsegments $\alpha_1, \dots, \alpha_n$ of lengths $l_1, \dots, l_n$, with the endpoints $x_0 = x_u, x_1, \dots, x_n = y_u$ such that the following hold:

[2]The parameters D and E of the ladder play no role in the proposition.

(1) *The K-qi sections $\Sigma_{x_i} \subset L = L_K(\alpha)$ through the points x_i for $i = 0, \dots, n$, are such that each pair $(\Sigma_{x_i}, \Sigma_{x_{i+1}})$, $i = 0, \dots, n-1$, bounds a (K, D, E)-subladder $\mathfrak{L}^i = \mathfrak{L}_{K,D,E}(\alpha_i) \subset \mathfrak{L}$. These subladders satisfy*

$$L^i \cap L^{i+1} = \Sigma_{x_i}.$$

(2) *Each ladder $\mathfrak{L}^i, i = 0, \dots, n-1$, contains a (K, C)-narrow carpet $\mathfrak{A}^i = \mathfrak{A}_K(\alpha'_i)$, where $\alpha'_i \subset \alpha_i$ contains x_i and*

$$0 \le l_i - \text{length}(\alpha'_i) \le M_K \le M_{\bar{K}},$$

if $i < n-1$, while $\alpha'_n = \alpha_n$.

(3) $N^{fib}_{C/2}(\mathcal{L}^i) \cap N^{fib}_{C/2}(\mathcal{L}^j) = \emptyset$ *provided that $|i - j| > 1$.*

(4) *In each ladder $\mathfrak{L}^i$, the pair of sections $(\Sigma_{x_i}, \Sigma_{x_{i+1}})$ is $B = B_{4.14}(K, C)$-cobounded, for $i = 0, \cdots, n-2$. Moreover, the projection of $\Sigma_{x_{i+1}}$ to Σ_{x_i} is uniformly close to the point x_{w_i}, where*

$$[x_{w_i} y_{w_i}]_{X_{w_i}} = A_{w_i}$$

is the narrow end of the carpet $\mathfrak{A}^i$.

Proof. We will orient the interval α from x_u to y_u and, will use the corresponding natural order on α.

Suppose $x_j \in \alpha$ have been chosen and $x_i < y$. To choose x_{i+1} consider the subset

$$\Omega_{i+1} := \{x \in \alpha, x > x_i : d_{X_b}(\Sigma_x \cap \mathcal{L}_b, \Sigma_{x_i} \cap \mathcal{L}_b) > C, \ \forall b \in V(\pi(\Sigma_x) \cap \pi(\Sigma_{x_i}))\}.$$

In other words, Ω_{i+1} consists of points $x > x_i$ in α such that the sections Σ_x and Σ_{x_i} are fiberwise C-separated.

Now there are two possibilities. If $\Omega_{i+1} = \emptyset$ then define $n = i+1$, $x_{i+1} = y_u$. This will conclude the induction.

Otherwise, pick $x_{i+1} \in \Omega_{i+1}$ such that

$$x_{i+1} - \inf(\Omega_{i+1}) < M_K/2.$$

According to Corollary 3.12, $\Sigma_{x_i}, \Sigma_{x_{i+1}}$ bound a (K, D, E)-subladder $\mathfrak{L}^i := \mathfrak{L}_{K,D,E}(\alpha_i) \subset \mathfrak{L}$, where $\alpha_i = [x_i x_{i+1}]_{X_u} \subset \alpha$. By the construction, $L^i \cap L^{i-1} = \Sigma_{x_i}$.

In order to construct a (K, C)-narrow carpet $\mathfrak{A}^i = \mathfrak{A}_K(\alpha'_i) \subset \mathfrak{L}^i$, we take a subsegment $\alpha'_i = [x_i x'_{i+1}]_{X_u} \subset \alpha_i = [x_i x_{i+1}]_{X_u}$ such that

$$x'_{i+1} \notin \Omega_{i+1}, d_{X_u}(x'_{i+1}, x_{i+1}) < M_K.$$

Since $x'_{i+1} \notin \Omega_{i+1}$, by the definition of Ω_{i+1}, the sections $\Sigma = \Sigma_{x_i}, \Sigma' = \Sigma_{x'_{i+1}}$ contain points $y, y' \in X_w$ (for some $w \in V(S)$), such that

$$d_{X_w}(y, y') \le C.$$

Among such vertices $v \in T$ we choose one which is closest to u (there might be several).

The K-sections Σ, Σ' contain subsections $\gamma_y, \gamma_{y'}$ (over the interval $[\![u, v]\!]$) connecting y, y' to x_i, x'_{i+1} respectively.

Since $M_K \le C$ and w was chosen to be closest to u, we have that for every vertex $t \in [\![u, v[\![$,

$$d_{X_t}(\gamma_y(t), \gamma_{y'}(t)) > C \ge M_K.$$

Thus, the sections γ_y, γ'_y form the top and the bottom of a C-narrow K-carpet $\mathfrak{A}_{K,C}(\alpha'_i)$ in $\mathfrak{L}^i$ with the narrow end $\beta_i = L_w \cap \mathfrak{A}^i$.

This proves Parts (1) and (2) of the proposition.

We next prove Part (3). Suppose that $i + 1 < j$. By the construction, for each vertex $v \in \pi(\mathcal{L}^i) \cap \pi(\mathcal{L}^j)$ the length of the subsegment in $\mathcal{L}_v$ between $\Sigma_{x_{i+1}}, \Sigma_{x_j}$ is $> C$. Hence, the minimal fiberwise distance between $\mathcal{L}^i \cap X_v, \mathcal{L}^j \cap X_v$ is $> C$ as well. Part 3 follows since $\mathcal{L}_v$ is a geodesic segment in X_v.

Finally to prove (4) we will use the description of the paths $c(x, y)$ constructed for the proof Theorem 4.8 which are uniform quasigeodesics in the subladder $L^i = L_K(\alpha_i)$.

First of all, we observe that, the difference in the lengths of α_i', α_i is at most M_K and α_i' bounds a (K, C)-narrow carpet $\mathfrak{A}^i$ in $\mathfrak{L}^i$. In other words, $\mathfrak{L}^i$ is a carpeted ladder and Theorem 4.8 applies in our case.

The carpet $\mathfrak{A}^i = (\pi : A^i \to [\![u, w_i]\!])$ is bounded by horizontal paths $\gamma_i \subset \Sigma_{x_i}, \gamma_i' \subset \Sigma_{x_{i+1}'}$ and the vertical paths α_i', β_i (where β_i has length $\leq C$).

Consider points $z_i \in \Sigma_{x_i}$ and $z_{i+1} \in \Sigma_{x_{i+1}}$. Since the fiberwise vertical separation between Σ_{x_i} and $\Sigma_{x_{i+1}}$ is $> C \geq M_{\bar{K}}$, we conclude that the path $c(z_i, z_{i+1})$ has to be of *type 2* in the terminology of the proof of Proposition 4.8.

In other words, $c(z_i, z_{i+1})$ is the concatenation of five (actually, four) subpaths. Two of these subpaths (containing z_i, z_{i+1} respectively, connect z_i to $\tilde{z}_i$, z_{i+1} to $\tilde{z}_{i+1}$ and are contained in, respectively, γ_i, γ_{i+1}. The point $\tilde{z}_{i+1}$ is within vertical distance $\leq M_{\bar{K}}$ from a point $\bar{z}_{i+1} \in top(\mathfrak{A}^i)$, while the point $\tilde{z}_i$ actually equals $\bar{z}_i$. The vertical geodesic

$$[\bar{z}_{i+1}\tilde{z}_{i+1}]_{L_{v_{i+1}}}, v_i = \pi(\tilde{z}_{i+1}),$$

is contained in $c(z_i, z_{i+1})$. The path $c(z_i, z_{i+1})$ contains one more subpath, namely,

$$c_{\mathfrak{A}^i}(\bar{z}_i, \bar{z}_{i+1}),$$

a path in the combing of A^i constructed in the proof of Proposition 4.1. By the construction of $\mathfrak{A}^i$, each vertex-space A_v of $\mathfrak{A}^i$ has length $\geq C \geq M_{\bar{K}} > M_K$. Thus, by the definition of $c_{\mathfrak{A}^i}(\bar{z}_i, \bar{z}_{i+1})$ in the proof of Proposition 4.1 this path is a concatenation of subpaths contained in the top and the bottom of $\mathfrak{A}^i$ and, *most importantly*, the narrow end $\beta_i = [x_{w_i} y_{w_i}]_{L_{w_i}}$.

In particular, each uniform quasigeodesic path $c(z_i, z_{i+1})$ contains the point $x_i^- := x_{w_i} \in A^i_{w_i} \cap \Sigma_{x_i}$.

According to Lemma 1.101, it follows that the nearest-point projections (taken in L^i) of $\Sigma_{x_{i+1}}$ to Σ_{x_i} is contained in the R-neighborhoods of x_{w_i}, where $R = R(K, C)$. Hence, by Corollary 1.143, the pair of sections Σ_i, Σ_{i+1} is $B(K, C)$-cobounded in L^i. This concludes the proof of the proposition. □

In Section 7.1 (when describing uniform quasigeodesics in ladders) we will need a bit more detailed information about the uniformly cobounded pair $(\Sigma_{x_i}, \Sigma_{x_{i+1}})$. Let $x_i^- \in \Sigma_{x_i}, x_i^+ \in \Sigma_{x_{i+1}}$ be a pair of points realizing the minimal distance between these subsets in the ladder L^i. According to Part (4) of the proposition, x_i^- is uniformly close (in terms of K and C) to the point x_{w_i}. We will also need to identify the other point, x_i^+, up to a uniformly bounded error.

Let $u_{i+1} \in [\![u, w_{i+1}]\!]$ benote the maximum (in the oriented interval $[\![u, w_{i+1}]\!]$) of the subset

$$\{t \in V([\![u, w_{i+1}]\!]) : d_{X_t}(\Sigma_{x_{i+1}} \cap X_t), A_t) \leq M_{\bar{K}}\}.$$

For each vertex $v \in [\![u, u_{i+1}]\!]$ we observe that the top-most point y_v' of the segment A_v divides L_v^i in two subsegments:

$$A_v \cup A_v', \quad A_v' = [y_v' y_v]_{X_v},$$

where

$$y_v = top(L_v^i).$$

The first part of the next lemma will be used in Section 7.1, while second part will be used in Section 8.3 (proof of Lemma 8.30).

LEMMA 4.15.

(1)
$$d(y_{u_{i+1}}, x_i^+) \le D_{4.15}(K, C).$$

(2) *For each v as above, the length of* A'_v *is* $\le R_{4.15}(K)$.

PROOF.

(1) Take a point $y \in \Sigma_{x_{i+1}} \cap X_v$. Then, as we noted in the proof of the last part of the proposition, since each path $c(x, y)$ connecting (any) $x \in \Sigma_{x_i}$ to y has type 2, it has to pass through a point of the segment $A'_{u_{i+1}}$, and the latter has length $\le M_{\bar{K}}$. Thus, $c(x, y)$ passes within distance $M_{\bar{K}}$ from $y_{u_{i+1}}$. It follows that $d(y_{u_{i+1}}, x_i^+)$ is uniformly bounded in terms of K and C.

(2) This part is an application of uniform flaring. We have two K-qi sections γ_0, γ_1 over the interval $J = [\![u, u_{i+1}]\!]$, defined by restricting $top(\mathfrak{A}^i)$ and Σ_{i+1}. The vertical separation between these sections at the endpoints of J is $\le M_{\bar{K}}$. Thus, by Corollary 2.53, the vertical separation between these sections elsewhere in the interval is $\le \tau_{2.53}(K, M_{\bar{K}})$. Hence, we can take
$$R_{4.15}(K) = \tau_{2.53}(K, M_{\bar{K}}).$$
□

THEOREM 4.16. *For each K and a hyperbolic tree spaces* $\mathfrak{X}$ *satisfying the uniform* $\kappa = \kappa_{4.7}(K)$*-flaring assumption as in Proposition* 4.8*, and arbitrary D and E, there exists* $\delta = \delta_{4.16}(K)$ *such that every* (K, D, E)*-ladder* $\mathfrak{L} = \mathfrak{L}_K(\alpha) \subset \mathfrak{X}$ *has* δ*-hyperbolic total space L with respect to the intrinsic metric of the ladder.*

PROOF. Set $C = M_{\bar{K}}$. By Proposition 4.14 we have a subdivision $x_0 = x_u, x_1, \ldots, x_n = y_u$ of the segment $\alpha = [x_u y_u]_{X_u}$. The K-qi sections $\Sigma_i := \Sigma_{x_i}$ in $\mathfrak{L}$ passing through x_i decompose the ladder $\mathfrak{L}$ into subladders $\mathfrak{L}^i := \mathfrak{L}_K([x_{i-1}x_i]_{X_u})$ containing (K, C)-narrow carpets $\mathfrak{A}^i$. Hence, by Proposition 4.8, each L^i (the total space of $\mathfrak{L}^i$) is $\delta_{4.8}(K, C)$-hyperbolic. By the construction,
$$L^i \cap L^{i+1} = \Sigma_i, i = 1, \ldots, n-1.$$
The subsets Σ_i are K-sections, hence, are $\lambda_{1.92}(\delta_{4.8}(K, C), K)$-quasiconvex in L^i, L^{i+1}. Furthermore, by Proposition 4.14, each pair of ladders $\mathfrak{L}^{i-1}, \mathfrak{L}^{i+1}$, $i \ge 1$, is $B_{4.14}(K, C)$-cobounded. Thus, arbitrary ladder $L = L_K(\alpha)$ is uniformly hyperbolic by Theorem 2.65.
□

CHAPTER 5

Hyperbolicity of flow-spaces

In this chapter we shall prove that the k-flow-spaces (for k in a suitable range) of each vertex-space $X_u \subset X$, are uniformly hyperbolic (with hyperbolicity constant depending on k) provided that $\mathfrak{X}$ is a hyperbolic tree of spaces satisfying a certain uniform flaring condition. The strategy of the proof is to show that:

(a) Every pair of points x, y in $Fl_k(X_u)$ belong to a common (K, D, E)-ladder $L_{x,y}$ (essentially contained in $Fl_k(X_u)$), where D is a sufficiently large number, K depends only on k, and E depends on k and D. This is done in Section 5.1. Actually, in Section 5.1 we prove a stronger result, the existence of *tripods of ladders* connecting points $x_1, x_2, x_3 \in Fl_k(X_u)$ such that in each vertex-space of $\mathfrak{Fl}_k(X_u)$ the three geodesics of these ladders form a geodesic tripod. Hyperbolicity of total spaces of such tripods of ladders is almost immediate; see Section 5.3.

(b) Since ladders are uniformly hyperbolic (as it was proven in the previous chapter), this appears to yield a preferred family of paths $c(x, y)$ connecting points of $Fl_k(X_u)$; these paths are obtained (from uniform quasigeodesics in ladders) by applying vertex-space-wise projections from the corresponding ladders to $Fl_k(X_u)$. Hyperbolicity of tripods of ladders should then yield the uniform slimness condition for the family of paths $c(x, y)$. The trouble, however, is that $L_{x,y}$ is far from canonical, and, thus, it is far from clear why the paths $c(x, y)$ satisfy the fellow-traveling condition. If different ladders $L^1_{x,y}, L^2_{x,y}$ were at uniformly bounded minimal distance from each other in each fiber-space where both ladders are nonempty, one could use hyperbolicity of the union of these two ladders. Unfortunately, it is unclear why there should be a uniform bound on such minimal distance. To resolve the problem, we use the construction of a *coarse projection* of the ladder $L^1_{x,y}$ to $L^2_{x,y}$ defined in Section 5.2. This projection is used in Proposition 5.18 to construct a uniformly hyperbolic subspace Z in X containing the two ladders. The coarse independence of the paths $c(x, y)$ on the choice of $L_{x,y}$ is then almost immediate; see Corollary 5.19.

This, in turn, will conclude the verification of hyperbolicity of flow-spaces, Theorem 5.17.

5.1. Ubiquity of ladders in $Fl_k(X_u)$

In this section we prove that for all (sufficiently large) D and k, and all $x, y \in Fl_k(X_u)$, there is a (K, D, E)-ladder $\mathfrak{L}$ containing x, y, where $K = K(k), E = 3k + D$. Furthermore, $\mathfrak{L}$ will be contained in the fiberwise $4\delta_0$-neighborhood of $\mathfrak{Fl}_K(X_u)$. We will actually prove a stronger result, about the existence of a *tripod* of ladders containing given three points in $Fl_K(X_u)$.

In Definition 5.1 and in what follows, i is taken modulo 3.

Definition 5.1. A (K, D, E)-*tripod of ladders* in $\mathfrak{X}$ is a semicontinuous (K, D, E)-family $\mathfrak{Y}$ over a subtree $S \subset T$, which is a union of three (K, D, E)-ladders $\mathfrak{L}^i = (L^i \to S_i)$, $i = 1, 2, 3$, such that:

(1) There exists a K-section Ξ defined over the subtree

$$S_{123} := S_1 \cap S_2 \cap S_3$$

and called the *center-section* of $\mathfrak{Y}$, such that for each $v \in V(S_{123})$ and $i \in \{1, 2, 3\}$, $\Xi(v) = top(\mathcal{L}^i) \cap X_v$.

(2) For each $v \in V(S_{123})$, $e \in E(S_{123})$, the vertex- and edge-space Y_v, Y_e of $\mathfrak{Y}$ is a δ_0-tripod in X_v, X_e

$$Y_v = \bigcup_{i=1}^{3}[x^i_v z_v]_{X_v}, Y_e = \bigcup_{i=1}^{3}[x^i_e z_e]_{X_e}, z_v = \Xi(v), L^i_v = [x^i_v z_v]_{X_v}, L^i_e = [x^i_e z_e]_{X_e}.$$

(3) If $v \in V(S_i) \setminus V(S_{123})$, then $L^{i+1}_v = L^{i-1}_v = \emptyset$ (two legs of the tripod are missing) and $L^i_v = L_v = [x_v y_v]_{X_v}$ (we, thus, omit the superscript i in this situation). The same applies to the edges.

We will refer to the union of bottoms of the ladders $\mathfrak{L}^i$ as the *bottom* (denoted $bot(\mathfrak{Y})$) of the tripod of ladders $\mathfrak{Y}$.

A tripod of ladders $\mathfrak{Y}$ is said to be *degenerate* if for some $i \in \{1, 2, 3\}$, $\mathfrak{L}^{i+1} = \mathfrak{L}^{i-1} = \Xi$ and, thus, $\mathfrak{Y}$ is reduced to a single ladder, $\mathfrak{L}^i$.

In the following proposition, $D_0 = D_{1.142}(\delta'_0, \lambda'_0)$,

$$C = 2(\lambda'_0 + 2\delta'_0 + D_{1.54}(\delta'_0, L'_0)) + C_{1.110}(\delta'_0, \lambda'_0),$$

$$D_1 := \max(D_0, C_{1.143}(\lambda'_0, \delta'_0, C)),$$

$$D = D_{5.2} := D_1 + \max(3\delta_0, 2\delta_0 + 2(\lambda'_0 + 2\delta'_0)). \tag{5.1}$$

Assume also that $k = r^\wedge = (15L'_0 r)^3 \geq K_0$ and r satisfies the inequality

$$r \geq r_1 = \max(2\lambda'_0 + 5\delta'_0, \lambda + 4\lambda'_0 + 8\delta'_0 + 5\delta_0),$$

where

$$\lambda = C_{1.128}(\delta'_0, \lambda'_0, L'_0).$$

In other words,

$$k \geq k_{5.2} = K_1 := \max(K_0, (15L'_0 r_1)^3). \tag{5.2}$$

Note that, in particular, $k \geq \lambda + 4\lambda'_0 + 8\delta'_0 + 5\delta_0$.

In the proposition we will be also using the function $\kappa' = K'_{3.17}(\kappa)$ defined in Lemma 3.17.

Proposition 5.2 (Existence of tripods of ladders). *Let $\mathfrak{X} = (\pi : X \to T)$ be a tree of spaces satisfying Axiom* **H1**. *Then for k and D as above, there exist constants*

$$K = K_{5.2}(k), E = E_{5.2}(k)$$

such that the following holds.

For points $x^i, i = 1, 2, 3$ in $\mathcal{F}l_k(X_u)$, we let $\gamma^i := \gamma_{x^i}$ denote k-sections in $Fl_k(X_u)$ over $[\![u, \pi(x^i)]\!]$, connecting x^i to X_u.

Then there exists a (K, D, E)-tripod of ladders $\mathfrak{Y} = \mathfrak{L}^1 \cup \mathfrak{L}^2 \cup \mathfrak{L}^3$, centered at u such that:

(1) *$\mathcal{Y} \subset N^{fib}_{5\delta_0}\mathcal{F}l_k(X_u)$, while $bot(\mathcal{Y}) \subset \mathcal{F}l_k(X_u)$.*

(2) *$\gamma^i \subset bot(\mathfrak{L}^i)$, $i = 1, 2, 3$, thus, $\gamma^i \subset bot(\mathfrak{Y})$.*

(3) *If, for some i, $\gamma^{i-1} = \gamma^{i+1}$, then the tripod of ladders $\mathfrak{Y}$ is degenerate and the section $\gamma^{i-1} = \gamma^{i+1}$ is contained in the center-section Ξ of $\mathfrak{Y}$.*
(4) *There exist (K, D, E)-ladders L^{ij} containing x^i, x^j, such that $top(L^{ij}) \subset bot(L^j)$, $bot(L^{ij}) \subset bot(L^j)$, and L^{ij} is contained in δ_0-fiberwise neighborhood of $\mathfrak{L}^i \cup \mathfrak{L}^j$.*

PROOF. We first note that, according to Lemma 3.28, given $x \in Fl_k(X_u)$, there exist a maximal K-section $\Sigma_x \subset Fl_k(X_u)$ through x, intersecting X_u and containing γ_x. Thus, we define the maximal k-sections $\Sigma_{x^i}, i = 1, 2, 3$ through points x^i, intersecting X_u, and containing γ^i (if it is given), otherwise, chosen arbitrarily. (Note that these sections Σ_{x^i} need not be disjoint and, in general, they have different domains in T.) In line with Part 3 of the proposition, if $\gamma^{i-1} = \gamma^{i+1}$, we require $\Sigma_{x^{i-1}} = \Sigma_{x^{i+1}}$. We define tripods $Y_v, v \in V(T)$ inductively, by induction on the distance from v to u.

As the base of induction, we define x^i_u as $\gamma^i(u)$ and L^i_u as a geodesic segment $[x^i_u z_u]_{X_u}$, where z_u is a δ_0-center of the geodesic triangle $\Delta_u = \Delta x^1_u x^2_u x^3_u$ in X_u.

We then proceed inductively as in the proof of Lemma 3.17 to which we refer the reader for the notation used below. Namely, assume that segments L^i_v are defined for vertices of the subtree $B_n \subset T$,

$$L^i_v = [x^i_v z_v]_{X_v}, v \in V(B_n), Y_v = L^1_v \cup L^2_v \cup L^3_v,$$

where z_v is a δ_0-center of the geodesic triangle $\Delta_v = \Delta x^1_v x^2_v x^3_v$. (In order to simplify the notation, we allow for empty vertex and edge-spaces in ladders.) We let d_{Y_v} denote the intrinsic path-metric on Y_v. Then the inclusion map $(Y_v, d_{Y_v}) \to X_{vw}$ is an $L'_0(2\delta_0)$-qi embedding for each edge $[v, w] \in E(T), v \in B_n$, directed away from u. We will be also assuming (inductively) that the extremities x^i_v of Y_v belong to $Q_v = Fl_k(X_u) \cap X_v$ and, thus, Y_v is contained in the $5\delta_0$-neighborhood of Q_v taken in X_v.

Next, we apply the modified projection $\bar{P}_{X_{vw},Y_v}$ to X_w. The image $\bar{Y}_v$ is the closed convex hull of $P_{X_{vw},Y_v}(X_w)$ in the tripod Y_v (with respect to the path-metric of Y_v). Specifically, if $\bar{Y}_v \neq \emptyset$, then it is the convex hull (taken in Y_v) of three points $\bar{x}^i_v, i = 1, 2, 3$, such that $\bar{x}^i_v$ is the nearest-point projection (taken in (Y_v, d_{Y_v})) of x^i_v to $\bar{Y}_v$.

Note that it is entirely possible for the center z_v not to belong to $\bar{Y}_v$, in which case $\bar{Y}_v$ is a segment contained in the relative interior of one of the legs of Y_v. It even can happen that $\bar{Y}_v$ is empty, if $Q_w = \emptyset$. Our next task is to analyze implications of the containment $\bar{Y}_v \subset L^i_v$.

LEMMA 5.3. *Suppose that $\bar{Y}_v$ is contained in L^i_v, $d_T(u, v) = n$ and $e = [v, w]$ is an edge of T oriented away from u. Then for $j = i \pm 1$ we have*

$$P_{X_{vw},L^j_v}(X_w) \subset B^e(z_v, C),$$

where $B^e(z_v, C)$ is the ball with respect to the metric of X_{vw} and $C = C_{5.3}$.

PROOF. According to Corollary 1.110, for the λ'_0-quasiconvex subsets $V = L^j_v, U = Y_v$ in X_{vw} we have

$$d_{X_{vw}}(P_{X_{vw},V}, P_{U,V} \circ P_{X_{vw},U}) \leq C_{1.110}(\delta'_0, \lambda'_0).$$

Here all the projections are taken with respect to the restriction of the metric on X_{vw}. Thus, we need to prove that $P_{U,V}(L^i_v)$ is uniformly close to the point z_v. This is obviously true for the intrinsic nearest-point projection $P'_{U,V}(L^i_v)$ (taken with respect to the metric d_{Y_v}), since $P'_{U,V}$ sends L^i_v to $\{z_v\}$. Therefore, we need to compare the intrinsic projection $Y_v \to L^j_v$ and the extrinsic projection, with respect to the metric of X_{vw}.

Take some $x \in L^i_v$ and let $\bar{x} = P_{X_{vw},V}(x)$. According to Lemma 1.105, the geodesic $\alpha^* = [x z_v]_{X_{vw}}$ passes within distance $\lambda'_0 + 2\delta'_0$ from the point $\bar{x}$. Since the segment $\alpha =$

$[xz_v]_{X_v}$ is an L'_0-quasigeodesic in X_{vw}, it follows that

$$\mathrm{Hd}_{X_{vw}}(\alpha, \alpha^*) \le D_{1.54}(\delta'_0, L'_0).$$

Hence, α^* contains a point y satisfying

$$d_{X_{vw}}(y, z_v) \le \lambda'_0 + 2\delta'_0 + D_{1.54}(\delta'_0, L'_0).$$

Since

$$d_{X_{vw}}(x, \bar{x}) \le d_{X_{vw}}(x, z_v),$$

we get

$$d_{X_{vw}}(\bar{x}, z_v) \le 2(\lambda'_0 + 2\delta'_0 + D_{1.54}(\delta'_0, L'_0)).$$

Combining this estimate with Corollary 1.110, and the hypothesis that $P_{X_{vw},U}(X_w) \subset L^i_v$, we conclude that for each point $q \in X_w$, $P_{X_{vw},U}(q) = x \in L^i_v$ and

$$\begin{aligned} d_{X_{vw}}(P_{X_{vw},L^j_v}(q), z_v) &\le C_{5.3} \\ &:= 2(\lambda'_0 + 2\delta'_0 + D_{1.54}(\delta'_0, L'_0)) + C_{1.110}(\delta'_0, \lambda'_0). \end{aligned}$$

□

Combining the lemma with Corollary 1.143, yields Corollary 5.4.

COROLLARY 5.4. *If $\bar{Y}_v$ is contained in L^i_v, then for $j = i \pm 1$, the pair (L^j_v, X_w) is $C_{1.143}(\lambda'_0, \delta'_0, C)$-cobounded in X_{vw}, where $C = C_{5.3}$.*

REMARK 5.5. The assumption $D \ge D_1$ made in the proposition ensures that $D \ge C_{1.143}(\lambda'_0, \delta'_0, C)$. Thus, $\bar{Y}_v \subset L^i_v$ implies that the pair (L^j_v, X_w) is D_1-cobounded in X_{vw}, hence, D-cobounded.

We now return to the construction of a family of tripods. Let $e = [v, w]$ be an edge directed away from u, $v \in B_n \subset T$, $w \notin B_n$. There are several things which can now happen, primarily depending on the coboundedness of Y_v and X_w, but also on intersections of the sections Σ_{x^i} with X_w.

CASE 1. Suppose that the tripod Y_v and X_w are D_1-cobounded (in X_{vw}) and all three sections Σ_{x^i} are disjoint from X_w. Then we set $Y_w = \emptyset$.

REMARK 5.6. Observe that if the pair (Y_v, X_w) is D_1-cobounded, so are the pairs (L^i_v, X_w), $i = 1, 2, 3$.

CASE 2. Suppose that the tripod Y_v and X_w are not D_1-cobounded. We will also assume that the tripod Y_v has "all its legs," i.e., $L^i_v \neq \emptyset$, $i = 1, 2, 3$. According to Lemma 1.128,

$$\bar{Y}_v \subset N_\lambda(P_{X_{vw},Y_v}(X_w)), \tag{5.3}$$

where, as before,

$$\lambda = C_{1.128}(\delta'_0, \lambda'_0, L'_0).$$

We now use the fact that $D_1 \ge D_0 = D_{1.142}(\delta'_0, \lambda'_0)$. Since Y_v and X_w are not D_1-cobounded, by Corollary 1.146,

$$P_{X_{vw},Y_v}(X_w) \subset N^e_{4\lambda'_0+8\delta'_0}(X_w) \cap Y_v,$$

hence,

$$\bar{Y}_v \subset N^e_{\lambda+4\lambda'_0+8\delta'_0}(X_w) \cap Y_v.$$

However, $Y_v \subset N^{fib}_{5\delta_0}(Q_v)$, by the inductive hypothesis. Hence, each point $x \in X_w$ within distance (in X_{vw})

$$\lambda + 4\lambda'_0 + 8\delta'_0 + 5\delta_0$$

from $\bar{Y}_v$, belongs to $Q'_w = N^e_r(Q_v) \cap X_w$, $r = k^\vee$ (see the definition of flow-spaces in Section 3.3). Since, by the assumption of the proposition,

$$k^\vee > \lambda + 4\lambda'_0 + 8\delta'_0 + 5\delta_0,$$

we see that

$$\bar{Y}_v \subset N^e_{\lambda+4\lambda'_0+8\delta'_0}(Q_w) \cap Y_v.$$

Since the extremities $\bar{x}^i_v$, $i = 1, 2, 3$, of the (possibly degenerate) tripod $\bar{Y}_v$ are at the distance $4\lambda'_0 + 8\delta'_0 + 5\delta_0$ from Q_w, we take points $\tilde{x}^i_w \in Q_w$ which are nearest-point projections of $\bar{x}^i_v$, $i = 1, 2, 3$, and, thus,

$$d_{X_{vw}}(\bar{x}^i_v, \tilde{x}^i_w) \le 4\lambda'_0 + 8\delta'_0 + 5\delta_0. \tag{5.4}$$

(If $\bar{x}^i_v = \bar{x}^j_v$ then $\tilde{x}^i_w = \tilde{x}^j_w$.) Similarly, if z_v belongs to $\bar{Y}_v$ then there exists $\tilde{z}_w \in Q_w$ satisfying

$$d_{X_{vw}}(z_v, \tilde{z}_w) \le \lambda + 4\lambda'_0 + 8\delta'_0 + 5\delta_0.$$

In this case, we define the tripod $\tilde{Y}_w \subset X_w$ centered at $\tilde{z}_w$ and with the legs $[\tilde{z}_w \tilde{x}^i_w]_{X_w}$.

The actual tripod Y_w, as we will see, is uniformly Hausdorff-close to $\tilde{Y}_w$. For now, we observe that, according to Lemma 1.55:

$$\mathrm{Hd}_{X_{vw}}(\tilde{Y}_w, \bar{Y}_v) \le D_{1.55}(\delta'_0, L'_0, 4\lambda'_0 + 8\delta'_0 + 5\delta_0). \tag{5.5}$$

Depending on the intersections $\Sigma_{x^i} \cap X_w$, the points $\tilde{x}^i_w$ might be the vertices of the tripod Y_w. Specifically, there are four subcases:

(a) If for some i, $\Sigma_{x^i} \cap X_w = \{x^i_w\}$, then we use the point x^i_w as one of the vertices of Δ_w. Thus,

$$d_{X_{vw}}(x^i_v, x^i_w) \le k$$

in this subcase.

(b) If for some i, $\bar{Y}_v$ is disjoint from (necessarily both) $L^{i\pm1}_v$ and $\Sigma_{x^{i+1}} \cap X_w = \emptyset$, $\Sigma_{x^{i-1}} \cap X_w = \emptyset$, we set $L^{i\pm1}_w = \emptyset$. Thus, in this subcase the tripod Y_w will be missing two legs. This degenerate tripod will be equal the oriented geodesic segment $L^i_w = L_w = [x_w y_w]_{X_w}$, where $x_w = \tilde{x}^{i\pm1}_w$ and y_w will be either $\tilde{x}^i_w$ (if $\Sigma_{x^i} \cap X_w = \emptyset$) or, as in subcase (a), $\Sigma_{x^i} \cap X_w = \{y_w\}$. In this situation, by the construction, for

$$\hat{x}_v := \bar{x}^{i\pm1}_v$$

$$d_{X_{vw}}(x_w, \hat{x}_v) \le 4\lambda'_0 + 8\delta'_0 + 5\delta_0,$$

while

$$d_{X_{vw}}(y_w, \hat{y}_v) \le \max(k, 4\lambda'_0 + 8\delta'_0 + 5\delta_0) = k.$$

Here $\hat{y}_v = \bar{x}^i_v$ (if $\Sigma_{x^i} \cap X_w = \emptyset$) or $\hat{y}_v = x^i_v$ (otherwise).

REMARK 5.7. In this subcase, due to our assumptions on D_1, both pairs $(L^{i\pm1}_v, X_w)$ will be D_1-cobounded; see Remark 5.5.

(c) If for some i, $\bar{Y}_v$ is disjoint from (necessarily both) $L^{i\pm1}_v$ and for exactly one element $j \in \{i \pm 1\}$, the section Σ_{x^j} intersects X_w, then we discard the point $\bar{x}^{i\pm1}_v$ and let $z_w = x^{i-1}_w = x^{i+1}_w$ be that point of intersection. We let x^i_w either be equal to the intersection point of Σ_{x^i} and X_w (if the intersection is nonempty) or equal to $\tilde{x}^i_w$. Thus,

$$d_{X_{vw}}(z_w, \bar{x}^j_v) \le k,$$

while either

$$d_{X_{vw}}(x^i_w, \hat{y}_v) \le \max(4\lambda'_0 + 8\delta'_0 + 5\delta_0, k) = k,$$

where, as in the subcase (b), $\hat{y}_v = \bar{x}^i_v$, or $\hat{y}_v = x^i_v$.

(d) In the "generic" case (i.e., when $z_v \in \bar{Y}_v$), for each i such that $\Sigma_{x^i} \cap X_w = \emptyset$, we set $x^i_w = \tilde{x}^i_w$. (Of course, if $\Sigma_{x^i} \cap X_w$ is nonempty, we use this intersection point as x^i_w; see subcase (a).) As above, we obtain:

$$d_{X_{vw}}(x^i_w, \{\bar{x}^i_v, x^i_v\}) \leq \max(4\lambda'_0 + 8\delta'_0 + 5\delta_0, k) = k, i = 1, 2, 3.$$

Except for subcase (b), we, thus, obtain three points x^1_w, x^2_w, x^3_w spanning a (possibly degenerate) geodesic triangle $\Delta_w = \Delta x^1_w x^2_w x^3_w \subset X_w$. We let z_w be a δ_0-center of this triangle. (Subcase (c) above does not cause trouble because the triangle Δ_w is degenerate and one of its sides equals the point z_w, which is, therefore, the center of Δ_w.) Accordingly, we define geodesic segments

$$L^i_w := [x^i_w z^i_w]_{X_w}$$

and the tripod $Y_w = L^1_w \cup L^2_w \cup L^3_w$.

Subcase (b) requires a separate discussion since the tripod Y_w is missing two out of its three legs. In this situation, by the definition of the point x_w,

$$d_{X_{vw}}(x_w, \bar{x}^{i\pm1}_v) \leq \lambda + 4\lambda'_0 + 8\delta'_0 + 5\delta_0 \leq k,$$

by the assumption on k made in the proposition.

CASE 3. We still assume that Y_v and Q_w are not D_1-cobounded, but consider the case that Y_v is degenerate and has only one leg, $L^i_v = L_v = [x_v y_v]_{X_v}$: The other two legs are empty. We treat this case exactly the same way as we treated the subcases (2b) and (2c) above: The tripod $\bar{Y}_v$ has empty intersection with the empty legs $L^{i\pm1}_v$ of Y_v. The points $x_w, y_w \in X_w$ define the oriented segment $L_w = [x_w y_w]_{X_w}$ and the points x_w, y_w are within distance k, respectively, from points $\hat{x}_v, \hat{y}_v$, where (as in subcase (2b)) $\hat{x}_v \in \{\bar{x}_v, x_v\}$, $\hat{y}_v \in \{\bar{y}_v, y_v\}$,

$$\bar{Y}_v = [\bar{x}_v \bar{y}_v]_{X_v} \subset L_v$$

By the definition, the points $\hat{x}_v, \hat{y}_v$ satisfy

$$x_v \leq \hat{x}_v \leq \hat{y}_v \leq y_v$$

in the oriented segment $[x_v y_v]_{X_v}$. (Compare Lemma 3.17(a3).)

This concludes the construction of the segments L^i_w. We just note that since $Q_w \subset X_w$ is $4\delta_0$-quasiconvex and $x^i_w \in Q_w$ for all i, we get:

$$Y_w \subset N^{fib}_{5\delta_0}(Q_w).$$

These are the inductive assumptions we made earlier. We set

$$\mathcal{L}^i := \bigcap_{v \in V(T)} L^i_v, i = 1, 2, 3.$$

We define a subtree of spaces $\mathfrak{Y} \subset \mathfrak{X}$ using the tripods Y_v, Y_e as, respectively, vertex and edge-sets. The incidence maps $Y_e \to Y_v$ will be compositions of restrictions of incidence maps of $\mathfrak{X}$ with nearest-point projections $X_v \to Y_v$.

We, are done with the induction but it remains to verify that each $\mathcal{L}^i$ satisfies the properties required by Lemma 3.17: This lemma is used to promote the unions of geodesics segments in vertex-spaces of X to the union of vertex-spaces of a ladder. We also have to show that $\mathfrak{Y}$ is a (κ_1, D_1, E_1)-semicontinuous family in $\mathfrak{X}$, as required by the definition of a tripod of ladders, for suitable constants κ_1, E_1.

As we observed in the discussion of subcases, points x^i_w satisfy

$$d_{X_{vw}}(\hat{x}^i_v, x^i_w) \leq k, \tag{5.6}$$

where $\hat{x}_v^i \in \{x_v^i, \bar{x}_v^i\}$.

We next turn our attention to the center z_w of Y_w. Except for the generic subcase (d) above, the tripod Y_w is degenerate and z_w is one of its extremities, i.e., equals to one of the points $\bar{x}_w^j$. Hence, apart from the generic case, as we observed while discussing nongeneric cases,

$$d_{X_{vw}}(z_w, \bar{x}_v^j) \leq k. \tag{5.7}$$

Note that the point z_w is within uniformly bounded (in terms of k) distance from z_v in subcase (2c) but might be quite far from z_v in the subcase (b). The next lemma allows us to control the position of z_w in the generic subcase (d).

LEMMA 5.8. *Suppose that we are in the generic subcase (d). Then*

$$d_{X_{vw}}(z_v, z_w) \leq C_{5.8}(k).$$

PROOF. We define the geodesic triangles (in X_v, X_w) $\hat{\Delta}_v := \Delta \hat{x}_v^1 \hat{x}_v^2 \hat{x}_v^3$. $\Delta_w := \Delta x_w^1 x_w^2 x_w^3$ and corresponding geodesic triangles in X_{vw}:

$$\hat{\Delta}_v^* := \Delta_{X_{vw}} \hat{x}_v^1 \hat{x}_v^2 \hat{x}_v^3, \quad \Delta_w^* := \Delta_{X_{vw}} x_w^1 x_w^2 x_w^3.$$

Then the points z_v, z_w are, respectively, a $3\delta_0$-center of $\hat{\Delta}_v$ and δ_0-center of Δ_w. Since the sides of $\hat{\Delta}_v$ are L_0'-quasigeodesics in X_{vw}, stability of quasigeodesics implies that z_v is within distance $3\delta_0 + D_{1.54}(\delta_0', L_0')$ from all three sides of $\hat{\Delta}_v^*$. Since the respective endpoints of the geodesic sides of the triangles $\hat{\Delta}_v^*, \Delta_w^*$ are within distance k in X, it follows that z_v is within distance $3\delta_0 + D_{1.54}(\delta_0', L_0') + \delta_0' + k$ from all the sides of Δ_w^*, i.e., z_v is a $3\delta_0 + D_{1.54}(\delta_0', L_0') + \delta_0' + k$-center of Δ_w^*.

Similarly, the point z_w is a $\delta_0 + D_{1.54}(\delta_0', L_0')$-center of the triangle Δ_w^* in X_{vw}. Thus, by Lemma 1.77,

$$d_{X_{vw}}(z_w, z_v) \leq D_{1.77}(\delta_0, 3\delta_0 + D_{1.54}(\delta_0', L_0') + \delta_0' + k).$$

Setting $C_{5.8}(k) := D_{1.77}(\delta_0, 3\delta_0 + D_{1.54}(\delta_0', L_0') + \delta_0' + k)$ concludes the proof. □

COROLLARY 5.9. *For every edge $e = [v, w]$ (pointing away from u), we have*

$$\mathrm{Hd}_{X_{vw}}([z_v \bar{x}_v^i]_{X_v}, L_w^i) \leq D_{5.9}(k) = D_{1.55}(\delta_0', L_0', C_{5.8}(k)),$$

unless $L_w^i = \emptyset$. In any case,

$$\mathrm{Hd}_{X_{vw}}(\bar{Y}_v, Y_w) \leq D_{5.9}(k).$$

PROOF. In generic subcase (d) the first claim is an immediate application of Lemma 1.55 and Lemma 5.8. In other subcases, both tripods $\bar{Y}_v, Y_w$ are degenerate, equal to geodesic segments whose respective endpoints are within distance k in X_{vw}, e.g., $Y_w = L_w^i$. Therefore, we similarly conclude that

$$\mathrm{Hd}_{X_{vw}}([z_v \bar{x}_v^i]_{X_v}, L_w^i) \leq D_{1.55}(\delta_0', L_0', k).$$

Observing that

$$k \leq C_{5.8}(k),$$

we obtain the first claim in other subcases as well. The second claim is an immediate corollary of the first one. □

We are now ready to verify that $\mathfrak{Y}$ satisfies axioms of a (K, D, E)-semicontinuous family and that each $\mathfrak{L}^i$ is the vertex set of a (κ_1, D_1, E_1)-ladder. The constants κ_1 and E_1 will be computed in the end of the proof of the proposition.

In line with the proof of Lemma 3.17, for the edges $e = [v, w]$ (oriented away from u) we define tripods $Y_e \subset X_e$ as

$$T_{z_e}(x_e^1 x_e^2 x_e^3)$$

where x_e^i is a nearest-point projection of x_w^i to X_e in X_{vw}, while z_e is a δ_0-center of the geodesic triangle $\Delta x_e^1 x_e^2 x_e^3 \subset X_e$. Thus, by the construction, each Y_v, Y_e is a δ_0-quasiconvex subset of the respective vertex and edge-space of $\mathfrak{X}$.

According to Corollary 5.9, every point $x \in Y_w$ is within distance

$$\kappa_1 := D_{5.9}(k) \tag{5.8}$$

from a point $y \in \bar{Y}_v \subset Y_v$. Thus, every point of $\mathcal{Y}$ is connected to X_u by a κ_1-qi section. Similarly, we get the inequality $\mathrm{Hd}_{X_{vw}}(Y_w, Y_e) \le \kappa_1$. Let's verify the inequality (3.2) (see Definition 3.1) for a suitable value of the parameter E, i.e., get a uniform bound on the Hausdorff distance between Y_w and the projection of Y_v to X_w. First of all, since

$$d_{X_{vw}}(Y_v, X_w) \le k = \max(4\lambda_0' + 8\delta_0' + 5\delta_0, k),$$

using Lemma 1.130 we obtain

$$\mathrm{Hd}_{X_{vw}}(P_{Y_v}(X_w), P_{X_w}(Y_v)) \le R_{1.130}(k, \lambda, \delta_0') = 2\lambda_0' + 3\delta_0' + k.$$

Thus, we need to estimate the Hausdorff distance between $P_{Y_v}(X_w)$ and Y_w. According to (5.3),

$$\mathrm{Hd}_{X_{vw}}(\bar{Y}_v, P_{Y_v}(X_w)) \le \lambda.$$

Combining these inequalities with Corollary 5.9, we get:

$$\mathrm{Hd}_{X_{vw}}(P_{X_w}(Y_v), Y_w) \le E_1 := D_{5.9}(k) + \lambda + (2\lambda_0' + 3\delta_0' + k). \tag{5.9}$$

For Axiom 4 of a semicontinuous family we observe that, by the construction, if $Y_w = \emptyset$ then Y_v and X_w are D_1-cobounded. We conclude:

LEMMA 5.10. *$\mathfrak{Y}$ is a (κ_1, D_1, E_1)-semicontinuous family containing sections $\gamma^1, \gamma^2, \gamma^3$ and $\mathcal{Y} \subset N^{fib}_{5\delta_0}(\mathcal{F}l_k(X_u))$.*

Next, consider the families of intervals $\mathcal{L}^i, i = 1, 2, 3$ ($\mathcal{L}^i$ is the union of geodesic intervals L_v^i, $v \in V(S)$). We will be verifying the conditions of Lemma 3.17 for each $\mathcal{L}^i$. The easiest thing to check is the first part of condition (a2) of the lemma, dealing with Property 4 of a semicontinuous family of spaces. Namely, by the definition of L_w^i, it is empty only when L_v^i and X_w are $D_2 = D_1$-cobounded in X_{vw}; see Remarks 5.6 and 5.7.

Next, consider the condition (a3) of the lemma. Suppose L_w^i is nonempty, equals the oriented segment $[x_w^i z_w]_{X_w}$. There are several cases to consider, for instance, suppose we are in generic subcase (2d). Then there exists a point $\hat{x}_v^i \in [x_v^i z_v]_{X_v} = L_v^i$ within distance k from x_w^i, while according to Lemma 5.8

$$d_{X_{vw}}(z_v, z_w) \le C_{5.8}(k).$$

Of course, in this case, in the oriented interval L_v^i, we have

$$x_v^i \le \hat{x}_v^i \le z_v \le z_v.$$

In all nongeneric cases, there are points $\bar{x}_v^{i\pm1}, \hat{x}_v^i \in L_v^i$ within distance k from z_w, x_w^i (subcase (2c)) or points $\hat{x}_v, \hat{y}_v \in L_v^i$ (or L_v) within distance k from x_w, y_w (subcase (2c) or case 3), and these points appear in the oriented interval L_v^i (or L_v) in the correct order.

This verifies condition (a3) of Lemma 3.17 with the constant

$$k_2 = \max(k, C_{5.8}(k)) = C_{5.8}(k) \tag{5.10}$$

playing the role of K in Lemma 3.17.

Lastly, we analyze the projection of L^i_v to X_w. As in case of the projection of Y_v to X_w, we have:

$$d_{X_{vw}}(L^i_v, X_w) \le k$$

and, thus,

$$\mathrm{Hd}_{X_{vw}}(P_{L^i_v}(X_w), P_{X_w}(L^i_v)) \le R_{1.130}(k, \lambda, \delta) = 2\lambda'_0 + 3\delta'_0 + k. \tag{5.11}$$

In other words, the projection of L^i_v to X_w is uniformly Hausdorff-close to the projection of X_w to L^i_v. Therefore, we analyze the latter using the arguments from the proof of Lemma 5.3. We have four projections:

$$P_{X_{vw},Y_v}, P_{X_{vw},L^i_v}, P_{Y_v,L^i_v}, P'_{Y_v,L^i_v},$$

where the first three are nearest-point projections with respect to the metric of X_{vw}, while the last one is the intrinsic nearest-point projection with respect to the metric of Y_v. We have, of course,

$$P'_{Y_v,L^i_v} \circ P_{X_{vw},Y_v}(X_w) = P_{X_{vw},Y_v}(X_w) \cap [\bar{x}^i_v z_v]_{X_v} \subset P'_{Y_v,L^i_v}(\bar{Y}_v) = [\bar{x}^i_v z_v]_{X_v}.$$

According to Lemma 1.128,

$$\mathrm{Hd}_{X_{vw}}(P_{X_{vw},Y_v}(X_w) \cap [\bar{x}^i_v z_v]_{X_v}, [\bar{x}^i_v z_v]_{X_v}) \le \lambda. \tag{5.12}$$

As in the proof of Lemma 5.3,

$$d_{X_{vw}}(P_{X_{vw},L^i_v}, P_{Y_v,L^i_v} \circ P_{X_{vw},Y_v}) \le C_{1.110}(\delta'_0, \lambda'_0),$$

while for $x \in Y_v$,

$$d_{X_{vw}}(P_{Y_v,L^i_v}(x), P'_{Y_v,L^i_v}(x)) \le 2(\lambda'_0 + 2\delta'_0 + D_{1.54}(\delta'_0, L'_0)).$$

Combining the two inequalities, we obtain that for each $q \in X_w$,

$$d_{X_{vw}}(P'_{Y_v,L^i_v} \circ P_{X_{vw},Y_v}(q), P_{X_{vw},L^i_v}(q)) \le C_{5.3} = C_{1.110}(\delta'_0, \lambda'_0) + 2(\lambda'_0 + 2\delta'_0 + D_{1.54}(\delta'_0, L'_0))$$

Thus, taking into account (5.12),

$$\mathrm{Hd}_{X_{vw}}(P_{X_{vw},L^i_v}(X_w), [\bar{x}^i_v z_v]_{X_v}) \le C_{5.3} + \lambda.$$

Combined with the inequality (5.11), we get:

$$\mathrm{Hd}_{X_{vw}}(P_{X_w}(L^i_v), [\bar{x}^i_v z_v]_{X_v}) \le R_{1.130}(k, \lambda, \delta'_0) + C_{5.3} + \lambda.$$

Recall that by Corollary 5.9

$$\mathrm{Hd}_{X_{vw}}([z_v \bar{x}^i_v]_{X_v}, L^i_w) \le D_{5.9}(k).$$

Thus,

$$\mathrm{Hd}_{X_{vw}}(P_{X_w}(L^i_v), L^i_w) \le E_2 := D_{5.9}(k) + R_{1.130}(k, \lambda, \delta'_0) + C_{5.3} + \lambda. \tag{5.13}$$

This concludes verification of the conditions of Lemma 3.17 and we obtain:

LEMMA 5.11. *For k_2 given by the equation (5.10), $\kappa_2 = k'_2 = K'_{3.17}(k_2)$, E_2 as in (5.13), and $\lambda = C_{1.128}(\delta'_0, \lambda'_0, L'_0)$, each $\mathcal{L}^i$ defined earlier is the vertex-set of a (κ_2, D_2, E_2)-ladder $\mathfrak{L}^i$ in $\mathfrak{X}$. Each ladder $\mathfrak{L}^i$ contains the section γ^i.*

This concludes the proof of part (i) of the proposition. We now prove part (ii). The goal, of course, is to verify the conditions of Lemma 3.17 for the family of segments $L_v^{ij}, v \in V(S)$, $j = i + 1$. We define the monotonic map

$$f_{ij} : L_v^i \cup L_v^j \to L_v^{ij} = [x_v^i x_v^j]_{X_v},$$

using Corollary 1.81. According to Lemma 5.11, the map moves each point distance $\leq 4\delta_0$. The points

$$x_v' := f_{ij}(\hat{x}_v^i), y_v' := f_{ij}(\hat{x}_v^j).$$

satisfy

$$x_v^i \leq x_v' \leq y_v' \leq x_v^j$$

in the oriented interval L_v^{ij}. Since

$$\max(d_{X_{vw}}(\hat{x}_v^i, x_w^i), d_{X_{vw}}(\hat{x}_v^j, x_w^j)) \leq k_2,$$

we obtain

$$\max(d_{X_{vw}}(x_v', x_w^i), d_{X_{vw}}(y_v', x_w^j)) \leq k_3 := k_2 + 4\delta_0. \tag{5.14}$$

Next: Since $L_v^{ij} \subset N_{2\delta_0}^{fib}(Y_v))$,

$$\operatorname{diam}(P_{X_w}(L_v^{ij})) \leq L_1' \cdot 3\delta_0 + \operatorname{diam}(P_{X_w}(Y_v)) \leq 3\delta_0 + D_1.$$

Since

$$\operatorname{Hd}_{X_v}(L_v^i \cup L_v^j, L_v^{ij}) \leq 2\delta_0, \tag{5.15}$$

$$d_{X_{vw}}(P_{L_v^i \cup L_v^j}, P_{L_v^{ij}}) \leq 2\delta_0 + 2(\lambda_0' + 2\delta_0').$$

Therefore,

$$\begin{aligned} \operatorname{diam}(P_{L_v^{ij}}(X_w)) &\leq \operatorname{diam}(P_{L_v^i \cup L_v^j}(X_w)) + 2\delta_0 + 2(\lambda_0' + 2\delta_0') && (5.16)\\ &\leq \operatorname{diam}(P_{Y_v}(X_w)) + 2\delta_0 + 2(\lambda_0' + 2\delta_0') && (5.17)\\ &\leq D_1 + 2\delta_0 + 2(\lambda_0' + 2\delta_0'). \end{aligned}$$

Therefore, for

$$D_3 := D_1 + \max(3\delta_0, 2\delta_0 + 2(\lambda_0' + 2\delta_0')),$$

for every boundary edge $e = [v, w]$ of S, the subsets $L_v^{ij}, X_w \subset X_{vw}$ are D_3-cobounded.

Lastly, we estimate the Hausdorff distance between the projection of L_v^{ij} to X_w and L_w^{ij} for every edge $e = [v, w] \in S$. We again use the inequality (5.15) and the coarse Lipschitz property of the projection P_{X_{vw}, X_w}:

$$\operatorname{Hd}(P_{X_w}(L_v^i \cup L_v^j), P_{X_w}(L_v^{ij})) \leq L_1' \cdot 3\delta_0.$$

Therefore, the inequality (5.13), implies that

$$\operatorname{Hd}(P_{X_w}(L_v^{ij}), L_w^{ij}) \leq E_3 := L_1' \cdot 5\delta_0 + E_2.$$

It follows (in view of Lemma 3.17) that $\mathfrak{L}^{ij}$ is the union of vertex-sets of a (κ_3, D_3, E_3)-ladder, $\kappa_3 = k_3'$. Taking

$$K := \max(\kappa_1, \kappa_2, \kappa_3), D := \max(D_1, D_3), E := \max(E_1, E_2, E_3),$$

concludes the proof of the proposition. □

COROLLARY 5.12. *For k, K, D, E as in the proposition, any two points $x, y \in \mathfrak{Fl}_k(X_u)$ belong to a (K, D, E)-ladder $\mathfrak{L}$ centered at u and contained in the fiberwise $4\delta_0$-neighborhood of $\mathfrak{Fl}_k(X_u)$. Furthermore, if we are given two k-leaves γ_x, γ_y in $Fl_k(X_u)$ connecting x, y to X_u, the ladder $\mathfrak{L}$ can be chosen to satisfy:*

$$\gamma_x \subset bot(\mathfrak{L}), \gamma_y \subset top(\mathfrak{L}).$$

5.2. Projection of ladders

In this section we discuss an important *projection* procedure which converts a pair of ladders $\mathfrak{L}^i = (\pi : L^i \to S_i), i = 1, 2$, with the common center u into a pair of subladders $\bar{\mathfrak{L}}^i \subset \mathfrak{L}^i, i = 1, 2$ (with the same center u, but possibly a different set of qi sections $\bar{\Sigma}_\bullet$), within uniformly bounded (fiberwise) Hausdorff distance from each other. This construction will be used in Section 5.4 to show the coarse independence of a combing path $c(x, y)$ in a ladder $L_{x,y}$ containing the given points x, y on the choice of a ladder $L_{x,y}$.

The intersection $\pi(\mathfrak{L}^1) \cap \pi(\mathfrak{L}^2) = S_1 \cap S_2$ is a subtree $S \subset T$. This subtree will contain (but, in general will be different from) the tree $\bar{S}$ which is the common base of the ladders $\pi : \bar{\mathfrak{L}}^i \to \bar{S}$. For each $v \in V(S)$ we let

$$\bar{L}_v^1 := \bar{P}_{X_v, L_v^1}(L_v^2) \subset L_v^1, \bar{L}_v^2 := \bar{P}_{X_v, L_v^2}(L_v^1) \subset L_v^2;$$

see Definition 1.124 for the definition of the modified fiberwise projection $\bar{P}$. In the definition of ladders $\bar{\mathfrak{L}}^i$ below, the segment $\bar{L}_v^i$ will equal the fiber $\bar{L}_v^i$ of $\bar{\mathfrak{L}}^i$, unless $v \notin V(\bar{S})$.

By Lemma 1.142 and Corollary 1.143 we have the dichotomy:

(i) Either the pair of geodesic segments $L_v^1, L_v^2 \subset X_v$ is $7\delta_0$-separated (i.e., $d_{X_v}(L_v^1, L_v^2) > 7\delta_0$), in which case this pair is $D_{1.142}(\delta, \delta) = 9\delta_0$-cobounded.

(ii) Or $d_{X_v}(L_v^1, L_v^2) \leq 7\delta_0$ and

$$\mathrm{Hd}_{X_v}(P_{L_v^1}(L_v^2), P_{L_v^2}(L_v^1)) \leq 12\delta_0.$$

According to Remark 1.127, in this case

$$\mathrm{Hd}_{X_v}(P_{L_v^1}(L_v^2), \bar{L}_v^1) \leq 4\delta_0$$

and

$$\mathrm{Hd}_{X_v}(P_{L_v^2}(L_v^1), \bar{L}_v^2) \leq 4\delta_0.$$

Combining the inequalities, we get that in the second case,

$$\mathrm{Hd}_{X_v}(\bar{L}_v^1, \bar{L}_v^2) \leq 20\delta_0. \tag{5.18}$$

Accordingly, for each vertex $v \in S$ such that the pair of geodesic segments $L_v^1, L_v^2 \subset X_v$ is $7\delta_0$-separated, we remove from $V(S)$ all the vertices (and edges) $w \neq v$ such that v is between u and w; we let $\bar{S}$ denote the subtree of S spanned by the remaining set of vertices of S. Note that if $v = u$, $\bar{S} = \{u\}$.

We define

$$\bar{\mathcal{L}}^i := \bigcup_{v \in V(\bar{S})} \bar{L}_v^i.$$

Thus, apart from the boundary vertices v of $\bar{S}$, the intervals $\bar{L}_v^1, \bar{L}_v^2$ are fiberwise Hausdorff $20\delta_0$-close, while for *some* boundary vertices v of $\bar{S}$, both $\bar{L}_v^1, \bar{L}_v^2$ have length $\leq 9\delta_0$. We will prove (Claim 5.15) that even for the boundary vertices of $\bar{S}$, the Hausdorff distance between the intervals $\bar{L}_v^1, \bar{L}_v^2$ is also uniformly bounded.

REMARK 5.13. By the construction, both $\bar{\mathcal{L}}^i$ contain the intersection $\mathcal{L}^1 \cap \mathcal{L}^2$.

Thus, we obtain subsets $\bar{\mathcal{L}}^i \subset \mathcal{X}$ which are unions of geodesic segments in vertex-spaces $X_v, v \in V(\bar{S})$. Our next goal it to prove that these subsets are unions of vertex-spaces of ladders in $\mathfrak{X}$.

Recall that L_1' is a coarse Lipschitz constant for the composition of the inclusion map $X_v \to X_{vw}$ with the nearest-point projection $P_{X_{vw},X_w} : X_{vw} \to X_w$, cf. Notation 2.6.4. Set

$$\epsilon = \epsilon_{5.14}(E) = E + (2E + 21\delta_0 L_1') + L_{1.102}(\delta_0, \delta_0)(2E + 21\delta_0 L_1' + 1),$$

$$k := 2(2K + \epsilon + 20\delta_0) + K + \epsilon + 20\delta_0,$$

$$\tilde{K} = \tilde{K}_{5.14}(K, E) = K'_{3.17}(k),$$

$$\tilde{E} := \tilde{E}_{5.14}(K, E) = 2k,$$

$$\tilde{D} := \tilde{D}_{5.14}(D) = \max(10\delta_0 \cdot L_1', D, C_{1.143}(\lambda_0', \delta_0', 21\delta_0 L_1')).$$

PROPOSITION 5.14 (Projections of pairs of ladders). *If $\mathfrak{L}^1, \mathfrak{L}^2$ are (K, D, E)-ladders centered at u, then $\bar{\mathfrak{L}}^i = (\pi : \bar{L}^i \to \bar{S}), i = 1, 2$, constructed above, are $(\tilde{K}, \tilde{D}, \tilde{E})$-ladders centered at u (where $\tilde{K} = \tilde{K}_{5.14}(K, E)$, $\tilde{D} = \tilde{D}_{5.14}(D)$, $\tilde{E} = \tilde{E}_{5.14}(K, E)$), such that $\bar{\mathcal{L}}^i = \bar{L}^i \cap \mathcal{X}$ is the union of vertex-spaces of $\bar{\mathfrak{L}}^i$.*

PROOF. As with tripods of ladders, we will prove the proposition by verifying the conditions of Lemma 3.17. Below, $e = [v, w]$ is an edge of T oriented away from u, with $v \in V(\bar{S})$.

(1) We first check part 1 of the condition (a2) in Lemma 3.17, i.e., that for every boundary edge e of $\bar{S}$, $e = [v, w] \notin E(\bar{S})$, $v \in V(\bar{S})$, both pairs of subsets $\bar{L}_v^i, X_w \subset X_{vw}$ are $\tilde{D}$-cobounded, $i = 1, 2$.

(i) It can happen that $e = [v, w] \notin E(\bar{S})$ because the pair $L_v^1, L_v^2 \subset X_v$ was $7\delta_0$-separated, i.e., both $\bar{L}_v^1, \bar{L}_v^2$ have length $\leq 9\delta_0$. Since the projection $P_{X_{vw},X_w} : X_{vw} \to X_w$ is L_1'-coarse Lipschitz, diameters of the projections $P_{X_{vw},X_w}(\bar{L}_v^i) \subset X_w$ are at most $10\delta_0 \cdot L_1' \leq \tilde{D}$. (Recall that $\delta_0 \geq 1$.)

(ii) If the pair $L_v^1, L_v^2 \subset X_v$ was not $7\delta_0$-separated then the Hausdorff distance in X_v between $\bar{L}_v^1, \bar{L}_v^2$ is $\leq 20\delta_0$. After swapping the labels of 1 and 2 we may assume that there is an edge $e = [v, w] \notin E(S_1)$. Then, because $\mathfrak{L}^1$ was a (K, D, E)-ladder, the pair L_v^1, X_w is D-cobounded in X_{vw}. The same, of course applies to the pair $\bar{L}_v^1, X_w$, since $\bar{L}_v^1 \subset L_v^1$. We need to get a coboundedness estimate for the pair $\bar{L}_v^2, X_w$. Since

$$\text{Hd}_{X_v}(\bar{\mathcal{L}}_v^1, \bar{\mathcal{L}}_v^2) \leq 20\delta_0 \tag{5.19}$$

and the projection $P_{X_{vw},X_w} : X_v \to X_w$ is coarse L_1'-Lipschitz, the diameter the projection of $\bar{L}_v^2$ to X_w is at most

$$L_1'(20\delta_0 + 1) \leq 21\delta_0 L_1' \leq \tilde{D}.$$

The estimate

$$\text{diam}_{X_{vw}}(P_{X_{vw},\bar{L}_v^2}(X_w)) \leq C_{1.143}(\lambda_0', \delta_0', 21\delta_0 L_1')$$

follows from Corollary 1.143.

This verifies Part (1) of the condition (a2) in Lemma 3.17.

(2) We assume now that $e = [v, w]$ is an edge of $\bar{S}$, in particular, $\bar{L}_v^1, \bar{L}_v^2$ are $20\delta_0$-Hausdorff close in X_v. Since $\mathfrak{L}^1, \mathfrak{L}^2$ satisfy the (K, D, E)-ladder axioms, for $P = P_{X_{vw},X_w}$, $P(L_v^i)$ and L_w^i are E-Hausdorff close in X_{vw}, $i = 1, 2$. Our goal is to estimate the Hausdorff distance between L_w^i and the projection of L_v^i to X_w $(i = 1, 2)$. In this part of the proof we will get only half of the estimate, we will get the other half in Part (4) of the proof.

Pick $x \in \bar{L}^1_v$. Then there exists $y \in \bar{L}^2_v$ such that $d_{X_v}(x,y) \le 20\delta_0$; we also have

$$d_{X_w}(P(x), L^1_w) \le E, \quad d_{X_w}(P(y), L^2_w) \le E.$$

Thus, there exist $x' \in L^1_w$ and $y' \in L^2_w$ such that

$$d_{X_w}(P(x), x') \le E, \quad d_{X_w}(P(y), y') \le E, \tag{5.20}$$

which in turn implies the inequality

$$d_{X_w}(x', y') \le 2E + 21\delta_0 L'_1. \tag{5.21}$$

We next estimate the distance $d_{X_w}(y', \bar{L}^2_w)$. Since the projection $P' : X_w \to L^2_w$ is $L_{1.102}(\delta_0, \delta_0)$-coarse Lipschitz, we have

$$d_{X_w}(P'(x'), y') \le L_{1.102}(\delta_0, \delta_0)(d_{X_w}(x', y') + 1) \le L_{1.102}(\delta_0, \delta_0)(2E + 21\delta_0 L'_1 + 1).$$

Since $x' \in L^1_w$, $P'(x') \in \bar{L}^2_w$, and we obtain:

$$d_{X_w}(y', \bar{L}^2_w) \le L_{1.102}(\delta_0, \delta_0)(2E + 21\delta_0 L'_1 + 1). \tag{5.22}$$

Switching the roles of x and y we similarly obtain:

$$d_{X_w}(x', \bar{L}^1_w) \le L_{1.102}(\delta_0, \delta_0)(2E + 21\delta_0 L'_1 + 1). \tag{5.23}$$

Combining the equations (5.20) and (5.23) we obtain

$$d_{X_w}(P(x), \bar{L}^1_w) \le E + L_{1.102}(\delta_0, \delta_0)(2E + 21\delta_0 L'_1 + 1)$$

and, similarly, combining (5.20) and (5.22) we obtain

$$d_{X_w}(P(y), \bar{L}^2_w) \le E + L_{1.102}(\delta_0, \delta_0)(2E + 21\delta_0 L'_1 + 1).$$

At the same time, combing the equations (5.20), (5.21) and (5.22) we get:

$$d_{X_w}(P(x), \bar{L}^2_w) \le E + (2E + 21\delta_0 L'_1) + L_{1.102}(\delta_0, \delta_0)(2E + 21\delta_0 L'_1 + 1) \le \tilde{E}.$$

Similarly,

$$d_{X_w}(P(y), \bar{L}^1_w) \le \epsilon = E + (2E + 21\delta_0 L'_1) + L_{1.102}(\delta_0, \delta_0)(2E + 21\delta_0 L'_1 + 1).$$

Thus, we proved:

$$P(\bar{L}^1_v) \cup P(\bar{L}^2_v) \subset N^e_\epsilon(\bar{L}^1_w) \cap N^e_\epsilon(\bar{L}^2_w),$$

which gives us half of the estimate on $\mathrm{Hd}_{X_{vw}}(P(\bar{L}^i_v), \bar{L}^i_w)$, but also gives an upper bound on the minimal distance between $\bar{L}^1_w$ and $\bar{L}^2_w$, something which was not a priori clear from the construction. We will derive the other half of the estimate on $\mathrm{Hd}_{X_{vw}}(P(\bar{L}^i_v), \bar{L}^i_w)$ in Part 4 of the proof. Before proceeding with Part 3 of the proof we establish:

Claim 5.15. For every vertex $w \in V(\bar{S})$

$$\mathrm{Hd}_{X_{vw}}(\bar{L}^1_w, \bar{L}^2_w) \le C_{5.15}(E) = \epsilon + 20\delta_0.$$

Proof. There are two cases to consider according to the definition of the projections $\bar{\mathcal{L}}^i$:

(i) Suppose that the pair of geodesic segments L^1_w, L^2_w is $7\delta_0$-separated, hence, each segment $\bar{L}^i_w, i = 1, 2$ has length $\le 9\delta_0$. As we observed the end of Part 2 of the proof, there exists $z \in X_w$ which lies in the intersection

$$N^e_\epsilon(\bar{L}^1_w) \cap N^e_\epsilon(\bar{L}^2_w).$$

Therefore,

$$\mathrm{Hd}_{X_{vw}}(\bar{L}^1_w, \bar{L}^2_w) \le \epsilon + 9\delta_0.$$

(ii) Suppose that the pair of geodesic segments L^1_w, L^2_w is not $9\delta_0$-cobounded. According to (5.18),

$$\mathrm{Hd}_{X_v}(\bar{\mathcal{L}}^1_v, \bar{\mathcal{L}}^2_v) \le 20\delta_0. \qquad \square$$

We now return to the proof of the proposition.

(3) We next verify the condition (a3) in Lemma 3.17.

Since $\mathfrak{L}^i, i = 1, 2$ were K-ladders, each $x_i \in \bar{L}^i_w$ is within distance K (measured in X_{vw}) from some point in L^i_v. Thus, setting $y_i := P_{X_{vw}, L^i_v}(x_i)$,

$$d_{X_{vw}}(x_i, y_i) \le 2K, i = 1, 2.$$

According to Claim 5.15, given $x_1 \in \bar{L}^1_w$ we can find $x_2 \in \bar{L}^2_w$ such that

$$d_{X_{vw}}(x_1, x_2) \le \epsilon + 20\delta_0.$$

Thus,

$$d_{X_{vw}}(y_1, y_2) \le 2K + \epsilon + 20\delta_0.$$

Let $\bar{y}_i$ denote a nearest-point projection of y_i to $L^{3-i}_v, i = 1, 2$. Then the above inequality implies that

$$d_{X_{vw}}(\bar{y}_i, y_i) \le 2(2K + \epsilon + 20\delta_0).$$

However, by the construction, $\bar{y}_i$ belongs to $\bar{L}^{3-i}_v$. Hence, by the above estimates:

$$d_{X_{vw}}(x_1, \bar{L}^1_v) \le d_{X_{vw}}(x_1, \bar{y}_2) \le k := 2(2K + \epsilon + 20\delta_0) + K + \epsilon + 20\delta_0.$$

Similarly, for each $x_2 \in \bar{L}^2_w$ we have

$$d_{X_{vw}}(x_2, \bar{L}^2_v) \le k = 2(2K + \epsilon + 20\delta_0) + K + \epsilon + 20\delta_0.$$

This verifies the condition (a3) in Lemma 3.17.

(4) By Part 3, each $x \in \bar{L}^i_w$ is within distance k from some $y \in \bar{L}^i_v, i = 1, 2$. Therefore,

$$d_{X_{vw}}(x, P_{X_{vw}, X_w}(y)) \le 2k.$$

In other words,

$$\bar{L}^i_w \subset N^e_{2k}(P_{X_{vw}, X_w}(\bar{L}^i_v)).$$

Combining this with the estimate in the end of Part 2, we obtain:

$$\mathrm{Hd}_{X_{vw}}(P(\bar{L}^i_v), \bar{L}^i_w) \le \max(2k, \epsilon) = 2k = \tilde{E}.$$

Since we defined $\tilde{E}$ to be $2k$, we are done with the proof of the proposition.

5.3. Hyperbolicity of tripods families

PROPOSITION 5.16 (Tripod families are hyperbolic). *Suppose that $\mathfrak{X}$ is a tree of hyperbolic spaces satisfying the uniform $\kappa_{4.7}(K)$-flaring condition. Then for each (K, D, E)-tripod family $\mathfrak{Y} = (\pi : Y \to S)$ in $\mathfrak{X}$, the total space Y is $\delta_{5.16}(K)$-hyperbolic.*

PROOF. The total space Y of $\mathfrak{Y}$ is the union of total spaces L^i of the ladders $\mathfrak{L}^i$, $i = 1, 2, 3$. The pairwise intersections of these ladders equal their triple intersection, namely, the K-qi section Ξ, which is, intrinsically, a tree. Thus, Y has a structure of a tree of spaces with the vertex-spaces L^i, $i = 1, 2, 3$, and Ξ. According to Theorem 4.16 each ladder L^i is $\delta_{4.16}(K)$-hyperbolic (this is where we need the $\kappa_{4.7}(K)$-flaring condition), the space Y is $\delta_{5.16}(K)$-hyperbolic according to Corollary 2.69. $\square$

5.4. Hyperbolicity of flow-spaces

In Theorem 5.17, we fix $k \geq k_{5.2}$, $K = K_{5.2}(k)$.

THEOREM 5.17. *Suppose that k, K are as above, $\mathfrak{X}$ is a tree of hyperbolic spaces satisfying the uniform $\kappa_{4.7}(K)$-flaring condition. Then there is a function $\delta = \delta_{5.17}(k)$ such that for each $u \in v(T)$, the flow space $Fl_k(X_u)$ is δ-hyperbolic.*

PROOF. According to Corollary 5.12, whenever $k \geq k_{5.2}$, $K = K_{5.2}(k)$, $D = D_{5.2}$, $E = E_{5.2}$, for any two points $x, y \in \mathcal{F}l_k(X_u)$ there exists a (K, D, E)-ladder $\mathfrak{L} = \mathfrak{L}_{x,y}$ centered at u and containing x, y, such that $\mathcal{L}$ is contained in the fiberwise $5\delta_0$-neighborhood of $\mathcal{F}l_k(X_u)$.

Recall that the total space $L_{x,y}$ of the ladder $\mathfrak{L}$ is $L_{3.4}(K, D, E, \delta_0)$-qi embedded in X. Define $c(x, y)$ to be a projection to $Fl_k(X_u)$ of a geodesic in $L_{x,y}$ connecting x to y. We note that the definition of $c(x, y)$ depends on the choice of $\mathfrak{L}_{xy}$ which is far from canonical. Our first task is to prove that different choices lead to uniformly Hausdorff-close paths.

PROPOSITION 5.18. *Let $\mathfrak{L}^1 = \mathfrak{L}^1_{x,y}$, $\mathfrak{L}^2 = \mathfrak{L}^2_{x,y}$ be (K, D, E)-ladders containing x, y. Then $L^1_{x,y} \cup L^2_{x,y}$ is contained in a $\delta_{5.18}(K, D, E)$-hyperbolic subspace Z in X.*

PROOF. We let $\bar{\mathfrak{L}}^i \subset \mathfrak{L}^i$ denote the $(\tilde{K}, \tilde{D}, \tilde{E})$-subladders obtained by the projection construction described in Section 5.2; see also Proposition 5.14. Note that $\tilde{K} \geq K, \tilde{E} \geq E, \tilde{D} \geq D$. Also, note that the subladders $\bar{\mathfrak{L}}^i$ are nonempty since they both contain x and y.

The subladders $\bar{L}^i$ have equal projection to T, which is a subtree $\bar{S} \subset S_1 \cap S_2$. As usual, we extend these ladders over the rest of the tree $S_1 \cup S_2$ by empty fibers. According to Claim 5.15, the ladders $\bar{\mathfrak{L}}^1, \bar{\mathfrak{L}}^2$ are fiberwise $C_{5.15}(E)$-Hausdorff close. Therefore, for each $v \in V(\bar{S})$ the union $\bar{L}^1_v \cup \bar{L}^2_v$ is $C_{5.15}(E) + \delta_0$-quasiconvex in X_v. For each vertex $v \in V(S)$ we set

$$Z^0_v := \mathrm{Hull}_{\delta_0}(\bar{L}^1_v \cup \bar{L}^2_v),$$

$$Z^i_v := L^i_v \cup Z^0_v, i = 1, 2,$$

and $Z_v := Z^1_v \cup Z^2_v$. Thus, Z^j_v $(j = 0, 1, 2)$ and Z_v are rectifiably connected $4\delta_0$-quasiconvex subsets of X_v (see Lemma 1.95). By (1.3),

$$\mathrm{Hd}_{X_v}(Z^0_v, \bar{L}^1_v \cup \bar{L}^2_v) \leq C_{5.15}(E) + 2\delta_0$$

and, hence,

$$\mathrm{Hd}_{X_v}(Z^0_v, \bar{L}^i_v) \leq 2C_{5.15}(E) + 2\delta_0, i = 1, 2. \tag{5.24}$$

Accordingly,

$$\mathrm{Hd}_{X_v}(Z^i_v, L^i_v) \leq 2C_{5.15}(E) + 2\delta_0, i = 1, 2.$$

We repeat the same construction (and estimates) for all edges $e \in E(\bar{S})$. We, thus, obtain four subtrees of spaces $\mathfrak{Z}^j$ $(j = 0, 1, 2)$ and $\mathfrak{Z}$ in $\mathfrak{X}$ whose vertex-spaces are, respectively Z^j_v $(j = 0, 1, 2)$ and Z_v, $v \in V(S)$. The total space Z of $\mathfrak{Z}$ contains both ladders L^1, L^2. We equip Z^j and Z with natural path-metrics d_{Z^j}, d_Z; the goal is to show that Z is uniformly hyperbolic. The space Z is the union of subsets Z^1, Z^2 whose intersection is Z^0. According to Corollary 3.4, the ladders $L^i, \bar{L}^i, i = 1, 2$ are $L_{3.4}(\tilde{K}, \tilde{D}, \tilde{E}, \delta_0)$-qi embedded in X. In view of (5.24), the inclusion $Z^0 \to Z$ is an $(L_{3.4}(\tilde{K}, \tilde{D}, \tilde{E}, \delta_0), 2C_{5.15}(E) + 2\delta_0)$-qi embedding.

Thus, $Z = Z^1 \cup Z^2$ satisfies the assumptions of Theorem 2.65 and, therefore, is δ-hyperbolic for some δ which depends only on K, D, E. □

Corollary 5.19. *Let $L^i_{x,y}, i = 1, 2$ be two (K, D, E)-ladders containing x, y and $c^i(x, y)$ be projections to $Fl_k(X_u)$ of geodesics $[xy]^i \subset L^i_{x,y}$. Then*

$$\mathrm{Hd}_X(c^1(x, y), c^2(x, y)) \le C_{5.19}(K, D, E).$$

Proof. The paths $c^i(x, y)$ are uniformly close to the geodesic segments $[xy]^i \subset L^i_{x,y}, i = 1, 2$; hence, it suffices to bound the Hausdorff distance between these segments. Since both $[xy]^i$ are contained in a $\delta_{5.19}(K, D, E)$-hyperbolic space (Z, d_Z) and are $L_{3.4}(K, D, E, \delta_0)$-quasigeodesics in Z with common endpoints, by Lemma 1.55, we get:

$$\mathrm{Hd}_X([xy]^1, [xy]^2) \le \mathrm{Hd}_Z([xy]^1, [xy]^2) \tag{5.25}$$

$$\le D_{1.55}(\delta_{5.19}(K, D, E), L_{3.4}(K, D, E, 0)). \tag{5.26}$$

The corollary follows. □

We now check that the family of paths c satisfies the conditions of Corollary 1.64, characterizing hyperbolic spaces; this will conclude the proof of the theorem.

Condition a1. This follows from the fact that $c(x, y)$ is within uniformly bounded distance from a geodesic in $L_{x,y}$ and that $L_{x,y}$ is $L_{3.4}(K', D, E, \delta_0)$-qi embedded in X.

Condition a2. Consider points $x, y, z \in \mathcal{F}l_k(X_u)$. According to Proposition 5.2, there exists a (K, D, E)-tripod family $\mathfrak{Y} = (\pi : Y \to S)$ contained the fiberwise $5\delta_0$-neighborhood of $\mathcal{F}l_k(X_u)$ and containing the points x, y, z. Moreover, according to Part (ii) of Proposition 5.2, there are (K, D, E)-ladders $\mathfrak{L}_{x,y}, \mathfrak{L}_{y,z}, \mathfrak{L}_{z,x}$ contained in the fiberwise δ_0-neighborhood of $\mathfrak{Y}$ and containing the respective pairs of points x, y, etc. By Proposition 5.16, Y is $\delta_{5.16}(K, D, E)$-hyperbolic. The paths $c(x, y), c(y, z), c(z, x)$ are uniformly close to geodesics $[xy]_{L_{x,y}}, [yz]_{L_{y,z}}, [zx]_{L_{z,x}}$, which are κ-quasigeodesics in Y, where κ depends only on K, D and E. Therefore, by the $\delta_{5.16}(K, D, E)$-hyperbolicity of Y, for

$$\epsilon = 2D_{1.54}(\delta_{5.16}(K, D, E), \kappa) + \delta_{5.16}(K, D, E)$$

we have

$$[xy]_{L_{x,y}} \subset N^Y_\epsilon([yz]_{L_{y,z}} \cup [zx]_{L_{z,x}}) \subset N_\epsilon([yz]_{L_{y,z}} \cup [zx]_{L_{z,x}}),$$

where the first neighborhood is taken in Y and the second is taken in X. Condition (a2) follows.

Lastly, by the construction, each $c(x, y)$ is uniformly close to an $L_{3.4}(K', D, E, \delta_0)$-quasigeodesic in X. □

In the next corollary illustrates an application of Theorem 5.17 to proving hyperbolicity of various subspaces of X. As in Theorem 5.17, we assume that $k \ge k_{5.2}$, but set $K := K_{5.2}(k^\wedge)$. We will use similar arguments in Section 6.1.2 to prove hyperbolicity of unions of pairs of flow-spaces. We refer to Definition 3.33 for the notion of a *generalized flow-space*, $Fl_k(\mathfrak{Q})$, used below.

Corollary 5.20 (Hyperbolicity of generalized flow-spaces). *Suppose that $\mathfrak{X}$ satisfies the uniform $\kappa_{4.7}(K)$-flaring condition. Then for every k-bundle $\mathfrak{Q} = (\pi : Q \to S) \subset \mathfrak{X}$, the k-flow space $Fl_k(\mathfrak{Q})$ is $\delta_{5.20}(k)$-hyperbolic.*

Proof. Pick $u \in V(S)$. Observe that every $x \in Fl_k(\mathfrak{Q})$ is connected to $Fl_k(\mathfrak{Q}) \cap X_u$ by a k-leaf γ_x. This leaf is contained in $Fl_{k^\wedge}(Q_u)$ (see Proposition 3.26(2)). Therefore,

$$Fl_k(\mathfrak{Q}) \subset Fl_{k^\wedge}(Q_u).$$

In view of the uniform flaring condition of the corollary, Theorem 5.17 applies to $Fl_{k^\wedge}(Q_u)$ and, hence, the latter is $\delta_{5.17}(k^\wedge)$-hyperbolic. According to Theorem 3.34, $Fl_k(\mathfrak{Q})$ is an $L_{3.34}(k,k)$-coarse Lipschitz retract of X. Hence $Fl_k(\mathfrak{Q})$ is $\lambda(k)$-quasiconvex in $Fl_{k^\wedge}(Q_u)$ and, therefore (in view of hyperbolicity of the latter), is $\delta_{5.20}(k)$-hyperbolic. □

CHAPTER 6

Hyperbolicity of trees of spaces: Putting everything together

In this chapter we finish the proof of Theorem 2.62, establishing hyperbolicity of trees of hyperbolic spaces, satisfying the uniform K-flaring condition for suitable values of K. The key is to show hyperbolicity of flow-spaces $Fl_K(X_J)$ for intervals $J \subset T$, Theorem 6.17. This is done in three steps:

STEP 1. Hyperbolicity of $Fl_K(X_J)$ for *special intervals* J (Theorem 6.14). This is the hardest part of the chapter, we deal with it in Section 6.1. An outline of this part of proof is given in the introduction to Section 6.1.

STEP 2. Hyperbolicity of $Fl_K(X_J)$, when J is the union of three special subintervals (Proposition 6.15).

STEP 3. Hyperbolicity of $Fl_K(X_J)$ for general intervals, which is done by subdividing J as the union of subintervals J_i, each of which is a union of (at most) three special subintervals, and then using quasiconvex chain-amalgamation (Theorem 2.65).

Once we are done with Theorem 6.17, applying quasiconvex amalgamation (Corollary 2.69) one more time, in Proposition 6.18 we will prove that flow-spaces $Fl_K(X_S)$ are uniformly hyperbolic, whenever S is a tripod in T. We then conclude the proof of Theorem 2.62 by appealing to Corollary 1.64 one last time, by constructing a slim combing in X via geodesics in flow-spaces of interval-spaces; see Section 6.3.

6.1. Hyperbolicity of flow-spaces of special interval-spaces

This section deals with Step 1 described in the introduction to the chapter. Recall that in Section 3.3.5 we defined an interval $J = [\![u, v]\!] \subset T$ to be *special* (more precisely, K-*special*) if one of its endpoints (say, u) has the property that $J \subset \pi(Fl_K(X_u))$. The main result of this section is Theorem 6.14, where we prove that the flow-space $Fl_K(X_J)$ of every special interval $J \subset T$ is uniformly hyperbolic with hyperbolicity constant depending only on K. The fact that this is true is not at all surprising since, assuming that X is hyperbolic, a uniform neighborhood of two intersecting uniformly quasiconvex subsets is uniformly qi embedded in X and, hence, is uniformly hyperbolic. However, at this stage we did not yet prove hyperbolicity of X and, furthermore, we are interested in describing uniform quasigeodesics in X. The most difficult part of the proof is to show that for each special interval J, a certain uniform (depending on K) neighborhood in X of the union $Fl_K(X_u) \cup Fl_K(X_v)$ is uniformly properly embedded in X and uniformly hyperbolic (Corollary 6.11). The idea is to:

(i) embed such union in a larger "modified" flow-space

$$Fl_K(X_u) \cup Fl_K(X_v) \cup Fl_R(\mathfrak{H})$$

(for a certain $R = R(K)$ and a metric bundle $\mathfrak{H}$ over the interval $J = [\![u, v]\!]$),

(ii) prove uniform hyperbolicity of a uniform neighborhood of this triple union using quasiconvex amalgamation (Theorem 2.65),
(iii) lastly, use the fact that a uniform neighborhood of the union of intersecting quasiconvex subsets of a hyperbolic space is again uniformly quasiconvex and hyperbolic.

We then use quasiconvex amalgamation to prove the same result for the union of three flow-spaces

$$Fl_K(X_u) \cup Fl_K(X_w) \cup Fl_K(X_v),$$

where each subinterval $[\![u, w]\!]$, $[\![w, v]\!]$ is special. Theorem 6.14 is then proven by verifying that the family of paths in $Fl_K(X_J)$ which are geodesics in pairwise unions $N_D(Fl_K(X_s) \cup Fl_K(X_t))$, $s, t \in V(J)$, satisfy the slim combing axioms from Corollary 1.64.

6.1.1. Proper embeddings of unions of pairs of intersecting flow-spaces. Assuming that $Fl_K(X_u) \cap X_v \neq \emptyset$, we will show that a certain uniform neighborhood of the union of two flow-spaces $Fl_K(X_u) \cup Fl_K(X_v)$ is uniformly properly embedded in X. We first deal with the following easier case when T is an interval, which we will identify with an interval $[0, n] \subset \mathbb{R}$, $n \in \mathbb{N}$, and that the vertex-set of T equals to the set of integer points in the interval. Recall that M_k is the parameter from the definition of uniform k-flaring (Definition 2.48).

LEMMA 6.1. *Suppose $\mathfrak{X} = (\pi : X \to T)$ is a tree of hyperbolic spaces, such that the tree T is an interval $T = [0, n]$, $K \geq K_0$ and $\mathfrak{X}$ satisfies the uniform k-flaring condition for $k = (L+1)^2 K$, where $L = L_{3.21}(K)$. Assume, moreover, that for vertices $u, v \in V(T)$, and a $4\delta_0$-quasiconvex subset $Q = Q_u \subset X_u$, we have $Fl_K(Q) \cap X_v \neq \emptyset$.*

Then the fiberwise M_k-neighborhood of $Fl_K(Q) \cup Fl_K(X_n)$ is uniformly properly embedded in X, with the distortion function depending only on K.

PROOF. It suffices to show that for each $D > 0$ and $x, y \in \mathcal{F}l_K(Q) \cup \mathcal{F}l_K(X_v)$ with $d(x, y) < D$, the intrinsic distance between x and y in the M_k-fiberwise neighborhood U of $Fl_K(Q) \cup Fl_K(X_v)$ is bounded by a constant depending on D, K only.

CLAIM 6.2. The statement of the lemma holds for $v = n$.

PROOF. Without loss of generality $x \in \mathcal{F}l_K(Q) \setminus \mathcal{F}l_K(X_n)$ and $y \in \mathcal{F}l_K(X_n) \setminus Fl_K(Q)$. (Otherwise, the claim follows from Theorem 3.21.) In the proof we will be repeatedly using Mitra's retractions ρ defined in Theorem 3.21.

REDUCTION 1. We first reduce to the case where x, y are in the same vertex-space and, moreover, $x = \rho_{Fl_K(Q)}(y)$.

Observe that, since T is an interval with an extremal vertex n, $y \in Fl_K(X_n)$ and $Fl_K(X_u) \cap X_n \neq \emptyset$, $\pi(y) \in \pi(Fl_K(Q))$. In particular, by the definition of Mitra's retraction $\rho_{Fl_K(Q)}$,

$$x' := \rho_{Fl_K(Q)}(y) \in X_{\pi(y)} \cap \mathcal{F}l_K(Q).$$

We also apply Mitra's retraction $\rho_{Fl_K(Q)} : X \to Fl_K(A)$ to xy, a geodesic in X connecting x to y. The image $\rho_{Fl_K(Q)}(xy)$ is a path of length $\leq D_1 := (D+1)L_{3.21}(K)$ in $Fl_K(Q)$ joining x to x'. Thus,

$$d_{Fl_K(Q)}(x, x') \leq D_1,$$

while $d(x', y) \leq d(x, y) + d(x', x) \leq D + D_1$. Hence, it is enough to bound the distance of x' and y in U in terms of $D + D_1$. This proves Reduction 1. Note that $\pi(x) \neq n$ since $x \in Fl_K(Q) \setminus Fl_K(X_n)$.

We, therefore, assume from now on that $j = \pi(x) = \pi(y)$ and $x = \rho_{Fl_K(Q)}(y)$ and $d(x, y) \le D$.

REDUCTION 2. We next reduce to the case when both x, y are connected by k-qi leaves to the same point of X_n, where k is as in the statement of lemma.

Let γ_y be a K-qi in $Fl_K(X_n)$ leaf joining y to some $z \in X_n$. In view of the assumption that $x = \rho_{Fl_K(Q)}(y)$, the path

$$\bar{\gamma}_x := \rho_{Fl_K(Q)}(\gamma_y) \subset Fl_K(Q)$$

connects x to $z_1 = \bar{\gamma}_x(n) = \rho_{Fl_K(Q)}(z) \in X_n$. For $y_1 = \rho_{Fl_K(X_n)}(x)$, the path

$$\gamma_{y_1} := \rho_{Fl_K(X_n)} \circ \rho_{Fl_K(Q)}(\gamma_y) \subset X_n$$

also connects y_1 to z_1. Since both Mitra's retractions that we used are $L = L_{3.21}(K)$-coarse Lipschitz, the paths $\bar{\gamma}_x, \gamma_{y_1}$ are k-qi leaves for $k = (L + 1)^2 K$.

Observe that since projections of $Fl_K(Q)$ and $Fl_K(X_n)$ to T both contain the interval $[\![j, n]\!]$, both Mitra's retractions restricted to the vertex-spaces $X_i, i \in [\![j, n]\!]$, amount to fiberwise nearest-point projections to respective flow-spaces. In particular, $y_1 \in X_j$.

We next estimate $d_{X_j}(y, y_1)$. Since $\rho_{Fl_K(X_n)}$ is L-coarse Lipschitz, $\rho_{Fl_K(X_n)}(y) = y$, $\rho_{Fl_K(X_n)}(x) = y_1$ and $d(x, y) \le D$, we obtain

$$d_{Fl_K(X_n)}(y, y_1) \le (L + 1)D.$$

At the same time, $d(x, y_1) \le d(x, y) + d(y, y_1) \le 2D$. Therefore, it suffices to prove the claim for the pair of points x, y_1: Both are connected to $z_1 \in X_n$ by k-leaves $\bar{\gamma}_x \subset Fl_K(Q), \gamma_{y_1} \subset Fl_K(X_n)$ respectively. This proves Reduction 2.

Thus, we consider the case of points $x, y \in X_j$ such that $d(x, y) \le D$, $x \in Fl_K(Q) \cap X_j, y \in Fl_K(X_n) \cap X_j$ and there exist k-qi leaves γ_x, γ_y in $Fl_K(Q)$, $Fl_K(X_n)$, respectively, connecting x, y to a point $z \in X_n$. Let $t \in [\![j, n]\!]$ be the minimal vertex such that

$$d_{X_t}(\gamma_x(t), \gamma_y(t)) \le M_k.$$

(Such t exists since $d_{X_n}(\gamma_x(n), \gamma_y(n)) = 0$.) Recall that the vertex-spaces of X are η_0-uniformly properly embedded in X; in particular, $d_{X_j}(x, y) \le \eta_0(D)$.

If $d_{X_j}(x, y) \le M_k$, we will be done by taking

$$D_{6.1}(K) := M_k,$$

since the intrinsic distance between x and y in $U := N^{fib}_{M_r}(Fl_K(Q) \cup Fl_K(X_n))$ would be $\le d_{X_j}(x, y) \le \eta_0(D)$.

Otherwise (if $d_{X_j}(x, y) > M_k$), by the uniform k-flaring condition, the length of the interval $[\![j, t]\!]$ is at most $\tau_{2.47}(k, \max(\eta_0(D), M_k))$. Therefore, the intrinsic distance between x, y in U is at most

$$k\tau_{2.47}(k, \max(\eta_0(D), M_k)) + M_k.$$

This concludes the proof of the claim. □

We return to the proof of the lemma and consider the general case, when v need not be equal to the extreme vertex n of $T = [\![0, n]\!]$. Let $x \in X_{u_0}, y \in X_{v_0}$ be points within distance D, $u_0, v_0 \in V(T)$. There are two cases to consider, depending on the order of the vertices v, u_0 in $[\![0, n]\!]$.

(1) Suppose that v lies in the interval $[\![u, u_0]\!]$ (or $[\![u_0, u]\!]$). Since $x \in Fl_K(X_u) \cap X_{u_0}$ Proposition 3.26 (specifically, (3.14) with $r = 0$) implies that $x \in Fl_K(X_v)$. Thus, x, y both belong to $Fl_K(X_v)$ and we conclude using the fact that flow-spaces are uniformly quasi-isometrically embedded in X.

(2) Suppose that v does not lie in the interval between u, u_0 in T. Without loss of generality (reversing the orientation on T if necessary), with respect to the order on the interval T,

$$\max(u, u_0) < v.$$

If $v_0 \le v$ then we can shorten the tree T replacing it with the subinterval $[\![0, v]\!]$ and reduce the problem to the one solved in the claim above. Thus, we can assume that

$$0 \le \max(u, u_0) < v < v_0 \le n.$$

In particular, $d_T(u_0, v_0) = d_T(u_0, v) + d_T(v, v_0)$. By projecting a geodesic xy in X to the tree T, we see that $d_T(u_0, v_0) \le D$ and, therefore,

$$\max\left(d_T(u_0, v), d_T(v, v_0)\right) \le D.$$

Hence, a K-qi leaf γ_y in $FL_K(X_v)$ connecting y to $y_1 \in X_v$ has length $\le KD$, which implies that

$$d(y, y_1) \le d_{FL_K(X_v)}(y, y_1) \le (K + 1)D.$$

Consider the subtree $S = [\![0, v]\!]$ in T. According to Lemma 2.18, the inclusion $X_S \to X$ is an $\eta = \eta_{2.18}$-uniformly proper map (where the function η depends only on the parameters of the tree of spaces $\mathfrak{X}$). In particular,

$$d_{X_S}(x, y_1) \le D_1 := \eta((K + 1)D).$$

Thus, it suffices to estimate the distance $d_{X_1}(x, y_1)$ in the fiberwise M_k-neighborhood of $Fl_K(Q) \cup FL_K(X_v) \cap X_S$. Since v is an extreme vertex of S, this is done in the claim above. The lemma follows. □

Lastly, we consider the case of a general tree T.

Proposition 6.3. *Suppose $\mathfrak{X} = (\pi : X \to T)$ is a tree of hyperbolic spaces, $K \ge K_0$ and $\mathfrak{X}$ satisfies the uniform k-flaring condition, where, as in the lemma,*

$$k = k_{6.3}(K) = (L + 1)^2 K, \quad L = L_{3.21}(K).$$

Assume, moreover, that for some $4\delta_0$-quasiconvex subset $Q = Q_u \subset X_u$, we have $Fl_K(Q) \cap X_v \neq \emptyset$. Then the M_k-neighborhood of $Y = Fl_K(Q) \cup Fl_K(X_v)$ in X is $\eta_{6.3,K}$-properly embedded in X.

Proof. Suppose

$$x \in Fl_K(X_u) \cap X_{u_0}, y \in Fl_K(X_v) \cap X_{v_0}$$

satisfy $d(x, y) \le D$. We need to show that there is a constant constant $D_1 > 0$ depending only on D and K such that the distance between x, y in $N_{M_k}(Y)$ is at most D_1.

If both x, y are either in $Fl_K(X_u)$ or $Fl_K(X_v)$ then certainly this is true since K-flows of vertex-spaces are uniformly quasi-isometrically embedded in X.

Suppose, therefore, that

$$x \in \mathcal{F}l_K(X_u) \setminus \mathcal{F}l_K(X_v) \quad \text{and} \quad y \in \mathcal{F}l_K(X_v) \setminus \mathcal{F}l_K(X_u) \tag{6.1}$$

If u, v span an edge in T then our assumptions on the location of x and y imply that the set of vertices $\{u_0 = \pi(x), v_0 = \pi(y), u, v\}$ is contained in a common interval in T and, hence, the claim follows from Lemma 6.1. Thus, we assume that $d_T(u, v) \ge 2$. More generally, we can assume that u_0, v_0, u, v do not belong to a common interval in T.

The same assumption (6.1) implies that the center of $\Delta u_0 uv \subset T$ is in $[\![u, v[\![$ and the center of $\Delta v_0 uv$ is in $]\!]u, v]\!]$, and if these centers are equal, then they are in $]\!]u, v[\![$.

The proof of the proposition is broken in three cases:

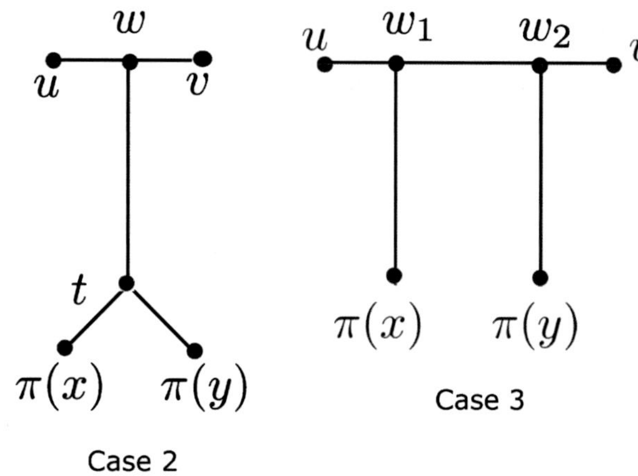

FIGURE 6.1. Cases 2 and 3

CASE 1. Suppose x, y are in the same vertex-space X_t; we let w denote the center of Δuvt. As noted before, $w \in]\!]u, v[\![$. Let $Q = Q_w = Fl_K(X_u) \cap X_w$; this is a $4\delta_0$-quasiconvex subset of X_w.

For the interval $S = [\![v, t]\!]$ consider the subtree of spaces $X_S \subset X$. The points x, y belong to the K-flow-spaces $Fl_K(Q), Fl_K(X_v)$ in X_S, and

$$\emptyset \neq Fl_K(X_u) \cap X_v \subset Fl_K(Q) \cap X_v.$$

We also have $d_{X_S}(x, y) \leq \eta(D)$, where $\eta = \eta_{2.18}$. Thus, the conclusion follows from Lemma 6.1 applied to the tree of spaces X_S over the interval S.

CASE 2. Suppose the triangles Δu_0uv and Δv_0uv have the common center w; see Figure 6.1. Let t denote the center of the triangle Δwu_0v_0. Without loss of generality, we may assume that $d_T(\pi(x), t) \leq d_T(\pi(x), \pi(y))$. Since $d(x, y) \leq D$, we also obtain $d_T(v_0, t) \leq D$. Let γ_x, γ_y denote, respectively, K-qi leaves in $Fl_K(X_u), Fl_K(X_v)$ connecting x, y to X_u, X_v. Then, for $x_1 = \gamma_x(t)$ and $y_1 = \gamma_y(t)$ we have

$$d_{Fl_K(Q)}(x, x_1) \leq KD \quad \text{and} \quad d_{Fl_K(X_v)}(y, y_1) \leq KD.$$

In particular $d(x_1, y_1) \leq (1 + 2K)D$ and x_1, y_1 belong to the same vertex-space X_t. This reduces the proof to that of Case 1.

CASE 3. Suppose the triangles Δu_0uv and Δv_0uv have distinct centers u_1 and v_1 respectively. These centers necessarily belong to the interval $[\![u, v]\!]$; see Figure 6.1. As in Case 2 we first we take two K-qi leaves γ_x, γ_y in $Fl_K(X_u), Fl_K(X_v)$, connecting x, y respectively to X_u and X_v. Since $d(x, y) \leq D$, we also have $d_T(u_0, u_1) \leq D, d(v_0, v_1) \leq D$.

We then replace x with $x_1 = \gamma_x(u_1)$ and replace y with $y_1 = \gamma_y(v_1)$, which are within distance KD from x, respectively y, in $Fl_K(X_u), Fl_K(X_v)$. Furthermore,

$$d(x_1, y_1) \leq D(1 + 2K).$$

Since x_1, y_1 belong to the subtree of spaces $X_{uv} = \pi^{-1}(uv)$, analogously to Case 1, we conclude the proof by applying Lemma 6.1. □

COROLLARY 6.4. *Suppose $\mathfrak{X} = (\pi : X \to T)$ is a tree of hyperbolic spaces, $K \geq K_0$ and $\mathfrak{X}$ satisfies the uniform k-flaring condition for*

$$k = k_{6.3}(K).$$

Assume also that for a $4\delta_0$-quasiconvex subset $Q = Q_u \subset X_u$, we have $Fl_K(Q) \cap X_v \neq \emptyset$.

Then for each $r \geq M_k$, the r-neighborhood of $Fl_K(Q) \cup Fl_K(X_v)$ in X is $\eta_{6.4,K,r}$-properly embedded in X.

6.1.2. Hyperbolicity of the union of two flow-spaces in the case of special intervals. The goal of this section is to prove that a uniform neighborhood of $Fl_K(X_u) \cup Fl_K(X_v)$ in X is uniformly hyperbolic, with the hyperbolicity constant depending only on K, provided that $Fl_K(X_u) \cap X_v \neq \emptyset$ and $\mathfrak{X}$ satisfies a suitable uniform flaring condition.

Here is the idea of the proof. Recall that $Fl_K(X_u)$ and $Fl_K(X_v)$ are (uniformly) hyperbolic and are (uniformly) qi embedded in X. If X were hyperbolic, it would follow (since $Fl_K(X_u) \cap Fl_K(X_v) \neq \emptyset$) that a uniform neighborhood of $Fl_K(X_u) \cup Fl_K(X_v)$ is uniformly hyperbolic. Hyperbolicity of X, of course, is not yet proven, so instead, we will find (see the proof of Corollary 6.4) a larger subset $U = U_{r_1}$ containing $Fl_K(X_u) \cup Fl_K(X_v)$, which is uniformly hyperbolic and uniformly properly embedded in X (Proposition 6.10). Thus, a suitable uniform neighborhood of $Fl_K(X_u) \cup Fl_K(X_v)$ in U is uniformly hyperbolic and uniformly properly embedded in U. From this (since U is uniformly properly embedded in X), we will conclude that a suitable uniform neighborhood of $Fl_K(X_u) \cup Fl_K(X_v)$ in X is also uniformly hyperbolic and uniformly properly embedded in X.

LEMMA 6.5. *For all $D \geq 0, K \geq K_0$ there is a constant $K_{6.5} = K_{6.5}(K) \geq K$ such that the following holds:*

Let $u, v \in T$ be vertices such $Fl_K(X_u) \cap X_v \neq \emptyset$. Then for each

$$x \in N_D(\mathcal{F}l_K(X_u)) \cap N_D(\mathcal{F}l_K(X_v)),$$

there is a vertex $t \in V(T)$ and a $K_{6.5}$-qi section Σ of $\pi : X \to T$ over the tripod (triangle) $S = \Delta tuv$ such that

$$x \in N_{3D'}(\Sigma \cap X_t),$$

where $D' = D + DL_{3.21}(K)$ and

$$\Sigma \subset Fl_K(X_u).$$

PROOF. Suppose that $y_1 \in Fl_K(X_u)$ and $y_2 \in Fl_K(X_v)$ satisfy $d(x, y_i) \leq D$. Let $\alpha_i = \pi([xy_i]_X)$, $i = 1, 2$. Let $T_1 = \pi(Fl_K(X_u))$ and $T_2 = \pi(Fl_K(X_v))$. Then we know that at least one of the α_i will contain a point $t \in T_1 \cap T_2$. Suppose that $t \in \alpha_1$. Let $x_t \in X_t \cap [xy_1]_X$. Let $x_1 = \rho_{Fl_K(X_u)}(x_t)$ and $x_2 = \rho_{Fl_K(X_v)}(x_t)$. As both flow spaces $Fl_K(X_u), Fl_K(X_v)$ intersect X_t, the corresponding Mitra's projections restricted to X_t are merely nearest point projections in X_t to $Fl_K(X_u) \cap X_t$ and $Fl_K(X_v) \cap X_t$ respectively. Hence $x_1 \in Fl_K(X_u) \cap X_t$ and $x_2 \in Fl_K(X_v) \cap X_t$. On the other hand, since the Mitra's projections are $L_{3.21}(K)$-coarse Lipschitz retractions, we have $d(y_i, x_i) \leq DL_{3.21}(K)$, $i = 1, 2$. Thus, $d(x, x_i) \leq D + DL_{3.21}(K) = D'$ for $i = 1, 2$.

Let γ_2 be a K-qi leaf in $Fl_K(X_v)$ connecting x_2 to X_v. We apply Mitra's projection $\rho = \rho_{Fl_K(X_u)}$ to γ_2; call the result γ_2'. Since, by the assumption of the lemma,

$$S \subset \pi(Fl_K(X_u)),$$

ρ restricted to X_S amounts to the fiberwise projection to $Fl_K(X_u)$. In particular, $\pi(\gamma_2') = [\![v, t]\!]$. Hence, γ_2' is a $K_{6.5} := KL_{3.21}(K)$-qi section over $[\![v, t]\!]$ whose image is contained in $Fl_K(X_u)$.

Let $w \in uv$ denote the center of the triangle S. Since γ_2' is contained in $Fl_K(X_u)$, the point $\gamma_2'(w)$ can be joined to X_u by a K-qi leaf γ_1' inside $Fl_K(X_u)$. Clearly, the union of these two qi leaves $\gamma_1' \cup \gamma_2'$ forms a $K_{6.5}$-qi section Σ over the tripod S.

Lastly, we have

$$d_{X_t}(x_2, \gamma_2'(t)) \leq d_{X_t}(x_1, x_2) \leq 2D',$$

$$d(x, x_2) \leq D', \quad d(x, \gamma_2'(t)) \leq 3D'.$$

The lemma follows. □

Set

$$k = K_{6.5}(K) \tag{6.2}$$

and define $\mathcal{S} = \mathcal{S}_{k,J}$, the set of all k-qi sections over the interval $J = [\![u, v]\!]$.

Assuming that $Fl_K(X_u) \cap X_v \neq \emptyset$ (which is the standing assumption of the previous and this section), $\mathcal{S}$ is nonempty since $k \geq K$ and we are assuming that

$$Fl_K(X_u) \cap X_v \neq \emptyset.$$

For each vertex $w \in V(J)$, let H_w denote the (fiberwise) δ_0-hull of the subset $\{\gamma(w) : \gamma \in \mathcal{S}\}$ in X_w. Define

$$\mathcal{H} := \bigcup_{w \in V(J)} H_w.$$

Each H_w, of course, is a $4\delta_0$-quasiconvex subset of X_w. Then Lemma 3.17(b) implies that $\mathcal{H}$ is the union of vertex-spaces of a $k' = K'_{3.17}(k)$-metric bundle $\mathfrak{H} = \mathfrak{H}_{k,J}$ over the interval J; see Definition 3.5.

As in Section 3.3.3, we define the generalized κ-flow-spaces $Fl_\kappa(\mathfrak{H})$ of the metric bundle $\mathfrak{H}$. From the definition, we recall that with each vertex $w \in V(J)$ we associate a subtree $T_w \subset T$ equal to the maximal subtree in T containing w and disjoint from all other vertices of the interval J.

Below we will frequently use the function

$$\kappa \mapsto \kappa^\wedge = (15L_0'\kappa)^3,$$

defined in (3.11).

Recall that $Fl_K(X_u) \cap X_v \neq \emptyset$.

Lemma 6.6. *For $D \geq 0, K \geq K_0$ set $k = K_{6.5}(K)$, $k' = K'_{3.17}(k)$. Then the following hold:*

(1) *Suppose that*

$$x \in N_D(\mathcal{F}l_K(X_u)) \cap N_D(\mathcal{F}l_K(X_v)).$$

Then there exists a vertex $t \in V(T)$ with

$$d(x, \mathcal{F}l_K(X_u) \cap \mathcal{F}l_K(X_v) \cap X_t) \leq D,$$

such that the center w of Δuvt satisfies

$$x \in N_{3D}(Fl_{k^\wedge}(H_w) \cap X_{T_w}).$$

In other words, we have

$$N_D(\mathcal{F}l_K(X_u)) \cap N_D(\mathcal{F}l_K(X_v)) \subset \bigcup_{w \in V([\![u,v]\!])} N_{3D}(Fl_{k^\wedge}(H_w) \cap X_{T_w}).$$

(2) *For each* $\kappa \geq k'$ *and each vertex* $w \in V([\![u,v]\!])$ *we have*

$$Fl_\kappa(H_w) \cap X_{T_w} \subset Fl_{\kappa^\wedge}(X_u) \cap Fl_{\kappa^\wedge}(X_v).$$

In other words,

$$Fl_\kappa(\mathfrak{H}) \subset Fl_{\kappa^\wedge}(X_u) \cap Fl_{\kappa^\wedge}(X_v).$$

PROOF.

(1) By Lemma 6.5 there is a $k = K_{6.5}(K)$-qi section Σ over a tripod $uv \cup wt$ such that

$$d(x, \Sigma) \leq 3D.$$

In particular, Σ restricted to $J = [\![u,v]\!]$ is a k-qi section over J hence, belongs to $\mathcal{S}_{k,J}$. Therefore, by the definition of $\mathcal{H}$, for the vertex $w \in V(J)$, the intersection $\Sigma \cap X_w$ belongs to H_w.

Since $\Sigma(w)$ is in H_w, the restriction of Σ to the interval $[\![t,w]\!]$ is a k-qi leaf connecting $\Sigma \cap X_t$ to H_w and, by Proposition 3.26(2),

$$\Sigma \cap X_t \in Fl_{k^\wedge}(H_w).$$

Thus,

$$x \in N_{3D}(Fl_{k^\wedge}(H_w)) \cap X_t.$$

This proves Part (1) for fiberwise neighborhoods. The proof for neighborhoods taken in X is identical and we omit it.

(2) Given $x \in \mathcal{F}l_\kappa(H_w) \cap X_{T_w}$, pick a κ-qi leaf γ_x in the flow-space $Fl_\kappa(H_w)$ connecting x to H_w.

Since $z = \gamma_x(w) \in H_w$, and $\mathfrak{H}$ is a k'-metric bundle over the interval $J = [\![u,v]\!]$, there exists a k'-qi leaf γ_z in this bundle (over $[\![w,v]\!]$) connecting z to X_v. Thus, since $\kappa \geq k'$, the point x can be connected to both X_u and X_v by κ-qi leaves. According to Proposition 3.26(2),

$$x \in Fl_{\kappa^\wedge}(X_u) \cap Fl_{\kappa^\wedge}(X_v). \qquad \square$$

For the rest of this section (until Corollary 6.4) we will be working under the following assumption (which is stronger than the one we had earlier):

- We assume that $X_v \cap Fl_K(X_u) \neq \emptyset, X_u \cap Fl_K(X_v) \neq \emptyset$.

REMARK 6.7.

(1) This assumption implies that for all vertices $w \in V([\![u,v]\!])$,

$$X_w \cap Fl_K(X_u) \neq \emptyset, X_w \cap Fl_K(X_v) \neq \emptyset.$$

(2) The stronger assumption we are now making is not too far from the condition that $X_v \cap Fl_K(X_u) \neq \emptyset$ made earlier, since

$$X_u \cap Fl_{K^\wedge}(X_v) \neq \emptyset;$$

see Proposition 3.26(2). We will be using this fact in the proof of Corollary 6.4.

We now fix some $K \geq K_0$, set, $k = K_{6.5}(K)$ and take some $R \geq k^\wedge$. For an interval $J = [\![u,v]\!] \subset T$, set $\mathfrak{H} := \mathfrak{H}_{k,J}$ and $Y^0 := Fl_R(\mathfrak{H})$. Since $\mathfrak{F}l_R(\mathfrak{H})$ is a generalized flow-space with the parameters $K_1 = k'$ and $K_2 = R \geq K_0$, Lemma 6.8 is a corollary of Theorem 3.34.

LEMMA 6.8. *The inclusion map* $Y^0 \to X$ *is a* $L_{3.34}(k', R)$*-qi embedding.*

We also define the unions $Y^1 = Fl_K(X_u) \cup Fl_R(\mathfrak{H})$, $Y^2 = Fl_K(X_v) \cup Fl_R(\mathfrak{H})$ and their neighborhoods $U^i_r := N_r(Y^i), i = 1, 2$, taken in X.

LEMMA 6.9. *For every* $r \geq 0$,

$$Fl_R(\mathfrak{H}) \subset U_r^1 \cap U_r^2 \subset N_{3r}(Fl_R(\mathfrak{H})),$$

i.e., the intersection is uniformly (in terms of r, R *and* K*) Hausdorff-close to* $Fl_R(\mathfrak{H})$.

PROOF. Consider $x \in U_r^1 \cap U_r^2$. Thus, there exist points $x_1 \in Fl_K(X_u) \cup Fl_R(\mathfrak{H})$, $x_2 \in Fl_K(X_v) \cup Fl_R(\mathfrak{H})$ at distance $\leq r$ from x. If one of these points is in $Fl_R(\mathfrak{H})$ then $d(x, Fl_R(\mathfrak{H})) \leq r$. Therefore, assume that $x_1 \in Fl_K(X_u)$, $x_2 \in Fl_K(X_v)$. By Lemma 6.6(1), $d(x, Fl_R(\mathfrak{H})) \leq 3r$, as required. □

Recall that $K \geq K_0$, $k = K_{6.5}(K)$. Set $R := k^\wedge$. In the next proposition, N' indicates a metric neighborhood of Y_1 or Y_2 taken inside $Fl_{R^\wedge}(X_u)$ or $Fl_{R^\wedge}(X_v)$ respectively. The most useful part of the proposition is (2): Part (1) is used only to prove (2).

PROPOSITION 6.10. *Assume that* $K \geq K_0$ *and* $\mathfrak{X}$ *satisfies the uniform* κ*-flaring condition for*

$$\kappa = \max(k_{6.3}(K), \kappa_{4.7}(R^\wedge)).$$

Then there exist $\delta'_{6.10} = \delta'_{6.10}(K, C)$, $\delta_{6.10} = \delta_{6.10}(K)$, $L'_{6.10} = L'_{6.10}(K, C)$, $C_{6.10} = C_{6.10}(K)$, *and a function* $\eta_{6.10} = \eta_{6.10,K}$ *such that the following hold.*

(1) *For each* $C \geq C_{6.10}$, *both* $U'_1 = N'_C(Y^1)$ *and* $U'_2 = N'_C(Y^2)$, *equipped with the induced path-metrics, are* $\delta'_{6.10}(K, C)$*-hyperbolic and* $L'_{6.10}(K, C)$*-qi embedded in* X.

(2) *For*

$$r := \max(C_{6.10}, M_{k_{6.3}(K)}),$$

the union

$$U_r := U_r^1 \cup U_r^2$$

equipped with the induced path-metric, is $\delta_{6.10}(K, R)$*-hyperbolic and* $\eta_{6.10}$*-uniformly properly embedded in* X.

PROOF.

(1) We will only prove the claim for U'_1 since the proof for U'_2 is obtained by relabelling. Recall that

$$Y^1 = Fl_K(X_u) \cup Y^0.$$

Since $R \geq k'$, the definition of Y^0 and Lemma 6.6 imply that

$$Y^0 = Fl_R(\mathfrak{H}) \subset Fl_{R^\wedge}(X_u).$$

Furthermore, since $R^\wedge \geq R \geq k \geq K$, $Fl_K(X_u) \subset Fl_{R^\wedge}(X_u)$. Thus,

$$Y^1 \subset Fl_{R^\wedge}(X_u).$$

Recall that we are assuming that $\mathfrak{X}$ satisfies the uniform $\kappa_{4.7}(R^\wedge)$-flaring condition. Therefore, Theorem 5.17 applies and the flow-space $Fl_{R^\wedge}(X_u)$ is $\delta = \delta_{5.17}(R^\wedge)$-hyperbolic.

By Lemma 6.8, Y^0 is $L_{3.34}(k', R)$-qi embedded in X, while $Fl_K(X_u)$ is $L_{3.22}(K)$-qi embedded in X according to Corollary 3.22.

Hence, for

$$L := 2\max(L_{3.34}(k', R), L_{3.22}(K))$$

and

$$\lambda = \lambda_{1.92}(\delta, L),$$

both Y^0 and $Fl_K(X_u)$ are λ-quasiconvex in $Fl_{R^\wedge}(X_u)$. Moreover, these subsets have nonempty intersection (containing at least H_u). Thus, their union is $\lambda + \delta$-quasiconvex in $Fl_{R^\wedge}(X_u)$.

By Lemma 1.98(1), since

$$C \geq C_{6.10} \geq \lambda + \delta,$$

the C-neighborhood U_1' of $Y^1 = Fl_K(X_u) \cup Y^0$ in $Fl_{R^\wedge}(X_u)$ (equipped with the induced path-metric) is $\delta + C$-quasiconvex in $Fl_{R^\wedge}(X_u)$.

Furthermore, by Part (2) of the same lemma, since C was taken to be $\geq 2\lambda + 4\delta$, the inclusion map $U_1' \to Fl_{R^\wedge}(X_u)$ is a $(1, 6\delta + C)$-quasi-isometric embedding, where U_1' is equipped with the induced path-metric. Thus, U_1' is $\delta_{1.56}(\delta, 6\delta + C)$-hyperbolic; see Lemma 1.56. Moreover, the inclusion map $U_1' \to X$ is an $L'_{6.10}$-qi embedding with

$$L'_{6.10} = L_{3.22}(R^\wedge)(6\delta + C).$$

This proves Part (1), where we use:

$$\delta = \delta_{5.17}(R^\wedge), L = 2\max(L_{3.34}(k', R), L_{3.22}(K)),$$
$$\lambda = \lambda_{1.92}(\delta, L), C_{6.10} = 2\lambda + 4\delta, \delta'_{6.10} = \delta_{1.56}(\delta, 6\delta + C).$$

(2) In the proof of Part (1) we were using hyperbolicity of the union of two quasiconvex subsets in a hyperbolic space. In Part (2), such ambient hyperbolic space is unavailable, so we will use hyperbolicity of *pairwise quasiconvex amalgams* of hyperbolic spaces (Theorem 2.65, quasiconvex amalgamation), to prove hyperbolicity of U_r. We will be using Part (1) with $C = r$. We have, of course,

$$Y_i \subset U_i' = N_r'(Y_i) \subset U_r^i = N_r(Y_i), i = 1, 2.$$

Since each inclusion map $U_i' \to U_r^i$ is an L-quasi-isometry,

$$L' = \max(L_{3.22}(K), r),$$

Lemma 1.56 implies that U_r^i is $\delta'_{1.56}(\delta'_{6.10}, L')$-hyperbolic.

Since Y^0 is $L_{3.34}(k', R)$-qi embedded in X, it is $\lambda_{1.92}(\delta'_{6.10}, L_{3.34}(k', R))$-quasiconvex in $U_r^i, i = 1, 2$. Since Y^0 is $3r$-Hausdorff close to $U_r^1 \cap U_r^2$ (Lemma 6.9), it follows that $U_r^1 \cap U_r^2$ is λ'-quasiconvex in both U_r^1, U_r^2 where,

$$\lambda' = 3r + \delta'_{6.10}\lambda_{1.92}(\delta'_{6.10}, L_{3.34}(k', R)).$$

Thus, we are in position to apply Theorem 2.65 (quasiconvex amalgams), and conclude that the union $U_r = U_r^1 \cup U_r^2$ is $\delta_{6.10}(K, D, R)$-hyperbolic. It remains to prove that U_r is uniformly properly embedded in X, thereby proving half of (2).

Let x_1, x_2 be points in U_r. We need to estimate $d_{U_r}(x_1, x_2)$ in terms of $d(x_1, x_2)$. It suffices to consider the case when $x_i \in U_r^i, i = 1, 2$, since U_1', U_2' are both $L'_{6.10}$-qi embedded in X and are within Hausdorff distance r from U_r^1, U_r^2 respectively. Let $y_i \in Y^i$ be points at distance $\leq r$ from $x_i, i = 1, 2$; $d(y_1, y_2) \leq d(x_1, x_2) + 2r$. Since

$$Y^1 = Fl_K(X_u) \cup Y^0, Y^2 = Fl_K(X_v) \cup Y^0$$

using the fact that Y^0 is $L_{3.34}(k', R)$-qi embedded in X, the problem reduces to the case when $y_1 \in Fl_K(X_u), y_2 \in Fl_K(X_v)$. Recall that according to Corollary 6.4 (applied to $Q_u = X_u$), since

$$r \geq M_{k_{6.3}(K)},$$

the r-neighborhood of the union $Fl_K(X_u) \cup Fl_K(X_v)$ is $\eta_{6.4,K}$-properly embedded in X. Therefore,

$$N_r(Fl_K(X_u) \cup Fl_K(X_v)) \subset N_r(Y^1) \cup N_r(Y^2) = U_r,$$

and we obtain

$$d_{U_r}(x_1, x_2) \leq d_{N_{M_k}(Fl_K(X_u) \cup Fl_K(X_v))} \leq \eta_{6.3,K}(d(x_1, x_2) + 2r).$$

This concludes the proof of Part (2) and, hence, of the proposition. □

We can now prove the main result of this section.

COROLLARY 6.11. *Suppose that $K \geq K_0$ and u, v are vertices in T such that $Fl_K(X_u) \cap X_v \neq \emptyset$. Set $k_{6.11} := K_{6.5}(K^\wedge)$, $R_{6.11} := k^\wedge_{6.11}$ and assume that $\mathfrak{X}$ satisfies the uniform κ-flaring condition for*

$$\kappa = \kappa_{6.11}(K) = \max(k_{6.3}(K^\wedge), \kappa_{4.7}(R^\wedge_{6.11})).$$

Then there exist $\delta = \delta_{6.11}(K)$, $D = D_{6.11}(K)$, and a function $\eta = \eta_{6.11,K}$ such that the following hold:

The D-neighborhood $N_D(Fl_K(X_u) \cup Fl_K(X_v))$ (with the induced path-metric) in X is δ-hyperbolic and η-properly embedded in X.

PROOF. According to Proposition 3.26, for $K_1 = K^\wedge$,

$$Fl_{K_1}(X_v) \cap X_u \neq \emptyset.$$

Of course, we still have

$$Fl_{K_1}(X_u) \cap X_v \supset Fl_K(X_u) \cap X_v \neq \emptyset.$$

Therefore, Proposition 6.10(2) applies and we get that for $r_1 = r(K_1)$ as in the proposition, $U = U_{r_1}$ is $\delta(K_1)$-hyperbolic and η_{K_1}-properly embedded in X.

Since both $Fl_K(X_u), Fl_K(X_v)$ are $L = L_{3.22}(K)$-qi embedded in X (hence, in U), they are $\lambda_1 = \lambda_{1.92}(\delta(K_1), L)$-quasiconvex in U (Lemma 1.92). Set

$$D := \max\left(2\lambda_1 + 4\delta(K_1), M_{k_{6.3}(K)}\right).$$

By Lemma 1.98, the D-neighborhood V of $Fl_K(X_u) \cup Fl_K(X_v)$ in U is $(2\lambda_1 + 5\delta(K_1))$-quasiconvex in U. By the same lemma, V (equipped with its path-metric) is $(10\delta(K_1) + 2\lambda_1)$-qi embedded in U. Hence, V is $\delta_{1.56}(\delta(K_1), 10\delta(K_1) + 2\lambda_1)$-hyperbolic (see Lemma 1.56).

Note that the Hausdorff distance in X between V (the D-neighborhood of $Fl_K(X_u) \cup Fl_K(X_v)$ in U) and $N_D(Fl_K(X_u) \cup Fl_K(X_v))$ (the D-neighborhood in X) is $\leq D$. Since U is η_{K_1}-properly embedded in X and the inclusion map $V \to U$ is a $(10\delta(K_1) + 2\lambda_1)$-qi embedding, the composition $V \to N_D(Fl_K(X_u) \cup Fl_K(X_v))$ is ζ-proper for

$$\zeta(t) = (10\delta(K_1) + 2\lambda_1)\eta(t) + (10\delta(K_1) + 2\lambda_1)^2.$$

Corollary 1.60 now implies that $N_D(Fl_K(X_u) \cup Fl_K(X_v))$ is $\delta_{6.11}(K)$-hyperbolic for

$$\delta_{6.11}(K) = \delta_{1.56}(\delta_{1.56}(\delta(K_1), 10\delta(K_1) + 2\lambda_1), \zeta(2D + 1)).$$

It remains to estimate the distortion of $N_D(Fl_K(X_u) \cup Fl_K(X_v))$ in X. Since $D \geq M_{k_{6.3}(K)}$, Corollary 6.4 applies and $Fl_K(X_u) \cup Fl_K(X_v)$ is $\eta_{6.4,K,D}$-properly embedded in X. Therefore, its D-neighborhood is η-properly embedded in X for

$$\eta_{6.11,K}(t) = \eta(t) = 2D + \eta_{6.4,K,D}(t + 2D).$$ □

REMARK 6.12. This corollary gives us the value of K_* in Theorem 2.62:

$$K_* = \max(k_{6.3}(K^\wedge), \kappa_{4.7}(R^\wedge_{6.11})), \tag{6.3}$$

where $K = K_0$.

6.1.3. Hyperbolicity of flow-spaces of special interval-spaces $Fl_K(X_J)$. In this section we conclude the proof of Theorem 6.14. We will also prove uniform hyperbolicity of flow-spaces $Fl_K(X_J)$, whenever $J \subset T$ is a union of three special intervals.

For the next proposition, set $D = D_{6.11}(K)$ and $\kappa = \kappa_{6.11}(K)$.

PROPOSITION 6.13. *Assume that $K \geq K_0$ and $\mathfrak{X}$ satisfies the uniform κ-flaring condition. Let v_0, v_1, v_2 be vertices of T such that $v_0 \in [\![v_1, v_2]\!]$. We will assume that each subinterval $J_i = [\![v_0, v_i]\!]$, $i = 1, 2$, is special. Then the D-neighborhood (taken in X)*

$$Z := N_D(Fl_K(X_u) \cup Fl_K(X_v) \cup Fl_K(X_w))$$

(equipped with the path-metric induced from X) is $\delta = \delta_{6.13}(K)$-hyperbolic and $\eta = \eta_{6.13,K}$-properly embedded in X.

PROOF. Define $Y_i = N_D(Fl_K(X_{v_i}))$, $Z_i := N_D(Y_0 \cup Y_i)$, $i = 0, 1, 2$. Thus, $Z = N_D(Y_1 \cup Y_2)$ and, furthermore, $N_D(Y_0) = Z_0 = Z_1 \cap Z_2$ and Z_0 separates Z_1, Z_2 in X: Every path c connecting a point of Z_1 to a point of Z_2, has to intersect Z_0; see Proposition 3.26(1).

1: $Z = Z_1 \cup Z_2$ **is hyperbolic.** The hypothesis of the proposition implies that both Z_1 and Z_2 satisfy the assumptions of Corollary 6.11. Thus, each Z_i is $\delta_{6.11}(K)$-hyperbolic and $\eta_{6.11,K}$-properly embedded in X. Since $Fl_K(X_{v_0})$ is $L_{3.22}(K)$-qi embedded in X, it is $\lambda = \lambda_{1.92}(\delta_{6.11}(K), L_{3.22}(K))$-quasiconvex in Z_1, Z_2. Hence, $Z_0 = N_D(Fl_K(X_{v_0}))$ is $\lambda + D + 2\delta_{6.11}(K)$-quasiconvex in Z_1, Z_2. Theorem 2.65 (for quasiconvex amalgams) now implies that Z is $\delta = \delta_{6.13}(K)$-hyperbolic.

2: $Z_1 \cup Z_2$ **is uniformly properly embedded in** X**.** The proof is similar to that of Proposition 6.10(2). Take points $x_1 \in Z_1, x_2 \in Z_2$. Then the separation property mentioned earlier, implies that each geodesic z_1z_2 in X has to intersect Z_0 at some z_0. In particular, $\max(d(z_1, z_0), d(z_0, z_2)) \leq d(z_1, z_2)$.

By Proposition 6.3, for $i = 1, 2$,

$$d_{Z_i}(z_i, z_0) \leq 2D + \eta_{6.3,K}(d(z_i, z_0)) \leq 2D + \eta_{6.3,K}(d(z_1, z_2)),$$

and, therefore,

$$d_Z(z_1, z_2) \leq \eta(d(z_1, z_2)) = \eta_{6.13,K}(d(z_1, z_2)) := 4D + 2\eta_{6.3,K}(d(z_1, z_2)).$$ □

Recall that for subtrees $S \subset T$, we defined flow-spaces $Fl_K(X_S)$; see (3.16). We again assume that $K \geq K_0$, $\mathfrak{X}$ satisfies the uniform $\kappa_{6.11}(K)$-flaring condition and set $D = D_{6.11}(K)$.

THEOREM 6.14. *The flow-space $Fl_K(X_J)$ of any special interval $J = [\![u, v]\!] \subset T$ (equipped with the intrinsic path-metric) is $\delta_{6.14}(K)$-hyperbolic.*

PROOF. It suffices to prove uniform hyperbolicity of the D-neighborhood of $Fl_K(X_J)$ in X (with the path-metric induced from X), the claim then will follow from the fact that $Fl_K(X_J)$ is uniformly qi-embedded in X (Proposition 3.32).

Note that, in view of the assumption of the theorem, for any two vertices $t, s \in J$, at least one of the intersections is nonempty:

$$Fl_K(X_t) \cap X_s \neq \emptyset, \quad \text{or} \quad Fl_K(X_s) \cap X_t \neq \emptyset$$

(depending on which distance $d(t, v)$, $d(s, v)$ is larger). Thus, any triple of vertices v_0, v_1, v_2 in J satisfies the assumptions of Proposition 6.13, and, hence,

$$Z = N_D\left(Fl_K(X_{v_0}) \cup Fl_K(X_{v_1}) \cup Fl_K(X_{v_2})\right)$$

is $\delta_{6.13}(K)$-hyperbolic and $\eta_{6.13,K}$-properly embedded in X. Since each flow-space $Fl_K(X_{v_i})$ is $L_{3.22}(K)$-qi embedded in X, it follows that the D-neighborhood of the union of any two of these flow-spaces is $\lambda(K)$-quasiconvex in Z.

We are now in position to apply Corollary 1.64. For each pair of points $x, y \in \mathcal{F}l_K(X_I)$ we define the path $c(x, y)$ in $N_D(Fl_K(X_J))$ to be a geodesic between x, y in

$$N_D\left(Fl_K(X_{\pi(x)}) \cup Fl_K(X_{\pi(y)})\right).$$

In view of the uniform proper embeddedness of this union (in X) and the uniform hyperbolicity of the triple unions as above, this family of paths in $Fl_K(X_J)$ satisfies axioms of Corollary 1.64. □

We now deal with Step 2 outlined in the introduction to this chapter. This step is a rather direct application of Theorem 6.14: We apply quasiconvex amalgamation of pairs twice to show that $Fl_K(X_J)$ is hyperbolic.

Proposition 6.15. *Suppose that J is an interval in T, that can be subdivided as a union of three special subintervals $J = J_1 \cup J_2 \cup J_3$,*

$$J_i = [\![v_i, v_{i+1}]\!], i = 1, 2, 3.$$

Then $Fl_K(X_J)$ (equipped with the intrinsic path-metric) is $\delta_{6.15}(K)$-hyperbolic.

Proof. A quick way to argue is to appeal to Corollary 2.69 since $Fl_K(X_J)$ has a structure of a hyperbolic tree of spaces with the base-tree consisting of four vertices and three edges (forming an interval of length 3), where the vertex-spaces are $Fl_K(X_{J_i})$ and the edge-spaces are $Fl_K(X_{v_i})$. We will give a more explicit proof following the proof of Corollary 2.69 since it will also provide us with a description of uniform quasigeodesics in $Fl_K(X_J)$.

We will use the quasiconvex amalgamation (see Section 2.6.2) *twice*:

(a) We have

$$Fl_K(X_{J_1}) \cap Fl_K(X_{J_2}) = Fl_K(X_{v_2}).$$

The intersection is $L_{3.22}(K)$-qi embedded in X, hence, in $Fl_K(X_I)$, $I = J_1 \cup J_2$. Since both $Fl_K(X_{J_1}), Fl_K(X_{J_2})$ are $\delta_{6.14}(K)$-hyperbolic, Theorem 2.65 (for quasiconvex amalgams) implies $\epsilon_{6.15}(K)$-hyperbolicity of their union $Fl_K(X_I)$.

(b) We have

$$Fl_K(X_J) = Fl_K(X_I) \cup Fl_K(X_{J_3}),$$

and

$$Fl_K(X_I) \cap Fl_K(X_{J_3}) = Fl_K(X_{v_3}).$$

The intersection is $L_{3.22}(K)$-qi embedded in X, hence, in $Fl_K(X_I)$, $I = J_1 \cup J_2$. Since both $Fl_K(X_I), Fl_K(X_{J_3})$ are δ-hyperbolic,

$$\delta = \max(\delta_{6.14}(K), \epsilon_{6.15}(K)),$$

Theorem 2.65 (for quasiconvex amalgams) implies $\delta_{6.15}(K)$-hyperbolicity of their union $Fl_K(X_J)$. □

The following result is never used afterward; we include the proof for the sake of completeness of the picture.

Corollary 6.16 (Hyperbolicity of the union of two flow-spaces: General case). *For each $K \geq K_0$, for $D = D_{6.11}(K)$, assuming that $\mathfrak{X}$ satisfies the uniform $\kappa_{6.11}(K)$-flaring condition, the following holds:*

If $u, v \in V(T)$ are such that $Fl_K(X_u) \cap Fl_K(X_v) \neq \emptyset$, then

$$N_D(Fl_K(X_u) \cup Fl_K(X_v))$$

is an $L_{6.16}(K)$*-qi embedded,* $\delta_{6.16}(K)$*-hyperbolic subspace of* X*.*

Proof. Since $Fl_K(X_u) \cap Fl_K(X_v) \neq \emptyset$, the interval $J = [\![u, v]\!]$ splits as the union of two special subintervals, J_1, J_2, there is a vertex w in the interval $I = [\![u, v]\!]$ such that $X_w \cap Fl_K(X_u) \neq \emptyset$ and $X_w \cap Fl_K(X_v) \neq \emptyset$ (see Lemma 3.40). Thus, $N_D(Fl_K(X_u) \cup Fl_K(X_v))$ is contained in the D-neighborhood of the $\delta_{6.15}(K)$-hyperbolic subspace, $Fl_K(X_I)$. The result now follows from Lemma 1.98 on unions of quasiconvex subsets of hyperbolic spaces, combined with Proposition 6.3. □

6.2. Hyperbolicity of flow-spaces of general interval-spaces

This section deals with Step 3 outlined in the introduction to this chapter. Recall that according to Proposition 3.32, for every subtree $S \subset T$ and $K \geq K_0$, the flow-space $Fl_K(X_S)$ is $L_{3.32}(K)$-qi embedded in X.

Theorem 6.17. *For every* $K \geq K_0$*, assuming that* $\mathfrak{X}$ *satisfies the uniform* $\kappa_{6.11}(K)$*-flaring condition, for each interval* $J = [\![u, v]\!] \subset T$*, the flow-space* $Fl_K(X_J)$ *is* $\delta_{6.17}(K)$*-hyperbolic.*

Proof. When $u = v$, the statement is established in Theorem 5.17. Therefore, we consider the case of nondegenerate intervals J.

We apply the Horizontal Subdivision Lemma (Lemma 3.44) and its corollary (Corollary 3.45) to subdivide the interval J into subintervals $J_0, \ldots, J_n$ such that:

(i) Each J_i is the union of three K-special subintervals.
(ii) Whenever $|i - j| \geq 2$, the flow-spaces $Fl_K(X_{J_i}), Fl_K(X_{J_j})$ are $L_{3.21}(K)$-Lipschitz cobounded in X, hence (taking restrictions of Mitra's projections), in $Fl_K(X_J)$ as well.

According to Corollary 6.15, each flow-space $Fl_K(X_{J_i})$ is $\delta_{6.15}(K)$-hyperbolic.

Finally, we consider the union

$$Fl_K(X_J) = \bigcup_{0 \leq i \leq n} Fl_K(X_{J_i}).$$

By Proposition 3.26(1),

$$Fl_K(X_{J_{i-1}}) \cap Fl_K(X_{J_i}) = Fl_K(X_{u_i}).$$

Since whenever $|i - j| \geq 2$, the flow-spaces $Fl_K(X_{J_i}), Fl_K(X_{J_j})$ are $L_{3.21}(K)$-Lipschitz cobounded in $Fl_K(X_J)$ and the consecutive intersections $Fl_K(X_{u_i})$ are $L_{3.21}(K)$-qi embedded in $Fl_K(X_J)$, Theorem 2.65 applies and the flow-space $Fl_K(X_J)$ is $\delta_{6.17}(K)$-hyperbolic. □

6.3. Conclusion of the proof

In this section we finish the proof of the main result of this book, Theorem 2.62. We first prove that flow-spaces $Fl_K(X_S)$ are uniformly hyperbolic, whenever S is a tripod in T (Corollary 2.69). We then conclude the proof of Theorem 2.62 by appealing to Corollary 1.64 one last time by constructing a slim combing in X via geodesics in flow-spaces of interval-spaces.

Proposition 6.18. *Assume that* $\mathfrak{X}$ *satisfies the uniform* $\kappa_{6.11}(K)$*-flaring condition. Suppose* $S = T_b u_1 u_2 u_3 \subset T$ *is a tripod*[1] *with the center* b *and three extremities* u_1, u_2, u_3*. Then for every* $K \geq K_0$*, the flow-space* $Fl_K(X_S)$ *is* $\delta_{6.18}(K)$*-hyperbolic.*

[1] See Definition 1.74.

PROOF. The proof is similar to that of Proposition 6.15. The tripod S is the union of three segments (legs) $J_i = [\![u_i, b]\!], i = 1, 2, 3$, whose pairwise intersections equal $\{b\}$. According to Theorem 6.17, each flow-space $Fl_K(X_{J_i})$ is $\delta_{6.17}(K)$-hyperbolic. The intersection

$$\bigcap_{i=1}^{3} Fl_K(X_{J_i}) = Fl_K(X_{J_m}) \cap Fl_K(X_{J_n}) = Fl_K(X_b),\ 1 \leq m < n \leq 3$$

(see Proposition 3.26(1)) is $\delta_{5.17}(K)$-hyperbolic (Theorem 5.17) and $L_{3.21}(K)$-qi embedded in X. Therefore, $Fl_K(X_S)$ has structure of a tripod $\mathfrak{Y}$ of hyperbolic spaces (see Remark 2.64(2))

$$Y = Fl_K(X_S) \to S' = T_b v_1 v_2 v_3,$$

where the tripod S' has four vertices (b, v_1, v_2, v_3) and three edges, all incident to the vertex b. Namely, the vertex-spaces of $\mathfrak{Y}$ are $Y_b = Fl_K(X_b)$, $Y_{v_i} = Fl_K(X_{J_i})$. The three edge-spaces are all isomorphic to Y_b and have natural incidence maps (inclusion maps) to the vertex-spaces. Now, the proposition follows from Corollary 2.69. □

We can now finish our proof of Theorem 2.62. We let $K = K_0$ and K_* be as Notation 2.6.4 and Remark 6.12 respectively (the constants $\delta'_0, \lambda'_0, L'_0$ used in Notation 2.6.4 are defined in Notation 2.6.4). We assume that the tree of spaces $\mathfrak{X}$ satisfies the uniform K_*-flaring condition.

We will once again apply Corollary 1.64, with $X_0 = \mathcal{X}$, the union of vertex-spaces in X. For each pair of vertices $u, v \in V(T)$, and points $x \in X_u$, $y \in X_v$, we define the path $c(x, y)$ in X to be the intrinsic geodesic in the flow-space $Y = Fl_K(X_{uv})$. In order to verify the assumptions of Corollary 1.64, we observe that condition (a1) follows from the fact that each Y is $L = L_{3.32}(K)$-qi embedded in X. We also conclude that each path $c(x, y)$ is an L-quasigeodesic in X. Condition (a2) is immediate from Proposition 6.18. □

COROLLARY 6.19. *Let $\mathfrak{X} = (\pi : X \to T)$ be a tree of hyperbolic spaces (satisfying Axiom **H**). Then the following conditions are equivalent:*

(1) *X is hyperbolic.*
(2) *$\mathfrak{X}$ satisfies the uniform κ-flaring condition for all $\kappa \in [K_0, K_*]$.*
(3) *$\mathfrak{X}$ satisfies the Bestvina-Feighn exponential flaring condition for all $\kappa \geq 1$.*
(4) *Carpets in $\mathfrak{X}$ are uniformly hyperbolic. More precisely, there exists a function $\delta(K, C)$, such that each (K, C)-narrow carpet in $\mathfrak{X}$ has $\delta(K, C)$-hyperbolic total space.*
(5) *Ladders in $\mathfrak{X}$ are uniformly hyperbolic. More precisely, there exists a function $\delta(K, D, E)$, such that each (K, D, E)-ladder in $\mathfrak{X}$ has $\delta(K, D, E)$-hyperbolic total space.*

PROOF. The implication $2 \Rightarrow 1$ is the content of Theorem 2.62. The converse implication $1 \Rightarrow 2$ is Lemma 2.50. In Proposition 2.60 we proved the implication $1 \Rightarrow 3$, while the implication $3 \Rightarrow 2$ is proven in Lemma 2.59. Thus, we obtain the equivalence of 1, 2, and 3.

In order to establish equivalence of 1 with 4 and 5, observe that the uniform flaring condition is defined in terms of separation properties of pairs of K-sections over intervals in T and every such pair of sections is contained in a carpet, while each carpet is contained in a ladder. Applying the implication $1 \Rightarrow 2$ to trees of spaces which are carpets and ladders, we conclude that uniform hyperbolicity of carpets/ladders implies the uniform κ-flaring condition for $\mathfrak{X}$ for all $\kappa \geq 1$. □

CHAPTER 7

Description of geodesics

7.1. Inductive description

Let $\pi : X \to T$ be a tree of hyperbolic spaces (satisfying Axiom **H**) with hyperbolic total space X. We can now give a description of geodesics in X, more precisely, of uniform quasigeodesics. This description is inductive/hierarchical. The basis of induction is the fact that each K-qi leaf in X (i.e., the image of a K-qi section of $\pi : X \to T$ over a geodesic segment in T) defines a uniform *horizontal* quasigeodesic in X. Such quasigeodesics present one type of building blocks of geodesics in X. The second building block consists of *vertical quasigeodesics in* X: Such quasigeodesics are certain intrinsic geodesics in vertex-spaces X_v of X. While X_v (typically) are not quasi-isometrically embedded in X, *some* geodesics in X_v nevertheless are uniform quasigeodesics in X, namely, ones satisfying the *small carpet condition*; see Proposition 7.2. We will see that general geodesics in X are uniformly Hausdorff-close to alternating concatenations of horizontal and vertical uniform quasigeodesics. In Section 7.3 we, furthermore, give a simpler description of uniform quasigeodesics in a more limited class of trees of spaces, namely, *acylindrical* trees of spaces.

There are several basic classes of subtrees of spaces in $\mathfrak{X}$, which are used to describe geodesics in X. All of these are special cases of *semicontinuous families* (of subsets of vertex-spaces) in $\mathfrak{X}$; see Chapter 3, all are uniformly quasiconvex subsets of X. Here is the list of these classes of subspaces, listed in order of increase of the complexity of their definitions:

- Carpets $\mathfrak{A} \subset \mathfrak{X}$;
- Metric bundles $\mathfrak{H} \subset \mathfrak{X}$;
- Ladders $\mathfrak{L} \subset \mathfrak{X}$;
- Flow-spaces of vertex-spaces $\mathfrak{Fl}_K(X_u) \subset \mathfrak{X}$;
- Flow-spaces $\mathfrak{Fl}_K(X_S)$, where $S \subset T$ is a subtree and $X_S = \pi^{-1}(S)$;
- Flow-spaces of metric bundles $\mathfrak{Fl}_K(\mathfrak{H}) \subset \mathfrak{X}$.

We first describe geodesics in carpets, then use those to describe geodesics in ladders, then use those to describe geodesics in flow-spaces of vertex-spaces. At the same time, flow-spaces of metric bundles $Fl_k(\mathfrak{H})$ are uniformly quasiconvex subsets of certain flow-spaces $Fl_K(X_u)$ (form some $K \geq k$) and, hence, we do not give a separate description of geodesics in the former. After Step I.4, our description of geodesics in X only uses geodesics in flow-spaces $Fl_K(X_u)$ as building blocks. A key feature of flow-spaces is that they are uniformly quasiconvex in X (unlike vertex-spaces themselves). Each flow-space is itself a tree of spaces, $Fl_K(X_u) \to S_u$, where S_u is a subtree in T. The intersection pattern of the subtrees S_u (encoded in the flow-incidence graph Γ_K) is discussed in Section 3.3.5; it will provide a guide for describing geodesics in X in terms of geodesics in flow-spaces $Fl_K(X_u)$.

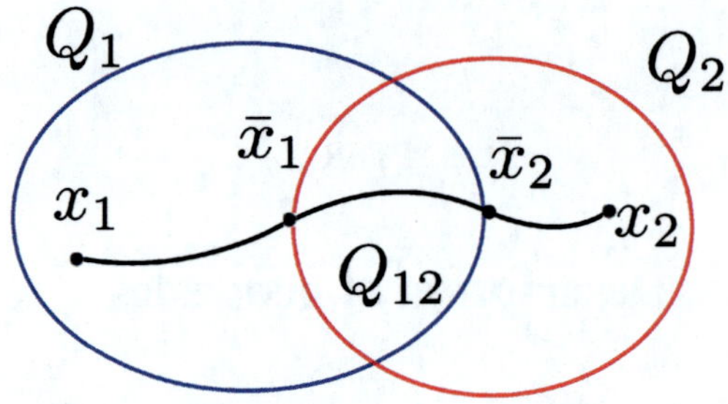

FIGURE 7.1. Geodesics in amalgams

We now describe (inductively) geodesics in X. Except for the two initial steps, the rest is a repetitive use of one of the following constructions.

(a) **Quasiconvex amalgams: Section 2.6.2.** Given two L-qi embedded subsets Q_1, Q_2 in a δ-hyperbolic space Y, with nonempty L-qi embedded intersection $Q_{12} = Q_1 \cap Q_2$, the union

$$Q = Q_1 \cup Q_2$$

is a *quasiconvex amalgam* of Q_1, Q_2.

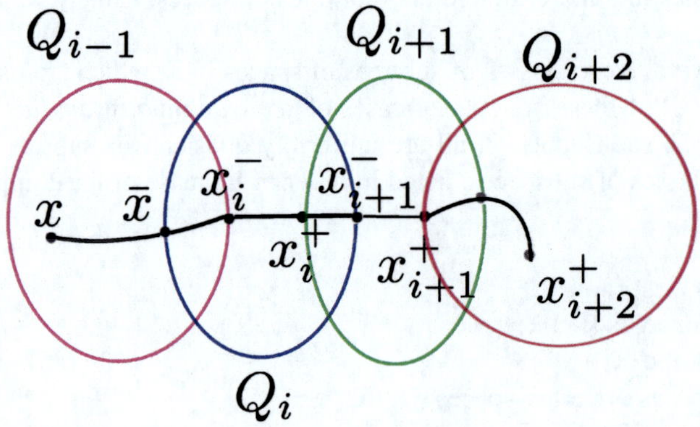

FIGURE 7.2. Geodesics in chain-amalgams

(b) **Pairwise cobounded quasiconvex chain-amalgamation.** We refer to Section 2.6.2 for a detailed definition and description of uniform quasigeodesics in unions $Q = Q_0 \cup Q_1 \cup \cdots \cup Q_n$ defining quasiconvex chain-amalgamation. Such amalgamations (with $n \geq 2$) will be used just in two instances in our proof. Briefly, paths c in Q are alternating concatenations of geodesics in $Q_{i,i+1}$ and Q_i connecting points $x_i^+, x_{i+1}^+ \in Q_{i,i+1}$ and $x_i^-, x_i^+ \in Q_i$ where latter pairs (up to a uniformly bounded error C) realize the minimal distance between $Q_{i-1,i}, Q_{i,i+1}$ in Q_i; see Figure 7.2.

We now begin the inductive description of uniform quasigeodesics. Regarding constants C, D, E, K appearing below: Ultimately, we will take $K = K_*$, $C = M_{\bar{K}}$, $D = D_{5.2}$,

$E = E_{5.2}$. However, for instance, the description of paths in carpets works for all $K \geq 1$ and all $C \geq 0$, etc.

Part 0: Geodesics in K-qi sections of $\mathfrak{X} = (\pi : X \to T)$. The basis for the entire description of uniform quasigeodesics in X is the fact that each K-qi leaf in X is a K-quasigeodesic in X.

Part I: Geodesics in flow-spaces of vertex-spaces $Fl_K(X_v)$.

These geodesics (or, rather, uniform quasigeodesics) are described in three steps.

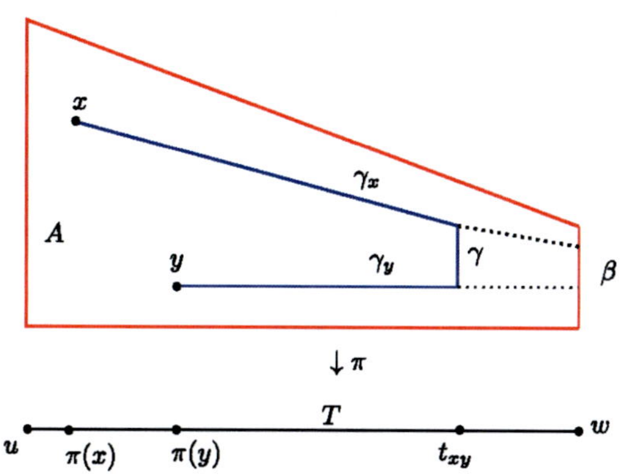

FIGURE 7.3. Geodesics in carpets

STEP I.1 (Quasigeodesics in carpets: Section 4.1, especially, Proposition 4.1). Let $\mathfrak{A} = (\pi : A \to J) \subset \mathfrak{X}$ be a (K, C)-carpet over an interval $J = [\![u, w]\!]$, such that the end A_w of $\mathfrak{A}$ over w is C-narrow (i.e., is a geodesic of length $\leq C$ in X_w).

Then for $x, y \in \mathcal{A}$ we consider K-sections γ_x, γ_y over subintervals $[\![\pi(x), w]\!]$, $[\![\pi(y), w]\!]$ in J. Without loss of generality, we may assume that $\pi(y) \in [\![\pi(x), w]\!]$.

Let $t_{xy} \in [\![w, \pi(y)]\!]$ denote the supremum of

$$\{t \in [\![w, u]\!] : d_{X_t}(\gamma_x(t), \gamma_y(t)) \leq M_K\}.$$

Then an $L_{4.1}(K, C)$-quasigeodesic $c(x, y)$ in A is defined as the concatenation of the section γ_x restricted to $[\![\pi(x), t]\!]$ with the vertical segment $\gamma = [\gamma_x(t)\gamma_y(t)]_{X_t}$, followed by the concatenation with the restriction of the section γ_y to the subinterval $[\![t, \pi(y)]\!]$; see Figure 7.3.

STEP I.2 (Quasigeodesics in carpeted ladders with narrow carpets: Section 4.2, especially Proposition 4.8). For a vertical geodesic segment $\alpha \subset X_u$, we consider a *carpeted ladder*, a (K, D, E)-ladder $\mathfrak{L} = \mathfrak{L}_K(\alpha) = (\pi : A \to J)$,which contains a (K, C)-carpet $\mathfrak{A} = \mathfrak{A}(\alpha')$, where $\alpha' \subset \alpha$ is a subsegment of length $\geq$ *length*$(\alpha) - M_{\bar{K}}$, where $\bar{K}$ is defined by

$$\bar{K} := K_{3.47}(\delta_0, K, K).$$

The definition of paths $c(x, y)$ connecting points $x, y \in \mathcal{L}$ is a two-part process.

Part a: Retraction ρ and paths c_x. In Section 3.4 we defined a retraction $\rho : L_K(\alpha) \to A$: This retraction is uniformly close to the nearest-point projection; see Remark 3.64. This retraction plays a critical role in the definition of our combing of $L_K(\alpha)$. Below is a review of the definition of the retraction in terms of the structure of the tree of spaces.

For a point $x \in \mathcal{L}_v$, let γ_x be a canonical K-leaf in $\mathcal{L}$ connecting x to α: Such leaves are a part of the definition of a ladder. Ultimately, which leaf one takes does not matter and the paths c_x change only a uniformly bounded amount if one makes a different choice. Let $t = t_x \in [\![u, v]\!]$ be the vertex farthest from u such that $\pi(x) \in J$ and there exists a point $\tilde{x} \in \gamma_x(t)$ for which

$$d_{X_t}(\tilde{x}, A_t) \leq M_{\bar{K}}.$$

(It is possible that $t = u$ and $\tilde{x} \in A_u = \alpha$.) Then define a path c_x connecting x to $\rho(x) = \bar{x}$ and equal to the concatenation

$$\gamma_{x,\tilde{x}} \star [\tilde{x}\bar{x}]_{X_t}.$$

Here $t = t_x$, and

$$\gamma_{x,\tilde{x}} = \gamma_x|_{[\![v,t]\!]},$$

is the subpath of γ_x connecting x to $\tilde{x}$, while $\bar{x} \in A_t$ is a nearest-point projection of $\tilde{x}$ to A_t in the vertex-space X_t. Since c_x is uniformly Hausdorff-close to $\gamma_{x,\tilde{x}}$, the point $\rho(x)$ essentially determines the path c_x.

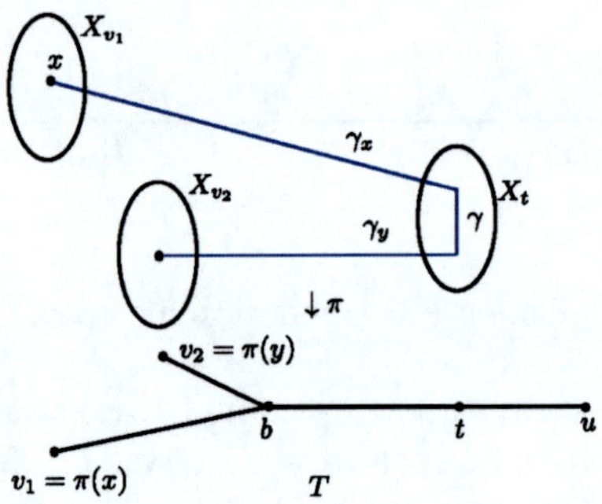

FIGURE 7.4. Geodesics in ladders: Type 1

Part b: Paths $c(x, y)$. For $x, y \in \mathcal{L}$ we let $b = b_{xy}$ be the center of the triangle $\triangle u\pi(x)\pi(y)$. The path $c(x, y)$ in $L_K(\alpha)$ connecting x to y is defined as follows.

Paths of type 1: There exists $t \in V([\![\pi(x), \pi(\bar{x})]\!] \cap [\![\pi(y), \pi(\bar{y})]\!]) \subset V([\![u, b]\!])$ such that

$$d_{X_t}(\gamma_x(t), \gamma_y(t)) \leq M_{\bar{K}},$$

i.e., the paths γ_x, γ_y "come sufficiently close" in some common vertex-space.

Then let $t = t_{x,y}$ be the maximal vertex in $[\![u, b]\!]$ with this property. Then define $c(x, y)$ to be the concatenation of the portions of γ_x and (the reverse of) γ_y over $[\![t, \pi(x)]\!]$ and $[\![t, \pi(y)]\!]$ respectively with the subsegment $\gamma \subset L_t$ joining their endpoints; see Figure 7.4.

Paths of type 2: Suppose type 1 does not happen. Then define $c(x, y)$ to be the concatenation of c_x and the reverse of c_y with a geodesic in A connecting $\rho(x)$ to $\rho(y)$; see Figure 7.5.

REMARK 7.1. The geodesic in A here is taken for granted; see Step I.1. In fact, instead of a geodesic one should take a path $c(\rho(x), \rho(y))$ in A defined in Step I.1. Such inductive arguments will be common in what follows and we will simply say "geodesic."

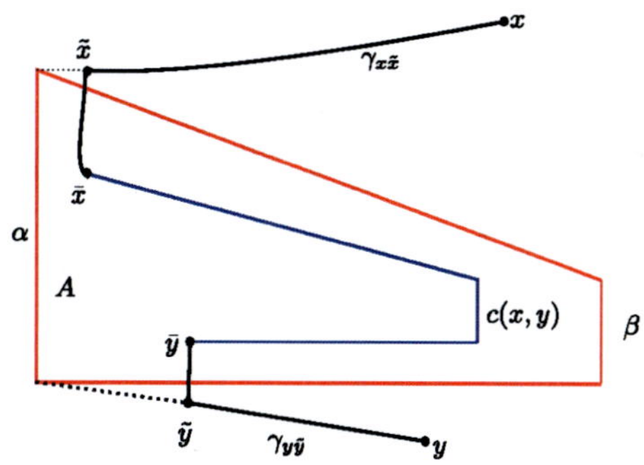

FIGURE 7.5. Geodesics in ladders: Type 2

One more thing of importance is that the vertical segments

$$[\tilde{x}\rho(x)]_{X_{u_x}}, \quad [\rho(y)\tilde{y}]_{X_{u_y}}$$

have length $\leq M_{\bar{K}}$, hence, for paths of both types we are essentially concatenating at most three paths, two of which are horizontal subpaths of c_x, c_y and one which lies in the carpet A. (The latter is also essentially a concatenation of two horizontal paths.)

STEP I.3 (Quasigeodesics in general (K, D, E)-ladders $\mathfrak{L}(\alpha)$: Section 4.3, especially Proposition 4.14). In this case, uniform quasigeodesics are constructed via *cobounded quasiconvex amalgamation*, amalgamating ladders $\mathfrak{L}_K(\alpha_1), \ldots, \mathfrak{L}_K(\alpha_{n+1})$ along uniformly pairwise cobounded uniformly quasiconvex subsets $\Sigma_1 = \Sigma_{p_1}, \ldots, \Sigma_{n-1} = \Sigma_{p_{n-1}}$, which are canonical K-qi sections of these ladders.

We consider a ladder $\mathfrak{L} = \mathfrak{L}_K(\alpha)$ of a vertical geodesic $\alpha = [pq]_{X_u}$ and a pair of points $x, x' \in L_K(\alpha)$, connected to the endpoints p, p' of α by K-qi leaves in $L_K(\alpha)$.

In Proposition 4.14 we prove the existence of a *Vertical Subdivision* of every vertical geodesic segment $\alpha \subset X_u$, $\alpha = [pp']_{X_u}$ into subsegments

$$\alpha_1 = [p_1 p_2]_{X_u}, \ldots, \alpha_n = [p_n p_{n+1}]_{X_u}, p_1 = p, p_{n+1} = p',$$

such that:

(a) For each i, the segment α_i defines a *carpeted ladder* $\mathfrak{L}^i = \mathfrak{L}_K(\alpha_i)$ which is a (K, D, E)-subladder in $\mathfrak{L}$ containing a (K, C)-narrow carpet $\mathfrak{A}_K(\alpha'_i)$ (for $C = M_{\bar{K}}$).
(b) The canonical sections Σ_i, Σ_{i+1} (in $\mathfrak{L}$) through the points p_i, p_{i+1} are, respectively, the bottom and the top sections of $\mathfrak{L}^i$, so that

$$L^{i-1} \cap L^i = \Sigma_i.$$

(c) The sections Σ_i, Σ_{i+1} are $B_{4.14}(K, C)$-cobounded unless $i = n$.
(d) For all i,

$$0 \leq l_i - \text{length}(\alpha'_i) \leq M_K \leq M_{\bar{K}}.$$

(e) The top and bottom sections $\Sigma_p, \Sigma_{p'}$ of $\mathfrak{L}$ pass through the points x, x'.

In particular, the ladder L is a quasiconvex amalgam of its subladders L^i with pairwise intersections $L^{i-1} \cap L^i = \Sigma_i$.

As a part of the proof of the vertical subdivision proposition (Proposition 4.14), we identify (up to a uniformly bounded error) the nearest points $x_i^- \in \Sigma_i, x_i^+ \in \Sigma_{i+1}$ between

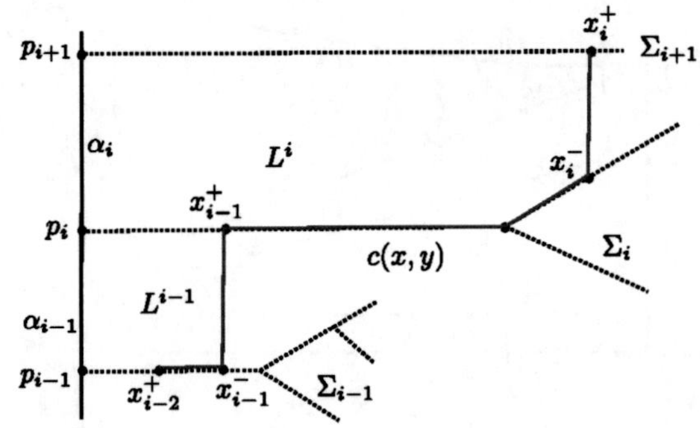

FIGURE 7.6. Geodesics in general ladders: Vertical subdivision

these cobounded subsets. Namely, by the construction, the ladder $\mathfrak{L}^i$ contains the carpet $\mathfrak{A}(\alpha_i')$ over an interval $[\![u, w_i]\!]$, with the narrow end

$$A_{w_i} = [x_{w_i} y_{w_i}]_{X_{w_i}}.$$

Then (up to a uniformly bounded error),

$$x_i^- = x_{w_i}.$$

The description of the point x_i^+ is more complicated; see Lemma 4.15.

The paths $c(x_i^-, x_i^+)$ are defined according to Step I.2 and the paths $c(x_i^+, x_{i+1}^-)$ are defined as in Part 0 (by using K-qi leaves in the K-qi section Σ_i). Lastly, according to Theorem 2.65,

$$c(x, x') = c(x, x_1^+) \star c(x_1^+, x_2^-) \star \cdots \star c(x_n^-, x_n^+) \star c(x_n^+, x'),$$

where $c(x, x_1^+)$, $c(x_n^+, x')$ are the uniformly quasigeodesic paths in carpeted ladders $\mathfrak{L}^1$ and $\mathfrak{L}^n$ respectively as defined in Step I.2.

Conclusion of Part I: Quasigeodesics in flow-spaces $Fl_k(X_u)$ of vertex-spaces. Any two points $x, x' \in Fl_k(X_u)$ belong to a common (K, D, E)-ladder $\mathfrak{L}_{x,x'} = \mathfrak{L}(\alpha)$, where α is a certain geodesic in X_u and K, D, E depend on k. The ladder itself is uniformly close to a uniformly quasiconvex subset of $Fl_k(X_u)$. Hence, Step I.3 yields a description of uniform quasigeodesics $c(x, x')$ connecting arbitrary points $x, x' \in Fl_k(X_u)$ (the path $c(x, x')$ is the projection to $Fl_k(X_u)$ of a geodesic in $L(\alpha)$ connecting x, x').

Part II: Connecting points in flow-spaces $Fl_k(X_J)$ for special intervals $J \subset T$. An interval $J = [\![u, v]\!] \subset T$ is said to be *special* (more precisely, *k-special*) if some $w \in \{u, v\}$ has the property that $J \subset \pi(Fl_k(X_w))$; such w is called a *special vertex* in J. The flow-space $Fl_k(X_J)$ then is called a *special flow-space* (more precisely, *k-special* flow-space). Similarly, an interval $J \subset T$ is *k-semispecial* if it is the union of two special intervals meeting at a vertex. Accordingly, for such J, the flow-space $Fl_k(X_J)$ is *k-semispecial.* Uniform quasigeodesics in $Fl_k(X_J)$ in this setting are described using pairwise quasiconvex amalgams of hyperbolic spaces.

STEP II.4 (Uniform quasigeodesics in special flow-spaces: Section 6.1.2, especially Proposition 6.10, Corollary 6.11). We assume that $J = [\![u, v]\!] \subset T$ is a K-special interval in T. The flow-spaces $Fl_K(X_u)$, $Fl_K(X_v)$ are uniformly quasiconvex in X, hence, a uniform D-neighborhood of this union in X is hyperbolic. The constant D is defined in Corollary

6.11. On this step, we describe uniform quasigeodesics in X connecting points of $Fl_K(X_u) \cup Fl_K(X_v)$ and uniformly close to such a union.

We first define a certain metric bundle $\mathfrak{H}$ over J. We then combine the generalized flow-space $Y_0 = Fl_R(\mathfrak{H})$ of this bundle (for some $R = R(K)$) with the flow-spaces $Fl_K(X_u)$, $Fl_K(X_v)$ to get "modified flow-spaces" $Y_1 = Fl_K(X_u) \cup Fl_R(\mathfrak{H})$, $Y_2 = Fl_K(X_v) \cup Fl_R(\mathfrak{H})$. The union

$$U = U_1 \cup U_2 = N_D(Y_1) \cup N_D(Y_2)$$

is uniformly qi embedded in X, while the intersection $U_1 \cap U_2$ is uniformly quasiconvex and uniformly Hausdorff-close to Y_0. Moreover, U_1, U_2 are uniformly Hausdorff-close to subsets in $Z_1 = Fl_{K_1}(Y_1), Z_2 = Fl_{K_1}(Y_2)$ for some (computable $K_1 \geq K$).

Hence, we are in the setting of the quasiconvex amalgamation of a pair (Section 2.6.2). To connect $x \in Fl_K(X_u), y \in Fl_K(X_v)$ by a uniform quasigeodesic uniformly close to $Fl_K(X_u) \cup Fl_K(X_v)$ we proceed as in Section 2.6.2.

We first project (in $Fl_{K_1}(Y_1), Fl_{K_1}(Y_2)$ respectively) the points x, y to points $\bar{x}, \bar{y} \in Y_0 = Fl_K(\mathfrak{H})$ (one can also use projections in U or in X, the difference will be uniformly bounded) and then connect $\bar{x}, \bar{y}$ in Y_0. The concatenations

$$[x\bar{x}]_{Z_1} \star [\bar{x}\bar{y}]_{Y_0} \star [\bar{y}y]_{Z_2}$$

are uniform quasigeodesics in X and are uniformly close to $Fl_K(X_u) \cup Fl_K(X_v)$.

STEP II.5 (Uniform quasigeodesics in semispecial flow-spaces: Proposition 6.13 and Theorem 6.14). Suppose now a semispecial interval $S = [\![u, v]\!] = I \cup J$, where both $I = [\![u, w]\!], J = [\![w, v]\!]$ are special intervals. Then for $s \in V(I)$ and $t \in V(J)$, uniform quasigeodesics in $Fl_K(X_S)$ connecting points $x \in Fl_K([\![s, w]\!]), y \in Fl_K([\![w, t]\!])$ are described as follows. The union

$$Fl_K([\![s, w]\!]) \cup Fl_K([\![w, t]\!])$$

is a quasiconvex amalgam over the flow-space $Fl_K(X_w)$. Therefore, according to the description of uniform quasigeodesics in quasiconvex amalgams, we first project (using nearest-point projections in $Z_1 = N_D(Fl_K(X_s) \cup Fl_K(X_w))$, resp., in $Z_2 = N_D(Fl_K(X_t) \cup Fl_K(X_w)))$ points x (resp., y) to $\bar{x} \in Fl_K(X_w)$ (resp., $\bar{y} \in Fl_K(X_w)$), and then take the concatenation

$$[x\bar{x}]_{Z_1} \star [\bar{x}\bar{y}]_{Fl_K(X_w)} \star [\bar{y}y]_{Z_2}.$$

The first and the last geodesics in this concatenation are from Step II.4, while the middle one is from the conclusion of Part I.

STEP II.6 (Uniform quasigeodesics in triple unions of special flow-spaces: Corollary 6.15). Suppose that J is an interval in T, that can be subdivided as a union of three special subintervals, $J = J_1 \cup J_2 \cup J_3$,

$$J_i = [\![v_i, v_{i+1}]\!], i = 1, 2, 3.$$

Set $I := J_1 \cup J_2$. Then geodesics in $Fl_K(X_J)$ are described by applying quasiconvex amalgamation of pairs *twice*: Once to

$$Fl_K(X_I) = Fl_K(X_{J_1 \cup J_2}),$$

which is the amalgam of $Fl_K(J_1), Fl_K(J_2)$ over $Fl_K(X_{v_2})$, and then once more, to

$$Fl_K(X_{I \cup J_3}),$$

which is the amalgam of $Fl_K(I), Fl_K(J_3)$ over $Fl_K(X_{v_3})$.

Part III: Connecting general points in X.

STEP III.7 (Horizontal subdivision). Any two points $x, y \in \mathcal{X}$ belong to the interval flow-space $Fl_K(X_J)$, where $J = [\![u, v]\!] \subset T$, $x \in X_u, y \in X_v$. Since the flow-space $Fl_K(X_J)$ is (uniformly) quasiconvex in X, it suffices to describe a uniform quasigeodesic in $Fl_K(X_J)$ connecting x to y. The key ingredient of this part is the *Horizontal Subdivision Lemma* (Lemma 3.44). This lemma gives a subdivision of the interval J into subintervals $J_i = [\![u_i, u_{i+1}]\!], i = 1, \dots, n$, such that:

(1) $Fl_K(J_i) \cap Fl_K(J_j) = \emptyset$ whenever $|i - j| \geq 2$.
(2) Each interval J_i is subdivided in three special subintervals.

This represents $Fl_K(X_J)$ as a (uniformly) pairwise cobounded quasiconvex chain with quasiconvex subsets $Q_i = Fl_K(X_{J_i}), i = 1, \dots., n$, whose consecutive intersections

$$Q_{i-1} \cap Q_i = Q_{i-1,i} = Fl_K(X_{u_i})$$

separate in the union $Fl_K(X_J)$ as required by a cobounded quasiconvex chain-amalgamation. Thus, uniform quasigeodesics in $Fl_K(X_J)$ are described according to Section 2.6.2; see the discussion of quasiconvex chain-amalgamation given early in this section.

This concludes our description of uniform quasigeodesics between points of X.

7.2. Characterization of vertical quasigeodesics

In this section we use the description of uniform quasigeodesics in X to characterize vertical geodesics in X (i.e., geodesics in vertex-spaces X_u) which are quasigeodesics in X. We assume that $\mathfrak{X} = (\pi : X \to T)$ is a tree of spaces satisfying the assumptions of Theorem 2.62, equivalently, a tree of hyperbolic spaces with hyperbolic total space X. Set $K = K_0$ and let $C = M_{\bar{K}}, D = D_{5.2}, E = E_{5.2}$.

Suppose that α is (finite or infinite) geodesic in X_u. We will say that α satisfies the *R-small carpet condition* if whenever $\alpha' \subset \alpha$ is a subsegment which bounds a (K, C)-narrow carpet $\mathfrak{A} = \mathfrak{A}(\alpha') \subset \mathfrak{X}$, $\mathfrak{A} = (\pi : A \to [\![u, w]\!])$, with fiberwise distances between $top(\mathfrak{A}), bot(\mathfrak{A})$ at least M_K, then we have

$$d_T(u, w) = \operatorname{length} \pi(A) \leq R.$$

In view of Corollary 2.39, for such α, the lengths of subsegments $\alpha' \subset \alpha$ bounding (K, C)-narrow carpets $\mathfrak{A}(\alpha')$, are uniformly bounded.

The main result of this section is that a vertical geodesic α satisfies the small carpet condition if and only if it is quasigeodesic in X. More precisely:

PROPOSITION 7.2.

(1) *Each vertical geodesic α satisfying the R-small carpet condition is an $L_{7.2}(R)$-quasigeodesic in X.*
(2) *If α fails the R-small carpet condition for all R, then α is not a quasigeodesic in X.*

PROOF.

(1) Let $\mathfrak{L} = \mathfrak{L}_K(\alpha)$ denote a (K, D, E)-ladder in X based on the segment α. Since $L_K(\alpha)$ is uniformly qi embedded in X (see Corollary 3.13), it suffices to show that α is a uniform quasigeodesic in $L_K(\alpha)$. We follow the description of uniform quasigeodesics geodesics in $L_K(\alpha)$ given in Step I.3 in the previous section. We subdivide the segment $\alpha = [pp']_{X_v} = [p_1 p_{n+1}]_{X_u}$ into subsegments

$$\alpha_i = [p_i p_{i+1}]_{X_u}, i = 1, \dots, n.$$

Each α_i (except for possibly $i = n$) contains a subsegment α_i' such that

$$M_K \leq \text{length}(\alpha_i) - M_K \leq \text{length}(\alpha_i')$$

and there exists a (K, C)-carpet $\mathfrak{A}^i = \mathfrak{A}_{(K,C)}(\alpha_i) \subset \mathfrak{L}$; the bottom and the top of $\mathfrak{L}^i$ are K-qi sections $\Sigma_i, \Sigma_{i+1} \subset \mathfrak{L}^i$ of a subladder $\mathfrak{L}^i \subset \mathfrak{L}$. Moreover, for each vertex t in the interval $\pi(A^i)$,

$$d_{X_t}(top(A^i)_t, bot(A^i)_t) \geq M_K.$$

For each i we mark the points $x_i^- \in \Sigma_i, x_i^+ \in \Sigma_{i+1}$ which (up to a uniformly bounded error) are the nearest points in L^i between these subsets and consider geodesics

$$\beta_i = [x_i^- x_i^+]_{L^i},$$

$$\gamma_i^- := \gamma_{x_{i-1}^+, x_i^-} = [x_{i-1}^+ x_i^-]_{\Sigma_i} \subset \Sigma_i,$$

$$\gamma_i^+ := \gamma_{x_i^+, x_{i+1}^-} = [x_i^+ x_{i+1}^-]_{\Sigma_{i+1}} \subset \Sigma_{i+1}.$$

The path $c(p, p')$ then is (up to a uniformly bounded error) equal the concatenation

$$c = \cdots \star \gamma_i^- \star \beta_i \star \gamma_i^+ \star \ldots.$$

Our goal is to show that the path c is uniformly Hausdorff-close to α. Because the lengths of projections $\pi(A_i)$ are uniformly bounded, the geodesics β_i are uniformly close to the vertical geodesics α_i. For the same reason, the paths $\gamma_i^\pm$ are uniformly short as well. Hence, each concatenation

$$\gamma_i^- \star \beta_i \star \gamma_i^+$$

is $r = r_{7.2}(R)$-Hausdorff close to the vertical geodesic segment α_i. It follows that $\text{Hd}(\alpha, c) \leq r$ as well. Now, the first statement of the proposition follows from Lemma 1.21.

(2) Suppose that α contains a sequence of subsegments α_i each bounding a (K, C)-carpet $\mathfrak{A}^i = \mathfrak{A}_{(K,C)}(\alpha_i) = (\pi : A^i \to J_i = [\![u, w_i]\!])$, such that

$$\lim_{i\to\infty} d_T(u, w_i) = \infty,$$

and for each vertex t in the interval J_i,

$$d_{X_t}(top(A^i)_t, bot(A^i)_t) \geq M_K.$$

Therefore, the concatenation c_i of the bottom of $\mathfrak{A}^i$, the narrow end β_i and the top of $\mathfrak{A}^i$ is an $L_{4.1}(K, C)$-quasigeodesic in A^i. Since

$$d(\alpha_i, \beta_i) \geq d_T(u, w_i),$$

the Hausdorff distances between α_i and c_i diverge to infinity. Morse lemma and hyperbolicity of X then imply that α_i cannot be uniform quasigeodesics in X. Therefore, α is not a quasigeodesic in X either. □

7.3. Visual boundary and geodesics in acylindrical trees of spaces

In this section we specialize our discussion of geodesics in trees of hyperbolic spaces to the case of (M, K, τ)-*acylindrical trees* of hyperbolic spaces satisfying Axiom **H**; see Definition 2.54; the constant K will be taken equal to $K = K_* = K_{2.62}(\delta_0, L_0)$ although, many arguments will go through for smaller values of K. Besides uniform quasigeodesics we will also describe the ideal boundary of X. In the group-theoretic setting, when X is the Cayley graph of the fundamental group of an acylindrical graph of hyperbolic groups with quasiconvex edge-subgroups, such description of the boundary is due to Dahmani, **[Dah03]** (who also gave a description in the relatively hyperbolic case).

LEMMA 7.3. *Suppose $\pi : X \to T$ is a (M, K, τ)-acylindrical tree of hyperbolic spaces satisfying Axiom **H** for some M and τ. Then (1) X is hyperbolic. (2) For subtrees $S \subset T$, the subspaces X_S are uniformly qi embedded in X.*

PROOF.

(1) Recall (see Section 2.5.2), that (M, K, τ)-acylindricity implies uniform K-flaring, hence, by Theorem 2.62, hyperbolicity of X.

(2) Let $v \in V(T)$. Then, by Proposition 7.2, every geodesic in X_v is a uniform quasigeodesic in X. Thus X_v is uniformly qi embedded in X. Consequently X_v is uniformly quasiconvex too. Then it follows that X_S is also uniformly quasiconvex in X for the following reason. Suppose a geodesic segment β of X joining a pair of points in X_S is not entirely contained in X_S. Then the closure β_w of each connected component of $\beta \setminus X_S$ joins two points of X_w for some $w \in S$. Since vertex-spaces X_w are uniformly quasiconvex in X, the geodesic β_w is uniformly close to X_w. It follows that the entire β is uniformly close to X_S.

Finally we know that X_S is uniformly properly embedded in X by Proposition 2.18. Hence, X_S is uniformly qi embedded in X by Lemma 1.102 and Lemma 1.16. □

Description of quasigeodesics. Suppose that $\mathfrak{X}$ is (M, K, τ)-acylindrical, $x \in X_u, y \in X_v$, $u, v \in V(T)$. Without loss of generality, $\tau \in \mathbb{N}$. Since, by the above lemma, $X_{[\![u,v]\!]}$ is uniformly qi embedded in X, it suffices to describe uniform quasigeodesics connecting x, y in $X_{[\![u,v]\!]}$. Hence, from now on, we will assume that the tree T is an interval $[\![u, v]\!]$. Let

$$t_0 = u, t_1, \dots, t_m = v$$

denote the consecutive vertices in the interval $[\![u, v]\!]$. Define $A_{i-1} := X_{e_i t_{i-1}} = f_{e_i t_{i-1}}(X_{e_i}) \subset X_{i-1}$ and $B_i := X_{e_i t_i} = f_{e_i t_i}(X_{e_i}) \subset X_{t_i}$ for $1 \le i \le m-1$. We inductively construct points $x_i^+, x_i^- \in X_{t_i}$ for $0 \le i \le m$ as follows.

Set $x_0^- = x$ and $x_0^+ = P_{X_{t_0}, A_0}(x)$. Now, suppose that $x_i^+, x_i^- \in X_{t_i}$ are already defined for some $i < m$. Then we define x_{i+1}^- to be an arbitrary point of $f_{e_{i+1} t_{i+1}}(f_{e_{i+1}, t_i}^{-1}(x_i^+))$ and $x_{i+1}^+ := P_{X_{t_{i+1}}, A_{i+1}}(x_{i+1}^-)$. We define $x_m^+ := y$.

Lastly, we define the path $\gamma(x, y)$ as the concatenation of the segments $[x_i^- x_i^+]_{X_{t_i}}$ for $0 \le i \le m$, and unit segments in $X_{t_i t_{i+1}}$ joining each pair x_i^+, x_{i+1}^- for $0 \le i \le m-1$.

PROPOSITION 7.4. *The path $\gamma(x, y)$ is a uniform quasigeodesic in X joining x, y. Moreover, x_m^- is uniformly close to the nearest point projection of x to X_v.*

PROOF. The proof is based on several lemmata. We first prove the claim for constants depending on m and then eliminate this dependence.

LEMMA 7.5.

(1) *The path $\gamma(x, y)$ is an $L_{7.5}(m)$-quasigeodesic.*
(2) *The distance between x_m^- and the projection of x to X_{t_m} inside $X_{[\![u,v]\!]}$ is bounded by $D_{7.5}(m)$.*

PROOF. In view of uniform quasiconvexity in $X = X_{[\![u,v]\!]}$ of vertex-spaces and of subintervals of spaces $X_{[\![t_i,t_j]\!]}$, the statements follow from, respectively, Parts (1) and (2) of Lemma 1.112. □

Below we set $R = R_{1.123}(\delta_0, \lambda_0) = \lambda_0 + 5\delta_0$ and note that $K \ge R + 1$.

LEMMA 7.6. *Each pair of vertex-spaces X_u, X_v satisfying $d_T(v, w) = m \ge \tau$, is $C_{7.6}(m)$-cobounded in X.*

PROOF. We start with a pair of points $x, z \in X_u$ and inductively project them to points $x_i^-, z_i^- \in X_{t_i}$, $1 \le i \le m$, using the notation above.

According to Lemma 1.123(1), for each i one of two things happens, where δ_0 is the hyperbolicity constant of X_{t_i} and λ_0 is the quasiconvexity constant of A_i in X_{t_i}:

(a) $d_{X_{t_i}}(x_i^+, z_i^+) \le D_{1.123}(\delta_0, \lambda_0) = 2\lambda_0 + 7\delta_0$.

(b) $[x_i^+ z_i^+]_{X_{t_i}} \subset N_R([x_i^- z_i^-]_{X_{t_i}})$, where, as above, $R = R_{1.123}(\delta_0, \lambda_0) = \lambda_0 + 5\delta_0$.

If (a) occurs for some i, then the distance between $x_j^-, z_j^- \in X_{t_j}$, $i+1 \le j \le m$ is uniformly bounded in terms of j, as the process of successive projections to next vertex-spaces is clearly uniformly Lipschitz. In particular, the distance between $x_m^-, z_m^- \in X_{t_m} = X_v$ is uniformly bounded as well, with bound depending on m. Suppose, therefore, that (b) holds for every i. Then for $k = R + 1$, there exists a pair of k-sections γ_0, γ_1 over $[\![u, v]\!]$ such that

$$\gamma_0(t_i), \gamma_1 \in [x_i^-, z_i^-]_{X_{t_i}}, i = 0, \dots, m.$$

Since, by the assumption, $K \ge k$, the acylindricity condition implies that $d_{X_v}(x_m^-, z_m^-) \le M$. Combining this with Lemma 7.5(2), we obtain a uniform (in terms of m) bound on the distance between the projections of x, z to X_v. □

We are now ready to prove the proposition. We let $N := \lfloor \frac{m}{\tau} \rfloor$ and define the subintervals

$$J_i = [\![v_i, v_{i+1}]\!], v_i = t_{i\tau}, i = 0, \dots, N,$$

in $[\![u, v]\!]$. These subintervals cover $[\![u, v]\!]$ except for the subinterval $J_{N+1} = [\![v_{N+1}, v]\!]$ whose length is $< \tau$. Set

$$x_i := x_{i\tau}^-, i = 0, \dots, N.$$

Since the subspaces X_{J_i} are uniformly quasiconvex in X and each pair $X_{J_{i-1}}, X_{J_{i+1}}$ is uniformly cobounded, it follows from Theorem 2.65 on quasiconvex chain-amalgams that the concatenation

$$[x_0 x_1] \star \cdots \star [x_N x_{N+1}] \star [x_{N+1} y]$$

is a uniform quasigeodesic in X. Furthermore, by Lemma 7.5(1), each path

$$\gamma(x_i, x_{i+1}), i = 0, \dots, N, \gamma(x_N, y),$$

is also a uniform quasigeodesic. Thus, the entire path $\gamma(x, y)$ is uniformly quasigeodesic as well. □

DEFINITION 7.7. We shall refer to the (uniform) quasigeodesics of the type described in Proposition 7.4 as *HV (horizontal-vertical) quasigeodesics* in what follows.

Up to a uniform error, these HV quasigeodesics describe all finite geodesics in X. Our next goal is to extend this description to the rays in X. We will do so under the extra assumption that X is a proper metric space. Of course, as before, we also suppose that $\pi : X \to T$ is a (M, K, τ)-acylindrical tree of hyperbolic spaces satisfying Axiom **H**.

Fix $v_0 \in V(T)$ and $x_0 \in X_{v_0}$. We will describe (quasi)geodesic rays in X starting from x_0. First of all, for every $v \in V(T)$ we fix, once and for all, an HV uniform quasigeodesic γ_v joining to x_0 to a point $x_v \in X_v$ where:

(1) x_v is uniformly close to a nearest point projection of x_0 to X_v as we obtained in Proposition 7.4;

(2) for each $w \in [\![v_0, v]\!]$ we have $\gamma_w \subset \gamma_v$.

One defines γ_v by induction on $d(v_0, v)$. Note that for each vertex $w \in [\![v_0, v]\!]$, the pre-image of w in γ_v under the projection π is an interval. In this situation, we will say that γ_v *projects monotonically* to $[\![v_0, v]\!]$.

Armed with this, we can now describe quasigeodesic rays in X.

Rays of type 1: Let $v \in V(T)$ and $\xi_v \in \partial_\infty X_v$. Let α_v be a geodesic ray in X_v joining x_v to ξ_v. Then $\rho_v := \gamma_v \star \alpha_v$ is a uniform quasigeodesic in X by Proposition 7.4.

Rays of type 2: On the other hand, suppose that $c = v_0\eta$ is a geodesic ray in T joining v_0 to $\eta \in \partial_\infty T$. Then the uniform quasigeodesic paths $\gamma_{c(n)}$ combine to form a uniform quasigeodesic ray γ_η in X which projects monotonically to c.

PROPOSITION 7.8. *Any (quasi)geodesic ray in X starting from x_0 is asymptotic to a quasigeodesic ray of exactly one of the above two types.*

PROOF. Suppose that ρ is a geodesic ray in X emanating from x_0. For each $n \in \mathbb{N}$ let $\alpha_n = \beta(x_0, \rho(n))$ be the arc-length parametrized HV quasigeodesic segment (discussed after Lemma 7.3), joining x_0 to $\rho(n)$. Since X is a proper metric space, the sequence of uniform quasigeodesic segments (α_n) subconverges to a (uniform) quasigeodesic ray, say α, asymptotic to $\rho(\infty)$.

Let $T_1 = \pi(\rho)$. By the stability of quasigeodesics it is clear that for all $m \geq n$, α_n is contained in a uniformly bounded neighborhood of α_m. Hence, the same is true for $\pi(\alpha_m), \pi(\alpha_n)$ for all $m \geq n$. We note that these are geodesic segments in T. It then follows that either T_1 is bounded or T_1 is a locally finite subtree of T containing a unique geodesic ray of T.

CASE 1. Suppose T_1 is bounded. There are two possibilities: Either T_1 is locally finite or it is not. If T_1 were locally finite then we can find a constant subsequence of the sequence of geodesic segments $\{\pi(\alpha_n)\}$. Clearly the corresponding subsequence converges to an HV quasigeodesic ray of type 1. Otherwise, suppose T_1 has a vertex v of infinite degree. In this case, we find a subsequence $\{\alpha_{n_k}\}$, say, such that for all k, α_{n_k} has an initial segment α'_{n_k}, say, joining x_0 to x_v by γ_v followed by a geodesic segment in X_v joining x_v to an edge space $X_{e_k v}$, say, of X_v. We note that these edge spaces are pairwise distinct. Finally, since X is a proper metric space, it follows that each vertex-space of X is a proper metric space and given $w \in V(T)$, $x \in X_w$ and $D \geq 0$, the ball $B(x; D)$ in X_w intersects only finitely many edge spaces of X_w. Thus, in this case, $\{d_{X_v}(x_v, X_{e_k v})\}$ is unbounded. Therefore, the geodesic segments $X_v \cap \alpha_{n_k}$ of X_v subconverges to a geodesic ray α_v, say. This produces a further subsequence of $\{\alpha_{n_k}\}$ that converges to the type 1 HV quasigeodesic ray which is simply the concatenation of γ_v and α_v.

CASE 2. Suppose T_1 is not bounded and let $c = v_0\eta$ be the geodesic ray in T_1. Let v_n be the projection of the endpoint of $\pi(\alpha_n)$ on c. Passing to subsequence, if necessary, we may assume that $\{d(v_0, v_n)\}$ is strictly increasing. It then follows that γ_{v_n} is an initial segment of α_n for all n and γ_{v_n} is an initial segment of γ_{v_m} for all $m \geq n$. From these it follows that α_n converge to an HV quasigeodesic ray of type 2. □

LEMMA 7.9. *Suppose $c : [0, \infty) \to T$ is a geodesic ray with $v_n = c(n), n \in \mathbb{N}$, and $c(\infty) = \eta$. Then every sequence $x_n \in X_{v_n}$ converges to $\gamma_\eta(\infty) \in \partial_\infty X$.*

PROOF. Let γ_n denote the concatenation of the path γ_{v_n} and the geodesic $[\![x_{v_n}, x_n]\!]_{X_{v_n}}$. These concatenations are uniform quasigeodesics in X. Clearly, the sequence $\{\gamma_n\}$ converges uniformly on compacts to γ_η. □

Our next goal is to describe the ideal boundary $\partial_\infty X$ in terms of ideal boundaries of vertex-spaces $\partial_\infty X_v$ and the ideal boundary of the tree T. Our description is similar to the one given by Dahmani, [**Dah03**, Section 2], in the setting of graphs of hyperbolic groups.

We set

$$\tilde{\mathcal{Z}} := \coprod_{v \in V(T)} \partial_\infty X_v$$

and define a relation $\sim$ on $\tilde{\mathcal{Z}}$ as follows:

If $\xi_u \in \partial_\infty X_u$ and $\xi_v \in \partial_\infty X_v$ then $\xi_u \sim \xi_v$ iff ξ_u belongs to the ideal boundary flow $Fl(\{\xi_v\})$ of ξ_v in $\partial_\infty X$; see Section 3.3.4. It follows from the discussion of ideal boundary flows in Section 3.3.4 that any two boundary flows are either equal or disjoint, hence, $\sim$ is an equivalence relation. We let $p : \tilde{\mathcal{Z}} \to \mathcal{Z} := \tilde{\mathcal{Z}}/_\sim$ be the quotient map. We will topologize $\mathcal{Z}$ later on. For now we note that, in general, this topology *is not* the quotient topology of the coproduct topology on $\tilde{\mathcal{Z}}$ (cf. Lemma 7.12), but the topology will restrict to the standard topology on each $\partial_\infty X_v$.

The space $\mathcal{Z}$ will be homeomorphic to one part of $\partial_\infty X$. The other part will be homeomorphic to $\partial_\infty T$ (with its natural topology of the ideal boundary of a tree). There are natural maps from both $\partial_\infty T$ and $\mathcal{Z}$ to $\partial_\infty X$ defined, respectively, as follows:

(1) $f_{\partial_\infty T} : \eta \mapsto \gamma_\eta(\infty)$, $\eta \in \partial_\infty T$.

(2) Since each X_v, $v \in V(T)$, is qi embedded in X, we have topological embeddings

$$q_v : \partial_\infty X_v \hookrightarrow \partial_\infty X.$$

By combining these embeddings, we obtain a map

$$q : \coprod_{v \in V(T)} \partial_\infty X_v \to \partial_\infty X.$$

Next, for two vertical geodesic rays α_u in X_u and α_v in X_v ($u, v \in V(T)$) we have equivalence of the following statements:

(a) $q_u(\alpha_u(\infty)) = q_v(\alpha_v(\infty))$.

(b) $\mathrm{Hd}(\alpha_u, \alpha_v) < \infty$.

(c) The boundary flow-space $Fl(\{\alpha_u(\infty)\})$ contains $\alpha_v(\infty)$.

While the equivalence of (a) and (b) is immediate, the equivalence of (b) and (c) follows from Lemma 3.35(2).

Thus, the maps q satisfies the property that $z_1 \sim z_2$ iff $q(z_1) = q(z_2)$. In particular, the map q descends to a map $f_T : \mathcal{Z} \to \partial_\infty X$. Combining the maps $f_{\partial_\infty T}, f_T$ we obtain a map

$$f : \partial_\infty T \sqcup \mathcal{Z} \to \partial_\infty X.$$

LEMMA 7.10. *The map f is a bijection.*

PROOF.

(1) That $f_{\mathcal{Z}}$ is injective is immediate from the equivalence of (a) and (c) above.

(2) If $\eta_1, \eta_2 \in \partial_\infty T$ are distinct, then the rays $v_0\eta_1, v_0\eta_2$ in T diverge, hence, their lifts $\gamma_{\eta_1}, \gamma_{\eta_2}$ in X diverge as well. Hence, $\gamma_{\eta_1}(\infty) \neq \gamma_{\eta_2}(\infty)$. It follows that $f_{\partial_\infty T}$ is injective.

(3) Proposition 7.8 directly implies that the images of $f_{\partial_\infty T}, f_T$ are disjoint and their union is the entire $\partial_\infty X$. ☐

We next topologize the disjoint union of $\mathcal{Z}$ and $\partial_\infty T$ by defining a basis of this topology as follows. Given a vertex $v \in T$, we let the *shadow* Sh_v of v in $V(T) \cup \partial_\infty T$ denote the set consisting of all $\xi \in \partial_\infty T$ and $w \in V(T)$ such that the rays $v_0\xi$ and segments v_0w contain v.

(A) Let η be a point in $\partial_\infty T$ and let v be a vertex in the ray $v_0\eta$. Then define

$$U_{v,\eta} := (Sh_v \cap \partial_\infty T) \sqcup \bigcup_{w \in Sh_v \cap V(T)} p(\partial_\infty X_v) \subset \partial_\infty T \sqcup \mathcal{Z}.$$

These subsets will form a basis of neighborhoods of η in $\partial_\infty T \sqcup \mathcal{Z}$.

(B) Let $[\zeta]$ be a point in $\mathcal{Z}$, $R > 0$. For each $\eta \in \partial_\infty T$ and $\zeta \in \partial_\infty X_v$ in the equivalence class $[\zeta]$ define the point $z_\eta \in X_v$ as x_v if the ray $v_0\eta$ does not contain v, and so that $[x_v z_\eta]_{X_v}$ is the maximal subsegment of γ_η contained in X_v. Similarly, define z_η for points $\eta \in \partial_\infty X_w$: If $[\![v_0, w]\!]$ is disjoint from v, then set $z_\eta := x_v$, otherwise it is the maximal subsegment of the HV ray $\rho_\eta = \gamma_w \star x_w\eta$ contained in X_v.

(B1) We define $U^1_{R,[\zeta]}$ as the set of points $\eta \in \partial_\infty T$ such that there exists a representative $\zeta \in \partial_\infty X_v$ of $[\zeta]$ for which $(\eta.z_v)_{x_v} > R$.

(B2) Similarly, define

$$U^2_{R,[\zeta]} := p(\{\eta \in \partial_\infty X_v : (\zeta.z_v)_{x_v} > R, \zeta \in [\zeta]\}).$$

Here in both (B1) and (B2) where the Gromov-product is taken in X_v. Set

$$U_{R,[\zeta]} := U^1_{R,[\zeta]} \cup U^2_{R,[\zeta]}.$$

We leave it to the reader to check the following properties satisfied by the collection of *basic subsets* $U_{v,\eta}$, $U_{R,[\zeta]}$ defined above:

LEMMA 7.11.

(1) $\eta \in U_{v,\eta}$, $[\zeta] \in U_{R,[\zeta]}$ *for every* η, ζ, *and* v, R *as above.*

(2) *For every two basis subsets* U', U'' *as above containing a point* $\xi \in \partial_\infty T \sqcup \mathcal{Z}$, *there exists another basis subset* U *containing* ξ *such that* $U \subset U' \cap U''$.

(3) *For any two distinct points* $\xi', \xi'' \in \partial_\infty T \sqcup \mathcal{Z}$, *there exist disjoint basic subsets* U', U'' *containing* ξ', ξ'' *respectively.*

It follows that the collection of basic sets $U_{v,\eta}$, $U_{R,\zeta}$ defines a Hausdorff topology τ on $\partial_\infty T \sqcup \mathcal{Z}$. From now on, we will equip $\partial_\infty T \sqcup \mathcal{Z}$ with this topology. We will see soon (Theorem 7.13) that the topology τ is compact and metrizable. The next lemma, where we use metrizability of $(\mathcal{Z}, \tau)$, describes some properties of $(\mathcal{Z}, \tau)$.

LEMMA 7.12.

(i) *For every sequence of distinct vertices* $v_n \in V(T)$ *the sequence of compacts* $p(\partial_\infty X_{v_n})$ *subconverges to a point in* $(\partial_\infty T \sqcup \mathcal{Z}, \tau)$.

(ii) *For every finite diameter subtree* $S \subset T$, *the restriction of* τ *to* $p(\coprod_{v \in V(S)} \partial_\infty X_v) \subset \mathcal{Z}$ *is compact.*

PROOF OF LEMMA 7.12(i). After extraction of a subsequence, there are two cases which may occur:

Case (i1) There is a vertex $v_0 \in V(S)$ such that all the edges $e_n = [v_0, w_n]$ in the segments $[\![v_0, v_n]\!]$ are pairwise distinct. Pick a base-point $x_0 \in X_{v_0}$. Then, again by properness of X, the sequence of distances from x_0 to $X_{e_n v_0}$ (in X_{v_0}) diverges to infinity. It follows that, after further extraction, the sequence of subsets $X_{e_n v_0}$ converges to a point $\xi \in \partial_\infty X_{v_0}$. By the description of the topology τ, the sequence (x_n) converges to $p(\xi) \in \mathcal{Z}$ as well.

Case (i2) The sequence (v_n) converges to a point $\eta \in \partial_\infty T$. Then, by the definition of the topology τ, every sequence (x_n) in X_{v_n} converges to η. □

Proof of Lemma 7.12(ii). Since $(\mathcal{Z}, \tau)$ is metrizable, it suffices to prove sequential compactness. Consider a sequence $x_n \in X_{v_n}$ where $v_n \in V(S)$. There are two cases to consider:

Case (ii1) Suppose that, after extraction, all the vertices v_n are distinct. Then, by Part (i) (Case i1), the sequence (x_n) subconverges to a point in $p(\partial_\infty X_v)$ for some $v \in V(S)$.

Case (ii2) The same vertex v appears in the sequence (v_n) infinitely many times. Then, after extraction, $v_n = v$ and the sequence (x_n) subconverges to a point in $\partial_\infty X_v$ (recall that X is assumed to be proper, which implies properness of X_v). □

We can now prove the main result of this section.

Theorem 7.13. *The map $f : \partial_\infty T \sqcup \mathcal{Z} \to \partial_\infty X$ is a homeomorphism.*

Proof. Since $\partial_\infty X$ is compact, $\partial_\infty T \sqcup \mathcal{Z}$ is Hausdorff and f is a bijection, it suffices to prove continuity of f^{-1}.

Case 1. Suppose that $\eta_n \in \partial_\infty T$ are such that the sequence $(\gamma_{\eta_n}(\infty))$ in $\partial_\infty X$ converges to $\gamma_\eta(\infty)$, $\eta \in \partial_\infty T$. This implies that there exists a constant D such that for every $R \geq 0$ and all sufficiently large n, the point $\gamma_\eta(R)$ lies in the D-neighborhood of γ_{η_n}. The same, therefore, holds for the rays $v_0\eta$, $v_0\eta_n$. That implies convergence $\eta_n \to \eta$.

Case 2. Suppose that $\xi_n \in \partial_\infty X_{w_n}$ is a sequence converging to $\xi = \rho_\eta(\infty)$. Then, for the same reason as in Case 1, the sequence (w_n) converges to η in $T \cup \partial_\infty T$. It then follows by the definition of a neighborhood basis at η in $\mathcal{Z} \cup \partial_\infty T$ that $\xi_n \to \eta$.

Case 3. Suppose that η_n is a sequence in $\partial_\infty T$ such that the corresponding sequence $\xi_n := \gamma_{\eta_n}(\infty)$ converges to some $\xi \in \partial_\infty X_v$. Then, after extraction, the sequence of HV (uniformly quasigeodesic) rays γ_{η_n} converges to an HV ray γ asymptotic to ξ. This limiting ray necessarily has the form of a concatenation $\rho_{\xi'}\gamma_w \star \alpha_w$, where α_w is a geodesic ray in a vertex-space X_w. Since the rays γ, γ_η are at finite Hausdorff distance from each other, it follows that $p(\xi) = p(\xi')$, i.e., $[\xi] = [\xi']$. Furthermore, let $[x_v z_n]_{X_v}$ is the maximal subinterval of γ_{η_n} contained in X_v. Then

$$\lim_{n\to\infty} (\xi'.z_n)_{x_v} = \infty.$$

Now, it follows from the definition of a neighborhood basis of $[\xi]$ in $\mathcal{Z} \sqcup \partial_\infty T$ (see the description of neighborhoods $U^1_{[\xi],R}$) that

$$\lim_{n\to\infty} \eta_n = [\xi]$$

in the topology of $\mathcal{Z} \sqcup \partial_\infty T$.

Case 4. The proof in the last case, when $\xi_n \in \partial_\infty X_{v_n}$ is a sequence converging to $\xi \in \partial_\infty X_v$ in $\partial_\infty X$ is similar to Case 3 and is left to the reader. □

CHAPTER 8

Cannon–Thurston maps

In this chapter, as an application of the description of uniform quasigeodesics in trees of hyperbolic spaces, we establish an existence theorem for *Cannon–Thurston* maps (CT-maps) between ideal boundaries of trees of hyperbolic spaces induced by the inclusion maps of subtrees of spaces, $X_S = Y \to X$, Theorem 8.13. The proof of this theorem occupies most of the chapter. Once this theorem is proven, we investigate the associated *Cannon–Thurston laminations*. In Section 8.7 we identify the CT-lamination $\Lambda(Y, X)$, while in Section 8.9 we relate $\Lambda(Y, X)$ to the collection of *ending laminations* of Y in X. We conclude the chapter with Section 8.11 where we discuss group-theoretic applications of our results on CT-maps and CT-laminations. In particular, in Section 8.11.1 we construct examples of undistorted surface subgroups of $PSL(2, \mathbb{C}) \times PSL(2, \mathbb{C})$ which are not Anosov. (The proof of nondistortion is an application of Theorem 8.21.)

8.1. Generalities on Cannon–Thurston maps

If X, Y are nonempty geodesic hyperbolic spaces and $f : Y \to X$ is a qi embedding, then f induces a (continuous) embedding of Gromov-boundaries, $\partial_\infty Y \to \partial_\infty X$: A sequence (y_n) in Y is a Gromov-sequence (see Section 1.13) if and only if $(f(y_n))$ is a Gromov-sequence in X (see, e.g., [**V05**, Theorem 5.38] or [**DK18**, Exercise 11.109]). The problem of existence of Cannon–Thurston maps concerns the existence of such an extension in the setting of *uniformly proper maps*.

The original motivation for Cannon–Thurston maps comes from the group theory: Given a hyperbolic subgroup H of a hyperbolic group G (with the inclusion map $\iota = \iota_{H,G}$) or, more generally, a homomorphism with finite kernel $\phi : H \to G$ between two hyperbolic groups, one says that *ϕ admits a Cannon–Thurston map* if there exists a ϕ-equivariant continuous map (a CT-map)

$$\partial_\infty \phi : \partial_\infty H \to \partial_\infty G.$$

Surprisingly, CT-maps for group homomorphisms exist quite often (see [**Mj14a, Mj14b, MP11, Mj16, Mj17, BR20**], etc.) but not always (see [**BR13**]). One of the earliest examples (justifying the terminology) of existence of CT-maps is due to Cannon and Thurston, [**CT07**]: If M is a compact hyperbolic 3-manifold fibered over the circle with the fiber F, then the natural embedding $\iota : \pi_1(F) \to \pi_1(M)$ admits a CT-map.

One can further generalize the setting of CT-maps to the nonequivariant one.

Definition 8.1. Let X, Y be geodesic Gromov-hyperbolic spaces and $f : Y \to X$ is a coarse Lipschitz map. Then f is said to *admit a CT-map* or *admits a CT-extension*, if there exists a map

$$\partial_\infty f : \partial_\infty Y \to \partial_\infty X$$

such that $f \cup \partial_\infty f : \bar{Y} = Y \cup \partial_\infty Y \to \bar{X} = X \cup \partial_\infty X$ is continuous at each point of $\partial_\infty Y$.

Remark 8.2. The continuity requirement and the fact that $\bar{X}$ is Hausdorff imply that if a CT-extension exists, then it is unique.

In order to connect to the group-theoretic situation, one takes X, Y to be Cayley graphs of the groups G and H respectively (where the finite generating set of G contains that of H), and $f : Y \to X$ the inclusion map induced by the inclusion $\phi = \iota : H \to G$. Since f is equivariant, the existence of a CT-extension of f would imply the existence of a continuous equivariant map $\partial_\infty \phi$.

In the setting when Y is a subspace of X with the induced path-metric and the identity embedding $f = \iota_{Y,X}$, we will use the notation $\partial_{Y,X}$ for the CT-map $\partial_\infty \iota_{Y,X}$ (if it exists). The same notation $\partial_{H,G}$ will be used for the CT-map $\partial_\infty \iota_{H,G}$, where $\iota_{H,G} : H \hookrightarrow G$ is the identity embedding of a hyperbolic subgroup H of a hyperbolic group G.

A criterion for the existence of CT-maps between hyperbolic metric spaces was established by Mahan Mitra in [**Mit98**, Lemma 2.1]. Recall that $(x.y)_p$ denotes the Gromov-product in a metric space.

THEOREM 8.3 (Mitra's criterion). *Let $f : Y \to X$ be a coarse Lipschitz proper map of proper geodesic hyperbolic metric spaces. Then a Cannon–Thurston map $\partial_\infty f : \partial_\infty Y \to \partial_\infty X$ exists if and only if for some (each) $y_0 \in Y$ the function*

$$t \mapsto \inf\{(f(y_1).f(y_2))_{f(y_0)} : y_1, y_2 \in Y \text{ are such that } (y_1.y_2)_{y_0} \geq t\}$$

is proper.

It is also instructive (in view of the relative version of the theorem, proven in Proposition 9.36) to restate the result using distances to geodesics (since these are uniformly close to Gromov-products):

THEOREM 8.4 (Mitra's criterion). *Let $f : Y \to X$ be a coarse Lipschitz proper map of proper geodesic hyperbolic metric spaces. Then a Cannon–Thurston map $\partial_\infty f : \partial_\infty Y \to \partial_\infty X$ exists if and only if for some (every) $y_0 \in Y$ there exists a function $C = C(y_0, D)$ such that:*

$$d_X(f(y_0), [f(y_1)f(y_2)]_X) \leq D \Rightarrow d_Y(y_0, [y_1y_2]_Y) \leq C$$

for all points $y_1, y_2 \in Y$ and geodesics connecting them and their images.

Half of Theorem 8.3 (most relevant for us) also holds for nonproper and nongeodesic hyperbolic spaces. The result was first proven in [**KS20**], we include a proof since it is quite simple and for the sake of completeness:

PROPOSITION 8.5. *Suppose that $f : Y \to X$ is a coarse Lipschitz map of hyperbolic spaces in the sense of Gromov, such that for each $p \in Y$ and each pair of sequences $y_n, y'_n \in Y$,*

$$\lim_{n\to\infty} (y_n.y'_n)_p = \infty \Rightarrow \lim_{n\to\infty} (f(y_n).f(y'_n))_{f(p)} = \infty.$$

Then f admits a Cannon–Thurston map.

PROOF. Note that the assumption in the proposition implies that the map f is metrically proper: If a sequence $y_n \in Y$ diverges to infinity in the sense that $d(p, y_n) \to \infty$, then the sequence $f(y_n)$ also diverges to infinity. To prove properness one takes a sequence $y'_n = y_n$.

The definition of the Cannon–Thurston extension of the map f is a natural one. The assumption in the proposition states that the image under f of each Gromov-sequence (y_n) in Y is also a Gromov-sequence in X. Thus, one defines the extension by the formula:

$$\partial_\infty f([y_n]) = [f(y_n)],$$

where (y_n) is a Gromov-sequence. A diagonal argument shows that this map is well defined and, hence, is partially continuous: If a sequence $y_n \in Y$ converges to $\zeta \in \partial_\infty Y$, then the sequence $f(y_n)$ converges to $\partial_\infty f(\zeta)$. From this, by the density of Y in $\bar{Y}$, it follows that $f \cup \partial_\infty f$ is continuous at $\partial_\infty Y$. □

REMARK 8.6. The converse to this proposition also holds if Y is proper: If a map $f : Y \to X$ admits a CT-extension, then for every pair of sequences $(y_n), (y'_n)$ in Y,

$$\lim_{n\to\infty} (y_n.y'_n)_p = \infty \Rightarrow \lim_{n\to\infty} (f(y_n).f(y'_n))_{f(p)} = \infty.$$

The reason is that such sequences $(y_n), (y'_n)$ have to subconverge to points $\xi, \xi' \in \partial_\infty Y$ (in view of properness of Y). Then, necessarily, $\xi = \xi'$. Since f admits a CT-extension to the points ξ, ξ', the sequences $(f(y_n)), (f(y'_n))$ have to converge to $\partial_\infty f(\xi)$, implying that

$$\lim_{n\to\infty} (f(y_n).f(y'_n))_{f(p)} = \infty.$$

In the next lemma we will use the notion of the *relative ideal boundary* $\partial_\infty(A, X)$ of a subset A of a hyperbolic space X; see Definition 1.83.

LEMMA 8.7. *Suppose that Y is a proper metric space, and $f : Y \to X$ admits a CT-extension. Then*

$$\partial_\infty f(\partial_\infty Y) = \partial_\infty(f(Y), X).$$

PROOF. By the continuity of the CT-extension, if a sequence $y_n \in Y$ converges to $\xi \in \partial_\infty Y$, then the sequence $(f(y_n))$ converges to $\partial_\infty f(\xi)$. Conversely, suppose that a sequence $x_n = f(y_n) \in f(Y)$ converges to a point $\eta \in \partial_\infty(f(Y), X)$. By the properness of Y, the sequence (y_n) subconverges to some $\xi \in \partial_\infty Y$ and, again by continuity of the CT-extension, $\partial_\infty f(\xi) = \eta$. □

LEMMA 8.8 (Functoriality of CT-maps). *If $f : X \to Y, g : Y \to Z$ are coarse Lipschitz maps which admit CT-extensions, then so does their composition and*

$$\partial_\infty(g \circ f) = \partial_\infty g \circ \partial_\infty f.$$

PROOF. Let (x_n) be a Gromov-sequence representing $\xi \in \partial_\infty X$. The continuity of the CT-extension at the point ξ implies that for $y_n = f(x_n)$ the sequence (y_n) in Y, is also a Gromov-sequence in Y representing the point $\eta = \partial_\infty f(\xi)$. Applying the same reasoning to the map g, we conclude that $(g(y_n))$ is a Gromov sequence in Z representing $\zeta = \partial_\infty g(\eta)$. Thus, $\partial_\infty g \circ \partial_\infty f$ defines the CT-extension of $g \circ f$. □

Since for δ-hyperbolic spaces in the sense of Rips, the Gromov-product is comparable to the distance to a suitable geodesic (see Lemma 1.47), Proposition 8.5 can be reformulated as

PROPOSITION 8.9. *Suppose that $f : Y \to X$ is a coarse Lipschitz map of hyperbolic spaces in the sense of Rips, such that for each pair of sequences $y_n, y'_n \in Y$,*

$$\lim_{n\to\infty} d_Y(p, [y_n y'_n]_Y) = \infty \Rightarrow \lim_{n\to\infty} d_X(f(p), [f(y_n) f(y'_n)]_X) = \infty.$$

Then f admits a Cannon–Thurston map.

The existence of a CT-map, of course, does not imply its injectivity, and the notion of a Cannon–Thurston lamination (introduced by Mitra in [**Mit97**]) is motivated by this lack of injectivity.

DEFINITION 8.10. Suppose that $f : Y \to X$ is a map of hyperbolic spaces which admits a CT-extension $\partial_\infty f$. The *Cannon–Thurston lamination* (the CT-lamination) of $f : Y \to X$ is the (closed) subset $\Lambda(f)$ of $\partial_\infty^{(2)} Y$ consisting of unordered pairs of distinct points $\{\xi, \eta\}$ such that $\partial_\infty f(\xi) = \partial_\infty f(\eta)$. In the case when Y is a subset of X and f is the inclusion map $Y \to X$, we will use the notation $\Lambda(Y, X)$ for the CT-lamination. A geodesic $\alpha \subset Y$ connecting points ξ, η with $\{\xi, \eta\} \in \Lambda(f)$, is called a *leaf* of the CT-lamination $\Lambda(f)$.

Note that, in view of the fact that the map $\partial_\infty f$ is continuous, the lamination $\Lambda(f)$ is a closed subset of $\partial^{(2)}_\infty Y$.

REMARK 8.11.

(1) The above definition of $\Lambda(f)$ requires existence of a CT-map. However, one can extend this definition to the general case as follows; see [**Mit99**, section 2], [**MR18**, Definition 3.1]. We say that a point $\{\xi, \xi'\} \in \partial^{(2)}_\infty Y$ belongs to $\Lambda(f)$ if there exist sequences (y_n), (y'_p) in Y converging to ξ, ξ' respectively, such that

$$\lim_{n\to\infty} (f(y_n).f(y'_n))_{f(p)} = \infty.$$

(2) As it was noted in [**MR18**], in the general setting, $\Lambda(f)$, *a priori* is not closed in $\partial^{(2)}_\infty Y$. If the map f admits a CT-extension, then the two definitions of $\Lambda(f)$ agree.

(3) If H is a hyperbolic subgroup of a hyperbolic group G and f is the inclusion map $H \to G$ then $\Lambda(f) = \emptyset$ if and only if H is quasiconvex in G; see [**Mit99**, Lemma 2.1].

Mitra proved in [**Mit98**] that if G is a hyperbolic group isomorphic to the fundamental group of a finite graph $\mathcal{G}$ of hyperbolic groups satisfying the conditions of the Bestvina-Feighn Combination Theorem, then for each vertex-group G_v of $\mathcal{G}$, the Cannon–Thurston map for the inclusion homomorphism $G_v \to G$ exists. More generally, he proves:

THEOREM 8.12. *If $X \to T$ is a tree of hyperbolic spaces with hyperbolic total space X, then for every vertex-space X_v the inclusion map $X_v \to X$ admits a CT-map.*

Later on, in the paper by Mj and Pal [**MP11**], this result was extended to the relatively hyperbolic setting; we will discuss the extension in the next chapter of the book.

Our goal is to generalize Theorem 8.12 to the case of fundamental groups of subgraphs of graphs of groups and, more generally, to inclusion maps $Y \to X$ of subtrees of spaces in a tree of hyperbolic spaces $\mathfrak{Y} \subset \mathfrak{X}$, satisfying the conditions of Theorem 2.62. The main result of this chapter is:

THEOREM 8.13. *Let $\mathfrak{X} = (\pi : X \to T)$ be a tree of hyperbolic spaces with hyperbolic total space X. Then for every subtree $S \subset T$, the inclusion map $Y = X_S \to X$ admits a CT-extension.*

The most difficult part of the proof is to relate, for points $x, y \in Y$, the geodesics $[xy]_X$ in X to the geodesics $[xy]_Y$ in Y. This is done in Section 8.2 in the form of a "cut-and-replace" theorem (Theorem 8.21). Once this theorem is established, the existence of a CT-map is an almost immediate consequence of Proposition 8.9 (see Theorem 8.48). The main tool in our proof of Theorem 8.21 is the description of geodesics in hyperbolic trees of spaces given in the previous chapter. As this description is inductive in nature (a 7-step process), the proof of the cut-and-replace theorem follows the same inductive process (but we will only need 6 steps). While our proof follows in main Mitra's proof in [**Mit98**], we have to deal with some substantial complications; in fact, we will derive Mitra's theorem (Theorem 8.12) as an easy application of the first part of our proof; see the end of Section 8.3.

In the proof of Theorem 8.21 we will be using the fact that $\mathfrak{X}$ satisfies the uniform K-flaring condition for all $K \geq 1$ (see Lemma 2.50). Of course, if $\mathfrak{X}$ satisfied the K-flaring condition, so does $\mathfrak{Y} \subset \mathfrak{X}$.

8.2. Cut-and-replace theorem

8.2.1. Definitions and notations. Suppose that $\mathfrak{X} = (\pi : X \to T)$ is a tree of spaces and $\mathfrak{Y} = (\pi : X_S = Y \to S)$ is a subtree of spaces in $\mathfrak{X}$. In the next section we shall prove that the inclusion $Y \to X$ admits a Cannon–Thurston extension $\partial_\infty Y \to \partial_\infty X$ provided that $\mathfrak{X}$ satisfies Axiom **H** and X is hyperbolic.

To prove this result we need to compare the Y-geodesics $[xy]_Y$ to X-geodesics $[xy]_X$ joining pairs points $x, y \in Y$. When the points are in the same vertex-space in Y, this is done by Mahan Mitra in [**Mit98**] by constructing ladders. In general, up to a uniformly bounded error, the relation between X-geodesics and Y-geodesics is given by a *cut-and-replace procedure* described below.

For each (continuous) path $c : I \subset \mathbb{R} \to X$ in X with $c(\partial I) \subset Y$, we define the following modification, a *cut-and-replace procedure*, transforming c to a new path $\hat{c} = c_S : I \to Y$.

Definition 8.14. For a closed subinterval $J = [s, s'] \subset I$, we say that the restriction $\zeta = c|_J$ is a *detour subpath* in c, if $c(\partial J) \subset Y$, while $c(J - \partial J)$ is disjoint from Y. Thus, the points $x = c(s), x' = c(s')$ belong to a common vertex-space $X_t \subset Y$. We then replace each detour subpath ζ in c by the corresponding X_t-geodesic $\hat{\zeta} = [xx']_{X_t}$, called a *replacement segment.*[1] We let $\hat{c} = c_S$ denote the resulting path $I \to Y$.

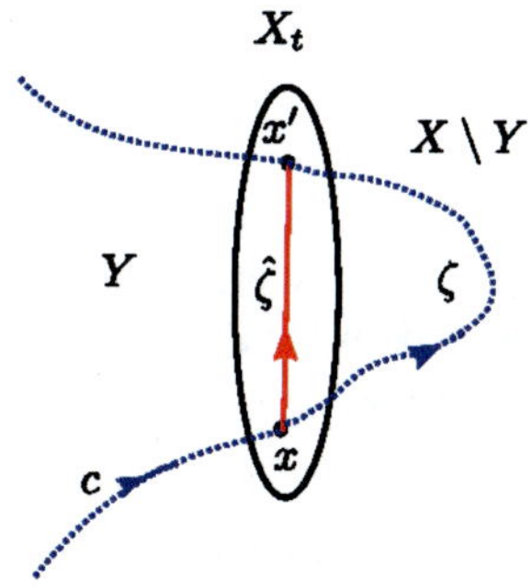

Figure 8.1. Detours and path-modification

Definition 8.15. Let ϕ be a continuous Λ-quasigeodesic in X with the endpoints in Y. We will say that ϕ is Λ'-*consistent* if $\hat{\phi}$ is a Λ'-quasigeodesic in Y. We will say that a pair of points $(y, y') \in Y^2$ is θ-*consistent*, where θ is a function $[1, \infty) \to [1, \infty)$, if every continuous Λ-quasigeodesic ϕ in X connecting y to y' is $\theta(\Lambda)$-consistent. We say that a subset of Y^2 is (uniformly) consistent if it consists of θ-consistent pairs for some function θ. We will say that a pair of points $(y, y') \in Y^2$ is Λ'-*weakly consistent* if y, y' are connected by *some* Λ'-consistent continuous quasigeodesic in X. Similarly, we will say that a subset of Y^2 is *uniformly weakly consistent* if it consists of Λ'-weakly consistent pairs of points for some Λ'.

While it is clear that every consistent pair of points is also Λ'-consistent for some Λ', it is far from clear if a subset of Y^2 consisting of Λ'-consistent pairs is uniformly

[1] While this X_t-geodesic is, in general, nonunique, if vertex-spaces are uniformly hyperbolic (which will be the case in all our examples) the ambiguity is uniformly bounded and we will ignore it.

consistent. The issue is that while any two uniform quasigeodesics ϕ, ϕ' connecting y to y' are uniformly close, we do not yet know if the same holds for $\hat{\phi}, \hat{\phi}'$. We will prove the claim, establishing equivalence of weak consistency and consistency, at the end of Part (I) of the proof of Theorem 8.21. Before that, we will be only proving consistency of specific quasigeodesics, namely slim combing paths $c(y, y')$ described in Section 7.1.

We will frequently use the following simple observation.

REMARK 8.16. Suppose that a pair $(x, y) \in Y^2$ in X is Λ'-consistent. Then every pair of points $x', y' \in N_r([xy]_X) \cap Y$ is θ'-consistent, where θ' depends only on θ' and r. In particular, perturbing the points x, y by a uniformly bounded amount, we do not lose consistency of the pair.

DEFINITION 8.17. Given two subsets Z', Z'' of a Gromov-hyperbolic space Z, we say that $y \in Z$ is an *R-transition point* between Z', Z'' in Z, if every geodesic in Z connecting Z', Z'' passes within distance R from y. We say that a finite sequence of points $z' = z_1, z_2, \dots, z_{n+1} = z''$ in a δ-hyperbolic space Z is *R-straight* if the geodesic $\gamma = z'z'' \subset Z$ passes through some points $x_i \in B(z_i, R)$ in the order $x_2, \dots, x_n$.

In relation to (sub)trees of spaces $Y \subset X$, we will talk about X-transition points and Y-transition points. We will say that $y \in Y$ is an R-transition point between $Z', Z'' \subset Y$ if it is R-transition point for Z', Z'' regarded as subsets in *both* Y and X.

When dealing with sets of finite sequences, we will refer to those as *uniformly straight* if they are R-straight for some uniform value of R. We use a similar terminology for transition points.

EXAMPLE 8.18. Suppose that Z, Z' are λ-quasiconvex and C-cobounded subsets in a δ-hyperbolic geodesic space X. Let β be a shortest geodesic between Z and Z'. Then every point of β is an R-transition point between Z and Z' with $R = R(C, \lambda, \delta)$. Conversely, every R-transition point between such Z, Z' is D-uniformly close to a point in β, where $D = D(C, B, \delta, R)$.

The importance of the concept of a transition point comes from another simple observation. Let $\mathfrak{X}$ be a tree of spaces and $\mathfrak{Y} \subset \mathfrak{X}$ be a subtree of spaces.

LEMMA 8.19. *Suppose that $Z_1, Z_2 \subset X$ and $y \in Y$ is an R-transition point between Z_1, Z_2. Assume that the set of pairs (z_i, y), $z_i \in Z_i$, $i = 1, 2$, is θ-consistent. Then the set of pairs $(z_1, z_2) \in Z_1 \times Z_2$ is θ'-consistent where $\theta' = \theta'(L, R)$.*

PROOF. Since we are dealing with quasigeodesics, we can as well consider L-quasigeodesics c in X connecting point $z_1 \in Z_1$ to $z_2 \in Z_2$ and passing through y (such exist due to the assumption that y is an R-transition points in X). Such c is a concatenation $c_1 \star c_2$, where c_1 connects z_1 to y. Thus, since $y \in Y$,

$$\hat{c} = \hat{c}_1 \star \hat{c}_2.$$

Each of the subpaths $\hat{c}_i$ is a $\theta(L)$-quasigeodesic by the consistency assumption for the pairs (z_i, y). We will estimate the quasi-isometry constant of $\hat{c}$ in Y. Let γ be a geodesic in Y connecting two points a, b in $\hat{c}$; the only interesting case to consider is when $a \in \hat{c}_1, b \in \hat{c}_2$. Let $\hat{c}(a, b)$ be the portion of $\hat{c}$ between a and b. Since y is an R-transition point in Y between z_1, z_2, the path γ has to pass within distance $r = r(R, L)$ from y. Subdividing γ as a concatenation $\gamma_1 \star \gamma_2$, where γ_1 connects z_1 to $x \in B(y, r)$, $i = 1, 2$, we see that

$$\text{length}(\hat{c}(a, b)) \le (\theta(L) + r)\, \text{length}(\gamma),$$

as required by a uniform quasigeodesic.[2] □

This lemma generalizes to the case of higher number of transition/concatenation points. Since the proof is similar, we leave it to the reader.

LEMMA 8.20. *Suppose that* $z' = z_1, z_2, \ldots, z_{n+1} = z''$ *is an R-straight sequence in both X and Y, and that each pair of points* $(z_i, z_{i+1}), i = 1, \ldots, n$, *is* θ*-consistent. Then the pair* (z', z'') *is* θ'*-consistent with the function* θ' *depending only on the hyperbolicity constant of Y and on* θ.

We are now ready to state the main technical result of this chapter.

THEOREM 8.21 (Cut-and-replace X-quasigeodesics to get Y-quasigeodesics). *Suppose that* $\mathfrak{X} = (\pi : X \to T)$ *is a tree of hyperbolic spaces satisfying the uniform* K_**-flaring condition,* $S \subset T$ *is a subtree and* $Y = X_S$. *Then* $Y \times Y$ *is* θ*-consistent with* θ *depending only on the parameters of* $\mathfrak{X}$.

We break the proof of Theorem 8.21 in three parts and each part in several steps, where we prove this theorem in special cases and then use these special cases to prove the general case in the last part.

8.3. Part I: Consistency of points in vertex flow-spaces

Suppose that $k \geq k_{5.2}$, $K = K_{5.2}(k)$, and that $\mathfrak{X}$ is a tree of hyperbolic spaces satisfying the uniform $\kappa_{4.7}(K)$-flaring condition. (As it was noted earlier, in Lemma 2.50, uniform κ-flaring holds for all κ if X is hyperbolic.) The main result of this section is Proposition 8.22.

PROPOSITION 8.22. *For every* $u \in V(S)$, *the set of pairs* $(y, y') \in Fl_{Y,k}(X_u) \times Fl_{Y,k}(X_u)$ *is* $\theta = \theta_{8.22,k}$*-consistent, with* θ *depending only on k and the parameters of* $\mathfrak{X}$.

PROOF. For most of the proof we will be only proving weak uniform consistency, working with the quasigeodesic paths $c = c_X(y, y')$ in the fiberwise $4\delta_0$-neighborhood of $\mathfrak{Fl}_{X,k}(X_u) \subset \mathfrak{Y}$ given by the slim combing of $\mathfrak{Fl}_{X,k}(X_u)$ described in Section 7.1.

According to Corollary 5.12, there exists a (K, D, E)-ladder $\mathfrak{L} = \mathfrak{L}_X(\alpha) \subset \mathfrak{Y}$ contained in the fiberwise $4\delta_0$-neighborhood of $\mathfrak{Fl}_{X,k}(X_u) \subset \mathfrak{Y}$, containing y (resp., y') in its bottom (resp., top), where K, D, E depend only on k and α is a geodesic in X_u. Recall that, by the very definition, $c_X(y, y') = c_{\mathfrak{L}}(y, y')$ is contained in the ladder $L_X(\alpha)$. Define $\mathfrak{L}_Y = \mathfrak{L} \cap \mathfrak{Y}$: It follows from the definition of a ladder that $\mathfrak{L}_Y$ is a (K, D, E)-ladder in $\mathfrak{Y}$. It also follows from the definition of the modification $c \mapsto \hat{c}$, that $\hat{c}$ is contained in L_Y (up to a uniformly bounded error which we will ignore).

There are several cases to consider, according to the construction of uniformly quasigeodesic paths $c_{\mathfrak{L}}(y, y')$, depending on the properties of the ladder $\mathfrak{L}$ and location of the points y, y'.

8.3.1. Part I.1: The points y, y' **belong to a** $(K, M_{\bar{K}})$**-narrow carpet** $\mathfrak{A} = \mathfrak{A}_X = (\pi : A_X \to [\![u, w]\!]) \subset \mathfrak{L}_X$**.** The carpet $\mathfrak{A}$ contains the "subcarpet" $\mathfrak{A}_Y = \mathfrak{A}_X \cap \mathfrak{Y}$, $[\![u, w']\!] = \pi(A_Y)$. Note that $\mathfrak{A}_Y$ is a (K, C)-carpet where C is the length of the "narrow end" $A_{w'}$ of $\mathfrak{A}_Y$, but we cannot bound C (from above) in terms of k. According to Corollary 3.63, we have the coarse $L_{3.63}$-Lipschitz retraction $\rho_{\mathfrak{A}} : X \to A_X$, where $L_{3.63}$ depends only on K, D, and E (hence, only on k). The restriction of this retraction to Y is a retraction to A_Y; in particular, A_Y is $L_{3.63}$-qi embedded in Y.

[2]Here and in what follows we repeatedly use the notation $\phi(p, q)$ for a subpath in a path ϕ between the points p, q in ϕ, see Section 1.2.1.

LEMMA 8.23. *The pairs (y, y') are $\Lambda_{8.23}$-weakly consistent, where $\Lambda_{8.23}$ depends only on k.*

PROOF. We let $\gamma_y, \gamma_{y'}$ denote the K-leaves in $\mathfrak{A}_X$ connecting, respectively, y, y' to points of A_w. Let $v \in [\![u, w]\!]$ be the infimum of all vertices t in $[\![u, w]\!]$ such that

$$d_{X_t}(\gamma_y(t), \gamma_{y'}(t)) \leq M_{\bar{K}}.$$

Then $c = c_{\mathfrak{A}}(y, y')$ is the concatenation of the subpath γ_y restricted to $[\![u, v]\!]$, followed by the vertical geodesic $[\gamma_y(v)\gamma_{y'}(v)]_{X_v}$ and then followed by $\gamma_{y'}$ restricted to $[\![v, \pi(y')]\!]$. The path $\hat{c}$ is a similar concatenation $c_1 \star c_2 \star c_3$ except v is replaced by the vertex v' which is the minimum of $\{v, w'\}$ in the oriented interval $[\![u, w]\!]$ (the paths c_1, c_3 are contained in $\gamma_y, \gamma_{y'}$ respectively and c_2 is contained in $A_{v'}$). But this path is exactly the path $c_{\mathfrak{A}_Y}(y, y')$ as defined in Step I.1 in Section 7.1, or in the proof of Proposition 4.1.

The quasigeodesic constant of $c_{\mathfrak{A}_Y}(y, y')$ *a priori* depends on both K and C. However, according to Remark 4.3, the dependence on C appears only in the proof of Lemma 4.2, establishing uniform bounds on distortion of paths $c_{\mathfrak{A}_Y}(y, y')$ in A_Y. It remains, therefore, to get a uniform distortion bound depending only on k. Take a pair of points $a, b \in \hat{c}$. There are several cases to consider depending on the location of the points a, b, we will treat just one since the rest are done by the same argument: We will assume that $a \in c_2, b \in c_3$. It suffices to bound the length of $\hat{c}(a, b)$ (between a, b) in terms of $d_{\mathfrak{A}}(a, b)$, equivalently, in terms of the length of $c(a, b)$ since the latter is a uniform (with quasigeodesic constant depending only on k) quasigeodesic in A_X. The latter path is a concatenation of $c(a, a')$ and $c(a', b)$, where a' is in $A_{v'}$ and $c(a', b) = c_3$; see Figure 8.2. We have

$$\text{length}(c_3) \leq \text{length}(c(a, b)),$$

while

$$\text{length}(c_2) = d_{A_{v'}}(a, a') \leq \eta(d_X(a, a')) \leq \eta(\text{length}(c(a, b))),$$

where $\eta = \eta_{2.18}$. Thus,

$$\text{length}(\hat{c}(a, b)) \leq \text{length}(c(a, b)) + \eta(\text{length}(c(a, b))),$$

thereby providing the required distortion bound depending only on k. □

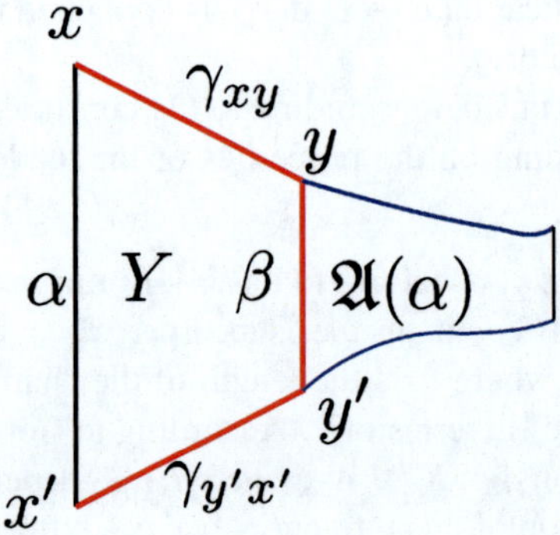

FIGURE 8.2

8.3.2. Part I.2: The points $y, y' \in Y$ belong to a carpeted X-ladder $\mathfrak{L} = \mathfrak{L}_X(\alpha)$. In this part of the proof we assume that $\mathfrak{L}$ contains a $(K, M_{\bar{K}})$-narrow carpet $\mathfrak{A} = \mathfrak{A}_X(\alpha')$, where α is a geodesic segment in X_u, $u \in S$, and $\alpha' \subset \alpha$ is a subsegment of length $\geq \text{length}(\alpha) - M_{\bar{K}}$. The path $c = c_{\mathfrak{L}}(y, y')$ can be of one of two types (see Section 7.1, Step I.2, for the definition of the types of slim combing paths in a ladder):

(1) If c is of type 1, then the assumption that y, y' are in Y and the definition of type 1 paths imply that $c = \hat{c}$, so there is nothing to prove.
(2) Suppose that c is of type 2. Then c is the concatenation $c_1 \star c_2 \star c_3$, where c_2 is contained in the carpet $\mathfrak{A}$, c_1, c_3 are contained in Y and, thus,

$$\hat{c} = c_1 \star \hat{c}_2 \star c_3.$$

Below we will use the same notation for the subcarpet $\mathfrak{A}_Y = \mathfrak{A} \cap \mathfrak{Y}$ as in Part I.1.

Lemma 8.24. *The pairs (y, y') are $\Lambda_{8.24}$-weakly consistent, where $\Lambda_{8.24}$ depends only on k.*

Proof. The paths c_1, c_3 are uniformly quasigeodesic in L_X (since c is) while $\hat{c}_2$ is uniformly quasigeodesic in A_Y according to Lemma 8.23. Since L_Y is uniformly hyperbolic, in order to prove the lemma, it suffices to verify that $\hat{c}$ is uniformly proper in L_Y (see Lemma 1.21). The proof is similar to the one in Lemma 8.23. We will prove uniform properness of $\hat{c}$ in L_X. As in the proof of Lemma 8.23, we only consider the most representative case, of points $a \in c_2, b \in c_3$: We need to bound $\text{length}(\hat{c}(a, b))$ in terms of $\text{length}(c(a, b))$. The path $c(a, b)$ is the concatenation $c(a, a') \star c_3(a', b)$, where a' is the concatenation point of c_2 and c_3. According to Lemma 8.23,

$$\begin{aligned}\text{length}(\hat{c}(a, a')) \leq \Lambda_{8.23} d_{\mathfrak{A}_Y}(a, a') &\leq \Lambda_{8.23}\eta_{2.18}(d_{\mathfrak{A}}(a, a')) \\ &\leq \Lambda_{8.23}\eta_{2.18}(\text{length}(c(a, b))),\end{aligned}$$

while

$$\text{length}(c_3)) \leq \text{length}(c(a, b)).$$

The lemma follows. □

8.3.3. Part I.3: General ladders. Suppose that $u \in S$, $\alpha = [pp']_{X_u} \subset Y$, $\mathfrak{L} = \mathfrak{L}_X(\alpha) \subset \mathfrak{X}$ is a K-ladder, $\mathfrak{L}_Y = \mathfrak{L} \cap \mathfrak{Y}$. Our goal is to prove uniform consistency of paths $c_{\mathfrak{L}}$ connecting points $y \in bot(\mathfrak{L}) \cap Y$, $y' \in top(\mathfrak{L}) \cap Y$. Recall that according to Proposition 4.14 (on vertical subdivision), we have a subdivision of α into subintervals $\alpha_i = [p_i p_{i+1}]_{X_u}$, subintervals $\alpha_i' \subset \alpha_i$, and a collection of K-qi sections Σ_i in $\mathfrak{L}$ through the points p_i dividing $\mathfrak{L}$ into subladders $\mathfrak{L}^i = \mathfrak{L}(\alpha_i)$ containing $(K, M_{\bar{K}})$-narrow carpets $\mathfrak{A}^i = \mathfrak{A}(\alpha_i')$. We also defined points $x_i^{\pm}$ in the sections $\Sigma_i^- = \Sigma_i, \Sigma_i^+ = \Sigma_{i+1}$ bounding $\mathfrak{L}^i$ such that the combing paths in L (connecting y, y') pass through the points $x_i^{\pm}$. Each section Σ_i is defined over some subtree $T_i \subset T$. The intersections $\Sigma_{i,Y} = \Sigma_i \cap Y$ project to subtrees $S_i \subset T_i$.

We next find uniform transition points in Y between the sections $\Sigma_{i,Y}$. We let $v_i^{\pm} = \pi(x_i^{\pm})$ and define vertices $w_i^{\pm}$ as nearest-point projections of $v_i^{\pm}$ to the subtrees S_i, S_{i+1}, where the projection is taken inside the trees T_i, T_{i+1} respectively. Set

$$y_i^{\pm} := \Sigma_{i,Y}^{\pm} \cap X_{w_i^{\pm}};$$

see Figure 8.3.

Since each path $c_{\mathfrak{L}}$ connecting *arbitrary* points $y^{\pm} \in \Sigma_i^{\pm}$ is a concatenation of paths in $\Sigma_i^{\pm}$ connecting $y^{\pm}$ and $x_i^{\pm}$ and of the path $c_{\mathfrak{L}^i}(x_i^-, x_i^+)$, we conclude that whenever $y^{\pm}$ are both in Y, the path $\hat{c}_{\mathfrak{L}}(y_i^-, y_i^+)$ passes through the points $y_i^{\pm}$. Since the above paths are uniformly quasigeodesic in the ladders L_Y^i (Part I.2), we see that the points $y_i^{\pm}$ are at

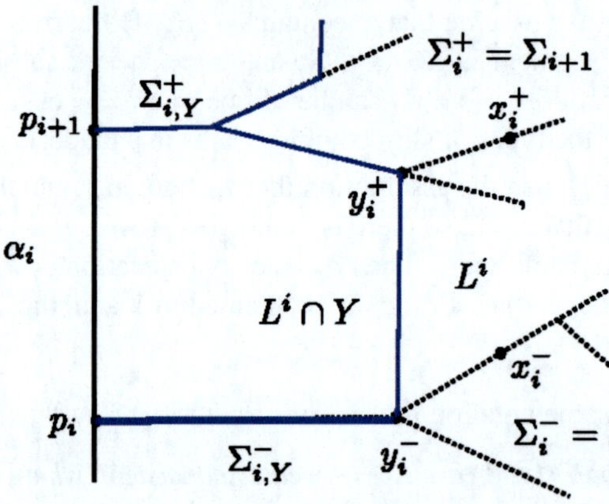

FIGURE 8.3. Ladder L^i and transition points

uniformly bounded (in terms of k) distance R from the nearest-point projections (in the ladder L^i_Y) of $\Sigma_i^- \cap Y$ to $\Sigma_i^+ \cap Y$ and vice versa, for all $i = 1, \dots, n-1$. For $i = n$, we have that the point y_n^- within distance R from the projection (in the ladder L^n_Y) of y' to the section $\Sigma_{n,Y} = \Sigma_n \cap Y$. Furthermore, by Part I.2, the paths $\hat{c}(y^+_{i-1}, y^+_i)$ are $\Lambda_{8.24}$-quasigeodesics in L^i_Y. Thus, by Theorem 2.65, the alternating concatenation of the paths $\hat{c}_{\mathfrak{L}}(y_i^-, y_i^+)$ in the ladders L^i_Y and of the paths $c_{\Sigma_{i,Y}}(y^+_{i-1}, y^-_i)$ in $\Sigma_{i,Y}$, is a $\Lambda_{8.3.3}$-quasigeodesic in L_Y connecting y to y'.

For each vertex v of $\pi(\mathfrak{A}^i)$ we break the geodesic segment $L^i_v \subset X_v$ as a concatenation of two subsegments: A^i_v (in the carpet $\mathfrak{A}^i$) and β^i_v.

LEMMA 8.25. *For $w = w_i^+ = \pi(y_i^+)$ the length of $\beta = \beta^i_w$ is $\le C_{8.25}(K)$.*

PROOF. If the length of β is $\le M_{\bar{K}}$ then we are done. Otherwise, let J be the largest subinterval in $\pi(A^i)$ containing v such that for all vertices $s \in J$ the length of the subinterval β^i_s is $> M_{\bar{K}}$. Since β^i_u has length $\le M_{\bar{K}}$ and for $v = v_i^+ = \pi(x_i^+)$ the length of β^i_v is also $\le M_{\bar{K}}$, uniform K-flaring implies that the length of J is $\le \tau = \tau(K, M_{\bar{K}}) - 2$ and, thus, by Lemma 2.38

$$\text{length}(\beta) \le C(k) := a^\tau (M_{\bar{K}} + b),$$

where $a = L_0'$ and $b = 2L_0'K$. □

At this point, if we knew that for all i the vertices $\pi(x_i^+), \pi(x^-_{i+1})$ are separated in T by the subtree $S_{i+1} = \pi(\Sigma_i^+)$, then the paths $c_{\Sigma_i}(x^+_{i-1}, x_i^-)$ would have to contain the subpaths $c_{\Sigma_{i,Y}}(y^+_{i-1}, y_i^-)$. This would imply that $\hat{c}$ is a concatenation of the uniform quasigeodesics

$$c_{\mathfrak{L}_Y}(y_i^-, y_i^+) \star c_{\Sigma_{i,Y}}(y_i^+, y^-_{i+1}), i = 1, 2, \dots,$$

and then we would be done with the proof of the proposition. However, this (the separation property) need not be the case. What we know, however, from the description of the points $x_i^\pm, y_i^\pm$, is that in the oriented interval $\pi(\mathfrak{A}_i)$ either

$$u \le \pi(y_i^+) \le \pi(y_i^-) \le \pi(x_i^+) \le \pi(x_i^-)$$

or

$$u \le \pi(y_i^+) = \pi(x_i^+) \le \pi(y_i^-) \le \pi(x_i^-).$$

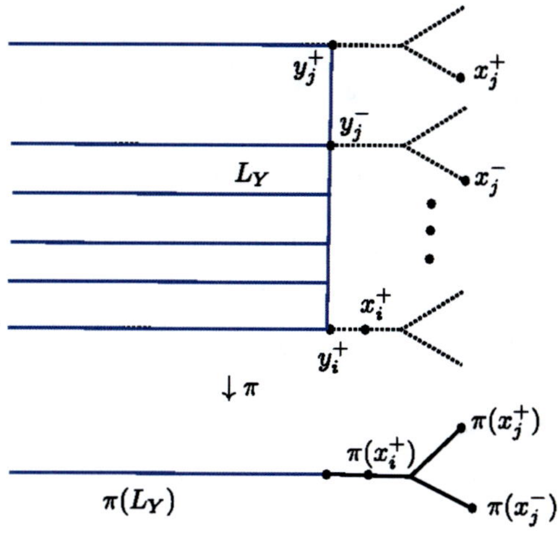

FIGURE 8.4. A clique $C = \{x_i^+, \dots, x_j^-, x_j^+\}$

Moreover, since $\mathfrak{L}, \mathfrak{L}_Y$ are ladders, $\pi(x_i^+), \pi(x_{i+1}^-)$ are separated in T by the subtree S_{i+1} if and only if they are separated in T by the subtree $\pi(L_Y) = \pi(\mathfrak{L}) \cap S$. In particular, for each i either $\pi(x_i^\pm)$ are in the same component of $T - \pi(L_Y)$ or x_i^+ lies in Y. In other words, we find that y_i^+ is an X-transition point between y, y' if either $y_i^+ = x_i^+$ or if $\pi(x_i^+)$ and $\pi(x_{i+1}^-)$ lie in distinct components of $T - \pi(\mathfrak{L}_Y)$. Thus, we group the points $x_i^\pm$ into *cliques* of the form

$$C = \{x_i^+, x_{i+1}^-, x_{i+1}^+, \dots, x_j^+\},$$

or

$$C = \{x_i^-, x_i^+, x_{i+1}^-, x_{i+1}^+, \dots, x_j^+\},$$

etc., consisting of maximal collections of consecutive points in c of the form $x_l^\pm$, whose projections to T are not separated by $\pi(L_Y)$. Here it is understood that if one of the two vertices s, t lies in a subtree $T' \subset T$, then s and t are separated by this subtree. We will denote a clique as

$$(x_i^\pm, \dots, x_j^\pm),$$

where $x_i^\pm, x_j^\pm$ are the first and the last element of a clique C, listed in the order of their appearance in the path $c = c(x, x')$. Elements of a clique have the property that their projections to T followed by the projection to $\pi(L_Y)$ equal to the same vertex $v = v_C$, and the corresponding points in Y

$$y_i^\pm, \dots, y_j^\pm$$

all belong to the same vertical geodesic segment L_v, $v = v_C$. Furthermore, for each $l \in [i, j]$

$$y_l^+ = y_{l+1}^-,$$

since they both belong to $L_v \cap \Sigma_{l+1}$.

The first and the last points $x_i^\pm, x_j^\pm$ of a clique C determine points $y_i^\pm, y_j^\pm \in L_v \cap c$ which break c as a concatenation of three subpaths $c_1 \star c_2 \star c_3$, where $c_2 = c(y_i^\pm, y_j^\pm)$. Thus, each clique defines a decomposition

$$\hat{c} = \hat{c}_1 \star \hat{c}_2 \star \hat{c}_3,$$

where $\hat{c}_2 = [y_i^{\pm} y_j^{\mp}]_{X_v} \subset L_v$. The latter path is a concatenation of the subsegments $[y_l^- y_l^+]_{X_v} \subset L_v$. Each of these subsegments, in turn, is a concatenation of two subsegments: A_v^l (the narrow end of the carpet $\mathfrak{A}_Y^l \subset \mathfrak{L}_Y^i$) and a subsegment β_v^l of length $\leq C(k)$; see Lemma 8.25. Since the segment A_v^l is a uniform quasigeodesic in L_Y^i (see Lemma 8.23), it follows that $[y_l^- y_l^+]_{X_v}$ is a uniform quasigeodesic in L_Y^i.

Thus, the entire path $\hat{c}$ is broken as an alternating concatenation of uniform Y-quasigeodesics $[y_l^- y_l^+]_{X_v}$ connecting $\Sigma_{l,Y}, \Sigma_{l+1,Y}$ and of (possibly degenerate) horizontal K-qi leaves connecting y_i^+, y_{i+1}^- inside $\Sigma_{l+1,Y}$. The points y_l^-, y_l^+, up to a uniformly bounded error, realize the shortest distance in L_Y^i between $\Sigma_{l,Y}, \Sigma_{l+1,Y}$.

We can now finish the proof of Proposition 8.22 for the slim combing paths c: The path $\hat{c}$ satisfies the conditions of Theorem 2.65 and, hence, is a uniform quasigeodesic in Y.

We conclude with Lemma 8.26.

Lemma 8.26. *The set of pairs $(y, y') \in Fl_k^Y(X_u) \times Fl_k^Y(X_u)$ is weakly special. More precisely, there exists $\Lambda_{8.26}(k)$ such that each slim combing path c in $N_{4\delta_0}^{fib} Fl_k^X(X_u)$ connecting y to y' satisfies the property that $\hat{c}$ is a $\Lambda_{8.26}(k)$-quasigeodesic in Y.*

Proof. Points y, y' belong to bottom/top of a k-ladder $L(\alpha)$ which is uniformly close to $Fl_k^Y(X_u)$. The combing path $c = c(y, y')$ in this ladder satisfies the property that $\hat{c}$ is a $\Lambda_{8.26}(k)$-quasigeodesic in Y. □

Lastly, we prove consistency for arbitrary uniform quasigeodesics ϕ connecting points of the given vertex-flow-space, i.e., uniform consistency of points in an arbitrary flow-space $Fl_k(X_u)$, i.e., prove Proposition 8.22 in full generality.

Since $Fl_k(X_u)$ is $\delta_{5.17}(k)$-hyperbolic, for each $\Lambda \geq 1$, each Λ-quasigeodesic ϕ in $Fl_k(X_u)$ connecting y, y' is within Hausdorff distance $D(k, \Lambda)$ from a combing path $c = c_L(y, y')$ contained in a ladder $\mathfrak{L} = \mathfrak{L}_X(\alpha) \subset N_{4\delta_0}^{fib} Fl_k(X_u)$, where $y, y' \in L_X(\alpha)$.

We will be using the notation from the proof of Lemma 8.26. Suppose that $x \in \mathcal{X}$ is a point in $c \subset L_K(\alpha), \alpha \subset X_u$, $x \notin L_Y$. Then x belongs to one of the subpaths $c(y_i^{\pm}, y_j^{\mp})$ determined by a clique C and $\pi(x)$ is a vertex in a subtree of T separated from $\pi(L_Y)$ by the vertex $v = v_C \in \pi(L_Y)$. In particular,

$$d(x, X_S) = d(x, X_v) \geq d(\pi(x), v)$$

and taking intersection with L_v of the canonical K-qi section $\Sigma_x \subset L_X(\alpha)$, we obtain a point $y'' \in L_v \subset L_Y$ within distance $Kd(x, X_S)$ from x. It follows that every point $z \in X_S$ within distance R from $x \in c$, lies within distance $(K + 1)R$ from a point $\hat{z} = y''$ in $\hat{c} \cap L_v$. In particular, each intersection point of ϕ with X_S is within distance $D(k, \Lambda) + (K + 1)R$ from a point in $\hat{c} \cap L_v$. Furthermore, by the construction, the map $z \mapsto \hat{z}$ is monotonic: If z_1 appears before z_2 in c, then $\hat{z}_1$ appears before $\hat{z}_2$ in $\hat{c}$. It now follows that for each L-quasigeodesic ϕ, the path $\hat{\phi}$ is $\hat{\Lambda}$-quasigeodesic in Y. □

This concludes Part I of the proof of Theorem 8.21.

The next result is an immediate corollary of Proposition 8.22:

Corollary 8.27. *Suppose that p, p' are points in a vertex-space $X_v \subset Y$ such that a Λ-quasigeodesic ϕ in X connecting p to p' intersects X_v only at its endpoints. Then the vertical geodesic $[pp']_{X_v}$ is a $\Lambda' = \Lambda'_{8.27}(\Lambda)$-quasigeodesic in Y.*

As another application of Part I, we also obtain a theorem which is essentially due to Mitra, [**Mit98**].

THEOREM 8.28. *There exists a constant $R = R(\Lambda)$ depending only on the tree of spaces $X \to T$ such that for every vertex $u \in V(T)$ and any pair of points $y, y' \in X_u$, and any Λ-quasigeodesic ϕ in X with the endpoints y, y', the intersection*

$$X_u \cap \phi$$

is contained in the R-neighborhood of the vertical geodesic $[yy']_{X_u}$.

PROOF. We will apply Proposition 8.22 to the subtree $S = \{u\}$. By the proposition, each intersection point $z \in X_u \cap \phi$ is within distance $D(k, \Lambda) + (K + 1)R$ from a point in the geodesic segment $L_u \subset [yy']_{X_u}$, where $K = K(k), R = R(k, \Lambda)$ and k can be taken to be uniform, say, $k = K_0$. □

COROLLARY 8.29. *There exists $R = R_{8.29}(r)$ such that for every vertex $v \in V(T)$ and any pair of points $p, q \in X_v$ and geodesics $\alpha = [pq]_{X_v}$, $\beta = [pq]_X$, we have*

$$N_r(X_v) \cap \beta \subset N_R(\alpha).$$

In particular, if $\alpha' = [xx']_{X_v} \subset \alpha = [xy]_{X_v}$ and $y \in N_r(\beta)$, $\beta = [xx']_X$, then $d(y, x') \le R$. (See Figure 8.5.)

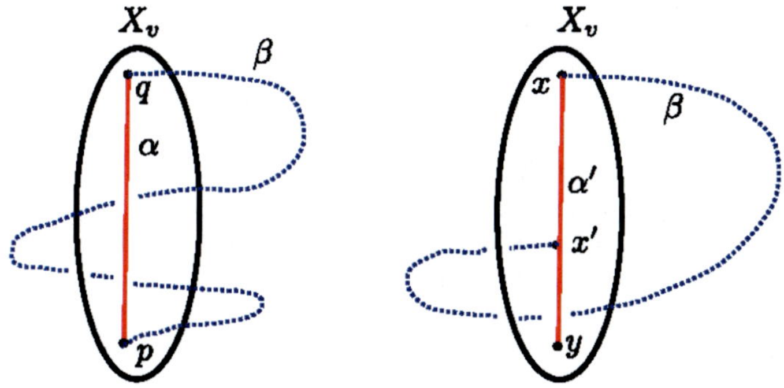

FIGURE 8.5

As another corollary, we obtain Mitra's theorem on the existence of CT-maps (Theorem 8.12).

Proof of Theorem 8.12. We will use Mitra's criterion for the inclusion map $f : X_u \to X$ (Proposition 8.9). We prove that f satisfies the assumption of Proposition 8.9 by arguing the contrapositive.

Fix a base-point $y \in Y$ and consider a sequence of vertical geodesics $\alpha_n = [x_n x_n']_{X_u}$. Assume that each X-geodesic $\beta_n = [x_n x_n']_X$ has nonempty intersection with the r-ball $B(y, r) \subset X$ for some fixed r. After replacing β_n with uniform quasigeodesics ϕ_n in X, we ensure that $y \in \phi_n$ for all n. Therefore, $y \in \hat{\phi}_n$ as well. Since the Hausdorff distance between $\hat{\phi}_n$ and α_n is uniformly bounded, the minimal distances between y and α_n are uniformly bounded too. Thus, the assumption of Proposition 8.9 is satisfied and, hence, the inclusion map $f : X_u \to X$ admits a CT-extension. □

In Part II of the proof of Theorem 8.21 we will need several technical results regarding projections to X-geodesics connecting points in X_u, $u \in V(S)$. These results occupy the rest of this section. We consider a (K, D, E)-ladder $\mathfrak{L} = \mathfrak{L}_X(\alpha)$, where $p, p' \in X_u$ and $\alpha = [pp']_{X_u}$. We will be investigating the nearest-point projection (taken in X) of points $y \in \alpha = [pp']_{X_u}$ to a Λ-quasigeodesic geodesic ϕ in X containing a detour path $\zeta = \phi(p, p')$ connecting the points p, p'. Most of the discussion deals with the case $\phi = \zeta$. Since the path ζ is uniformly Hausdorff-close to the combing path $c = c_X(p, p')$, in order to understand the projection of y to ζ, it suffices to analyze the projection of y to c (see Corollary 1.108).

LEMMA 8.30. *Suppose that p, p' belong to a common vertex-space X_u. Let $\{\alpha_i\}$ be a vertical subdivision of $\alpha = [pp']_{X_u}$. Then:*

(1) *For each $y \in \alpha_i = [p_i p_{i+1}]_{X_u}$ the projection $\bar{y} = P_{X,c}(y) \in c = c_X(p, p')$ is uniformly close to a point $\bar{y}'$ in the subladder $L^i = L_K(\alpha_i)$ determined by α_i:*

$$d_X(\bar{y}, \bar{y}') \leq C = C_{8.30}(K, D, E, \Lambda).$$

(2) *The point $\bar{y}'$ can be chosen to lie in the carpet $A^i = A(\alpha'_i) \subset L^i$, where $\alpha'_i \subset \alpha_i$ is as in Proposition 4.14.*

(3) *The point $\bar{y}' \in A^i$ can be chosen so that $y, \bar{y}'$ are connected by a (canonical in $A^i \subset L^i$) K-qi leaf contained in A^i.*

PROOF.

(1) Connect y to a point $z \in c \cap L^i$ by a geodesic $[yz]_X$. Since L^i is λ-quasiconvex in X, $[yz]_X$ lies in the λ-neighborhood of L^i. On the other hand, by Lemma 1.105, $[yz]_X$, as any geodesic connecting y to c, has to pass uniformly close to $\bar{y}$, namely, within distance $\lambda' + 3\delta_X$, where λ' is the quasiconvexity constant (in X) of the path c. It follows that $\bar{y}$ lies at a distance $C = \lambda + \lambda' + 3\delta_X$ from a point $\bar{y}' \in L^i$.

(2) By Part (1), $\bar{y} \in c$ lies within distance C from a point $\bar{y}' \in L^i$. If $\bar{y} \notin L^i$, take the smallest subpath in c connecting $\bar{y}$ to a point $z \in L^i$. Then (by the construction of the path (c) the point z realizes (up to a uniform additive error) the minimal distance from $\bar{y}$ to L^i. Hence, by replacing C with another uniform constant C', we can assume that $\bar{y}' \in c_i := c \cap L^i$, $d_X(\bar{y}, \bar{y}') \leq C'$.

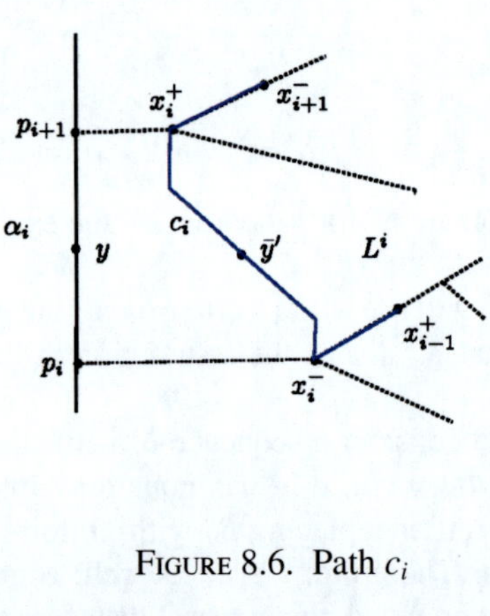

FIGURE 8.6. Path c_i

Note that c_i is the concatenation

$$c(x_{i-1}^+, x_i^-) \star c(x_i^-, x_i^+) \star c(x_i^+, x_{i+1}^-);$$

see Figure 8.6. The middle subpath lies in A^i (except for a vertical subpath of length $\leq M_{\bar{K}}$); therefore, there are just two cases we have to consider:

(a) $\bar{y}' \in c(x^+_{i-1}, x^-_i) \setminus A^i$. Then, by the description of combing paths in L^i (see Section 7.1, Step I.2), the path $c(\bar{y}', y)$ contains a subpath contained in $bot(L^i) \cap c(x^+_{i-1}, x^-_i) \subset c_i \subset c$ and connecting $\bar{y}'$ to a point A^i. Since $\bar{y}'$ is uniformly close to the projection of y to c (and since $c(\bar{y}', y)$ is uniformly quasigeodesic), the length of that subpath of $c(\bar{y}', y)$ is uniformly bounded and, hence, $\bar{y}'$ is uniformly close to a point in $c_i \cap A^i$, as required by the second statement of the lemma.

(b) $\bar{y}' \in c(x^+_i, x^-_{i+1}) \setminus A^i$; see Figure 8.7. The proof is similar to Case (a): The path $c(\bar{y}', y)$ starts with a subpath $c_{\bar{y}'} \subset L^i$ connecting $\bar{y}'$ to A^i (again; see Section 7.1, Step I.2). Since the length of the part of $c_{\bar{y}'}$ contained in c_i has to be uniformly bounded (as in Case (a)), by changing the location of $\bar{y}' \in c_i \cap top(L^i)$ by a uniformly bounded amount, we can assume that for

$$b = \text{center}(\Delta u \pi(x^+_i)\pi(x^-_{i+1})) = \text{center}(\Delta u w_i \pi(x^-_{i+1})),$$

the vertex $v = \pi(\bar{y})$ lies in the interval

$$[\![b, \pi(x^+_i)]\!] \subset [\![u, \pi(x^+_i)]\!],$$

where $\pi(A^i) = [\![u, w_i]\!]$. Therefore, by Lemma 4.15(2), the vertical distance from $\bar{y}'$ to the top of A^i_v is bounded by $R_{4.15}(K)$. This concludes the proof of (2).

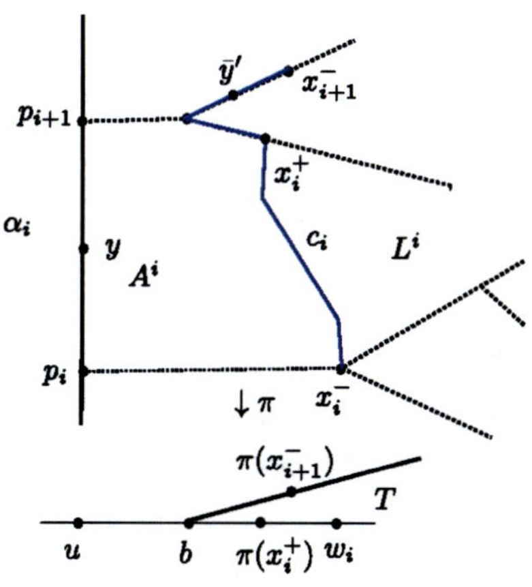

FIGURE 8.7. Path c_i

(3) Thus, we assume that $\bar{y}' \in c_i \cap A^i$. If $y \in \alpha_i \setminus \alpha'_i$, then $\bar{y}' \in top(A^i)$ and $c(\bar{y}', x^+_i)$ has uniformly bounded length and, without loss of generality, $c(\bar{y}', x^+_i)$ is a vertical segment of length $\leq M_{\bar{K}}$. The canonical K-section $\Sigma_y \subset L^i$ crosses the vertical interval $c(\bar{y}', x^+_i)$ at some point $\bar{y}''$ and $d(\bar{y}', \bar{y}'') \leq M_{\bar{K}}$. Thus, from now on, we assume that $y \in \alpha'_i \subset A^i$.

The intersection $c_i \cap A^i$ consists of a path in $top(A^i)$, the narrow end of A^i (which has length $\leq M_{\bar{K}}$) and of a path contained in $bot(A^i)$ which equals

$$c(x^+_{i-1}, x^-_i) \cap bot(A^i).$$

Since every point in the narrow end of A^i is connected to each point of α'_i by a K-qi section, we have two cases to consider:

(a)

$$\bar{y}' \in c(x^+_{i-1}, x^-_i) \cap bot(A^i).$$

By the definition of the path

$$c(\bar{y}', y),$$

it is a concatenation of a horizontal subpath contained in $c(x_{i-1}^{+}, x_i^{-}) \cap bot(A^i)$, followed by a vertical path of length $\leq M_{\bar{K}}$ and, then by a K-qi leaf

$$\gamma_{\bar{y}'', y} \subset A^i,$$

connecting $\bar{y}''$ to y. (Note that $\bar{y}''$ does not, in general, lie in c, it is just within distance $\leq M_{\bar{K}}$ from a point in c.) The first horizontal subpath is entirely contained in $c_i \subset c$, hence, has to have uniformly bounded length. Thus, $\bar{y}'$ is uniformly close to the point $\bar{y}'' \in A^i$ connected to y by a K-qi section. This concludes the proof in case (a).

(b)

$$\bar{y}' \in c(x_{i-1}^{+}, x_i^{-}) \cap top(A^i).$$

This case is similar to (a): The path $c(\bar{y}', y)$ is a concatenation of a subpath contained in c, followed by a short vertical subpath and, then, by a K-qi leaf connecting a point $\bar{y}'' \in A^i$ to y. The first two subpaths are uniformly short, hence, we are done. □

COROLLARY 8.31. *If $d(\bar{y}, p) \leq R$, then*

$$d_X(y, p) \leq R' = R'_{8.31}(R, K, C),$$

where $C = C_{8.30}$.

PROOF. Taking $\bar{y}'$ is as in Part (3) of Lemma 8.30, we obtain

$$d_T(u, \pi(\bar{y}')) \leq d_X(p, \bar{y}') \leq R + C.$$

Since $y, \bar{y}'$ are connected by a K-qi leaf over the interval $[\![u, \pi(\bar{y}')]\!]$,

$$d_X(y, \bar{y}') \leq K d_T(u, \pi(\bar{y}')) \leq K(R + C),$$

which, in turn, implies the inequality $d_X(y, p) \leq (K+1)(R+C)$. □

Lastly, we turn to the projection of $y \in [pp']_{X_u}$ to a Λ-quasigeodesic ϕ (with endpoints in Y) containing $\zeta = \phi(p, p')$ as a detour subarc:

LEMMA 8.32.

$$d(P_{X,\phi}(y), \zeta) \leq D_{8.32} = D_{8.32}(K, D, E, \Lambda).$$

PROOF. The result is an application of Corollaries 1.110 and 8.31. The subsets $U = \phi, V = \zeta$ are $D_{1.54}(\delta_X, \Lambda)$-quasiconvex in X. Assume that $\bar{y} := P_{X,\phi}(y)$ is not in the arc ζ. In view of the Morse lemma (Lemma 1.54), for each $z \in \zeta$ the geodesic $[\bar{y}z]_X$ passes within distance $D_{1.54}(\delta_X, \Lambda)$ from p or from p' (depending which component of $\phi \setminus \zeta$ the point $\bar{y}$ belongs to). We will assume that it is p rather than p'. Therefore, by Lemma 1.105(i), the projection of $\bar{y}$ is within distance $2D_{1.54}(\delta_X, \Lambda)$ from p or from p'. But, according to Corollary 1.110,

$$d_X(P_{U,V} \circ P_{X,U}(y), P_{X,V}(y)) \leq C_{1.110}(\delta_X, D_{1.54}(\delta_X, \Lambda)).$$

Hence,

$$d_X(P_V(y), p) \leq R := C_{1.110}(\delta_X, D_{1.54}(\delta_X, \Lambda)) + 2D_{1.54}(\delta_X, \Lambda).$$

Applying Corollary 8.31, we get

$$d_X(y, p) \leq R'_{8.31}(D),$$

and, therefore, $d_X(\bar{y}, p) \leq D_{8.32} := 2R'_{8.31}(R)$. □

COROLLARY 8.33.

$$d(P_{X,\phi}(y), P_{X,\zeta}(y)) \le D_{8.33}(K, D, E, \Lambda).$$

PROOF. If $P_{X,\phi}(y) \in \zeta$ then $P_{X,\phi}(y) = P_{X,\zeta}(y)$. Assume, therefore, that $P_{X,\phi}(y) \notin \zeta$. According to lemma, $P_{X,\phi}(y)$ is within distance $D_{8.33}$ from a point $q \in \zeta$. Therefore,

$$d_X(y, q) \le d(y, \zeta) + D_{8.33}.$$

By Corollary 1.107,

$$d_X(q, P_{X,\zeta}(y)) \le D_{8.33} + 2\lambda + 4\delta_X,$$

where $\lambda = D_{1.54}(\delta_X, \Lambda)$ is the quasiconvexity constant of $\zeta \subset X$. □

8.4. Part II: Consistency in semispecial flow-spaces

8.4.1. Part II.4: Projections in special flow-spaces. In this part of the proof (which is the most difficult part of Section 8.4) we are assuming that vertices $u, v \in V(S)$ define a K-special interval $J = [\![u, v]\!] \subset S$. In order to simplify the notation we set

$$F_w^X = Fl_K^X(X_w), \quad F_w^Y = Fl_K^Y(X_w),$$

for vertices $w \in V(S)$. Observe that $F_v^X \cap X_u \neq \emptyset \iff F_v^Y \cap X_u \neq \emptyset$ and $F_u^X \cap X_v \neq \emptyset \iff F_u^Y \cap X_v \neq \emptyset$, i.e., it does not matter in what space (X or Y) our interval J is special. In the proofs below, it does not matter which of the above intersections is nonempty. In this section we will relate the X-projection $\bar{x}$ of $x \in F_v^Y$ to F_u^X and the Y-projection $\bar{\bar{x}}$ of x to F_u^Y. At the same time, we will establish uniform consistency of some classes of pairs $(x, x') \in F_v^Y \times F_u^Y$. These results will be the key for proving uniform consistency of pairs of points in semispecial flow-spaces, which will be done in the next section. (This will apply, of course, to points in special flow-spaces as well.) After altering $\bar{x}$ and $\bar{\bar{x}}$ by uniformly bounded distance, we can assume that these points belong to vertex-spaces of X and Y respectively.

Recall (see Lemma 1.105) that the concatenation

$$\phi = [x\bar{x}]_X \star [\bar{x}x']_X$$

is a Λ-quasigeodesic in X, where Λ depends only on the quasiconvexity constant λ_X of F_u^X and the hyperbolicity constant of X (i.e., only on the parameters of $\mathfrak{X}$ and K). If $\bar{x} \notin Y$, we let $\zeta = \phi(\hat{x}, \hat{x}')$ denote the detour subpath in ϕ containing $\bar{x}$ and connecting points $\hat{x}, \hat{x}'$ which belong to the same vertex-space $X_t \subset Y$. In the case $\bar{x} \in Y \cap X_t$, we declare ζ to be degenerate, equal to the singleton $\{\bar{x}\}$, and, accordingly, set

$$\hat{x} = \hat{x}' = \bar{x}.$$

REMARK 8.34. Even if $\bar{x}$ is far away from Y, the points $\hat{x}, \hat{x}'$ depend quite a bit on the quasigeodesic ϕ (i.e., on the choice of geodesics $[x\bar{x}]_X, [\bar{x}x']_X$) connecting x, x'. The notation $\hat{x}, \hat{x}'$, therefore, is ambiguous and, in truth, should contain the symbol ϕ, which we omit for ease of notation.

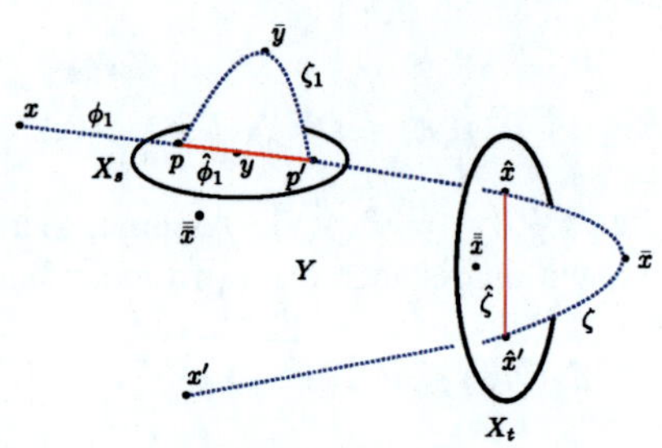

FIGURE 8.8. Two projections

Figure 8.9 is a picture of the detour at the vertex-space X_s in more detail.

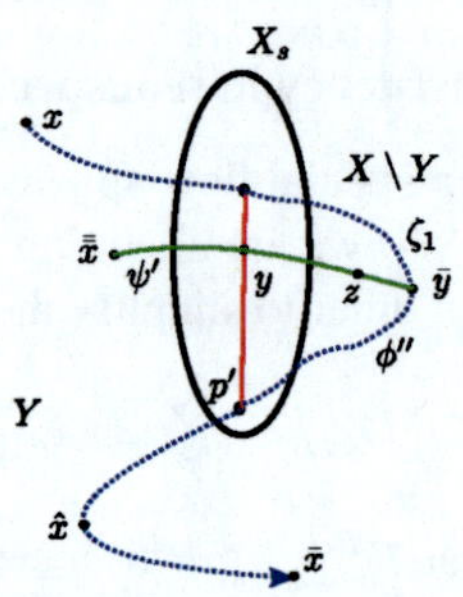

FIGURE 8.9. Two projections: Detour ζ_1

The key result of Part II.4 is Lemma 8.35.

LEMMA 8.35. *The point $\bar{\bar{x}}$ lies within distance $R_{8.35}(K, \Lambda)$ from a point of the segment $\hat{\zeta} = [\hat{x}\hat{x}']_{X_t}$, no matter what x' and ϕ are.*

PROOF. Consider the subpath $\phi' = \phi(x, \hat{x}')$ connecting x to $\hat{x}'$ and passing through $\hat{x}$:

$$\phi' = \phi_1 \star \zeta.$$

(The subpath ϕ_1 is geodesic since it is contained in $[x\bar{x}]_{X}$.); see Figure 8.8. Then

$$\hat{\phi}' = \hat{\phi}_1 \star \hat{\zeta}.$$

We already know (by Proposition 8.22 and Corollary 8.27) that $\hat{\phi}_1$ and $\hat{\zeta}$ are uniform Y-quasigeodesics, but we do not yet know that their concatenation is, since uniform consistency of the pairs $(x, \hat{x}')$ is not yet known, except in some special cases. We do know, however, the uniform consistency of the pair $(x, \bar{\bar{x}})$: Since $F^Y_v \cap F^Y_u \neq \emptyset$, Lemma 1.130 implies that the point $\bar{\bar{x}}$ lies in the $2\lambda_Y + 3\delta_Y$-neighborhood of F^Y_v, where λ_Y is the quasiconvexity constant of flow-spaces $Fl^Y_w \subset Y$. Therefore, the uniform consistency of the pairs $(x, \bar{\bar{x}})$ follows from Proposition 8.22.

Since the geodesic $[x\hat{x}']_Y$ passes uniformly close to the projection $\bar{\bar{x}}$ of x to F^Y_u, by the δ_Y-slimness of the geodesic triangle

$$\Delta_Y x\hat{x}\hat{x}' \subset Y,$$

either $\hat{\phi}_1$ or $\hat{\zeta}$ passes within the distance $R = R(K)$ from $\bar{\bar{x}}$. If it is $\hat{\zeta}$, we are done. Consider, therefore, the case where $\hat{\phi}_1$ passes within distance R from $\bar{\bar{x}}$. The path $\hat{\phi}_1$ is a concatenation of subarcs $\phi_1 \cap Y$ with vertical geodesics $\hat{\zeta}_i$, where each ζ_i is a detour subarc in ϕ_1.

Suppose first that $\phi_1 \cap Y$ passes within distance R from $\bar{\bar{x}} \in F^Y_u$. Since ϕ_1 is geodesic in X and $\bar{x}$ is the nearest-point projection of x to F^X_u, we obtain:

$$d(\bar{\bar{x}}, \bar{x}) \le 2R.$$

In particular, $d(\bar{x}, X_t) \le 2R$ as well (since $[\bar{x}\bar{\bar{x}}]_X$ has to cross X_t). By Corollary 8.29, the intersection

$$B(\bar{x}, 2R) \cap X_t$$

is uniformly close to a subset of $\hat{\zeta}$ and, hence, we are done, in this case as well.

It remains to analyze the harder case, where for one of the detour subarcs $\zeta_1 \subset \phi_1$, the path $\hat{\zeta}_1$ passes within distance R from $\bar{\bar{x}}$. Let X_s denote the vertex-space of Y containing the endpoints p, p' of ζ_1 and $y \in \hat{\zeta}_1 \subset X_s$ a point within distance R from $\bar{\bar{x}}$. Let $\bar{y}$ denote the nearest-point projection of y to ζ_1.

The point $\bar{y}$ lies within distance $D_{8.32}$ from the projection of y to ϕ_1 since the latter contains ζ_1; see Lemma 8.32. The concatenation $\psi = \psi_1 \star [\bar{y}y]_X$ (where ψ_1 is the subpath of ϕ_1 between x and $\bar{y}$) is a uniform quasigeodesic in X connecting x to the point y within distance R from $\bar{\bar{x}}$.

Since F^X_v is λ_X-quasiconvex in X, $[y\bar{\bar{x}}]_X$ has length $\le R$ and ψ is a uniform quasigeodesic in X, the path $\psi' = \psi \star [y\bar{\bar{x}}]_X$ (connecting x to $\bar{\bar{x}}$) lies within a uniform neighborhood of F^X_v. Since $\bar{\bar{x}}$ is in the $2\lambda_Y + 3\delta_Y$-neighborhood of $F^Y_v \subset F^X_v$, by Lemma 1.105(i) the path ψ' (connecting $x \in F^X_v$ to $\bar{\bar{x}}$ which is uniformly close to F^X_v) passes, at some point $z \in \psi'$, uniformly close to $\bar{x}$.

Where could this point z be? If z lies in ψ_1, then, since $\bar{y}$ is between z and $\bar{x}$ in ϕ, the length of the entire subpath ϕ'' of ϕ between $\bar{x}$ and $\bar{y}$ is uniformly bounded. Similarly, if z lies in $[y\bar{\bar{x}}]_X$ then, since the concatenation of ϕ'' with $[\bar{y}y]_X$ is also a uniform quasigeodesic, the length of ϕ'' is uniformly bounded as well. The path ϕ'' includes a subpath (contained in ζ_1) between $\bar{y}$ and p', and we conclude that the distance between $\bar{y}$ and p' is uniformly bounded by some constant D. Therefore, by Corollary 8.31, the distance between y and p' is $\le D' = D'_{8.31}(D)$, implying a uniform upper bound on the distance from $\bar{\bar{x}}$ to $\hat{x} \in \phi''$. Therefore, in this case again, we see that $\bar{\bar{x}}$ is uniformly close to a point (namely, $\hat{x}$) in the segment

$$\hat{\zeta} = [\hat{x}\hat{x}']_{X_t}.$$

Of course, in this case the distance between $\bar{\bar{x}}$ and $\bar{x}$ is also uniformly bounded as well. □

In fact, we can pin down the location of a point $y \in \hat{\zeta}$ within distance $R_{8.35}(K)$ from $\bar{\bar{x}}$ a bit further. Namely, the set of points $\hat{x}'$ in the lemma is precisely the $4\delta_0$-quasiconvex subset $Q_t \subset X_t$ equal to the intersection

$$X_t \cap Fl^X_u.$$

Since $\bar{\bar{x}}$ is uniformly close to a point in each of the geodesics $[\hat{x}\hat{x}']_{X_t}$, $\hat{x}' \in Q_t$, by applying Corollary 1.109 we conclude that $\bar{\bar{x}}$ is uniformly close to a point in the geodesic segment $[\hat{x}y]_{X_t}$, where y is the projection (taken in X_t) of $\hat{x}$ to Q_t.

At the same time, since $\bar{\bar{x}}$ belongs to F^Y_u, y lies in the intersection of X_t with the $R_{8.35}(K)$-neighborhood of F^Y_u (the neighborhood is taken in X). Thus, by Lemma 3.29, the point y lies in

$$N^{fib}_D(Q_t) \subset X_t,$$

where $D = D_{3.29}(R, K)$. We conclude:

Corollary 8.36. *The point $\bar{\bar{x}}$ is uniformly close to the projection $\tilde{x}$ (taken in X_t) of $\hat{x}$ to $Q_t = X_t \cap Fl_K^X(X_u)$.*

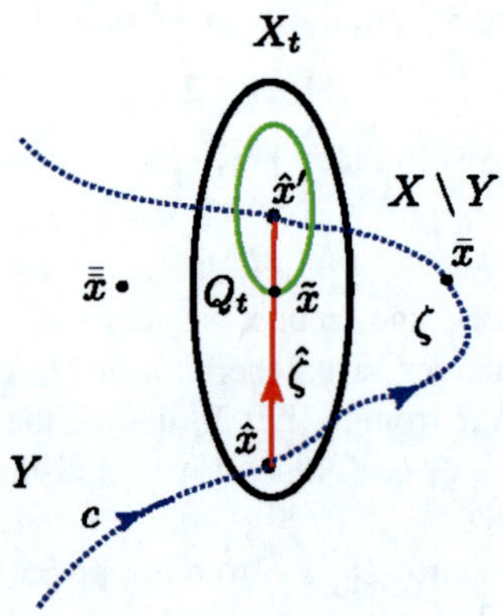

Figure 8.10. Location of $\bar{\bar{x}}$

Note that, while the point $\hat{x}$ depends heavily on the choice of the path ϕ connecting $x, x' \in F_u^Y$, the point $\tilde{x}$ is canonical (up to a uniform error, depending only on K).

Another observation relating the position of the points $\bar{\bar{x}}$ and $\bar{x}$ is that, setting $y := \tilde{x}$, if $\bar{y}$ denotes the X-projection of y to the detour arc ζ (connecting $\hat{x}$ and $\hat{x}'$), then $\bar{x}$ is uniformly close to a point in the subarc $\zeta(\hat{x}, \bar{y}) \subset \zeta$: Otherwise, as in the proof of Lemma 8.35, the concatenation $[y\bar{y}]_X \star \zeta(\bar{y}, \hat{x})$ is a uniform X-quasigeodesic. Therefore, it has to pass uniformly close to the point $\bar{x}$. At the same time, the concatenation $[y\bar{y}]_X \star \zeta(\bar{y}, \hat{x}')$ is also a uniform X-quasigeodesic. Thus, $\bar{x}$ would have to be within uniformly bounded distance from both $[y\bar{y}]_X$ and $\zeta(\bar{y}, \hat{x}')$, which means that $\bar{x}$ is uniformly close to $\bar{y}$. We, therefore, proved (see Figure 8.11).

Lemma 8.37. *$d_X(\bar{x}, \zeta(\hat{x}, \bar{y})) \le C_{8.37}(K)$.*

Corollary 8.38. *If $d_X(\bar{x}, Y) \le r$, then $d_X(\bar{x}, \bar{\bar{x}}) \le \bar{r} = \bar{r}_{8.38}(K, r)$.*

Proof. Choose $x' = y = \tilde{x}$ as above. As before, the concatenation $[x\bar{x}]_X \star [\bar{x}y]_X$ is a Λ-quasigeodesic in X for Λ depending only on K. Take a point $p \in X_t$ within distance r from $\bar{x}$. Then, according to Lemma 3.29,

$$d(p, F_u \cap X_t) \le D_{3.29}(r, K).$$

At the same time, by Corollary 8.29, the point p can be chosen to lie in the $R_{8.29}(r)$-neighborhood of the segment $[\hat{x}y]_{X_t}$. Since $y = \tilde{x}$ was the projection of $\hat{x}$ to $F_u \cap X_t$ (taken in X_t), it follows that the distance from p to y is uniformly bounded. □

We are now ready to prove (using the notation introduced above, with the point $y := \tilde{x}$ uniformly close to $\bar{\bar{x}}$):

Lemma 8.39. *For all K-special intervals $J = [\![u, v]\!]$ in S, and all pairs $(x, x') \in F_v^Y \times F_u^Y$:*

(1) *The points $\hat{x}$ are uniform transition points between $x, \hat{x}'$.*
(2) *The pairs $(x, \hat{x}')$ are uniformly consistent.*

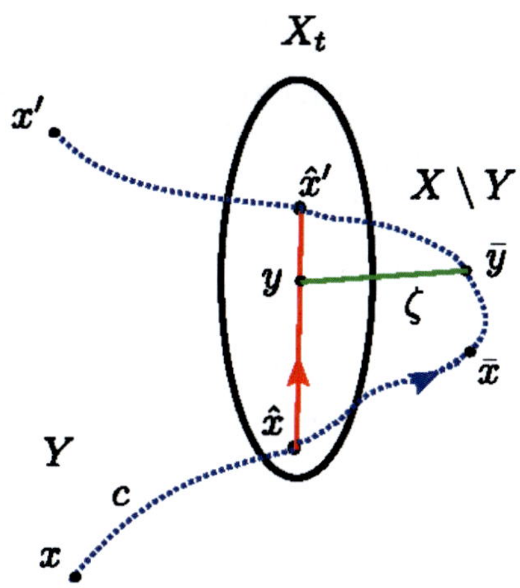

FIGURE 8.11. Location of $\bar{x}$ in the detour path

PROOF.

(1) By the construction, the point $\hat{x}$ belongs to the $\Lambda = \Lambda(K)$-quasigeodesic $\phi = [x\bar{x}]_X \star [\bar{x}\hat{x}']_X$ connecting x to $\hat{x}'$, implying that $\hat{x}$ is a uniform X-transition point between $x, \hat{x}'$. To prove the Y-transition property, note that the concatenation

$$[xy]_Y \star [y\hat{x}']_{X_t}$$

is a uniform Y-quasigeodesic (since y is uniformly close to the nearest-point projection in Y of x to F^Y_u and $[y\hat{x}']_{X_t}$ is a uniform quasigeodesic in F^Y_u). Thus, we only have to prove that $[xy]_Y$ passes uniformly close to $\hat{x}$. As we noted earlier, the pairs (x, y) are uniformly consistent (since $x \in F^Y_v$ and y lies in a uniform neighborhood of F^Y_v). The concatenation

$$\psi = [y\bar{y}]_X \star [\bar{y}x]_X$$

is a uniform quasigeodesic in X. The path $\hat{\psi} = [y\hat{x}]_{X_t} \star \hat{\psi}(\hat{x}, x)$ passes through $\hat{x}$ and is a uniform Y-quasigeodesic (by the uniform consistency of the pair (x, y)). The same, therefore, holds for the geodesic $[yx]_Y$. This implies that $\hat{x}$ is a uniform transition point between x and $\hat{x}'$.

(2) The pairs $(x, \hat{x})$, $(\hat{x}, \hat{x}')$ are uniformly consistent according to Proposition 8.22, because the first is in $F^Y_v \times F^Y_v$ and the second is in X_t^2. Since $\hat{x}$ is a uniform transition point between $x, \hat{x}'$ according to Part 1, the pairs $(x, \hat{x}')$ are uniformly consistent (see Lemma 8.19). □

This concludes Part II.4 of the proof of Theorem 8.21.

8.4.2. Part II.5: Pairs in semispecial flow-spaces. Consider a K-semispecial interval $J = J_1 \cup J_2$ which is a union of two K-special intervals J_1, J_2 meeting only at a common endpoint w. We will prove uniform consistency of pairs of points in $Fl^Y_K(X_J)$: This will also apply to the case of special intervals J since would arbitrarily subdivide it into two subintervals. Observe that it suffices to prove uniform consistency of pairs of points

$x_i \in F^Y_{v_i} = Fl^Y_K(X_{v_i})$, $v_i \in V(J_i)$, $i = 1, 2$. Indeed, if $v_i, i = 1, 2$ are both in, say, J_1, then we subdivide the interval J_1 further, to subintervals J_1', J_1'' containing v_1, v_2 respectively.

PROPOSITION 8.40. *Each pair (x_1, x_2) as above is $\theta_{8.40,K}$-consistent.*

PROOF. For $i = 1, 2$ consider the point

$$\bar{x}_i = P_{X, F^X_w}(x_i),$$

which is the nearest-point projection of x_i (taken in X) to the flow-space $F^X_w = Fl^X_K(X_w)$.

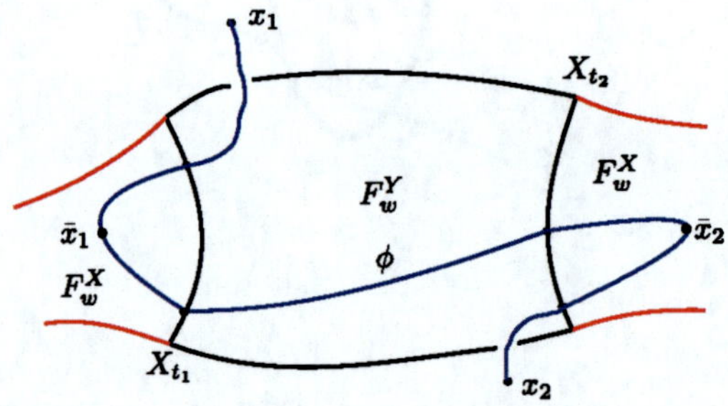

FIGURE 8.12. Path ϕ from x_1 to x_2

LEMMA 8.41. *The points $\bar{x}_1, \bar{x}_2$ cannot lie in the same component of $X - Y$.*

PROOF. Assume that both $\bar{x}_1, \bar{x}_2$ belong to $X - Y$; in particular, $\pi(\bar{x}_i) \neq w, i = 1, 2$. Then the vertex w cannot separate $w_i = \pi(\bar{x}_i)$ from v_i, $i = 1, 2$: If w were to separate these vertices, then the geodesic $[x_i \bar{x}_i]_X$ would cross into X_w before reaching $\bar{x}_i$ and, thus, $\bar{x}_i$ would not be a nearest point to x_i in F^X_w. In particular, the geodesics $[\![v_i, w_i]\!]$ lie in distinct components of $T - \{w\}$ and, hence, w_1, w_2 are separated by $w \in S$. Thus, $\bar{x}_1, \bar{x}_2$ cannot lie in the same connected component of $X - Y$. □

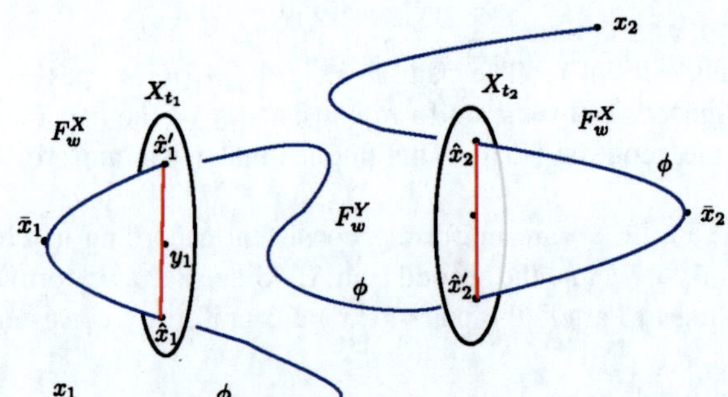

FIGURE 8.13. Transition points on the path ϕ from x_1 to x_2

As we proved in Section 2.6.2, the concatenations

$$\phi = [x_1 \bar{x}_1]_X \star [\bar{x}_1 \bar{x}_2]_X \star [\bar{x}_2 x_2]_X,$$

are Λ-quasigeodesics, where $\Lambda = \Lambda(K)$. As in the previous section, we consider detour subpaths $\zeta_i \subset \phi$ containing $\bar{x}_i$,

$$\zeta_i = \phi(\hat{x}_i, \hat{x}'_i),$$

where $\hat{x}_i, \hat{x}'_i \in X_{t_i}, i = 1, 2$. (See Figure 8.13.) By Lemma 8.41, these detour subpaths have to be disjoint, except, possibly at their endpoints. The points $\hat{x}'_i$ belong to the middle portion $[\bar{x}_1\bar{x}_2]_X$ of ϕ. We mark points

$$y_i \in [\hat{x}_i \hat{x}'_i]_{X_{t_i}}, i = 1, 2$$

which are uniformly close to the projections $\bar{\bar{x}}_i = P_{Y,F^Y_w}(x_i)$; see Lemma 8.35.

LEMMA 8.42. *The points $\hat{x}'_i$ are uniform transition points between x_1 and x_2, more precisely, the sequences*

$$x_1, \hat{x}'_1, \hat{x}'_2, x_2$$

are uniformly straight in both X and in Y.

PROOF. The uniform straightness in X follows from the fact that the above sequence appears in the Λ-quasigeodesic ϕ in X (in the correct order). To prove the uniform straightness in Y, note that the concatenation

$$\psi = [x_1 y_1]_Y \star [y_1 y_2]_Y \star [y_2 x_2]_Y,$$

is a uniform quasigeodesic in Y, since each y_i is uniformly close to the projection of x_i to the λ_Y-quasiconvex subset $F^Y_w \subset Y$ (again; see Section 2.6.2). Thus, it suffices to show that $\hat{x}'_1, \hat{x}'_2$ are uniform Y-transition points between y_1, y_2. As in the previous section, consider the nearest-point projections $\bar{y}_i \in \zeta_i$ of y_i. According to Lemma 8.32, there exist points

$$\tilde{y}_i \in [\bar{x}_i, \hat{x}'_i]_X, i = 1, 2,$$

within distance $C_{8.37}(K)$ from $\bar{y}_i$. The points $\bar{y}_i$ (and, hence, $\tilde{y}_i$) are uniformly close to the projections of y_i to the quasigeodesic ϕ (Lemma 8.32). Hence, the concatenation

$$[y_1\tilde{y}_1]_X \star [\tilde{y}_1\tilde{y}_2]_X \star [\tilde{y}_2 y_2]_X$$

is a uniform quasigeodesic in X, where $[\tilde{y}_1\tilde{y}_2]_X \subset [\bar{x}_1\bar{x}_2]_X$. The middle geodesic in this concatenation passes through the points $\hat{x}'_1, \hat{x}'_2$ (in this order), implying the uniform straightness of the sequence

$$y_1, \hat{x}'_1, \hat{x}'_2, y_2$$

in Y, as claimed. □

Now, we can finish the Part II.5 of the proof of Theorem 8.21. The sequence

$$x_1, \hat{x}'_1, \hat{x}'_2, x_2$$

is uniformly straight in both X and in Y. The pair

$$(\hat{x}'_1, \hat{x}'_2) \in F^Y_w \times F^Y_w$$

is uniformly consistent according to Proposition 8.22, while both pairs

$$(x_1, \hat{x}'_1), (\hat{x}'_2, x_2)$$

are uniformly consistent by Lemma 8.39(2). Therefore, by Lemma 8.20, the pair (x_1, x_2) is θ-consistent for some function $\theta = \theta_{8.40,K}$. □

This concludes Part II of the proof.

8.5. Part III: Consistency in the general case

Suppose that $x, y \in \mathcal{Y}$ belong to vertex-spaces X_u, X_v, respectively, $u, v \in V(S)$. Fix $K = K_0$. Using Lemma 3.44 and Theorem 6.17, we obtain a *horizontal subdivision* of the interval $J = [\![u, v]\!]$, into subintervals $J_i = [\![u_i, u_{i+1}]\!], i = 1, \dots, n$, such that the pairs of distinct flow-spaces $F_i^X := Fl_K^X(X_{u_i})$ have disjoint projections to T and, hence, are $C = C_{6.17}$-cobounded, unless, possibly $i = n$; $u = u_1, v = u_{n+1}$. (The same, of course also holds for the Y-flows.) As in Lemma 3.44, we also define vertices $u_i'', u_{i+1}' \in J_i$ such that

(1) u_i', u_i span an edge e_i in T (except, possibly, for $i = n + 1$, in which case we can have $d_T(u_{n+1}', u_{n+1}) \le 1$).

(2) Each subinterval $[\![u_i', u_i]\!], [\![u_i, u_i'']\!], [\![u_i'', u_{i+1}']\!]$ is K-special. In particular, the subinterval $J_i' = [\![u_i, u_{i+1}']\!]$ is semispecial.

For each i we define pairs of points $(x_i'', x_{i+1}') \in \mathcal{F}_i^X \times \mathcal{F}_{i+1}^X$ realizing the minimal distance between the subsets $\mathcal{F}_i^X, \mathcal{F}_{i+1}^X$ of X unless $i = n$ and $u_{n+1}' = u_{n+1}$, in which case we take x_n'' to be the projection of y to $\mathcal{F}_n^X$.

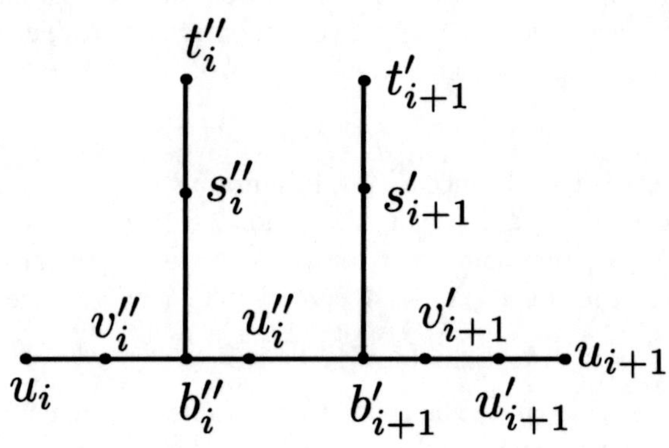

FIGURE 8.14. Projections to J

For each i define the vertices $t_i' := \pi(x_i')$, $t_i'' := \pi(x_i'')$ and let s_i', s_i'' denote their respective projections to the subtree $S \subset T$.

LEMMA 8.43. *Let b_i' (resp., b_i'', v_i', v_i'') denote the projection to J of the vertices t_i' (resp., t_i'', $\pi(y_i')$, $\pi(y_i'')$). Then:*

(1) *The projections b_i', v_i' lie in the subinterval $]\!]u_{i-1}'', u_i']\!]$.*

(2) *The projections b_i'', v_i'' lie in the subinterval $[\![u_i, u_i'']\!]$.*

PROOF. We will prove the claim for the vertex v_i', since the rest is proven by the same argument.

First of all,

$$\pi(F_i^Y) \subset \pi(F_i^X)$$

and, by the definition of the horizontal subdivision of the interval J, $\pi(F_i^X)$ intersects J in a subinterval of $]\!]u_{i-1}'', u_i'']\!]$. Thus, v_i' belongs to $]\!]u_{i-1}'', u_i'']\!]$. Suppose, for the sake of a contradiction, that v_i' is in the interval $[\![u_i, u_i'']\!]$. Then each geodesic connecting y_{i-1}'' to y_i' goes through the edge-space X_e of the edge

$$e = [u_i', u_i],$$

before reaching x_i'. But then, this geodesic also passes through the subset

$$X_{eu_i'} \subset X_{u_i'}.$$

Since $K \geq 1$, the entire subset $X_{eu_i'}$ is contained in the flow-space $Fl_K^Y(X_{u_i})$. Hence, y_i' cannot possibly be the Y-projection of y_i'' to F_i^Y. A contradiction. □

COROLLARY 8.44. *The pairs* (y_i'', y_{i+1}') *are uniformly consistent.*

PROOF. Let v_i'', v_{i+1}' denote the projections of $\pi(y_i'')$, $\pi(y_{i+1}')$ to J. By the lemma, both v_i'', v_{i+1}' lie in the interval $J_i' = [\![u_i, u_{i+1}']\!]$, which is semispecial. Moreover, $y_i'' \in Fl_K^Y(X_{u_i})$ and $y_i' \in Fl_K^Y(X_{u_{i+1}'})$. Thus,

$$(y_i'', y_{i+1}') \in Fl_K^Y(X_{J_i'})$$

and, hence, the statement is a special case of Proposition 8.40. □

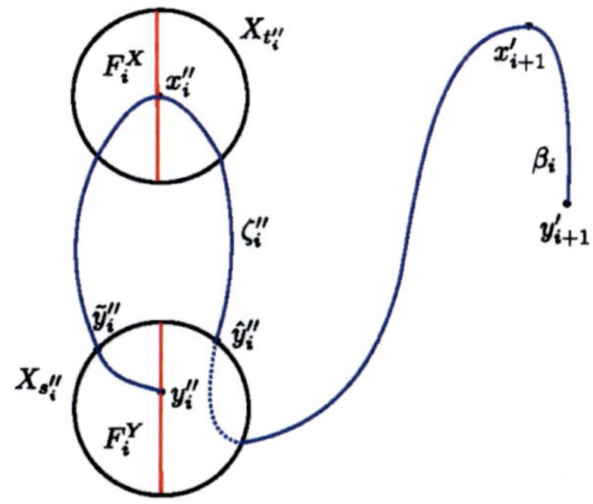

FIGURE 8.15. Transition points

Our next task is to relate the points x_i', y_i' and also relate the points x_i'', y_i''. Since y_i' is in Y, the segment $[y_i'x_i']_X$ has to cross the vertex-space $X_{s_i'}$; the same holds for the segment $[y_i''x_i'']_X$ and $X_{s_i''}$. We define the points $\tilde{y}_i' \in [y_i'x_i']_X$, $\tilde{y}_i'' \in [y_i''x_i'']_X$ by the condition that the subsegments

$$[\tilde{y}_i'x_i']_X \subset [y_i'x_i']_X$$

and

$$[\tilde{y}_i''x_i'']_X \subset [y_i''x_i'']_X$$

are the smallest subsegments terminating in x_i', x_i'', with the property that $\tilde{y}_i' \in X_{s_i'}$, $\tilde{y}_i'' \in X_{s_i''}$. (See Figure 8.15.) Thus,

$$Y \cap [\tilde{y}_i'x_i']_X \setminus \{\tilde{y}_i'\} = \emptyset,$$

and, similarly, for $[\tilde{y}_i''x_i'']_X$.

Similarly, define points $\hat{y}_i'', \hat{y}_{i+1}' \in [x_i''x_{i+1}']_X$ such that

$$[\hat{y}_i'x_i']_X \subset [x_{i-1}''x_i']_X, [\hat{y}_i''x_i'']_X \subset [x_{i+1}'x_i'']_X,$$

are the smallest subsegments terminating in x_i', x_i'' with $\hat{y}_i' \in X_{s_i'}$, $\hat{y}_i'' \in X_{s_i''}$. Observe that the concatenations

$$\beta_i := [y_i''x_i'']_X \star [x_i''x_{i+1}']_X \star [x_{i+1}'y_{i+1}']_X$$

are uniform quasigeodesics in X since $x_i'' \in F_i^X, x_{i+1}' \in F_{i+1}^X$ realize the minimal distance between these two uniformly quasiconvex subsets of X and $y_i'' \in F_i^Y \subset F_i^X, y_{i+1}' \in F_{i+1}^Y \subset F_{i+1}^X$. Both

$$\zeta_i' = [\hat{y}_i' x_i']_X \cup [x_i' \tilde{y}_i']_X, \zeta_i'' = [\hat{y}_i'' x_i'']_X \cup [x_i'' \tilde{y}_i'']_X$$

are detour subpaths in β_i, containing the points x_i', x_i'' respectively.

LEMMA 8.45.

(1) *The pairs* $(\hat{y}_i'', \hat{y}_{i+1}') \in Y^2$ *are uniformly consistent.*
(2) *The distances* $d(y_i', \tilde{y}_i')$, $d(y_i'', \tilde{y}_i'')$ *are uniformly bounded.*

PROOF.

(1) By the previous corollary, the pair (y_i'', y_{i+1}') is uniformly consistent. In particular, the path

$$\hat{\beta}_i = [\widehat{y_i'' \tilde{y}_i''}]_X \star [\tilde{y}_i'' \hat{y}_i'']_{X_{s_i''}} \star [\widehat{\hat{y}_i'' \hat{y}_{i+1}'}]_X \star [\hat{y}_{i+1}' \tilde{y}_{i+1}']_{X_{s_{i+1}'}} \star [\widehat{\tilde{y}_{i+1}' y_{i+1}'}]_X$$

is uniformly quasigeodesic. Thus, the geodesic $[y_i'' y_{i+1}']_Y$ passes within uniform distance D from the points $\tilde{y}_i'', \hat{y}_i'', \hat{y}_{i+1}', \tilde{y}_{i+1}'$ (in this order). Taking into account uniform consistency of the pairs (y_i'', y_{i+1}') and Remark 8.16, we conclude uniform consistency of the pairs $(\hat{y}_i'', \hat{y}_{i+1}')$.

(2) We will estimate the second distance, $d(y_i'', \tilde{y}_i'')$, since the proof for the first one is similar. Since F_i^X is λ_X-quasiconvex in X, and $(x_i', y_i') \in F_i^X \times F_i^X$, the point $\tilde{y}_i'$ lies within distance λ_X from a point q in F_i^X. The point q might not be in $F_i^Y = F_i^X \cap Y$, but it is within distance $K\lambda_X$ from $F_i^X \cap X_{s_i'} \subset F_\lambda^Y$. In Part 1 we observed that the geodesic $[y_i'' y_{i+1}']_Y$ passes (at some point p) within uniform distance D from the point $\tilde{y}_i''$. Thus, we found a point $p \in [y_i'' y_{i+1}']_Y$ within distance $D + (K+1)\lambda$ from F_i^Y. It follows that

$$d_Y(p, y_i'') \leq D + (K+1)\lambda_X$$

and, hence,

$$d_Y(\tilde{y}_i'', y_i'') \leq 2D + (K+1)\lambda_X. \qquad \square$$

REMARK 8.46. While it is not needed for our purposes, one can prove similarly to Lemma 8.35, that the points y_i', y_i'' are uniformly close to the projections (taken in the vertex-spaces $X_{s_i'}, X_{s_i''}$ respectively) of the points $\hat{y}_i', \hat{y}_i''$ to the $4\delta_0$-quasiconvex subsets

$$F_i^Y \cap X_{s_i'}, F_i^Y \cap X_{s_i''}$$

respectively. This provides a description (up to a uniform error) of the points y_i', y_i'' in terms of x_{i-1}'', x_i', x_i''.

LEMMA 8.47. *The pairs* $(\hat{y}_i', \hat{y}_i'')$ *are uniformly consistent.*

PROOF. Each interval

$$I_i = [\![s_i', s_i'']\!] \subset T$$

is contained $[\![t_i', t_i'']\!]$ and intersects J along the interval $[\![b_i', b_i'']\!]$, which, in turn, contains the vertex u_i; see Lemma 8.43. Since F_i^Y has nonempty intersection with both $X_{s_i'}, X_{s_i''}$, it follows that the interval I_i is semispecial (it is the union of special subintervals $[\![s_i', u_i]\!]$, $[\![s_i'', u_i]\!]$). Thus,

$$y_i', y_i'' \in X_{I_i} \subset Fl_K^Y(X_{I_i})$$

and, hence, the pair $(\hat{y}_i', \hat{y}_i'')$ is uniformly consistent by Proposition 8.40. $\square$

We now can finish the proof of Theorem 8.21. For each i, the points $\hat{y}''_i, \hat{y}'_{i+1}$ are both X and Y-transition points between, respectively, x''_i, x'_{i+1} and y''_i, y'_{i+1}. The sequence

$$x, \hat{y}''_1, \hat{y}'_2, \hat{y}''_2, \ldots, \hat{y}'_n, \hat{y}''_n, \hat{y}'_{n+1}, y$$

is uniformly straight in both X and in Y and the consecutive pairs of points in this sequence are uniformly consistent. Now, the uniform consistency of the pairs (x, y) follows from Lemma 8.20.

8.6. The existence of CT-maps for subtrees of spaces

We finally are ready to prove the main result of this chapter.

THEOREM 8.48. *Let $\mathfrak{X} = (\pi : X \to T)$ be a tree of hyperbolic spaces with hyperbolic total space X and let $\mathfrak{Y} = (\pi : Y \to S)$ be a subtree of spaces in $\mathfrak{X}$, where $S \subset T$ is a subtree and $Y = \pi^{-1}(S)$. Then the inclusion map $Y \to X$ admits a CT-extension.*

PROOF. We will derive this result from Theorem 8.21. Fix $p \in Y$ and suppose that $(y_n), (y'_n)$ are sequences in Y such that

$$\lim_{n\to\infty} (y_n . y'_n)^Y_p = \infty,$$

where the superscript Y refers to the Gromov-product of the intrinsic path-metric of Y. Equivalently, for the geodesic $\beta_n = [y_n y'_n]_Y \subset Y$ we have

$$\lim_{n\to\infty} d(p, \beta_n) = \infty.$$

For the sake of contradiction, assume that the corresponding X-geodesics $\alpha_n = [y_n y'_n]_X \subset X$ all pass through a ball $B_X(y, R) \subset X$ for some fixed R (which is equivalent to saying that the sequence $(y_n, y'_n)^X_p$ does not diverge to infinity), i.e., there exist points

$$q_n \in \alpha_n \cap B_X(p, R).$$

By Theorem 8.21, each $\hat{\alpha}_n$ is a Λ-quasigeodesic in Y connecting y_n, y'_n, for some Λ independent of n. By the Morse Lemma (Lemma 1.54),

$$\mathrm{Hd}_Y(\beta_n, \hat{\alpha}_n) \le D = D(\delta_Y, \Lambda).$$

There are two cases which may occur:

(a) $q_n \in Y$. Then (by the definition of $\hat{\alpha}_n$) $q_n \in \hat{\alpha}_n$, hence,

$$d_X(p, \beta_n) \le D + R$$

for all n, which is a contradiction.

(b) $q_n \notin Y$. Let $\zeta_n = [z_n z'_n]_X \subset \alpha_n$ be a detour subpath in α_n containing q_n; the endpoints z_n, z'_n of ζ_n belong to a vertex-space $X_{v_n} \subset Y$. Since $p \in Y$, $q_n \notin Y$, the vertex-space X_{v_n} separates p from q_n and, hence, the geodesic $[pq_n]_X$ has to pass through X_{v_n} at some point $p_n \in X_{v_n}$. We, obviously, have

$$d_X(p, p_n) \le R.$$

Thus, by Corollary 8.29, the points p_n are all uniformly close (within distance r depending only on the parameters of $\mathfrak{X}$) to the geodesic $[z_n z'_n]_{X_{v_n}} \subset \hat{\alpha}_n$. Therefore,

$$d_X(p, \beta_n) \le D + R + r,$$

which is again a contradiction. □

8.7. Fibers of CT-maps

Let $\mathfrak{X} = (\pi : X \to T)$ be a tree of hyperbolic spaces with hyperbolic total space X. According to Theorem 8.48, for every subtree $S \subset T$ and $Y = X_S$, the inclusion map

$$f_{Y,X} : Y = X_S \to X$$

admits a Cannon–Thurston extension

$$\partial_{Y,X} := \partial_\infty f_{Y,X} : \partial_\infty Y \to \partial_\infty X.$$

For the rest of the chapter, we will be working under the extra assumptions that X is a proper metric space, i.e., that an abstract tree of spaces $\mathfrak{X}$ admits a proper total space. Note that if some edge-spaces are nondiscrete, the total space X defined in the proof of Theorem 2.15 need not be proper even if $V(T) = \{v, w\}$ and $X_v, X_w, X_{[v,w]}$ are proper. However, this X will be proper (a locally finite metric graph) under the following conditions:

(1) Each vertex-space X_v is a locally-finite graph (with the standard graph-metric).
(2) All edge-spaces X_e are discrete and $X_{ev} = f_{ev}(X_e) \subset V(X_v)$ for each edge e and incident vertex v.
(3) Each finite subset of each vertex-space X_v meets only finitely many subsets of the form X_{ev}, where e are edges incident to v.

Proposition 8.49. *Suppose that $\mathfrak{X}$ has a proper total space X. Then there exists a number r, depending only on $\mathfrak{X}$ and X, for which the following holds. For each $Y = X_S \subset X$, if $\xi_\pm \in \partial_\infty Y$ are two distinct points such that $\partial_{Y,X}(\xi_+) = \partial_{Y,X}(\xi_-)$, then there exists a vertex $v \in S$ and two ideal boundary points $\xi'_\pm \in \partial_\infty X_v$ such that:*

(1) $\partial_{X_v,Y}(\xi'_\pm) = \xi_\pm$.
(2) The Y-geodesic $\eta : \mathbb{R} \to Y$, connecting ξ_+, ξ_-, is contained in the r-neighborhood of a vertex-space $X_v, v \in S$, and, moreover, is r-Hausdorff-close to a vertical geodesic α in X_v.

Proof. Set $x_n := \eta(n)$, $n \in \mathbb{Z}$. Define subarcs of the geodesic η by

$$\eta_n := [x_{-n}x_n]_Y, n \in \mathbb{N},$$

and geodesics $\beta_n = [x_{-n}x_n]_X$.

Since $\partial_{Y,X}(\xi_+) = \partial_{Y,X}(\xi_-)$, we have that

$$\lim_{n\to\infty} d_X(x_0, \beta_n) = \infty.$$

Thus, x_0 cannot be uniformly close to $\beta_n \cap Y$. At the same time, since x_0 belongs to each η_n (which, by Theorem 8.21, is uniformly close to the Λ-quasigeodesic $\hat{\beta}_n$ in Y), the point x_0 lies in the C-neighborhood of one of the replacement arcs $\hat{\zeta}_n = [p_n q_n]_{X_{v_n}} \subset \hat{\eta}_n$, where Λ and C depend only on the parameters of $\mathfrak{X}$. In particular, $d_X(x_0, X_{v_n}) \le C$ for all n. Since X is assumed to be proper, there are only finitely many vertex-spaces in X which can intersect $B(x_0, C)$. It follows that there is an infinite subset $M \subset \mathbb{N}$ and a vertex $v \in S$ such that for all $m \in M$, $v_m = v$. At the same time, since the distances $d(x_0, \beta_n)$ diverge to infinity, we also have

$$\lim_{n\to\infty} d(x_0, p_n) = \infty, \quad \lim_{n\to\infty} d(x_0, q_n) = \infty.$$

Hence, after passing to a further subsequence, the sequences $(p_m)_{m\in M}, (q_m)_{m\in M}$, diverge in X_v to two ideal boundary points $\xi'_\pm$.

By continuity at infinity of the CT-map $\partial_{X_v,Y} : \partial_\infty X_v \to \partial_\infty Y$, it follows that

$$\partial_{X_v,Y}(\xi'_\pm) = \xi_\pm.$$

After passing to a subsequence once more, we can assume that the sequence of geodesics $\hat{\zeta}_n \subset X_v$ converges to a complete geodesic $\zeta \subset X_v$ asymptotic to the points $\xi'_\pm$. Since geodesics $\hat{\zeta}_n$ are Λ-quasigeodesics in Y, so is their limit α. Since the Y-geodesic η is also asymptotic to $\xi_\pm$, it follows that η, α are within Hausdorff distance $r := 2D_{1.54}(\delta_Y, \Lambda)$, which depends only on the parameters of $\mathfrak{X}$. It also follows that η is contained in $N_r(X_v)$. □

In the following addendum to this proposition we describe more precisely (up to a uniform error) the nature of geodesic segments $\beta_{-m,n}$ connecting the points x_{-m}, x_n, $n, m \in \mathbb{N}$, in the setting of the theorem. Since the points x_{-m}, x_n belong to the r-neighborhood of X_v, we will be considering instead of $\beta_{-m,n}$ the uniform quasigeodesics $c_{-m,n}$ (from the slim combing of X described in Section 7.1) connecting points y_{-m}, y_n, $m, n \in \mathbb{N}$, where $y_i := \alpha(i), i \in \mathbb{Z}$. We will see that the path $c_{-m,n}$ first diverges away from X_v (in the metric of X) at a linear speed and then converges back to X_v at a linear speed. We refer the reader to Section 7.1 (step I.1) for the description of uniform quasigeodesics connecting points in narrow carpets used in the proof of the next proposition.

Remark 8.50. In what follows, we will frequently use the following notation. Let $\mathfrak{A} = \mathfrak{A}(\alpha)$ be a (K, C)-narrow carpet with the narrow end $\beta \subset X_w$, bottom and top sections γ_-, γ_+. Then we define the path $c = c_{\mathfrak{A}}$ as the concatenation

$$\gamma_- \star \beta \star \gamma_+$$

connecting the bottom and the top points of the segment $\alpha \subset X_u$. Such paths will be uniform (with qi constants depending on K and C) quasigeodesics in $\mathfrak{A}$ and, hence, in X, as long as for each vertex $v \in X_v$, the length of A_v is $\geq M_{\bar{K}}$; see Section 4.1 or Section 7.1, Step I.1.

Proposition 8.51.

(1) *For all $n > 0, m > 0$, the segment $\alpha_{-m,n} = [y_{-m}y_n]_{X_v}$ bounds a certain (K, C)-narrow carpet $\mathfrak{A}_{-m,n} = \mathfrak{A}_K(\alpha_{-m,n})$ in X over the interval $J_{-m,n} = [\![v, w_{-m,n}]\!]$. Here K and C depend only on the parameters of $\mathfrak{X}$.*

(2) *For $n > 0, m > 0$, the uniform quasigeodesic $c_{-m,n}$ in $A_{m,n} \subset X$ connecting points y_{-m}, y_n is the concatenation*

$$\gamma_{-m} \star \beta_{-m,n} \star \gamma_n,$$

where $\beta_{-m,n}$ is a vertical geodesic in $X_{w_{-m,n}}$ of length $\leq C$ and γ_{-m}, γ_n are K-qi sections over the interval $J_{-m,n} = [\![v, w_{-m,n}]\!]$.

Proof. Set $K := K_0$ (defined in Notation 2.6.4). Consider maximal K-qi sections Σ_{-m}, Σ_n in X through the points y_{-m}, y_n. If, for some n_0, m_0, these sections have vertical separation $\geq M_K$ everywhere, then they are uniformly cobounded in X; see 7.1. In this situation, for all $n \geq n_0, m \geq m_0$, the path $c_{-m,n}$ (and, hence, the geodesic $\beta_{-m,n} = [x_{-m}x_n]_X$) has to come uniformly close to a pair of points $x^- \in \Sigma_{-m_0}, x^+ \in \Sigma_{n_0}$, contradicting the assumption that

$$\lim_{n\to\infty, m\to\infty} d(y_0, \beta_{-m,n}) = \infty.$$

Thus, for each pair (m, n) there is a vertex $w_{m,n}$ and a pair of K-qi leaves $\gamma_n \subset \Sigma_n, \gamma_{-m} \subset \Sigma_{-m}$ over an interval $J_{m,n} = [\![v, w_{m,n}]\!]$, with vertical separation $\geq M_K$, such that

$$d_{X_{w_{m,n}}}(\gamma_{-m}(w_{m,n}), \gamma_n(w_{m,n})) \leq M_K.$$

The uniform quasigeodesic $c_{-m,n}$ connecting points y_{-m}, y_n then is defined as the concatenation

$$\gamma_{-m} \star \beta_{-m,n} \star \gamma_n, \quad \beta_{-m,n} = [\gamma_{-m}(w_{-m,n})\gamma_n(w_{-m,n})]_{X_{w_{-m,n}}}.$$

(This argument is similar to the proof of Proposition 7.2.) □

We can now give a complete description of pairs of distinct points in the fibers of the CT-map $\partial_{Y,X}$. Recall that the metric space X is assumed to be proper.

THEOREM 8.52. *There are constants K, C depend only on the parameters of $\mathfrak{X}$ and a function $D = D(k)$ such that the following hold:*

(1) *Suppose that $\xi^{\pm}$ are distinct points in $\partial_\infty Y$ such that $\partial_{Y,X}(\xi^-) = \partial_{Y,X}(\xi^+)$. Then there exists a vertex-space $X_u \subset Y$ and a complete geodesic $\alpha : \mathbb{R} \to X_u$, which is a uniform quasigeodesic in Y asymptotic to $\xi^{\pm}$, such that the intervals $\alpha_{-m,n} = [\alpha(-m)\alpha(n)]_{X_u} \subset \alpha$, bound (K, C)-narrow carpets $\mathfrak{A}(\alpha_{-m,n})$ in X for all $m > 0, n > 0$.*
(2) *Conversely, if X_u is a vertex-space of $\mathfrak{Y}$, $\alpha \subset X_u$ is a complete geodesic asymptotic to distinct points $\xi^{\pm} \in \partial_\infty Y$, such that each subinterval $\alpha_{-m,n}$ as above bounds a (K, C)-narrow carpet $\mathfrak{A}(\alpha_{-m,n})$ in X, then $\partial_{Y,X}(\xi^-) = \partial_{Y,X}(\xi^+)$.*
(3) *Suppose that X_u is a vertex-space of $\mathfrak{Y}$, $\alpha \subset X_u$ is a complete geodesic asymptotic to distinct points $\xi^{\pm} \in \partial_\infty Y$. Then $\partial_{Y,X}(\xi^-) \neq \partial_{Y,X}(\xi^+)$ if and only if for some (equivalently, every) $k \geq 1$, there exist points $x, y \in \alpha$ and maximal k-qi sections Σ_x, Σ_y over, possibly different, subtrees T_x, T_y in T through the points x, y such that the vertical separation between Σ_x, Σ_y over every vertex of $T_x \cap T_y$ is $\geq D$.*

PROOF. We take $K = K_0$ and $C = M_{\bar{K}}$.

(1) The first part of the theorem is the content of Propositions 8.49 and 8.51.

(2) Since the lengths of the intervals $\alpha_{-m,n} = [y_{-m}y_n]_{X_u}$ diverge to ∞ as $m \to \infty, n \to \infty$, for all sufficiently large m, n, without loss of generality, we may assume that the vertical separation between the top and the bottom of each carpet $\mathfrak{A}(\alpha_{-m,n})$ is $\geq M_K$. (Otherwise, since $C = M_{\bar{K}} \geq M_K$, we take a smaller (K, C)-subcarpet $\mathfrak{A}'(\alpha_{-m,n}) \subset \mathfrak{A}(\alpha_{-m,n})$ containing no vertical intervals of length $< M_K$.) Now, just as in proof of Proposition 7.2, each carpet $\mathfrak{A}(\alpha_{-m,n})$ defines a uniform X-quasigeodesic $c_{-m,n}$ connecting y_{-m} to y_n and

$$d_X(y_0, c_{-m,n}) \geq \text{length}(\pi(A(\alpha_{-m,n}))).$$

Thus,

$$\lim_{m\to\infty, n\to\infty} d_X(x_0, [y_{-m}y_n]_X) = \infty$$

and, therefore, $\partial_{Y,X}(\xi^-) = \partial_{Y,X}(\xi^+)$.

(3) This part of the proof is similar to that of Proposition 8.49. We choose $D = M_k$.

(a) Suppose that sections Σ_x, Σ_y exist. After reparameterizing α, we can assume that $x = \alpha(-m_0), y = \alpha(n_0), m_0 > 0, n_0 > 0$. There is a ladder $\mathfrak{L}_{X,k}(\alpha) \subset \mathfrak{X}$ containing sections Σ_x, Σ_y. Consider the combing paths $c_{-m,n} = c(\alpha(-m), \alpha(n))$ in the ladder $\mathfrak{L}_{X,k}(\alpha)$ for $n \geq n_0, m \geq m_0$. These paths have to go through both sections Σ_x, Σ_y and, hence, pass uniformly close to a pair of points x', y' (independent of m, n) in these sections realizing the minimal fiberwise distance between Σ_x, Σ_y (the sections are cobounded in $L_{X,k}(\alpha)$; see Section 7.1). Thus, for $y_0 = \alpha(0)$, the minimal distances $d_X(y_0, c_{-m,n})$ are uniformly bounded (from above), hence,

$$\limsup_{m,n\to\infty} (\alpha(-m).\alpha(n))_{y_0} < \infty,$$

and, therefore, $\partial_{Y,X}(\xi^-) \neq \partial_{Y,X}(\xi^+)$.

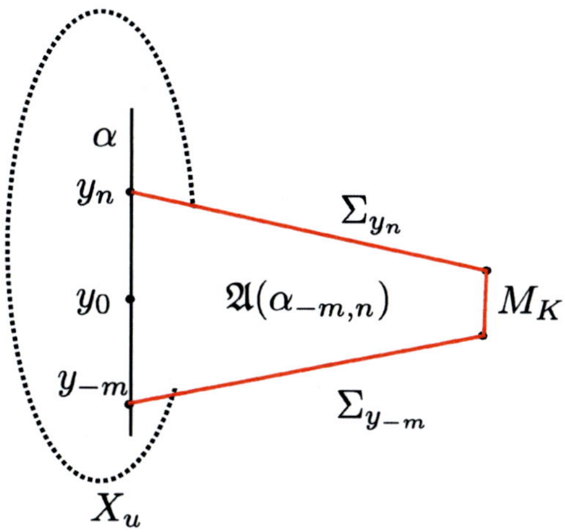

FIGURE 8.16. The carpet $\mathfrak{A}(\alpha_{-m,n})$

(b) Suppose that the points x, y do not exist. Take arbitrary maximal k-qi sections $\Sigma_{y_{-m}}, \Sigma_{y_n}$ in X through the points $y_{-m} = \alpha(-m), y_n = \alpha(n)$. Then for all $m \geq 0, n \geq 0$ the minimal vertical separation between $\Sigma_{y_{-m}}, \Sigma_{y_n}$ is $\leq M_k$ and, hence, each interval $\alpha_{-m,n} = [y_{-m}y_n]_{X_u} \subset \alpha$ bounds a (k, M_k)-narrow carpet $\mathfrak{A}(\alpha_{-m,n})$ in X. According to Part 2, then $\partial_{Y,X}(\xi^-) = \partial_{Y,X}(\xi^+)$. □

REMARK 8.53. Note that in Part (2) of the theorem we can bound from above lengths of the intervals $[\![u, w_{m,n}]\!] = \pi(\mathfrak{A}(\alpha_{-m,n}))$ in terms of lengths of the segments $\alpha_{-m,n}$: In view of the exponential flaring satisfied by $\mathfrak{X}$ (see Lemma 2.59 and Proposition 2.60) there exists a constant λ (depending only on K and the parameters of $\mathfrak{X}$) such that

$$\text{length}(\alpha_{-m,n}) \geq \lambda^t C, \quad t = d_T(u, w_{m,n}).$$

Thus,

$$d_T(u, w_{m,n}) \leq \tau_{m,n} := \log_\lambda(\text{length}(\alpha_{-m,n})) - \log_\lambda(C).$$

As an application of Part (1) of the theorem we obtain Corollary 8.54.

COROLLARY 8.54. *Suppose that γ is a geodesic in Y whose projection to T is unbounded. Then γ is not a leaf of the CT-map $\partial_{Y,X}$.*

8.8. Boundary flows and CT laminations

In this section we will be using the notion of *ideal boundary flows* Fl_t (where t are vertices and edges of the tree T) which are defined and discussed in Section 3.3.4.

Before proving the next proposition we will need two definitions which will be discussed in much greater detail in Section 8.9. Recall that $\partial_\infty(Z, X)$ denotes the *limit set* (relative ideal boundary) of a subset $Z \subset X$. Recall also that $\Lambda(Y, X)$ denotes the *Cannon–Thurston lamination* of a hyperbolic subspace Y in a hyperbolic space X; see Definition 8.10.

DEFINITION 8.55. Consider a point $\xi \in \partial_\infty T$ and a geodesic ray $v\xi$ in T joining a vertex $v \in T$ to ξ. Define the *ξ-relative ideal boundary*

$$\partial^{\xi}_{\infty}(X_v, X) := \{\eta \in \partial_\infty(X_v, X) : \exists \text{ a qi section } \gamma \text{ over } v\xi,\ \gamma(\infty) = \eta\}$$

and

$$\Lambda^{\xi}(X_v, X) = \{\{z^-, z^+\} \in \Lambda(X_v, X) : \partial_{X_v,X}(z^{\pm}) \in \partial^{\xi}_{\infty}(X_v, X)\}.$$

Note that, by the definition of $\Lambda(X_v, X)$, $\partial_{X_v,X}(z^+) = \partial_{X_v,X}(z^-)$.

While the definition of $\Lambda^{\xi}(X_v, X)$ at this point looks rather unmotivated, in the next section we will prove that it equals the the *ξ-ending lamination* $\Lambda(X_v, X_{v\xi})$. Examples of points in $\partial^{\xi}_{\infty}(X_v, X)$ are given by points $\gamma(\infty)$ for qi sections γ that are limits (as $n \to \infty$) of bottom sections of carpets $\mathfrak{A}(\alpha_{-m_0,n})$ appearing in the proof of Theorem 8.52.

PROPOSITION 8.56. *Fix a point $\xi \in \partial_\infty T$ and a pair of distinct points $z^+, z^+ \in \partial_\infty X_v$ such that $\{z^+, z^+\} \in \Lambda^{\xi}(X_v, X)$. Let $L_v = \alpha \subset X_v$ be a bi-infinite geodesic asymptotic to the points $z^{\pm}$.*

(1) *For all vertices and edges t in the ray $v\xi$ we have*

$$Fl_t(\{z^{\pm}\}) = \{z_t^{\pm}\} \neq \emptyset.$$

(2) *For each vertex/edge t in the ray $v\xi$ let L_t be a bi-infinite geodesic in X_t connecting the points $z_t^{\pm}$. Then the collection of such geodesics forms the union of vertex/edge sets of a metric bundle $\mathfrak{L} = (\pi : L \to v\xi) \subset \mathfrak{X}$ over $v\xi$, which is also a K'-ladder for some K'.*

(3) *There are constants K_1, C_1 such that for each n, the segment $\alpha_{-n,n} = [\alpha(-n)\alpha(n)]_{X_v}$ bounds a (K_1, C_1)-narrow carpet $\mathfrak{B}^n \subset \mathfrak{L}$, where Λ is the ladder from (2).*

(4) *Every qi section over $v\xi$ contained in L is asymptotic to $z = \partial_{X_v,X}(z^-) = \partial_{X_v,X}(z^+) \in \partial_\infty X$. In particular, any two such qi sections are at a finite Hausdorff distance from each other.*

(5) *$\partial_\infty L$ is the singleton $\{z\}$, in particular, $\partial_\infty(L, X) = \{z\}$.*

PROOF. We let $\mathfrak{Y} = (\pi : Y \to v\xi)$ denote the restriction of $\mathfrak{X}$ to the ray $v\xi$. We will need Lemma 8.57.

LEMMA 8.57. $\partial_{X_v,Y}(z^+) = \partial_{X_v,Y}(z^-)$.

PROOF. We consider a sequence of (K, C)-narrow carpets $\mathfrak{A}^n = \mathfrak{A}(\alpha_{-n,n}) = (\pi : A^n \to [\![v, w_n]\!])$ in $\mathfrak{X}$ given by Theorem 8.52(1) and bounded by the intervals

$$\alpha_{-n,n} = [\alpha(-n)\alpha(n)]_{X_u} \subset \alpha.$$

Set

$$[\![v, v_n]\!] = v\xi \cap [\![v, w_n]\!].$$

(i) Let us first verify that $\lim_{n\to\infty} v_n = \xi$. Suppose not. After passing to a subsequence, the sequence (v_n) would have to be constant. For sufficiently large each n, there is a point x_n in the narrow end of $\mathfrak{A}^n$ and a K-section γ_{x_n} in A^n connecting x_n to $\gamma(v)$. The concatenation $\gamma_{x_n} \star \gamma$ is then a uniform quasigeodesic in X (since v_n is fixed). But then $\lim_{n\to\infty} x_n \neq \gamma(\infty)$ in $\partial_\infty X$, contradicting the assumption that $\{z^+, z^-\} \in \Lambda^{\xi}(X_v, X)$.

(ii) The carpets $\mathfrak{A}^n$ define uniform quasigeodesics $c_{-n,n}$ in X connecting points $\alpha(\pm n)$ (see the proof of Theorem 8.52). Applying the cut-and-replace procedure to the paths $c_{-n,n}$ with respect to the subtree of spaces $Y \subset X$, we obtain uniform quasigeodesics $\hat{c}_{-n,n}$ in Y which project to the intervals $[\![v, v_n]\!]$. Since $\lim_{n\to\infty} d_T(v, v_n) = \infty$, it follows that

$$d_Y(\alpha(0), \hat{c}_{-n,n}) = \infty,$$

implying that $\partial_{X_v,Y}(z^+) = \partial_{X_v,Y}(z^-)$. □

We now begin the proof of the proposition. In view of the lemma, we can replace $\mathfrak{X}$ with $\mathfrak{Y}$.

(1) Since $\partial_{X_v,Y}(z^+) = \partial_{X_v,Y}(z^-)$, Theorem 8.52(1) implies that for each $n > 0$ the interval

$$\alpha_{-n,n} = [\alpha(-n)\alpha(n)]_{X_u} \subset \alpha$$

bounds a (K, C)-narrow carpet $\mathfrak{A}^n = \mathfrak{A}(\alpha_{-n,n}) = (\pi : A^n \to [\![v, w_n]\!])$ in $\mathfrak{Y}$ for some K, C depending only on the parameters of $\mathfrak{X}$, $w_n \in v\xi$. Since for each vertex $t \in v\xi$

$$\mathrm{Hd}_Y(A^n_v, A^n_t) \le Kd_T(v, t),$$

it follows that the entire geodesic L_v is contained in the $Kd_T(v, t)$-neighborhood of the vertex-space X_t. Lemma 3.35 now implies that

$$Fl_t(\{z^\pm\}) = \{z^\pm_t\} \neq \emptyset.$$

This proves (1).

(2) By the construction, for each pair of vertices $s, t \in v\xi$ with $d_T(s, t) = 1$, the geodesics L_s, L_t are asymptotic to the same pair of points in X_{st}. Therefore, $\mathrm{Hd}(L_s, L_t) \le D$ for some D depending only on L'_0 and the hyperbolicity constant δ'_0 of X_{st}. The same applies to the geodesic $L_e \subset X_e$, $e = [s, t]$. This proves that the union of L_t, $t \in V(v\xi)$, L_e, $e \in E(v\xi)$ forms a metric bundle $\mathfrak{L}$ in $\mathfrak{Y}$. This metric bundle has structure of a K'-ladder according to Lemma 3.17.

(3) As noted in the proof of (1), we already have the carpets $\mathfrak{A}^n$ in $\mathfrak{Y}$ bounded by the segments $\alpha_{-n,n}$. The trouble is that these carpets need not be contained in the ladder $\mathfrak{L}$. However, according to Theorem 3.3, there is a uniformly coarse Lipschitz projection $\nu : \mathfrak{Y} \to \mathfrak{L}$ which, for every vertex $t \in v\xi$, equals to the restriction of the nearest-point projection P_{X_t,L_t}. Taking the corresponding modified projection $\bar{P}_{X_t,L_t}(A^n_t)$ (see Definition 1.124) we then obtain a collection of subsegments $B^n_t \subset L_t$ satisfying axioms of a (K_1, C_1)-narrow carpet over $[\![v, w_n]\!]$, which we denote $\mathfrak{B}^n$.

(4) Suppose that γ is a k-qi section of $\mathfrak{L}$ over the ray $v\xi$, $p = \gamma(v) \in \alpha$. Then p belongs to the segment $\alpha_{-n,n}$ for all sufficiently large $n \ge n_p$. We let c_n denote the uniform quasigeodesic in Y connecting the endpoints of $\alpha_{-n,n}$ and equal to the concatenation of the two horizontal boundary sections of $\mathfrak{B}^n$ along with its narrow end (the vertical geodesic segment in X_{w_n}). Since both sequences $(\alpha(n))$, $(\alpha(-n))$ converge to the same point $z \in \partial_\infty Y$, it follows that the sequence of quasigeodesics (c_n) also converges to z. For each $n \ge n_p$, there exists a vertex $v_n \in [\![v, w_n]\!]$ such that

$$d_{X_{v_n}}(\gamma(v_n), x_n) \le K_1 + K',$$

where x_n lies in c_n. Clearly,

$$\lim_{n\to\infty} d_Y(\alpha(0), x_n) = \infty.$$

In particular, the sequence (x_n) converges to $\gamma(\infty)$. Since the sequence (x_n) also converges to z, we obtain $\gamma(\infty) = z$.

(5) The proof of this part is similar to that of (4). Consider a geodesic ray β in L, $\beta(0) = \alpha(0) = p$. Then for each n, there is a point x_n in β within distance K' from the path c_n in $\mathfrak{B}^n$ defined as in the proof of (4). Again, $d(p, x_n) \to \infty$ and, hence, $\beta(\infty) = \lim_{n\to\infty} c_n = z$. □

We now can relate boundary flows to CT-laminations.

PROPOSITION 8.58. *Suppose that* $\mathfrak{X} = (\pi : X \to T)$ *is a tree of hyperbolic spaces with hyperbolic total space* X, $\mathfrak{Y} = (\pi : Y \to S) \subset \mathfrak{X}$ *is a subtree of spaces and* $v \in V(S)$ *satisfies the following conditions:*

(1) $\partial_\infty X_w \subset Fl(\partial_\infty X_v)$ *for each vertex* $w \in V(S)$.
(2) $\partial_{X_v,X}(\partial_\infty X_v) = \partial_\infty X$.

Then the CT-map $\partial_{X_v,Y} : \partial_\infty X_v \to \partial_\infty Y$ *is also surjective.*

PROOF. We claim that each $z \in \partial_\infty Y$ belongs to $\partial_{X_v,Y}(X_v)$. Since $\partial_{X_v,X}$ is surjective, there is $z_1 \in \partial_\infty X_v$ such that $\partial_{X_v,X}(z_1) = \partial_{Y,X} \circ \partial_{X_v,Y}(z_1) = \partial_{Y,X}(z)$. Thus, for $z' = \partial_{X_v,Y}(z_1) \in \partial_\infty Y$, $\partial_{Y,X}(z') = \partial_{Y,X}(z)$.

If $z' = z$, then we are done. If not, then by the description of the fibers of the CT-map $\partial_{Y,X}$ given in Proposition 8.49, there exists a vertex-space $X_w \subset Y$ such that a geodesic $\beta = zz' \subset Y$ asymptotic to z, z' is Hausdorff-close to a geodesic $\alpha \subset X_w$. In particular, $z \in \partial_\infty(X_w, Y)$. The first assumption of the proposition implies that $z \in \partial_\infty(X_v, Y)$. Since $\partial_\infty(X_v, Y) = \partial_{X_v,Y}(\partial_\infty X_v)$ (see Lemma 8.7), the claim follows. □

8.9. Cannon–Thurston lamination and ending laminations

In this section we shall significantly expand on Theorem 8.52(1); many of our results are generalizations of the ones proven by Mitra in [**Mit97**]. Throughout this section we will assume that $\mathfrak{X} = (\pi : X \to T)$ is a tree of hyperbolic spaces with proper hyperbolic total space X.

To motivate the discussion, we recall Thurston's notion of the *ending laminations* in the setting of hyperbolic 3-manifolds. (We refer the reader for a detailed overview of end-invariants of hyperbolic 3-manifolds to Minsky's surveys [**Min03b**] and [**Min03a**].) For simplicity of the discussion, we consider a noncompact complete connected hyperbolic 3-manifold M with finitely generated fundamental group that does not split as a nontrivial free product and such that M has positive injectivity radius. The manifold M contains a (unique up to isotopy) compact submanifold with smooth boundary M_c (the *compact core* of M), such that the complement $M \setminus \mathrm{int}(M_c)$ is homeomorphic to $\partial M_c \times \mathbb{R}_+$.

REMARK 8.59. The existence of a compact core was first proven by P. Scott in [**Sco73a, Sco73b**] in the context of topological 3-manifolds with finitely generated fundamental groups. The uniqueness, up to an isotopy, again, in the topological context was proven in [**MMS85**]. Tameness of the ends of M, i.e., the existence of a product decomposition of $M \setminus \mathrm{int}(M_c)$ was proven primarily in the work of F. Bonahon, [**Bon86**], (who was inspired by the earlier work of W. Thurston, [**Thu80**]), in combination with the paper by Freedman, Hass and Scott, [**FHS83**]. See also [**Kap09**, Theorems 1.26, 1.27] and [**Kap09**, Chapter 14].

The group $G = \pi_1(M_c) \cong \pi_1(M)$ is hyperbolic (see [**Kap09**, Theorem 19.8]). In view of the Loop Theorem, [**Hem04**], the assumption that G does not split as a free product implies that for each surface component $S \subset \partial M_c$, the inclusion map $S \to M_c$ is π_1-injective.

Each component $E = S \times \mathbb{R}_+$ of $M \setminus M_c$ is an *end* of M; the surface S is a component of ∂M_c; it is a compact surface which admits a hyperbolic metric (which we fix from now on). For each end $E = S \times \mathbb{R}_+$ one defines an *ending lamination* $\lambda = \lambda(E)$ of E, which is a certain nonempty compact subset of S, equal to a disjoint union of complete geodesics in S. Lifting λ to the universal covering space of S, one obtains a $\pi_1(S)$-invariant closed subset of $\tilde{S} \cong \mathbb{H}^2$ equal to a disjoint union of geodesics. Each geodesic β in $\tilde{\lambda}$ is uniquely determined by an unordered pair $\{\xi^+, \xi^-\} \in \partial_\infty^{(2)}\mathbb{H}^2$, the ideal boundary points such that $\beta = \xi^-\xi^+$. We, thus, identity λ with a $\pi_1(S)$-invariant closed subset of $\partial_\infty^{(2)}\mathbb{H}^2$ consisting of such pairs. Consider a component $\tilde{E}$ of the pre-image of E in $\mathbb{H}^3$ (the universal covering space of M); the boundary surface $\tilde{S}$ of $\tilde{E}$ is a copy of the universal covering space of S. It was proven by Minsky [**Min94**] (under the above assumption on the injectivity radius of M) and Mj [**Mj17**] in full generality, that each inclusion map $\tilde{S} \to \tilde{E}$ has a CT-map; these CT-maps combine in a CT-map for the inclusion $\tilde{M}_c \to \tilde{M} = \mathbb{H}^3$ (where $\tilde{M}_c$ is the universal covering space of M_c equipped with the pull-back Riemannian metric). Each ending lamination $\lambda(E)$ is $\pi_1(S)$-equivariantly homeomorphic to the CT-lamination $\Lambda(\tilde{S}, \tilde{E})$ and the union of G-orbits of these laminations in $\partial_\infty \tilde{M}_c$ is the CT-lamination $\Lambda(\tilde{M}_c, \mathbb{H}^3)$.

We now relate this discussion to trees of spaces. For each end E, the space $\tilde{E}$ (with its intrinsic Riemannian path-metric) is $\pi_1(E)$-equivariantly quasi-isometric to the total space of a certain metric bundle $\mathfrak{X}_E = (E \to \mathbb{R}_+)$, with vertex and edge-spaces isometric to the hyperbolic plane. (This bundle structure is implicit in [**Min94**]. It is obtained via pull-back of the universal bundles over the Teichmüller spaces of boundary surfaces of M_c.) Putting these spaces together, we obtain a tree of spaces $\mathfrak{X} = (X \to T)$ (on which $G = \pi_1(M)$ is acting) which has a distinguished vertex v fixed by G. The total space X of $\mathfrak{X}$ is isometric to $\mathbb{H}^3$. The tree T is a union of geodesic rays; the intersection of any two distinct rays in this collection is the vertex v. Thus, the G-orbits of ending laminations $\lambda(E)$ in $\partial_\infty^{(2)} G$ can be described as CT-laminations

$$\Lambda(X_v, X_{v\xi}) \subset \partial_\infty^{(2)} X_v,$$

where, ξ are the ideal boundary points of T and $X_{v\xi}$ is the total space of the pull-back of $\mathfrak{X}$ to the ray $v\xi$ in T. The result stated above, relating ending laminations of M with the CT-lamination $\Lambda(\tilde{M}_c, \mathbb{H}^3)$ can then be restated as:

$$\Lambda(X_v, X) = \bigcup_{\xi \in \partial_\infty T} \Lambda(X_v, X_{v\xi}).$$

In the context of general trees of hyperbolic spaces, points at infinity $\xi \in \partial_\infty T$ play the role of ends of the hyperbolic manifolds and, accordingly, ending laminations are defined as CT-laminations $\Lambda(X_v, X_{v\xi})$. The main goal of this section is to prove an analogue (actually, a sharper version) of the above equality in the setting of more general trees of spaces, Theorem 8.63. In particular, we will also prove that for each $\xi \in \partial_\infty T$, the ending lamination $\Lambda(X_v, X_{v\xi})$ equals the subset $\Lambda^\xi(X_v, X) \subset \Lambda(X_v, X)$ defined in the previous section. This alternative interpretation of the ending lamination $\Lambda(X_v, X_{v\xi})$ will be used in several places, e.g., the proof of Theorem 8.63, Parts (4) and (5) and proof of Proposition 8.67.

The results below are motivated by similar results obtained in [**KS20**, section 6.2]; our notation and proofs are similar (see also [**Mit97, Bow13**]).

LEMMA 8.60. *Suppose that $\mathfrak{X} = (\pi : X \to T)$ is a tree of hyperbolic spaces with proper hyperbolic total space X. We fix a vertex $v \in V(T)$ and $K \geq 1$.*

(1) *For every $\xi \in \partial_\infty T$, there is a geodesic ray ρ joining v to ξ.*
(2) *There is K_1 depending on K and the parameters of $\mathfrak{X}$ such that the following holds. Let (w_n) be a sequence of points in $V(T) \cup \partial_\infty T$ and let (γ_n) be a sequence of K-qi sections of $\mathfrak{X}$ of over the geodesic vw_n. Suppose that the sequence $(\gamma_n(v))$ belongs to a bounded subset B of X and the sequence $(\gamma_n(w_n))$ converges to a point $\eta \in \partial_\infty X$. Then the sequence (w_n) converges to a point $\xi \in \partial_\infty T$ and there is a K_1-qi section γ over the geodesic $v\xi$ such that $\eta = \gamma(\infty)$ and $\gamma(v) \in B$.*

PROOF.

(1) Let (w_n) be a Gromov-sequence of vertices in T representing the point ξ. Then

$$\lim_{m,n\to\infty} d_T(v, w_m w_n) = \infty.$$

It follows that the union of geodesic segments $vw_n, n \in \mathbb{N}$, is a locally finite subtree $S \subset T$. Therefore, the sequence of segments vw_n subconverges to a geodesic ray ρ in S emanating from v. In order to prove that ρ joins v to ξ we note that for each $m \in \mathbb{N}$ and all sufficiently large n, the points $t_m = \rho(m)$ satisfy

$$d_T(v, t_m w_n) = d_T(v, t_m) = m.$$

Thus, the sequence (t_m) is a Gromov-sequence equivalent to (w_n).

(2) Pick a base-point x_0 in a bounded subset $B \subset X$ of diameter D containing all the points $\gamma_n(v)$, e.g., we can take $x_0 = \gamma_1(v)$.

Since the sequence $(\gamma_n(w_n))$ converges $\eta \in \partial_\infty X$, we have

$$\lim_{n\to\infty} d_T(v, w_n) = \infty.$$

Moreover, since the sequence of geodesic segments $\gamma_i^* = [x_0\gamma_n(w_n)]_X$ *coarsely converges* to a geodesic ray $x_0\xi$ (see [**DK18**, Definition 8.32]), there is a constant C (depending only on K, D and the hyperbolicity constant of X) such that for each R, there is a number n_0 such that for all $m, n \geq n_0$ the Hausdorff-distance between $\gamma_m \cap B(x_0, R)$ and $\gamma_n \cap B(x_0, R)$ is $\leq C$. Since γ_i are sections over geodesic segments vw_i in T, it follows that

$$\lim_{i\to\infty} \sup\{R : vw_m \cap B(v, R) = vw_n \cap B(v, R)\ \forall m, n \geq i\} = \infty.$$

In particular, (w_n) is a Gromov-sequence in T converging to some $\xi \in \partial_\infty T$ and by Part 1 of the lemma, the sequence of segments (vw_n) converges to the ray $v\xi$.

Furthermore, in view of properness of X, the sequence of K-qi sections (γ_n) subconverges to a K-qi section γ over the geodesic ray $v\xi$ in T (this is a coarse version of the Arzela–Ascoli theorem, cf. [**DK18**, Proposition 8.34]). □

Fix $K \geq K_0$, pick a vertex $v \in V(T)$ and let $\mathfrak{Y} = (\pi : Y \to S)$ be a (K, D, E, λ)-semicontinuous family in $\mathfrak{X}$ relative to a vertex $v \in S \subset T$; see Definiton 3.1. The following proposition is motivated by the results of [**KS20**] and [**Bow13**, Proposition 8.2].

PROPOSITION 8.61. *We have*

$$\partial_\infty Y = U := \partial_\infty(X_v, Y) \cup \left(\bigcup_{\xi\in\partial_\infty S} \partial^\xi_\infty(X_v, X) \right).$$

PROOF. Recall that, according to Theorem 3.3, Y is qi embedded in X. Thus, we will identify $\partial_\infty Y$ with a subset of $\partial_\infty X$. Since U is obviously contained in $\partial_\infty Y$, we only have to prove that every point $z \in \partial_\infty Y$ lies in U. Fix a base-point $x \in X_v$. Suppose that $x_n \in Y_{v_n}$ is a sequence of points converging to z. Let γ_n be a K-qi section in Y over vv_n, joining x_n to $y_n \in Y_v$.

(i) Suppose first that (y_n) is a bounded sequence. Then by Lemma 8.60(2), the sequence (v_n) converges to some $\xi \in \partial_\infty T_v$ and the sequence (γ_n) coarsely converges to a K-qi section γ in Y over the ray $v\xi$, so that $x_n \to \gamma(\infty)$. Thus $z = \lim_{n\to\infty} x_n \in \partial^\xi_\infty(X_v, X) \subset U$ in this case.

(ii) Consider now the case when (y_n) is an unbounded sequence. After extraction, we can assume that (y_n) converges to some $z' \in \partial_\infty(X_v, X) = \partial_\infty(X_v, Y)$ (since Y is quasiconvex in X). We claim that $z' = z$. Since each γ_n is a K-qi section, it suffices to show that $d(x, \gamma_n) \to \infty$. Suppose that the sequence $d(x, \gamma_n)$ is bounded, and $p_n = \gamma_n(v_n)$, $v_n \in V([\![v, \pi(y_n)]\!])$, is a sequence such that $d_X(x, p_n) \le C$ for all n. Since $d_T(v, v_n) = d_T(\pi(x), \pi(p_n)) \le d_X(x, p_n) \le C$, it follows that

$$d_X(y_n, p_n) \le CK, d_X(x, y_n) \le C + CK,$$

contradicting the assumption that the sequence (y_n) is unbounded. Thus,

$$z = z' \in \partial_\infty(X_v, Y) \subset U. \qquad \square$$

COROLLARY 8.62. *For any ladder $\mathfrak{L} = (\pi : L \to \pi(L))$ in $\mathfrak{X}$, centered at a vertex $v \in V(T)$, we have*

$$\partial_\infty(L, X) = \bigcup_{\xi \in \partial_\infty \pi(L)} \partial^\xi_\infty(L_v, X).$$

PROOF. Since L_v is a finite geodesic segment, $\partial_\infty(L_v, L) = \emptyset$, and, thus, the corollary is an immediate consequence of Proposition 8.61. $\square$

We now return to the discussion of properties of ending laminations $\Lambda(X_v, X_{x\xi})$ and their relation to the CT-laminations $\Lambda(X_v, X)$ and their subsets $\Lambda^\xi(X_v, X)$ defined in the previous section (see Definition 8.55).

THEOREM 8.63 (Properties of ending laminations). *Suppose that $\mathfrak{X} = (\pi : X \to T)$ is a tree of hyperbolic spaces with hyperbolic and proper total space X. There exists K depending only of the parameters of $\mathfrak{X}$ and the hyperbolicity constant of X such that the following hold:*

(1) *Let v be a vertex of T and let $\alpha : \mathbb{R} \to X_v$ be a complete geodesic in X_v such that $\{\alpha(-\infty), \alpha(\infty)\} \in \Lambda(X_v, X)$. Then for both $z \in \{\alpha(\pm\infty)\}$, there exists point $\xi \in \partial_\infty T_z$ such that for each $p \in \alpha$, there is a K-qi section γ over $v\xi$ satisfying $\gamma(v) = p$ and $\gamma(\infty) = \partial_{X_v,X}(\alpha(\infty))$. In other words, for each point $\{z^-, z^+\} \in \Lambda(X_v, X)$, there exists $\xi \in \partial_\infty T$ such that*

$$\partial_{X_v,X}(z^\pm) \in \Lambda^\xi(X_v, X)$$

and, thus,

$$\Lambda(X_v, X) = \bigcup_{\xi \in \partial_\infty T_v} \Lambda^\xi(X_v, X).$$

(2) *For each $\xi \in \partial_\infty T$ we have*

$$\Lambda^\xi(X_v, X) = \Lambda(X_v, X_{v\xi}) = \Lambda(X_v, Fl_K(X_v) \cap X_{v\xi}).$$

(3) *Each $\Lambda^\xi(X_v, X)$ is a closed subset of $\partial^{(2)}_\infty X_v$.*

(4) *Suppose $\xi_1 \neq \xi_2 \in \partial_\infty T$, and α_1, α_2 complete geodesics in X_v such that $\{z_i^-, z_i^+\} = \{\alpha_i(-\infty), \alpha_i(\infty)\} \in \Lambda^{\xi_i}(X_v, X)$, $i = 1, 2$. Then the subsets $\{z_1^-, z_1^+\}$, $\{z_2^-, z_2^+\}$ of $\partial_\infty X_v$ are disjoint. In particular, the ending laminations $\Lambda^{\xi_1}(X_v, X)$, $\Lambda^{\xi_2}(X_v, X)$ are disjoint and the point ξ in (1) is uniquely determined by $\{z^-, z^+\} \in \Lambda(X_v, X)$.*

(5) *Ending laminations Λ^ξ depend upper semicontinuously[3] on ξ: Suppose that $\xi_n \to \xi$ in $\partial_\infty T_v$, $\{z_n^+, z_n^-\} \in \Lambda^{\xi_n}(X_v, X)$ and $\{z_n^+, z_n^-\} \to \{z^+, z^-\} \in \partial^{(2)}_\infty X_v$. Then $\{z^+, z^-\} \in \Lambda^\xi(X_v, X)$.*

(6) *If $\xi_1 \neq \xi_2 \in \partial_\infty T_v$, then any two leaves α^i of $\Lambda^{\xi_i}(X_v, X)$, $i = 1, 2$, are uniformly cobounded in X_v. Namely, given $D > 0$ there exists $R = R(D)$ (independent of α^1, α^2 but possibly depending on ξ_1, ξ_2) such that $\alpha^1 \cap N_D(\alpha^2)$ has diameter $\le R$.*

PROOF.

[3]It is shown in [**DKT16**, Section 7] that in general Λ^ξ does not depend continuously on ξ.

(1) The proof follows the argument in the proof of Proposition 8.56(1) (or Part (3) of that proposition). By Theorem 8.52(1), for each $n > 0$ the interval

$$\alpha_{-n,n} = [\alpha(-n)\alpha(n)]_{X_u} \subset \alpha$$

bounds a (K, C)-narrow carpet $\mathfrak{A}^n = \mathfrak{A}(\alpha_{-n,n})$ in X, which, in turn, defines a uniform quasigeodesic $c_n = c(\mathfrak{A}^n)$ in $\mathfrak{A}(\alpha_{-n,n})$ connecting the points $\alpha(-n), \alpha(n)$. Take any point $p \in \alpha$. Then for all $n \geq n_p$, p belongs to the segment $\alpha_{-n,n}$. Since $\mathfrak{A}^n$ is a K-metric bundle, there exists a K-section γ^n over the segment $[\![v, w_n]\!] = \pi(A^n)$. The endpoint $x_n = \gamma^n(w_n)$ belongs to c_n, which implies that the sequence (x_n) converges to the limit point $\partial_{X_v,X}(\alpha(-\infty)) = \partial_{X_v,X}(\alpha(\infty))$. Then the existence of the point ξ follows from Lemma 8.60(2): Proposition 8.56(1) implies that $\xi \in \partial_\infty T_z$, $z = \alpha(\pm\infty)$. The rest of the assertions of Part (1) follow immediately from the definition of $\Lambda^\xi(X_v, X)$.

(2) The inclusion $\Lambda^\xi(X_v, X) \subset \Lambda(X_v, X_{v\xi})$ is clear from the definition of $\Lambda^\xi(X_v, X)$. The opposite inclusion is a direct consequence of Part (1) of the theorem. The inclusion $\Lambda(X_v, Fl_K(X_v) \cap X_{v\xi}) \subset \Lambda(X_v, X_{v\xi})$ for every $K \geq K_0$ is clear from the fact that $Fl_K(X_v)$ is quasiconvex in X. Suppose that $\{z^-, z^+\} \in \Lambda(X_v, X_{v\xi})$. Proposition 8.56(1) implies that for each vertex $t \in v\xi$, $Fl_t(\{z^\pm\}) \neq \emptyset$. By Lemma 3.37(1), there exists K such that for each vertex/edge t in $v\xi$ there exists a bi-infinite geodesic $L_t \subset Fl_K(X_v)$ asymptotic to the points $Fl_t(\{z^\pm\})$. By Proposition 8.56(2), these geodesics form a ladder $\mathfrak{L} = (\pi : L \to v\xi)$ in X. By Part 5 of the same proposition, $\partial_\infty L$ is a singleton, which implies

$$\partial_{X_v, Fl_K(X_v)\cap X_{v\xi}}(z^+) = \partial_{X_v, Fl_K(X_v)\cap X_{v\xi}}(z^-).$$

In other words,

$$\{z^-, z^+\} \in \Lambda(X_v, Fl_K(X_v) \cap X_{v\xi}).$$

(3) In view of Part (2), the assertion follows from the fact that the CT-lamination $\Lambda(X_v, X_{v\xi})$ is closed in $\partial^{(2)}_\infty X_v$.

(4) By Proposition 8.56(2) and (4), there are quasi-isometric sections γ_i over $v\xi_i$ asymptotic to $\partial_{X_v,X}(z_i^\pm)$, $i = 1, 2$. Since $\xi_1 \neq \xi_2$, $\text{Hd}(v\xi_1, v\xi_2) = \infty$, which, in turn, implies that $\text{Hd}(\gamma_1, \gamma_2) = \infty$. Thus $\gamma_1(\infty) \neq \gamma_2(\infty)$, whence $\{z_1^-, z_1^+\} \cap \{z_2^-, z_2^+\} = \emptyset$.

(5) Since $\partial_{X_v,X}(z_n^+) = \partial_{X_v,X}(z_n^-)$ and $\partial_{X_v,X}$ is continuous, we have $\partial_{X_v,X}(z^+) = \partial_{X_v,X}(z^-)$. Thus $\{z^+, z^-\} \in \Lambda(X_v, X)$. Since $z^+ \neq z^-$, geodesics α_n in X_v connecting the points $z_n^\pm$ all intersect a certain bounded subset of X_v. Hence, we can parameterize these geodesics so that the sequence $(\alpha_n(0))$ is bounded in X_v. In view of properness of X_v, by the Arzela–Ascoli theorem, the sequence of geodesics α_n subconverges to a geodesic α in X. This geodesic is necessarily asymptotic to the points $z^\pm$; see e.g., [**DK18**, Theorem 11.104]. Since $\{z_n^+, z_n^-\}$ belongs to $\Lambda^\xi(X_v, X)$, there exist uniform qi sections γ_n over $v\xi_n$ connecting $\alpha_n(0)$ to $\partial_{X_v,X}(z^\pm)$. By the continuity of the CT-map, $z_n \to z$ implies that $\partial_{X_v,X}(z_n) \to \partial_{X_v,X}(z)$. Accordingly, $\gamma_n(\infty) \to \partial_{X_v,X}(z)$. Hence, by Lemma 8.60(2) there is a qi section γ over $v\xi$ such that $\gamma(\infty) = \partial_{X_v,X}(z^\pm)$, which implies that $\{z^+, z^-\}$ belongs to $\Lambda^\xi(X_v, X)$.

(6) Note that this statement is a strengthening of Part (4) since that part is equivalent to the statement that the leaves α^1, α^2 are cobounded in X_v. Note also that, since vertex-spaces of $\mathfrak{X}$ are uniformly properly embedded in X, the following two properties are equivalent for subsets $Y^1, Y^2 \subset X_v$:

(i) There exists a function $R_v(D)$ such that $\text{diam}_{X_v}(Y^1 \cap N_D^{X_v}(Y^2)) \leq R_v(D)$.

(ii) There exists a function $R(D)$ such that $\text{diam}_X(Y^1 \cap N_D(Y^2)) \leq R(D)$.

Set $\alpha^i(\pm\infty) = z_i^\pm$, $i = 1, 2$. Let $x_i^\pm \in \alpha_i$, $i = 1, 2$, be points such that

$$d_{X_v}(x_1^\pm, x_2^\pm) \leq D. \tag{8.1}$$

Our goal is to get an upper bound (in terms of D) on the distances $d_{X_v}(x_i^+, x_i^-)$, $i = 1, 2$.

(a) We first consider the special case when the rays $v\xi_1, v\xi_2$ intersect only at the vertex v. We have subtrees of spaces $\mathfrak{Y}^i = (\pi : X_{v\xi_i} \to v\xi_i)$ in X, $i = 1, 2$. Since for $i = 1, 2$,

$$\{z_i^-, z_i^+\} \in \Lambda^{\xi_i}(X_v, X) = \Lambda(X_v, X_{v\xi_i}),$$

there is a sequence of (K, C)-narrow carpets

$$\mathfrak{A}^{i,n} = \mathfrak{A}(\alpha^i_{-n,n}) \subset \mathfrak{Y}^i, n \in \mathbb{N},$$

where $\alpha^i_{-n,n}$ is the subinterval in α^i between $\alpha^i(-n), \alpha^i(n)$. In particular, for all sufficiently large n, $x_i^\pm \in \alpha^i_{-n,n}$, $i = 1, 2$. Connect $x_i^\pm$ to the narrow end of $\mathfrak{A}^{i,n}$ by a K-section $\gamma_i^\pm$ in $\mathfrak{A}^{i,n}$, $i = 1, 2$. Since $d_{X_v}(x_1^\pm, x_2^\pm) \le D$, both concatenations

$$\phi^- := \gamma_1^- \star [x_1^- x_2^-]_{X_v} \star \gamma_2^-, \phi^+ := \gamma_1^+ \star [x_1^+ x_2^+]_{X_v} \star \gamma_2^+$$

are k-quasigeodesics in X with k depending only on K and D. The respective endpoints of these quasigeodesics are at most C-apart from each other. It follows that $d_{X_v}(x_i^+, x_i^-) \le R = R(k, \delta_X)$, cf. Lemmata 1.55 and 2.50.

(b) We now consider the general case: The rays $v\xi_1, v\xi_2$ intersect along a finite subinterval vw (this subinterval is finite since $\xi_1 \ne \xi_2$). Since $\{z_i^-, z_i^+\} \in \Lambda^{\xi_i}(X_v, X)$, there exist vertical geodesics β^1, β^2 in X_w within uniformly bounded (in terms of $d_T(v, w)$) Hausdorff distance from α^1, α^2 respectively; see Proposition 8.56. By Part (a), the geodesics β^1, β^2 are uniformly cobounded in X_w. It follows that α^1, α^2 are uniformly cobounded as well. □

Corollary 8.64. *For each vertex $v \in T$ and*

$$z \in \partial_\infty(X_v, X) \setminus (\bigcup_{\xi \in \partial_\infty T} \partial^\xi_\infty(X_v, X)),$$

the pre-image $\partial^{-1}_{X_v,X}(z)$ is a singleton.

Proof. By Lemma 8.7, $\partial_\infty(X_v, X) = \partial_{X_v,X}(\partial_\infty X_v)$. Thus, all we need is to show that $|\partial^{-1}_{X_v,X}(z)| \le 1$. As we noted in the previous section, for each $\xi \in \partial_\infty T$,

$$\partial_{X_v,X}(\Lambda^\xi(X_v, X)) \subset \partial^\xi_\infty(X_v, X).$$

Hence, $z \notin \partial_{X_v,X}(\Lambda^\xi(X_v, X))$ for any $\xi \in \partial_\infty T$. However, according to Theorem 8.63(1),

$$\Lambda(X_v, X) = \bigcup_{\xi \in \partial_\infty T_v} \Lambda^\xi(X_v, X),$$

which means that there is no $\{z^+, z^-\} \in \Lambda(X_v, X)$ satisfying $\partial_{X_v,X}(z^\pm) = z$. Thus, $\partial^{-1}_{X_v,X}(z)$ contains at most one point. □

Corollary 8.65.

(1) *If for each $\xi \in \partial_\infty T$, $\Lambda^\xi(X_v, X) = \emptyset$, then $\Lambda(X_v, X) = \emptyset$, i.e., the CT-map $\partial_{X_v,X}$ is 1-1.*

(2) *If $z^\pm \in \partial_\infty X_v$ are distinct points such that $\partial_{X_v,X}(z^+) = \partial_{X_v,X}(z^-)$, then there exists $\xi \in \partial_\infty T$ such that for every vertex $w \in v\xi$, $Fl_w(\{z^\pm\}) \ne \emptyset$.*

Proof.

(1) The first claim is a direct consequence of Theorem 8.63(1).

(2) Since $\partial_{X_v,X}(z^+) = \partial_{X_v,X}(z^-)$, $\{z^+, z^-\} \in \Lambda(X_v, X)$. By Theorem 8.63(1), there exists $\xi \in \partial_\infty T$ such that $\{z^+, z^-\} \in \Lambda^\xi(X_v, X)$. Now the claim follows from Proposition 8.56(1). □

8.10. Conical limit points in trees of hyperbolic spaces

In this section we consider trees of hyperbolic spaces $\mathfrak{X} = (\pi : X \to T)$ with proper total space X and discuss the relation between conicality for limit points of subtrees of spaces $Y \subset X$ and the CT-maps $\partial_{Y,X}$. Namely, identifying $\Lambda(Y, X)$ with a subset $\Sigma(Y, X)$ of $\partial_\infty Y$ equal

$$\bigcup_{\{z^+,z^-\}\in\Lambda(Y,X)} \{z^+, z^-\},$$

we'll see that $\partial_{Y,X}(\Sigma(Y, X))$ is disjoint from the conical limit set of Y in $\partial_\infty X$.

The next definition is motivated by the notion of conical limit points of group actions on hyperbolic spaces; see Definition 1.138, as well as [**DK18**, Definition 11.93 and Section 11.13.4].

Definition 8.66. Suppose X is an arbitrary hyperbolic geodesic metric space and $Y \subset X$. Then a point $\xi \in \partial_\infty(Y, X) \subset \partial_\infty X$ is called a *conical limit point* of Y if for some (any) (quasi)geodesic $\alpha \subset X$ asymptotic to ξ there is $R > 0$ and a sequence of points $\{y_n\}$ in $N_R(\alpha) \cap Y$ converging to ξ. The set of conical limit points of Y is called the *conical limit set* of Y in $\partial_\infty X$.

Thus, if Y is an orbit Gx of an isometric proper action $G \curvearrowright X$, then ξ is a conical limit point of Y if and only if it is a conical limit point of the G-action on X.

Proposition 8.67. *Suppose $\mathfrak{X} = (\pi : X \to T)$ is a tree of hyperbolic spaces with proper and hyperbolic total space, and $\mathfrak{Y} = (\pi : Y \to S) \subset \mathfrak{X}$ is a subtree of spaces. Let $\partial_{Y,X} : \partial_\infty Y \to \partial_\infty X$ be the CT-map. If $\eta \in \partial_\infty(Y, X)$ is a conical limit point of Y, then $|\partial_{Y,X}^{-1}(\eta)| = 1$.*

Proof. If not, then there are distinct points $z_\pm \in \partial_\infty Y$ such that $\partial_{Y,X}(z_-) = \partial_{Y,X}(z_+) = \eta$. Consider a geodesic β in Y asymptotic to the points $z_\pm$. By Proposition 8.49, there exists a vertex-space $X_v \subset Y$ and a complete geodesic $\alpha \subset X_v$ such that $\mathrm{Hd}(\alpha, \beta) < \infty$. Let $z'_\pm = \alpha(\pm\infty)$. It follows that $\partial_{X_v,X}(z'_\pm) = \eta$. By Theorem 8.63(1), there is a point $\xi \in \partial_\infty T$ and a qi section γ over the ray $v\xi$, such that $\gamma(\infty) = \eta$. We claim that $\xi \in \partial_\infty T \setminus \partial_\infty S$. If not, then the ray $v\xi$ is contained in the subtree S. By Theorem 8.63(2),

$$\{z'_-, z'_+\} \in \Lambda^\xi(X_v, X) = \Lambda(X_v, X_{v\xi}) \subset \Lambda(X_v, Y).$$

But then $\partial_{X_v,Y}(z'_\pm) = z_\pm$ and $z_+ = z_-$, contradicting our assumption that the points $z_\pm$ are distinct. It then follows that $\lim_{n\to\infty} d_X(\gamma(n), Y) = \infty$. Since $\gamma(\infty) = \eta$ and γ is a quasigeodesic in X, this contradicts the hypothesis that η is a conical limit point of Y and proves the proposition. □

Remark 8.68. This proposition is a geometric counterpart of the following group-theoretic result: If H is a hyperbolic subgroup of a hyperbolic group G, and the CT-map $\partial_{H,G}$ exists, and a limit point z of H in G is conical, then $|\partial_{H,G}^{-1}(z)| = 1$. The converse to this implication is false; see [**JKLO16**].

The next conjecture would show that there always is a large number of limit points of H for which the converse (in the remark) actually holds.

Conjecture 8.69. *Suppose that $H < G$ is a hyperbolic subgroup of a hyperbolic group, the CT-map $\partial_{H,G}$ exists, but H is not quasiconvex in G. Then there is a continuum of (nonconical) limit points of H in G whose pre-images under $\partial_{H,G}$ are not singletons.*

Here is some background on the conjecture. First of all, if *all* limit points of H have singletons as pre-images under $\partial_{H,G}$ (equivalently, $\partial_{H,G}$ is injective), then all limit points of H are conical and, hence, H is quasiconvex in G. In view of the equivariance of $\partial_{H,G}$, once there is one limit point whose pre-image is not a singleton, there are countably infinitely many. The question then becomes of the actual cardinality of the set of such points. It is easy to see, for instance, that if G is the fundamental group of a compact hyperbolic 3-manifold fibering over the circle and H is the fundamental group of the fiber, then the cardinality is that of continuum. More generally, the following theorem was proven in **[KL19]**.

Suppose that X is a Hadamard manifold of pinched negative curvature and Γ is a discrete isometry group of X. Then the followings are equivalent:

(1) Γ is geometrically infinite.
(2) There exists a sequence of closed geodesics in M which escapes every compact subset of M.
(3) The set of nonconical limit points of Γ has cardinality of continuum.

At this point, it is very unclear if this result also holds in the setting of subgroups of hyperbolic groups. However, it is not unreasonable to expect that, indeed failure of quasi-convexity implies continuum of nonconical limit points. The part about pre-images under CT-maps in Conjecture 8.69 would further strengthen the analogy with surface subgroups of hyperbolic 3-manifold groups.

8.11. Group-theoretic applications

In this section we collect group-theoretic applications of our existence results for CT-maps.

8.11.1. Maps to products and examples of undistorted subgroups in $PSL(2,\mathbb{C})\times PSL(2,\mathbb{C})$. Set $H := PSL(2,\mathbb{C})$ and $G := H\times H$. We equip G with a left-invariant Riemannian metric and the corresponding left-invariant distance function d_G. A finitely generated subgroup $\Gamma < G$ is said to be *undistorted* if the inclusion map

$$(\Gamma, d_\Gamma) \to (G, d_G)$$

is a qi embedding, where d_Γ is a word metric on Γ. Since G acts properly, isometrically and transitively on $\mathbb{H}^3\times\mathbb{H}^3$, a subgroup $\Gamma < G$ is undistorted if and only if for (some/every) point $x\in\mathbb{H}^3\times\mathbb{H}^3$ the orbit map $\gamma\mapsto\gamma x$ is a qi embedding of Γ (with its word metric) into $\mathbb{H}^3\times\mathbb{H}^3$.

An element $h\in H$ is called *parabolic* if it has precisely one fixed point in the Riemann sphere. An element $g=(h_1,h_2)\in G$ is called *semisimple* if neither component h_1 nor h_2 is a parabolic element of H. We will not attempt to define here Anosov subgroups of G; it suffices to say that each Anosov subgroup $\Gamma < G$ is Gromov-hyperbolic and for one of the factors $H_\pm$ of $G = H\times H = H_+\times H_-$, the projection of Γ to $H_\pm$ has finite kernel and convex-cocompact image. Moreover, each Anosov subgroup is undistorted in G. We refer to the reader to **[GW12, KL18a, KLP17]** for the detailed definitions. O. Guichard constructed in **[Gui04]** (see also **[GGKW17]**) an example of an undistorted non-Anosov free subgroup $\Gamma < G$. The subgroup in his example contained nonsemisimple elements. (Its projections to both factors were geometrically finite with parabolic elements, we refer the reader to **[Bow95]** for definitions of geometric finiteness.) Let S be a closed connected oriented hyperbolic surface with the fundamental group π.

THEOREM 8.70. *There exists an undistorted subgroup* $\Gamma < G$ *isomorphic to* π*, such that every element of* Γ *is semisimple, but* Γ *is not Anosov.*

PROOF. Let c be a complete geodesic in the Teichmüller space $T(S)$ of S, such that the projection of c to the moduli space of S is bounded. For instance, c can be taken to be the unique invariant geodesic (*axis*) in $T(S)$ of a pseudo-Anosov homeomorphism h of S. The asymptotics of c in positive/negative directions are described by two transversal geodesic laminations $\lambda^{\pm}$ on S, called *ending laminations*: Such laminations contain no closed geodesics and each component of $S \setminus \lambda^{\pm}$ is simply-connected (see [**Kla18**]). In the example where c is the axis of a pseudo-Anosov homeomorphism h, the laminations $\lambda^{\pm}$ are stable/unstable laminations of h (see, e.g., [**CB88**]).

Take $\lambda^{\pm}$, the ending laminations of a pseudo-Anosov homeomorphism of S or, more generally, any two *transversal* ending geodesic laminations on S. There exist discrete embeddings $\rho_{\pm} : \pi \to H$ such that the image $\Gamma_{\pm}$ of each $\rho_{\pm}$ is a singly-degenerate subgroup of H without parabolic elements such that $\lambda^{\pm}$ is the *ending lamination* of the geometrically infinite end[4] $E^{\pm}$ of the hyperbolic manifold $M^{\pm} = \mathbb{H}^3/\Gamma_{\pm}$. Furthermore, there exists a discrete embedding $\rho_0 : \pi \to H$ such that the group $\Gamma_0 = \rho_0(\pi)$ is *doubly-degenerate group*, whose quotient manifold $M_0 = \mathbb{H}^3/\Gamma_0$ has two ends $E_0^{\pm}$ with the ending laminations $\lambda^{\pm}$. We refer the reader to [**Ohs09**] for proofs of more general existence theorems of this type (which are generalizations of Thurston's double limit theorem). The ends $E^{\pm}$ are bi-Lipschitz homeomorphic to the ends $E_0^{\pm}$ of the manifold M_0 (see [**Min94**] or [**BCM12**] for more general results). The manifolds $M_0, M^{\pm}$ have injectivity radii bounded from below and $M^{\pm}$ has structure of a metric bundle over $\mathbb{R}$, whose fibers are uniformly bi-Lipschitz to the surface S. For instance, in the case when $\lambda^{\pm}$ are stable/unstable laminations of a pseudo-Anosov homeomorphism h, the manifold M_0 is isometric to a cyclic covering space of the mapping torus of h, equipped with the unique hyperbolic metric.

We then obtain a discrete and faithful representation

$$\rho : \pi \to \Gamma < G = H \times H, \rho(\gamma) = (\rho_+(\gamma), \rho_-(\gamma)).$$

By the construction, the image of each element of π is a semisimple element of G. Since the projections $\Gamma_{\pm}$ of Γ to the factors $H_{\pm}$ of $H \times H$ are geometrically infinite, the representation ρ is not Anosov. It remains to prove that Γ is undistorted in G, i.e., that the map ρ is a qi embedding. This qi embedding condition can be reformulated as follows. Consider the closed convex hulls $C^{\pm} \subset \mathbb{H}^3$ of the limit sets of the subgroups $\Gamma_{\pm} < H$. The group π acts properly discontinuously, isometrically and cocompactly on the boundaries of these convex hulls. Accordingly, we obtain quasi-isometries

$$\mathbb{H}^2 \to \partial C^{\pm},$$

where the targets are equipped with intrinsic path-metrics. The quotient manifolds $C^{\pm}/\Gamma_{\pm}$ are isometric to the ends $E^{\pm} \subset M^{\pm}$. Since $C^{\pm}$ are isometrically embedded in $\mathbb{H}^3$, Γ is qi embedded in G if and only if the map

$$f : \mathbb{H}^2 \to \mathbb{H}^3 \times \mathbb{H}^3$$

given by the composition of the isometries $f_{\pm} : \mathbb{H}^2 \to \partial C^{\pm}$ with the inclusion maps $\partial C^{\pm} \to C^{\pm} \to \mathbb{H}^3$, is a qi embedding. Since the ends $E^{\pm}$ are bi-Lipschitz homeomorphic to the ends $E_0^{\pm}$ of the manifold M_0, f is a qi embedding if and only if the following holds:

For some (every) Γ_0-invariant embedded simply-connected hypersurface $\Sigma \subset \mathbb{H}^3$ separating $\mathbb{H}^3$ into components $\Sigma^{\pm}$ (equipped with the induced path-metrics), the inclusion

[4] As it is customary in 3-dimensional topology we will be conflating ends and their neighborhoods.

maps $\Sigma \to \Sigma^{\pm}$ combine to a qi embedding

$$\Sigma \to \Sigma^{-} \times \Sigma^{+}.$$

Since $\mathbb{H}^3$ has a Γ-invariant structure of a metric bundle $\mathfrak{X} = (\pi : X \to T = \mathbb{R})$ with fibers uniformly qi to $\mathbb{H}^2$, we just need to prove that for some (every) vertex $v \in T$ and the ideal boundary points $\xi_{\pm}$ of $T = \mathbb{R}$, the inclusion maps $X_v \to X_{T_{\pm}} = X_{v\xi_{\pm}}$ combine to a qi embedding

$$\Phi : X_v \to X_{T_+} \times X_{T_-},$$

where $T_{\pm} = v\xi_{\pm}$, a half-line. We will prove that Φ is indeed a qi embedding (and even more) below, Proposition 8.71.

Suppose $\mathfrak{X} = (\pi : X \to T)$ is a tree of hyperbolic metric spaces with hyperbolic total space X and let $v \in V(T)$ be a vertex of finite degree $n \geq 2$, with edges $e_1, \ldots, e_n$ incident to v. For each $i = 1, \ldots, n$, let T_i denote the subtree in T which is the union of subintervals of the form $[\![v, w]\!]$, containing the edge e_i. We then obtain the subtrees of spaces $\mathfrak{X}_{T_i} = (\pi : X_{T_i} \to T_i)$ in $\mathfrak{X}$. For each i, we let $f_i : X_v \to X_{T_i}$ denote the inclusion map. We equip the product $Q = \prod_{1 \leq i \leq n} X_{T_i}$ with the ℓ_1-metric

$$d_Q(p, q) = \sum_{i=1}^{n} d_{X_{T_i}}(p_i, q_i), \quad p = (p_1, \ldots, p_n), q = (q_1, \ldots, q_n).$$

(One can also use the ℓ_2-metric, the *product* metric: The two metrics are qi to each other.)

In what follows, we take $K = K_*$, $D = D_{5.2}$, $E = E_{5.2}$, depending on the parameters of the tree of spaces $\mathfrak{X}$.

The next proposition is a generalization of a result from [**KS20**], where it was proven in the case when $\mathfrak{X}$ is a metric bundle:

PROPOSITION 8.71. *Under the above assumptions, the diagonal map*

$$\Phi : X_v \to Q = \prod_{1 \leq i \leq n} X_{T_i}, \quad x \mapsto (f_i(x)),$$

is a qi embedding.

PROOF. We note that the inclusion maps $X_v \to X_{T_i}$ are all 1-Lipschitz. Hence, the diagonal map Φ is n-Lipschitz. The proof of the proposition is divided in two cases.

CASE 1. Suppose that $n = 2$. Consider a pair of points $x, y \in X_v$ and let $\mathfrak{L} = \mathfrak{L}(\alpha) = \{\pi : L \to \pi(L)\}$ be a (K, D, E)-ladder centered at v, with $\alpha = [xy]_{X_v} = L_v$. For $i = 1, 2$, we have the (K, D, E)-ladders $\mathfrak{L}^i = \{L^i \to \pi(L) \cap T_i\}$ in $\mathfrak{X}_{T_i}$, obtained by pull-back of the ladder $\mathfrak{L}$ to the subtree of spaces $\mathfrak{X}_{T_i}$.

Let $c = c_L(x, y)$ be a combing path in L connecting x to y. We let $\hat{c}_1, \hat{c}_2$ be the paths in L_1, L_2 respectively, obtained from c via the cut-and-replace procedure with respect to the inclusions $L_1 \to L, L_2 \to L$ (see Definition 8.14). Note that, since L^i is a ladder in X_{T_i}, it is qi embedded in X_{T_i}. Moreover, according to Theorem 8.21, both $\hat{c}_1, \hat{c}_2$ are (uniform) κ-quasigeodesics in X_{T_i}. (Actually, this fact is established in Part I of the proof of Theorem 8.21.)

We will now estimate the length of α from above in terms of the distance between $\Phi(x), \Phi(y)$ in Q. We claim that the segment α is contained in the union $\hat{c}_1 \cup \hat{c}_2$. Indeed, by the definition of combing paths $c = c_L$ in L, there exists a finite monotonic sequence $x_0 = x, x_1, \ldots, x_m = y$ in α such that c is the concatenation of paths $c(x_i, x_{i+1})$ between points x_i, x_{i+1}, such that (after switching the roles of L_1, L_2 if necessary), $c(x_i, x_{i+1})$ is contained in L_1 for odd i and is contained in L_2 for even i. Now, it follows from the definition of the cut-and-replace procedure that $[x_i x_{i+1}]_{X_v} \subset \alpha$ is contained in $\hat{c}_1$ for each odd i and is contained in $\hat{c}_2$ for even i.

Thus,

$$d_{X_v}(x,y) = \text{length}(\alpha) \le \text{length}(\hat{c}_1) + \text{length}(\hat{c}_2) \le$$
$$(\kappa+1)\left(d_{X_{T_1}}(x,y) + d_{X_{T_2}}(x,y)\right) = 2(\kappa+1)d(\Phi(x),\Phi(y)).$$

It follows that Φ is a qi embedding.

CASE 2. Suppose that $n \ge 3$. Consider two points $x, y \in X_v$. Observe that for $p = (x,x), q = (y,y) \in X_{T_1} \times X_{T_2}$, we have

$$d_Q((\underbrace{x,\ldots,x}_{n \text{ times}}),(\underbrace{y,\ldots,y}_{n \text{ times}})) \ge d_{X_{T_1}\times X_{T_2}}(p,q).$$

Therefore, Case 1 implies that the diagonal embedding $\Phi : X_v \to Q$ is a qi embedding.

This concludes the proof of the theorem as well. □

QUESTION 8.72. In the example given in this theorem, is the subgroup $\Gamma < G$ a coarse Lipschitz retract of G?

Note that Anosov subgroups of semisimple Lie groups are coarse Lipschitz retracts; see [**KL18b**].

8.11.2. CT-maps for hyperbolic graphs of groups. In this section, $\mathcal{G}'$ is a finite graph of hyperbolic groups satisfying Axiom **H**, with underlying connected graph Γ' and Bass–Serre tree T'. We will also assume that the group $G' = \pi_1(\mathcal{G}')$ is hyperbolic.

We first prove the existence of CT-maps for some classes of hyperbolic subgroups $G < G'$.

PROPOSITION 8.73. *Suppose that $\Gamma \subset \Gamma'$ is a connected subgraph and $\mathcal{G} \subset \mathcal{G}'$ is the subgraph of groups obtained by restricting $\mathcal{G}'$ to Γ (see Section 2.1), with $G = \pi_1(\mathcal{G})$. Then the subgroup $G < G'$ admits a CT-map $\partial_\infty G \to \partial_\infty G'$.*

PROOF. Let T, T' denote the Bass–Serre trees of $\mathcal{G}, \mathcal{G}'$. The embedding of graphs of groups $\mathcal{G} \hookrightarrow \mathcal{G}'$ induces a G-equivariant embedding $T \hookrightarrow T'$. Since the subgraph of groups $\mathcal{G} \subset \mathcal{G}'$ is obtained by the restriction, for each vertex v and edge e of the subtree T', the stabilizer of v (resp., e) in G equals its stabilizer in G'. Thus, the tree of spaces $\mathfrak{X} = (\pi : X \to T)$ corresponding to the graph of groups $\mathcal{G}$ is obtained as the pull-back of the tree of spaces $\mathfrak{X}' = (\pi : X' \to T'), X = X'_T$. Since the groups G, G' are naturally quasi-isometric to the spaces X, X' (via respective orbit maps) the existence of a CT-extension for the embedding $G' \to G$ is equivalent to that of the embedding $X' \to X$. Since the existence of a CT-map for the inclusion $X' \hookrightarrow X$ is the content of Theorem 8.13, the proposition follows. □

The next theorem shows that one does not need to restrict to subgraphs of $\mathcal{G}'$ to obtain subgroups with CT-maps.

THEOREM 8.74. *Assume that $G < G' = \pi_1(\mathcal{G}')$ is a subgroup preserving a subtree $T \subset T'$ such that the quotient graph T/G is finite and that the vertex and edge stabilizers of this action on T are quasiconvex in the respective subgroups of G': $G_v < G'_v$ and $G_e < G'_e$ are quasiconvex for all $v \in V(T), e \in E(T)$. Then the subgroup G is hyperbolic and the inclusion map $G \to G'$ admits a CT-map.*

PROOF. We will use Proposition 2.30: As in the proof of the proposition we observe that the G-action on T defines a graph-of-groups decomposition of G: $\pi_1(\mathcal{G}) = G$, and the graph of groups $\mathcal{G}$ satisfies Axiom **H** (in view of the quasiconvexity assumptions in the theorem). We let $\mathfrak{X} = (\pi : X \to T)$ and $\mathfrak{X}' = (\pi : X' \to T')$ denote the trees of spaces corresponding to the graphs of groups $\mathcal{G}, \mathcal{G}'$ respectively.

Since G' is hyperbolic, so is X' and, hence, $\mathfrak{X}'$ satisfies the uniformly proper flaring condition. We have a G-equivariant relatively retractive morphism of trees of spaces $\mathfrak{X} \to \mathfrak{X}'$ ($h : X \to X'$, over the inclusion $T \to T'$). The uniformly proper flaring condition for $\mathfrak{X}'$ then implies the uniformly proper flaring condition for $\mathfrak{X}$, hence, X and, thus, G, is also hyperbolic. We let $\mathfrak{Y} = (\pi : X'_T \to T)$ denote the restriction of the tree of spaces $\mathfrak{X}'$ to T. The quasiconvexity assumption for the subgroups $G_v < G'_v, G_e < G'_e$, $v \in V(T), e \in E(T)$, implies that for each $v \in V(T)$, and edge $e = [v, w] \in E(T)$, the G_v-orbit of X'_{ev} is *locally finite* in X'_v; see Lemma 1.7. Thus, Proposition 2.30 implies that the map $h : X \to Y = X'_T$ is a qi embedding. According to Theorem 8.13, the inclusion $Y \to X'$ admits a CT-map $\partial_{Y,X'}$. Composing it with the boundary map of the qi embedding $X \to Y$, we obtain a CT-map $\partial_\infty h$ for the map $h : X \to X'$. Since G acts geometrically on X and G' acts geometrically on X' (see Section 1.3 for the definition and Lemma 1.32), we conclude from the existence of $\partial_{Y,X'}$ the existence of a CT-map for the subgroup $G < G'$. □

EXAMPLE 8.75. Let $G' = F\star_\varphi$ be a hyperbolic group which is the descending HNN extension of a finitely generated free group F via an injective endomorphism $\varphi : F \to F$. Then $G = F\star_{\varphi^n}$ is a hyperbolic subgroup of G' and the embedding $G \to G'$ admits a CT-map.

For a boundary vertex v of a subtree $T \subset T'$, we let $T'(v) \subset T'$ denote the maximal subtree of T' containing v and disjoint from the rest of the vertices of T. Thus, if g is an automorphism of T fixing v and preserving T, it preserves the subtree $T'(v)$ as well.

THEOREM 8.76. *Assume that $G < G'$ are as in Theorem* 8.74 *and that for each boundary vertex v of T in T', the stabilizer $G_v < G$ acts k-acylindrically on the subtree $T'(v) \subset T'$. Then G is a quasiconvex subgroup of G'.*

PROOF. According to Remark 8.11(2), in order to prove the quasiconvexity of G in G' it suffices to show that the CT-map $\partial_{G,G'}$ is injective, i.e., that the CT-lamination $\Lambda(G, G')$ is empty. For the sake of contradiction, suppose that $\Lambda(G, G') \neq \emptyset$. As we observed in the proof of Theorem 8.74, the action of G on the space $Y = X'_T$ is quasiconvex. Since $\Lambda(G, G') \neq \emptyset$, follows that there exists a pair of distinct limit points $z_\pm$ of G in $\partial_\infty Y$ with equal images under $\partial_{Y,X'}$. By Proposition 8.49, there is a bi-infinite vertical geodesic $\alpha \subset X'_v$ (for some $v \in V(T)$) which is a quasigeodesic in Y, such that $z_\pm = \alpha(\pm\infty)$.

LEMMA 8.77. *Suppose that ρ is a geodesic ray in a vertex-space X'_v, $v \in T$, which is also a quasigeodesic ray in Y, such that $z = \rho(\infty)$ is a limit point of G in $\partial_\infty Y$. Then z is a (conical) limit point of the action of G_v on X'_v.*

PROOF. Since the G-action on Y is quasiconvex, the limit point z is a conical limit point (see Definition 1.138). Thus, there is a sequence $g_i \in G$ and a constant r such that for $x = \rho(0)$, $d(g_i x, \rho) \leq r$, and $\lim_{i\to\infty} g_i(x) = z$. At the same time, the G-orbit of X'_v in Y is locally finite (since for $g \in G$, $gX'_v = X'_{gv}$ and each compact in X intersects only finitely many vertex-spaces). Since $g_i(x) \in N_r(X'_v)$, Proposition 1.137 implies that $g_i(x) \in N_R(G_v x)$ for some R independent of i. Hence, for $h_i \in G_v$ such that $d(g_i(x), h_i(x)) \leq R$, we obtain $\lim_{i\to\infty} h_i(x) = z$ in X'_v. □

We now return to the proof of the theorem. By the lemma, the points $z_\pm$ are limit points of the G_v-action on X'_v. Since $\partial_{Y,X'}(z_-) = \partial_{Y,X'}(z_+)$ and $z_\pm \in \partial_\infty X'_v \subset \partial_\infty Y$ are limit points of the G_v-action on X'_v, Theorem 8.63(1) implies that there is a point $\xi \in \partial_\infty T' \setminus \partial_\infty T$ such that $\{z_-, z_+\} \in \Lambda^\xi(X'_v, X')$. By Proposition 8.56(2), for each vertex $w \in V(v\xi)$, $Fl_w(\{z_\pm\}) \neq \emptyset$. Thus, according to Lemma 3.35, for each $x \in X_v$ and vertex $w \in v\xi$, the pair of subsets $G_v x, X'_w$ is not cobounded in X_{vw}. Lemma 2.36 then implies that the G_v-stabilizer of the interval $J_w = [\![v, w]\!]$ is infinite for each $w \in V(v\xi)$. Since $\xi \in \partial_\infty T' \setminus \partial_\infty T$, the intersection of the ray $v\xi$ with the subtree T is a finite interval $[\![v, v']\!]$. The vertex v' is a boundary vertex of T in T'. Let G_{J_w} denote the G-stabilizer of the interval J. Consider a vertex $w \in v\xi$ such that $v' \in V([\![v, w]\!])$ and, moreover, $d_T(v', w) > k$. Then the infinite subgroup $G_J \cap G_v$ fixes the interval $[\![v', w]\!] \subset T'(v')$ of length $> k$, contradicting the hypothesis that the group $G_{v'}$ acts k-acylindrically on the subtree $T'(v')$. □

REMARK 8.78. Assume that $G < G'$ are as in Theorem 8.74.

(1) The subgroup G is at most exponentially distorted in G' since Y is at most exponentially distorted in X' (Corollary 2.20) and the orbit map $o_y : G \to Gy \subset Y$ is a qi embedding.
(2) In the setting of Theorem 8.74, we can drop the assumption of hyperbolicity for G', but assume that for each boundary vertex $v \in T$ and the boundary edge $e = [v, w]$ the stabilizer G_e is finite. Then G is a coarse Lipschitz retract of G', since for each boundary edge e the projection of G_v to G'_e is uniformly bounded, cf. Theorem 2.22 and Proposition 2.30.
(3) In [**BR13**] Baker and Riley construct examples of finitely generated free subgroups G in certain hyperbolic groups G' such that the CT-maps for the inclusions $G \to G'$ do not exist. Their groups G' free-by-cyclic, hence, are isomorphic to fundamental groups of graphs of groups $\mathcal{G}'$ satisfying Axiom **H**. However, in their examples, the intersections of G with vertex/edge subgroups of $\mathcal{G}'$ are not finitely generated.

The next result is a direct group-theoretic application of Proposition 8.49 regarding the nature of CT-laminations for subgraphs of groups:

COROLLARY 8.79. *Suppose that $G < G'$ are as in Proposition 8.73 and $\xi^\pm \in \partial_\infty G$ are distinct points which have the same image in $\partial_\infty G'$ under the CT-map $\partial_\infty G \to \partial_\infty G'$. Then there exists a vertex $v \in T$ and a pair of points $\xi_v^\pm \in \partial_\infty G_v$ such that:*

(1) $\partial_{G_v,G}(\xi_v^\pm) = \xi^\pm$.
(2) *Geodesics in G_v connecting $\xi_v^\pm$ are uniform quasigeodesics in G.*

Here is another application, this time of Theorem 8.52. Consider a hyperbolic group H, an automorphism $f : H \to H$ and the semidirect product $G = H \rtimes_f \mathbb{Z}$. The next result describes when a geodesic in H is a leaf of the CT-lamination $\Lambda(H, G)$. For the formulation of the result we will use the notion of *pseudo-orbits* of the automorphism f, Definition 2.72.

COROLLARY 8.80. *A geodesic α in the Cayley graph of H is a leaf of $\Lambda(H, G)$ if and only if the following holds for some numbers $K \geq 1$ and $C \geq 0$:*

There exist K-pseudo-orbits $(y_i^\pm)$, of $h_{\pm n} = \alpha(\pm n) \in H$ under the automorphism f which approach each other within distance C. More precisely, there exist $i = i(n)$ such that

$$d_H(y_i^+, y_i^-) \leq C.$$

PROOF. We let $\mathfrak{X} = (\pi : X \to \mathfrak{T} = \mathbb{R})$ denote the tree of spaces corresponding to the graph-of-groups structure on G given by the HNN-extension of H via the automorphism f.

Suppose that $\mathfrak{A}^n = (\pi : A^n \to [\![0, w_n]\!])$, $w_n = i = i(n)$, is a (K, C)-narrow carpet in G bounded by $[h_{-n}h_n]_{X_0}$ and $\gamma_n^{\pm}$ denote the K-qi sections corresponding to the top/bottom of the carpet $\mathfrak{A}^n$. By the definition of a (K, C)-narrow carpet,

$$d_{t^iH}(\gamma_n^-(i), \gamma_n^+(i)) \le C.$$

As it was explained in Section 2.7, for each n, the sequences $y_j^{\pm} = \gamma_n^{\pm}(j)$ are precisely the partial pseudo-orbits of f in H through the points $h_{\pm n}$. Thus, the claim is a direct consequence of Theorem 8.52. □

REMARK 8.81. One can show that α as above is a leaf of $\Lambda(H, G)$ if and only if there are sequences $h_{\pm n}$ in H converging to $\alpha(\pm\infty)$ (but not necessarily of the form $\alpha(\pm n)$), such that the f-orbits of $h_{\pm n}$ approach each other within distance C, where C is a uniform constant depending only on the group H, its generating set, and f.

8.11.3. Miscellaneous results. The next proposition is a partial converse to Proposition 8.67.

PROPOSITION 8.82. *Suppose $\mathcal{G}$ is a finite graph of hyperbolic groups satisfying Axiom **H** and $G = \pi_1(\mathcal{G})$ is hyperbolic. Suppose $X \to T$ is the tree of spaces associated to this graph of groups. If v is a vertex of T and $z \in \partial_\infty X_v$ is such that the subtree $T_z = \pi(Fl(\{z\}))$ contains no geodesic rays, then $\partial_{X_v,X}(z)$ is a conical limit point of X_v in X.*

PROOF. By the definition, $T_z \subset T$ is the subtree whose vertex set consists of those vertices $w \in V(T)$ for which $Fl_w(z) \neq \emptyset$.

LEMMA 8.83. *The subtree T_z is finite.*

PROOF. Since T_z contains no rays, it suffices to prove that the tree T_z is locally finite. Consider a vertex $w \in T_z$ and the collection of edges $e_i, i \in I$, in T_z incident to w. Let $Fl_w(\{z\}) = \{z'\}$, $z' \in \partial_\infty X_w$. Then, according to the definition of the boundary flow in Section 3.3.4, for each edge e_i we have $Fl_{e_i}(\{z'\}) \neq \emptyset$. It follows that $z' \in \partial_\infty(G_{e_i}, G_w)$ for each $i \in I$. Since $\mathcal{G}$ is a finite graph of groups, there are only finitely many G_w-conjugacy classes of edge-stabilizers $G_{e_i} < G_w$. At the same time,

$$z' \in \bigcap_{i \in I} \partial_\infty(G_{e_i}, G_w) \neq \emptyset,$$

hence (since each subgroup G_{e_i} is quasiconvex in G_w) each intersection $G_{e_i} \cap G_{e_j}$ is an infinite subgroup of G_w; see, e.g., [**GMRS98**, Lemma 2.6]. The main theorem in [**GMRS98**] states that *quasiconvex subgroups of hyperbolic groups have finite width.* Without defining *width of subgroups* here, we only note that, as a consequence of this finiteness theorem, if $H_i, i \in I$, is a collection of pairwise distinct quasiconvex subgroups of a hyperbolic group H which belong to finitely many H-conjugacy classes and $|H_i \cap H_j| = \infty$ for all $i, j \in I$, then I is finite. Applying this result in our setting, with the ambient hyperbolic group H equal G_w and quasiconvex subgroups H_i equal to the edge-subgroups G_{e_i}, we conclude that the set I is finite. Thus, T_z is a locally finite tree and, hence, is finite. □

We can now prove the proposition. In the proof it will be convenient to assume that each edge-space X_e of $\mathfrak{X}$ is discrete, cf. introduction to Section 8.7. Let β be a ray in X asymptotic to the point $\partial_{X_v,X}(z)$. Suppose for a moment that the intersection $\beta \cap X_{T_z}$ is bounded. Then there exists a boundary edge $e = [v', w]$ of T_z, $w \notin V(T_z)$, $v' \in V(T_z)$, such that an unbounded subray β' of β projects to the subtree $T_{w,z} \subset T$ which is the maximal subtree of T containing w and disjoint from T_z. (Here we are using the discreteness assumption on the edge-spaces of $\mathfrak{X}$.) Let K be as in Proposition 8.56 and let $\alpha \subset X_v$ be a geodesic ray asymptotic to z.

Since $w \notin T_z = \pi(Fl(\{z\}))$, the intersection $Fl_K(\alpha) \cap X_w$ is bounded, cf. Lemma 3.35. Recall that the flow-space $Fl_K(\alpha)$ is quasiconvex in X. Hence, we will identify $\partial_\infty Fl_K(\alpha)$ with a subset of $\partial_\infty X$. We claim that

$$\partial_{\alpha, Fl_K(\alpha)}(z) \neq \beta(\infty) \in \partial_\infty Fl_K(\alpha).$$

Indeed, the assumption that $\pi(\beta') \subset T_{w,z}$ implies that each geodesic in $Fl_K(\alpha)$ connecting points of α to that of β' has to pass through $Fl_K(\alpha) \cap X_w$, i.e., within distance D from $p = \alpha(0)$, where $D = \mathrm{Hd}(\{p\}, Fl_K(\alpha) \cap X_w)$. But this means that the sequences $(\alpha(n))$, $(\beta(n))$ cannot define the same ideal boundary point of $\partial_\infty Fl_K(\alpha)$. Quasiconvexity of $Fl_K(\alpha) \subset X$ implies that $\partial_{X_v,X}(z) \neq \beta(\infty)$, which is a contradiction. Thus, β contains an unbounded sequence of points (x_n) contained in $Y := Fl_K(\alpha) \cap X_{T_z}$. Since (by the lemma) the subtree T_z is finite, the subset Y is Hausdorff-close to the ray $\alpha \subset X_v$. In other words, the point $\partial_{X_v,X}(z) = \beta(\infty)$ is a conical limit point of X_v. □

We conclude the chapter with a proposition that deals with the case of *nonhyperbolic* graphs of hyperbolic groups and relates this lack of hyperbolicity to various notions discussed earlier, such as boundary flow-spaces and unbounded sequences of carpets.

PROPOSITION 8.84. *Suppose $\mathcal{G}$ is a finite graph of hyperbolic groups satisfying Axiom **H**; let $\mathfrak{X} = (\pi : X \to T)$ denote the corresponding tree of metric spaces. If $G = \pi_1(\mathcal{G})$ is not hyperbolic, then there is a vertex $v \in V(T)$ such that $\pi(Fl(\partial_\infty X_v))$ contains a geodesic ray.*

PROOF. Let G be the fundamental group of $\mathcal{G}$. By Corollary 2.53(2) there is a constant $D > 0$, a sequence of intervals $I_n = [\![-t_n, t_n]\!] \subset T$ of length at least n and a sequence (Π^n) of pairs of κ-qi sections (γ_0^n, γ_1^n) over I_n ($n \in \mathbb{N}$), such that $\Pi^n_{max} \leq D$, but $\Pi^n_0 \to \infty$. Since the G-action is cofinite on T and cocompact on X, and the map $\pi : X \to T$ is G-equivariant, after extraction, we may assume that for each n the midpoint of I_n is a fixed vertex $v \in V(T)$ and the midpoint of $[\gamma_0^n(0)\gamma_1^n(0)]_{X_v}$ is within unit distance from a fixed point $x \in X_v$. After passing to a further subsequence, we may assume that the sequence of segments $[\gamma_0^n(0)\gamma_1^n(0)]_{X_v}$ converges to a complete geodesic in X_v. We note that for each n the vertical geodesic segments $[\gamma_0^n(t)\gamma_1^n(t)]_{X_t}, t \in V(I_n)$, form a K-metric bundle over I_n (with $K = \kappa'$; see Lemma 3.17) and hence there is a K-qi section γ^n over I_n passing through $B(x, 2)$ and contained in this metric bundle. Hence, after passing to a further subsequence, we may assume that the sequence of sections (γ^n) converges to a complete quasigeodesic γ in X and the sequence I_n converges of complete geodesic I in T such that γ is a qi section over I, cf. the proof of Lemma 8.60(2). Lastly, Lemma 3.35(3) implies that $I \subset \pi(Fl(\partial_\infty X_v))$. □

CHAPTER 9

Cannon–Thurston maps for relatively hyperbolic spaces

The goal of this chapter is to generalize the results in Chapter 8 (primarily, the existence of CT-maps and some basic facts about CT-laminations) in the context of *relatively hyperbolic spaces*. Such a generalization was achieved in [**MP11**] for the inclusion maps of vertex-spaces.

Relatively hyperbolic spaces are generalizations of the following object: Start with the classical real-hyperbolic space $\mathbb{H}^n$ and remove from it a collection of open horoballs $\mathring{B}_i, i \in I$, such that any two distinct horoballs are ϵ-separated for some uniform $\epsilon > 0$. The result is a pair $(Y, \mathcal{H})$, where Y is the result of removal of horoballs (a *truncated hyperbolic space*) and $\mathcal{H}$ is the collection of boundary spheres H_i of the horoballs B_i; these are the *peripheral horospheres*. The pair $(Y, \mathcal{H})$ is a basic example of a relatively hyperbolic space. Attaching hyperbolic cones H_i^h to each horosphere H_i results in a geodesic metric space quasi-isometric to $\mathbb{H}^n$, it is the *hyperbolification* of $(Y, \mathcal{H})$. In parallel, one defines the *electrification* Y^ℓ of $(Y, \mathcal{H})$ by attaching to Y cones along each horosphere H_i. The interplay between the geometry of Y (with the induced path-metric), Y^h and Y^ℓ is one of the main tools of the theory of relatively hyperbolic spaces. In Sections 9.1, 9.2 we cover basic definitions of the theory of relatively hyperbolic spaces, one (of the many!) alternative definition and tools of the theory. We also define Cannon–Thurston maps between *morphisms* of relatively hyperbolic spaces.

In Section 9.3 we introduce the notion of *trees of relatively hyperbolic spaces* (still denoted $\mathfrak{X}$), over simplicial trees T. The key difference with the notion of ordinary trees of spaces is that the incidence maps f_{ev} are required to be morphisms of relatively hyperbolic spaces which are *relative qi embeddings*. There are two ways to *electrify* the total space X of a tree of relatively hyperbolic spaces: One is to electrify each vertex, edge-space, and incidence map, and then define the total space X^ℓ of the resulting tree of hyperbolic spaces $\mathfrak{X}^\ell$. The other way is to form trees of spaces over subtrees in T, whose vertex and edge-spaces are peripheral horospheres; the total spaces of these trees of horospheres are *peripheral subspaces* in X. Then one forms a space $\widehat{X}$ by coning-off these peripheral subspaces in X. Similarly, there are two ways to *hyperbolify* X. One is to hyperbolify each vertex and edge-space of X; this results in a tree of hyperbolic spaces with the total space X^h. The other way is to hyperbolify X by attaching hyperbolic cones along each peripheral subspace of X; the result is a space X^P.

Following Mj and Reeves we then define a *relative flaring condition* for trees of relatively hyperbolic spaces which combines flaring for $\mathfrak{X}^\ell$ with another condition in terms of apexes of cones in the vertex and edge-spaces of $\mathfrak{X}^\ell$. According to a theorem by Mj and Reeves (which we *do not reprove* in this book), under the relative flaring condition, the space X^P is hyperbolic.

We are now ready to state the main results of the chapter:

(1) In Theorem 9.47 we prove an existence theorem for CT-maps in the relative setting: For each subtree $S \subset T$, the inclusion map $X_S^P \to X^P$ admits a CT extension $\partial_{X_S^P, X^P}$. Here X_S is the total space of a subtree of relatively hyperbolic spaces induced from $\mathfrak{X}$ by the inclusion $S \to T$.

This theorem generalizes an earlier result by Mj and Pal who proved it in the case when S is a single vertex. The proof of this theorem occupies a part of Section 9.3 and Sections 9.3.2, 9.4. Our proof depends critically on the results of Chapter 8 establishing existence of CT-maps in the case of trees of hyperbolic spaces.

(2) In Section 9.5 (the last section of the chapter) we prove Theorem 9.60. This theorem establishing certain basic properties of CT-laminations for the maps $\partial_{X_S^P, X^P}$, which are analogous to those in Theorem 8.52 in the absolute case. Mainly, Theorem 9.60 proves that the leaves of the CT-lamination for the map $\partial_{X_S^P, X^P}$ come from leaves of CT-laminations of maps $\partial_{X_v^P, X^P}$, for various vertices in T. Unlike Chapter 8 (Section 8.9), we do not relate CT-laminations in the relative setting to ending laminations.

9.1. Relative hyperbolicity

9.1.1. Relative hyperbolicity in the sense of Gromov. We refer the reader to the papers by Farb and Bowditch [**Far98, Bow12**] for the background on the theory of relatively hyperbolic spaces. Briefly, a *relatively hyperbolic space* is a pair $(Y, \mathcal{H})$ consisting of a geodesic metric space (Y, d) together with a collection $\mathcal{H} = \{H_i : i \in I\}$ of *peripheral subspaces*, which are nonempty subsets of Y satisfying certain conditions discussed below and, in an alternative form, in Section 9.2.1. While this is not always required for relatively hyperbolic spaces, we will assume that each $H_i \in \mathcal{H}$ is rectifiably connected and the inclusion maps $(H_i, d_{H_i}) \to (Y, d_Y)$ are uniformly proper, where d_{H_i} are the intrinsic path-metrics on H_i. Given such a pair $(Y, \mathcal{H})$, one defines two new metric spaces:

(1) The *extended hyperbolic space* $(Y^h, d^h) = (Y^h, d_{Y^h}) = \mathcal{G}(Y, \mathcal{H})$ (or the *horoballification* of $(Y, \mathcal{H})$), which is a path-metric space obtained by attaching along each H_i its *hyperbolic cone* H_i^h defined in Section 1.11; the latter are (intrinsically) *uniformly hyperbolic horoballs*, with H_i the boundary horosphere in H_i^h. Recall that each H_i^h has unique ideal boundary point, called the *foot-point* (or the *ideal center*) of H_i^h, $\xi(H_i^h)$.

(2) The *electric space* $(Y^\ell, d^\ell) = \mathcal{E}(Y, \mathcal{H})$ (the *electrification* of $(Y, \mathcal{H})$), obtained by coning off each H_i, i.e., attaching to Y along each H_i the cone $H_i^\ell = C(a_i, H_i)$ with the apex $a_i = a(H_i^\ell)$ within distance $1/2$ from each point in H_i. We refer the reader to Section 1.7 for the precise definition of the metric spaces $C(a_i, H_i)$. Here we recall only that each point $x \in H_i$ is connected to a_i by a canonical geodesic segment of length $1/2$, called a *radial line-segment*. The set of apexes of these cones is called the *cone-locus* of Y^ℓ and denoted $a(Y^\ell)$.

We will use the notation $\mathring{H}_i^h$ (the *open peripheral horoball*) and $\mathring{H}_i^\ell$ (the *open peripheral cone*) for the complements

$$H_i^h \setminus H_i, \quad H_i^\ell \setminus H_i$$

respectively.

DEFINITION 9.1 (Relative hyperbolicity in the sense of Gromov). A pair $(Y, \mathcal{H})$ is called *relatively hyperbolic in the sense of Gromov (GRH)* if the metric space (Y^h, d^h) is hyperbolic.

If $(Y, \mathcal{H})$ is GRH, then the electric space (Y^ℓ, d^ℓ) is hyperbolic. This is a standard fact, usually attributed to Farb, **[Far98]**: Although the proofs given in his paper are only in the setting of manifolds of negative curvature, they go through in greater generality. We will give a proof in Section 9.2 with the side-benefit of relating quasigeodesics in (Y^h, d^h) to those in (Y^ℓ, d^ℓ).

REMARK 9.2. There are many characterizations of relatively hyperbolic spaces and groups; see **[Bum05, Osi06, DS05, Hru10, Bow12, Far98, Ger09, GP15, GP16, Yam04]**. In this book we discuss just two definitions of relative hyperbolicity, Gromov's and Farb's.

9.1.2. Extrinsic geometry of the peripheral horoballs. Throughout this section we will assume that $(Y, \mathcal{H})$ is GRH with the hyperbolicity constant δ. Our goal is to discuss the *extrinsic* geometry of the peripheral horoballs H_i^h. Among other things, we will prove that they are uniformly quasiconvex and uniformly pairwise cobounded.

LEMMA 9.3.

(1) *The subsets H_i^h are $\lambda_{9.3}(\delta)$-quasiconvex and $L_{9.3}(\delta)$-qi embedded in Y.*
(2) *Y is uniformly properly embedded in Y^h with distortion function depending only on δ.*

PROOF.

(1) We observe that, by the definition of the metric on H_i^h, for each point $x = (z, t) \in H_i^h$ the distance from x to H_i equals $\log(t)$. It follows that for any two points $x_1 = (z, t_1), x_2 = (z, t_2) \in H_i^h$,

$$d_{Y^h}(x_1, x_2) = |\log(t_1/t_2)|$$

and the vertical segment in H_i^h between x_1, x_2 is isometrically embedded in Y^h. In particular, the vertical rays in H_i^h are isometrically embedded in Y^h. Since any two such rays $\rho_1(t), \rho_2(t)$ converge as $t \to \infty$, it follows that the horoballs H_i^h are uniformly quasiconvex in Y^h. The fact that the peripheral subspaces H_i are uniformly properly embedded in Y implies that the horoballs H_i^h are uniformly properly embedded in Y^h. Combined with the uniform quasiconvexity, we obtain that these horoballs are uniformly qi embedded in Y^h.

(2) Lastly, uniform properness of the inclusion maps $H_i \to H_i^h$ (see Proposition 1.71) implies that Y is uniformly properly embedded in Y^h as well. □

Recall that a closed subset C of a geodesic metric space X is called *strictly convex*; see, e.g., **[BH99]**, if for any two points $x, y \in C$, every geodesic $xy \subset X$ is contained in the interior of C, except maybe for its endpoints. For instance, closed balls and horoballs in the classical hyperbolic space are strictly convex, while a closed hyperbolic half-space is not. The next lemma establishes a form of coarse strict convexity of peripheral horoballs in the context of GRH spaces.

LEMMA 9.4. *There exist $L = L_{9.4}(K, r, \delta)$ and $R = R_{9.4}(K, r, \delta)$ satisfying the following properties. Suppose that $\beta : [a, b] \to Y^h$ is a continuous K-quasigeodesic in Y^h connecting points $z = \beta(a), y = \beta(b)$.*

(1) *Suppose that the image of β is entirely contained in $N_r(H)$ for some $H \in \mathcal{H}$. Then $d(y, z) \leq R = R_{9.4}(K, r, \delta)$.*

(2) *Suppose that the points* $z = \beta(a), y = \beta(b)$ *both belong to* $N_r(H^h) \setminus \mathring{H}^h$. *Then either* $b - a \leq 2L$ *or there exist* $a' \in [a, a + L]$ *and* $b' \in [b - L, b]$ *such that* $z' = \beta(a') \in H$, $y' = \beta(b') \in H$ *and the subpath* $\beta(z', y')$ *in* $\beta(z, y)$ *is entirely contained in* $\mathring{H}^h$, *except for the endpoints* z', y'.

PROOF.

(1) Let $\bar{y}, \bar{z} \in H^h$ be the images of y, z, respectively, under the projection P_{Y^h, H^h}. Thus, $d(y, \bar{y}) \leq r, d(z, \bar{z}) \leq r$. Let $c = c(\bar{z}, \bar{y})$ denote the combing path in H^h connecting $\bar{z}$ to $\bar{y}$, as defined in the proof of Proposition 1.68. By the version of stability of quasigeodesics in hyperbolic spaces (Lemma 1.55), the Hausdorff distance between the images of β and c is at most $D_{1.55}(\delta, \max(k, K), r)$, where k is the bound on quasi-isometry constant of c given in Remark 1.69.

We will consider the case when $\bar{x}, \bar{y}$ are both in H and leave the other cases to the reader as the proofs are similar. According to the description of the combing paths in H^h, for each constant $D \geq 0$ if the path c is contained in the D-neighborhood of H, then $A = d_H(\bar{z}, \bar{y}) \leq e^D$.

In our case, c is contained in the $D = (D_{1.55}(\delta, \max(k, K), r) + r)$-neighborhood of H and, hence, we conclude that $d_H(\bar{z}, \bar{y}) \leq e^D$. It follows that

$$d(y, z) \leq R_{9.4}(K, r, \delta) := e^D + 2r.$$

(2) We follow the arguments of Part 1, define points $\bar{y}, \bar{z}$ and the path c connecting them, setting now $D := D_{1.55}(\delta, \max(k, K), r)$, an upper bound on the Hausdorff distance between β and c. If $d_H(\bar{z}, \bar{y}) > e^D$, then there are points $\bar{y}', \bar{z}'$ within distance D from $\bar{y}, \bar{z}$ respectively, such that the subpath $c(\bar{z}, \bar{y})$ is disjoint from the D-neighborhood of H, apart from the endpoints. Thus, there are points y'', z'' in the image of β which are within distance D from, respectively, $\bar{y}', \bar{z}'$, such that the subpath $\beta(z'', y'')$ is entirely contained in $\mathring{H}^h$, except, possibly, the endpoints. We then take the points y', z' on the subpaths $\beta(y, y''), \beta(z, z'')$ where these paths cross into H^h and such that $\beta(z', y')$ is entirely contained in $\mathring{H}^h$, except for the endpoints x', y'. The lemma follows. □

DEFINITION 9.5. In what follows, we will use the notation $[a', b']$ for the maximal subinterval in $[a, b]$ satisfying the properties in Part 2 of the lemma.

COROLLARY 9.6. *Suppose that* $y, z \in N_r(H^h)$ *and the geodesic* $[yz]_{Y^h}$ *is disjoint from* $\mathring{H}^h$. *Then* $d(y, z) \leq 2L_{9.4}(1, r, \delta)$.

In the next corollary we use the notation introduced in Definition 9.5:

COROLLARY 9.7. *Suppose that* H_i, H_j *are distinct elements of* $\mathcal{H}$, $\beta : [0, T] \to Y$ *is a* K-*quasigeodesic*, $[s_i, t_i], [s_j, t_j]$ *are subintervals of length* $> 2L_{9.4}(K, r, \delta)$ *in* $[0, T]$ *such that* $\beta(s_i) \in H_i, \beta(t_i) \in H_i, \beta(s_j) \in H_j, \beta(t_j) \in H_j$. *Then* $[s_i', t_i'] \cap [s_j', t_j'] = \emptyset$.

LEMMA 9.8. *Consider three points,* $x \in H^h, z \in H$, *and* $y \in Y^h$, *and* K-*quasigeodesics* $\alpha = \alpha(z, x), \beta = \beta(y, z)$ *in* Y^h *connecting* z *to* x *and* y *to* z *respectively, such that* β *is disjoint from* $\mathring{H}^h$ *and* α *is contained in* H^h. *(See Figure 9.1.) Then the concatenation* $\beta \star \alpha$ *is an* $L_{9.8}(K, \delta)$-*quasigeodesic in* Y^h.

PROOF. We set $D := D_{1.54}(\delta, K), R := R_{9.4}(1, D, \delta)+1$. Note that $D > 2\delta$; see Lemma 1.54.

In view of stability of quasigeodesics, it suffices to get an upper bound on the qi constant of the concatenation $\beta^* \star \alpha^*$, where $\alpha^* = [zx]_{Y^h}, \beta^* = [yz]_{Y^h}$ (geodesics which are

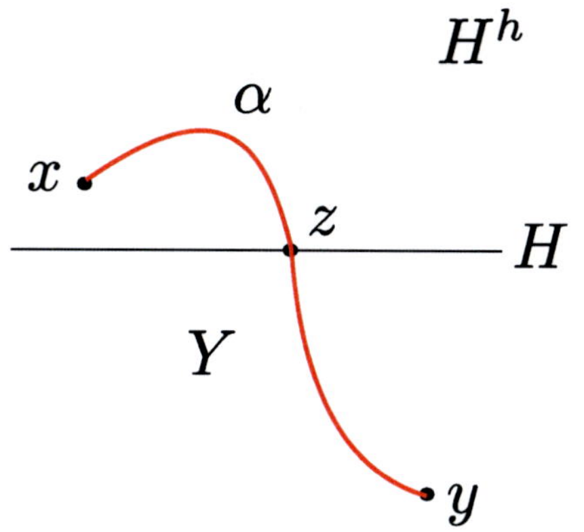

FIGURE 9.1

D-Hausdorff close to α, β, respectively, where $D = D_{1.54}(\delta, K)$). By the assumption of the corollary, β^* is contained in $N_D(Y^h \setminus H^h)$. There are two cases to consider:

(a) $d_{Y^h}(y, z) < R$ or $d_{Y^h}(x, z) < R$. Then the claim follows immediately: A concatenation of a uniform quasigeodesic with a uniformly bounded quasigeodesic is again a uniform quasigeodesic.

(b) $R \leq \min(d(x, z), d(y, z))$. Then, by Lemma 9.4(1), there exists a point $z' \in \beta^*$ at the distance $R' \leq R$ from z such that $z' \notin N_D(H^h)$. Take the point $x' \in \alpha^* \subset H^h$ at the same distance R' from z. Then

$$d(z', x') \geq D > 2\delta.$$

By Lemma 1.79, the concatenation $\beta^* \star \alpha^*$ is then an $L_{1.79}(R', \delta)$-quasigeodesic in Y^h. □

In the next lemma we prove a generalization of this result, for concatenations of three uniform quasigeodesics.

LEMMA 9.9. *Suppose we have three K-quasigeodesics $\beta(x_i, y_i)$, $\beta(y_i, y_j)$ and $\beta(y_j, x_j)$ in Y^h such that the paths $\beta(x_k, y_k)$ are contained in the distinct peripheral horoballs H_k^h, and $\beta(y_i, y_j) \cap \mathring{H}_k^h = \emptyset$, $k = i, j$. (See Figure 9.2.) Then the concatenation*

$$\beta(x_i, y_i) \star \beta(y_i, y_j) \star \beta(y_j, x_j)$$

is again a (uniform) $L_{9.9}(K, \delta)$-quasigeodesic in Y^h.

PROOF. We first consider the concatenation $\beta(y_i, y_j) \star \beta(y_j, x_j)$ and observe that it satisfies the assumptions of Lemma 9.8 with the peripheral horoball H^h equal H_j^h. Thus, $\beta = \beta(y_i, y_j) \star \beta(y_j, x_j)$ is a $K_1 = L_{9.8}(K, \delta)$-quasigeodesic in Y^h. Then we consider the concatenation $\alpha \star \beta$, where $\alpha = \beta(x_i, y_i)$. This concatenation again satisfies the assumptions of Lemma 9.8 with the peripheral horoball H^h equal H_i^h. The lemma follows. □

We now analyze the nearest-point projection to quasigeodesics of the type described in Lemma 9.8:

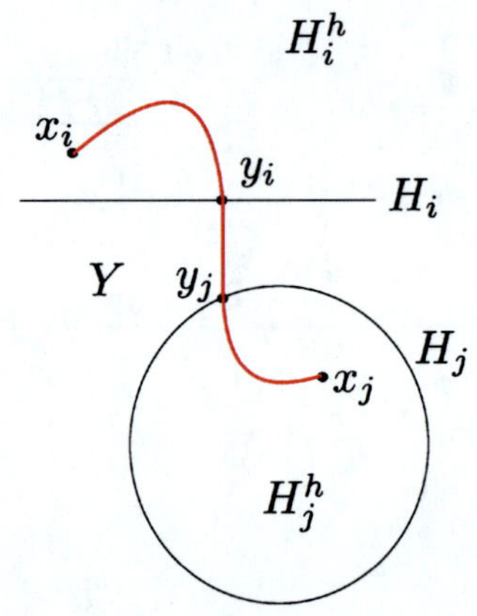

FIGURE 9.2

LEMMA 9.10. *Consider a quasigeodesic $\gamma = \beta \star \alpha$ as in Lemma 9.8. Then for each point $p \in H^h$ the nearest-point projection $\bar{p} = P_\gamma(z)$ satisfies:*

$$d(\bar{p}, \alpha) \le C_{9.10}(K).$$

PROOF. We continue with the notation of Lemma 9.8. Suppose that $\bar{p} \in \beta$. Since γ is a K-quasigeodesic, it is $\lambda = \lambda_{1.92}(\delta, K)$-quasiconvex; hence (see Lemma 1.105), the geodesic pz in Y^h passes within distance $\le D = \lambda + 2\delta$ from $\bar{p}$. In particular, for

$$r = D + \lambda_{9.3}(\delta),$$

$d(\bar{p}, H^h) \le r$, i.e., $\bar{p} \in N_r(H)$. We also have $p \in H$ and $\beta \cap \mathring{H} = \emptyset$. Hence, according to Lemma 9.4(2) the distance between $\bar{p}$ and z is $\le 2K$. □

LEMMA 9.11. *The peripheral horoballs $H_i^h, i \in I$, are uniformly pairwise cobounded in Y^h.*

PROOF. Consider two distinct peripheral horoballs H_i^h, H_j^h and suppose that $z_i \in H_i, z_j \in H_j$ are within distance c from each other in Y^h. The points z_i, z_j are connected by a path β of length $\le K = 4e^c$ in Y. Hence, β is a K-quasigeodesic in Y^h. Let ρ_i, ρ_j denote the vertical geodesic rays in H_i^h, H_j^h emanating from z_i, z_j respectively. Then, by Lemma 9.9, the concatenation α of the rays ρ_i, ρ_j and the path β is an $L = L_{9.9}(K, \delta)$-quasigeodesic in Y^h.

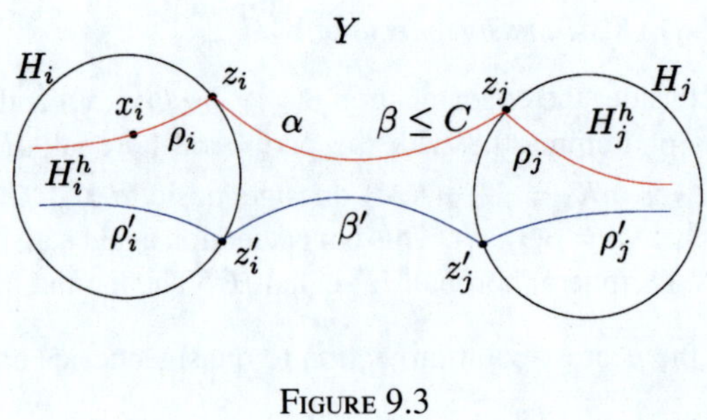

FIGURE 9.3

Suppose now that we have four points $z_i, z_i' \in H_i, z_j, z_j' \in H_j$ satisfying $d_{Y^h}(z_i, z_j) \le c$, $d_{Y^h}(z_i', z_j') \le c$. We then, as above, form two L-quasigeodesic lines α, α' in Y^h passing through the points z_i, z_j, and z_i', z_j' respectively and asymptotic to the centers $\xi(H_i^h)$, $\xi(H_j^h)$ of our peripheral horoballs. Thus, the δ-hyperbolicity of Y^h, and the fact that α, α' are L-quasigeodesics in Y^h asymptotic to the same pairs of points at infinity, imply that the Hausdorff distance between these quasigeodesics is $\le D = D_{1.87}(L, \delta)$; see Lemma 1.87.

We claim a uniform upper bound on the distance between z_i, z_i'. By Lemma 9.10, there exists a point $x_i \in \rho_i$ within distance $C = C_{9.10}(L)$ from the nearest-point projection $\bar{z}_i'$ of z_i to α'. Hence,

$$d(z_i', x_i) \le D' = D + C.$$

By the upper bound $\mathrm{Hd}(\alpha, \alpha') \le D$, in view of Lemma 9.10, there exists a point $x_i \in \rho_i$ within distance $D' = D'_{9.10}(L, D)$ from z_i'. Since z_i is the point in H_i closest to x_i, it follows that $d(x_i, z_i) \le D'$ as well. Combining the inequalities, we get $d(z_i, z_i') \le 2D'$. Similarly, $d(z_j, z_j') \le 2D'$. Uniform quasiconvexity of the horoballs H_i^h, H_j^h combined with Proposition 1.140 now implies that these horoballs are uniformly cobounded. □

LEMMA 9.12. *Every point $\xi \in \partial_\infty Y^h$ is the limit of a sequence (z_n) in Y, unless ξ is the center of a horoball H^h with bounded horosphere H.*

PROOF. There are two cases to consider.

(1) $\xi = \xi(H^h)$ is the center of a peripheral horoball H^h. Fix a base-point $y_0 \in H$ and the vertical ray ρ in H^h emanating from y_0. Since H is unbounded, take a sequence $z_n \in H$ which diverges to infinity, i.e., $D_n = d_H(y_0, z_n) \to \infty$. We claim that (z_n) converges to ξ. Indeed, for $y_n := (z_n, \log(D_n)) \in H^h$, the combing path $c(y_0, y_n)$ in H^h contains the subsegment of length $\log(D_n)$ in ρ. Hence, $\lim_{n\to\infty} z_n = \lim_{n\to\infty} \rho(\log(D_n)) = \xi$.

(2) ξ is not the center of any peripheral horoball. Then each geodesic ray $\rho = y\xi$ will cross into Y along a sequence z_n which diverges to infinity. Thus, (z_n) converges to ξ. □

9.1.3. Electrification and hyperbolization of quasigeodesics. In this section we describe two procedures of converting paths in Y^h to paths in Y^ℓ and vice-versa.

We first describe the procedure of *electrification* of continuous hyperbolic quasigeodesics. In view of Corollary 9.7, for each continuous K-quasigeodesic $\beta : [0, T] \to Y^h$ we obtain a maximal collection of maximal pairwise disjoint subintervals $[s_1', t_1'], \ldots, [s_n', t_n']$ in $[0, T]$ satisfying the conclusion of Corollary 9.7 with respect to certain peripheral subspaces $H_{j_1}, \ldots, H_{j_n} \in \mathcal{H}$. We then perform the following *electrification* procedure on β:

For each subinterval $[s_i', t_i']$ we replace the restriction of β to $[s_i', t_i']$ with the concatenation of two geodesic segments (of length $1/2$ each) in $H_{j_i}^\ell$ connecting $\beta(s_i'), \beta(t_i')$ to the apex $a(H_{j_i}^\ell)$ of the cone $H_{j_i}^\ell$. In the special case when we have to deal with the subintervals $[0, t_1']$ and/or $[s_n', T']$ (i.e, $\beta(0)$ or $\beta(T)$ belongs to one of the open horoballs $\mathring{H}_j^h$), we replace $\beta|_{[0,t_1']}$ (resp., $\beta|_{[s_n',T]}$) with the unit length geodesic in H_j^ℓ connecting the apex $a(H_j^\ell)$ to $\beta(t_1')$ (resp., $\beta(s_n')$). We let β^ℓ denote the resulting path in X^ℓ. Lastly, each subpath of β whose domain has length $\le 2L(K)$ and which connects points of H_i in H_i^h, is replaced by a geodesic in H_i connecting the same points. Lengths of such geodesics are uniformly bounded by some constant $E = E(K)$ since horospheres H_i are uniformly properly embedded in the horoballs H_i^h. The resulting map $\beta^\ell = \mathcal{E}_{\mathcal{P}}(\beta)$ will be called the *electrification* of β.

We note that β^ℓ visits each cone-point $a(H^\ell_{j_i}), i = 1, \ldots, n$, exactly once; more precisely, β^ℓ is *tight* in the following sense.

DEFINITION 9.13. A continuous path γ in Y^ℓ is *tight* if for each cone H^ℓ_i, $i \in I$, the pre-image $\gamma^{-1}(\mathring{H}^\ell_i)$ is a (possibly empty) interval and the restriction of γ to this interval is 1-1.

REMARK 9.14.

(1) The commonly used name for tight paths is *paths without backtracking*; we find this terminology cumbersome.
(2) The continuity assumption here is simply a matter of convenience and electrification can be defined for general K-quasigeodesics by using s_i, t_i such that $\beta(s_i), \beta(t_i)$ are within distance K from the horosphere H_i.
(3) The electrification procedure described in **[DM17]** is similar, except they replace by geodesics in H^ℓ_i the subsegments in β with endpoint in H_i at distance > 1. This construction, while technically simpler, does not result in tight paths, which we find undesirable.

More generally, we define tightness of paths in Y^h as follows:

DEFINITION 9.15. We say that a continuous path β in Y^h is *tight* (relative to $\mathcal{H}$) if for each peripheral horoball $H \in \mathcal{H}$, the pre-image $\beta^{-1}(\mathring{H}^h)$ is an open interval (possibly empty).

It is clear that for each tight path β, its electrification β^ℓ is a tight path in Y^ℓ.

We next use electrification to compare distances in Y^h and in Y^ℓ:

LEMMA 9.16. *For any two pair of points $x, y \in Y$, $d^\ell(x, y) \leq d_{Y^h}(x, y)$, i.e., the inclusion map $(Y, d_{Y^h}) \to (Y^\ell, d^\ell)$ is 1-Lipschitz.*

PROOF. Recall that by the definition of the metric on the hyperbolic cones H^h_i, if x, y belong to the same peripheral subspace H_i and $d_{Y^h}(x, y) \leq 1$, then $d_Y(x, y) \leq 1$. It follows that if β is a geodesic in Y^h connecting the points $x, y \in Y$ then the length of β^ℓ is at most the length of β. □

Hyperbolization of electric geodesics. Conversely, given any (continuous) path β in Y^ℓ, we define its *hyperbolization* β^h by replacing each subpath β_H of β connecting $x, y \in H \in \mathcal{H}$ and contained in $\mathring{H}^\ell$ except for its endpoints, with the combing path $c(x, y)$ in H^h; see Section 1.11. Recall that the paths $c(x, y)$ are uniform quasigeodesics in H^h (with respect to the intrinsic metric d^h on H^h). It is clear that if β was tight, so is β^h.

In the case when β is a quasigeodesic in Y^ℓ, the path β^h is called an *electro-ambient quasigeodesic*; see **[DM17]**. This construction yields a collection of paths in Y^h connecting points in Y. In Section 9.2.1 we describe (uniform) electro-ambient quasigeodesics connecting arbitrary pairs of points in Y^h.

The next result appears in **[DM17**, Lemma 2.15**]**.

LEMMA 9.17. *For tight*[1] *uniform quasigeodesics β in Y^ℓ connecting points of Y, the paths β^h are uniform quasigeodesics in Y^h. More precisely, there exists a function $L = L_{9.17}(K)$ such that if β is a tight K-quasigeodesic in Y^ℓ connecting points of Y, its hyperbolization β^h is an L-quasigeodesic in Y^h.*

An alternative and detailed proof was given by A. Pal and A. Kumar Singh in **[PKS15]**; we will discuss this further in Section 9.2.1.

[1] This assumption was forgotten in **[DM17]**.

9.2. Hyperbolicity of the electric space

In this section we will prove that for each relatively hyperbolic space $(Y, \mathcal{H})$, the space Y^ℓ is hyperbolic and describe uniform quasigeodesics in this space. The key result of this section is:

PROPOSITION 9.18. *Electrifications α^ℓ of uniform (continuous) quasigeodesics α in Y^h connecting points of Y, are uniformly proper in Y^ℓ.*

PROOF. Let $\alpha : J = [0, T] \to Y^h$ be a continuous K-quasigeodesic in Y^h connecting points $x, y \in Y$. Our goal is to prove that the length of the domain of α^ℓ is bounded in terms of the distance $M := d^\ell(x, y)$. This will imply uniform properness of the maps α^ℓ for all continuous K-quasigeodesics α in Y^h.

Let β be a geodesic in Y^ℓ connecting x to y and the length of β is $\leq M$. In particular, β goes through at most N cones H_i^ℓ where N depends only on M; these cones correspond to the horoballs $B_1 = H_{i_1}^h, \ldots, B_n = H_{i_n}^h$ in Y^h, $n \leq N$. It also follows that the length of each subsegment of β between two distinct horoballs is at most M. By renumbering the horoballs, we can also label these subsegments $\beta_1, \ldots, \beta_{n-1}$ so that β_i connects B_i to B_{i+1}. We also label β_0 the subsegment of β between x and B_1 and label β_n the subsegment of β between B_n and y. Set

$$B := \beta_0 \cup B_1 \cup \ldots . \cup B_n \cup \beta_n.$$

In particular, the subset $B \subset Y^h$ is $\lambda = \lambda(M)$-quasiconvex in Y^h. According to Lemma 1.142, there exists D and ϵ such that for each component C of $\alpha^{-1}(Y^h - N_D(B))$ the projection of $\alpha(C)$ to B has diameter $\leq \epsilon$. (Here ϵ, D depend on δ, K and λ, and $D = D_{1.142}(\delta, \lambda')$, where λ' is the maximum of λ and the quasiconvexity constant of the images of K-quasigeodesics in Y^h.) We now decompose the interval $J = [0, T]$ according to the mutual position of points $\alpha(t), t \in J$, with respect to the horoballs B_i and the segments β_i.

Let $J(\beta_j)$ denote the maximal subinterval in J such that the images of the endpoints of $J(\beta_j)$ under α belong to $N_D(\beta_j)$. Similarly, we define subintervals $J(B_j)$.

(1) Since α is a K-quasigeodesic and β_j has length $\leq M$, the length of each $J(\beta_j)$ is $\leq K(2D + M + 1)$.

(2) Similarly, Lemma 9.4 implies that the length of each $J(B_j) \setminus \alpha^{-1}(B_j)$ is at most $2L_{9.4}(K+D, 0, \delta)$. Recall that by the definition of α^ℓ, whenever $J(B_j)\setminus\alpha^{-1}(B_j)$ is nonempty, the interval $J(B_j)$ contributes the length $\leq 2+2L_{9.4}(K+D, 0, \delta)$ to the length of the domain of α^ℓ.

(3) It now remains to estimate the length of

$$J' := J \setminus \bigcup_{i=0}^{n+1} (J(\beta_i) \cup J(B_i)).$$

Let $C = [s, t]$ be a component of this complement. Then both points $\alpha(s), \alpha(t)$ belong to $\partial N_D(B)$ and the diameter of the projection of $\alpha(C)$ to B is $\leq \epsilon$. Hence, the distance in Y^h between $\alpha(s), \alpha(t)$ is at most $2D + \epsilon$. It follows that C has length $\leq K(2D + \epsilon + 1)$. Since the number of components C is $\leq 2N + 2$, it follows that the total length of J' is at most

$$K(2D + \epsilon + 1)(2N + 2).$$

Combining these estimates we conclude that the domain of α^ℓ has length at most

$$(M + 1)K(2D + M + 1) + (2 + 2L_{9.4}(K + D, 0, \delta))M + K(2D + \epsilon + 1)(2N + 2).$$

Uniform properness of the maps α^ℓ follows. □

As an application of this result we prove Theorem 9.19.

THEOREM 9.19.

(1) *If* $(Y, \mathcal{H})$ *is relatively hyperbolic, then* Y^ℓ *is hyperbolic.*
(2) *Moreover, continuous* K*-quasigeodesics* α *in* Y^h *yield* $k_{9.19}(K)$*-quasigeodesics* α^ℓ *in* Y^ℓ.

PROOF. We will verify that the conditions of Corollary 1.64 are met. Namely, we will check that the combing of Y^h by continuous K-quasigeodesics α results in a thin combing of Y^ℓ by paths α^ℓ connecting points of the subset $Y^\ell_0 \subset Y^\ell$ which is the union of Y and the set $a(Y^\ell)$ of apexes a_i of the cones H^ℓ_i, $i \in I$. We already know that the paths α^ℓ are uniformly proper (Property 1 in Corollary 1.64). Let us verify Property 2. Consider a triple of K-quasigeodesics $\alpha_{x,y}, \alpha_{y,z}, \alpha_{z,x}$ connecting points x, y, z in Y^h. Since Y^h is δ-hyperbolic,

$$\alpha_{x,y} \subset N_{3D_{1.54}(\delta,K)+\delta}(\alpha_{y,z} \cup \alpha_{z,x}).$$

Let u be a point of $\alpha_{x,y} \cap Y$ and suppose that v is a point in $\alpha_{y,z}$ at a distance

$$d_{Y^h}(u, v) \le 3D_{1.54}(\delta, K) + \delta$$

from u. If v happens to be in Y, then Lemma 9.16 implies that

$$d^\ell(u, v) \le 3D_{1.54}(\delta, K) + \delta$$

as well. If $v = \alpha_{y,z}(t)$ belongs to an open horoball $\mathring{H}^h_i$ but t does not lie in one of the subintervals $[s_i, t_i]$ in the domain of the path $\alpha_{y,z}$ which are coned-off when we define $\alpha^\ell_{y,z}$, then the distance in Y^h from $\alpha_{y,z}(t)$ to $\alpha_{y,z} \cap H_i$ is at most $2(K + 1)L_{9.4}(K, 0, \delta)$ and, thus, $d^\ell(u, \alpha_{y,z}) \le 2(K + 1)L_{9.4}(K, 0, \delta)$ as well. Lastly, if t is one of the intervals $[s_i, t_i]$ then p_i is in $\alpha_{y,z}$ and

$$d^\ell(u, p_i) \le d(u, v) + 1 \le 3D_{1.54}(\delta, K) + \delta + 1.$$

Finally, each point of $\alpha^\ell_{x,y}$ lies within unit distance (as measured in Y^ℓ) from a point of $\alpha^\ell_{x,y} \cap Y$ and, therefore, for

$$R = 3D_{1.54}(\delta, K) + \delta + 1 + 2(K + 1)L_{9.4}(K, 0, \delta),$$

the path $\alpha^\ell_{x,y}$ is contained in the R-neighborhood (in Y^ℓ) of the union $\alpha^\ell_{y,z} \cup \alpha^\ell_{z,x}$. □

The following consequence of the theorem appears in [**Kla99**, Proposition 4.3]; see also [**Far98**, Lemma 4.5 and Lemmas 4.8, 4.9] in the setting of manifolds of negative curvature:

COROLLARY 9.20. *For* $x, y \in Y$ *let* $\alpha = [xy]_{Y^\ell}$ *and* $\beta = [xy]_{Y^h}$. *Then the Hausdorff distance in* Y *between* $\alpha \cap Y$ *and* $\beta \cap Y$ *is uniformly bounded.*

PROOF.

(1) Take $z \in \alpha \cap Y$. Since α^h is a uniform k-quasigeodesic in Y^h connecting the end-points of β, the k-quasigeodesics α^h and β are D-Hausdorff-close in Y^h, $D = D_{1.54}(\delta, k)$. Thus, there exists $w \in \beta$ within distance D from z. If w happens to be in Y, we are done. Suppose, therefore, that $w \in \mathring{H}$ for some $H \in \mathcal{H}$. Since w is within distance D (as measured in Y^h) from Y, and β is a geodesic in Y^h, it follows that there exists a point $v \in \beta \cap H$ within distance $C = D + D'$ from w, where $D' = D_{1.54}(\delta, K)$ and K is the qi constant of the combing paths in H^h. Since $d_{Y^h}(v, z) \le D + C$, it follows that $d_Y(v, z)$ is also uniformly bounded by a uniform constant E, as (Y, d) is uniformly properly embedded in (Y^h, d^h); see Lemma 9.3(2). Hence,

$$\alpha \cap Y \subset N^Y_E(\beta \cap Y).$$

(2) The proof of the opposite inclusion is similar, swapping the roles of α^h and β and is left to the reader. □

We also record another application of Theorem 9.19.

COROLLARY 9.21. *There exists a constant $L = L_{9.21}(\delta)$ such that every two points x, y in X are connected by a tight L-quasigeodesic in X^ℓ. Namely, given a geodesic $c = [xy]_{X^h}$ in X^h, its electrification c^ℓ is a uniform tight quasigeodesic in X^ℓ.*

We define a coarse projection $q = q_Y : Y^h \to Y^\ell$ as follows.

(1) The map q is the identity on Y.
(2) For each peripheral subspace H_i, $q(\mathring{H}_i^h) = \{a(H_i^\ell)\}$, the apex of the cone over H_i.

Lemma 9.22 is immediate.

LEMMA 9.22. *The map q is $(1, 1)$-coarse Lipschitz.*

The next result is a direct corollary of Theorem 9.19(2).

COROLLARY 9.23. *There exists a function $\lambda \mapsto \hat{\lambda}$ such that if $Z \subset Y^h$ is a λ-quasiconvex subset, then $q(Z)$ is $\hat{\lambda}$-quasiconvex in Y^ℓ.*

LEMMA 9.24. *The projection q coarsely commutes with nearest-point projections. More precisely, there is $D = D_{9.24}(\delta, \lambda)$, where δ is the hyperbolicity constant of Y^h, such that for each λ-quasiconvex subset $Z \subset Y^h$,*

$$d(P_{Y^\ell, q(Z)}, q \circ P_{Y^h, Z}) \le D.$$

PROOF. For $y \in Y^h$, up to a uniformly bounded error, the projection $\bar{y} = P_{Y^h,Z}(y)$ of y to Z is defined by the property that for each $z \in Z$ the concatenation $\beta = y\bar{y} \star \bar{y}z = \beta_1 \star \beta_2$ is a uniform quasigeodesic (see Lemma 1.105). Suppose first that $\bar{y} \in Y$. Theorem 9.19(2) implies that the concatenation $\beta^\ell = \beta_1^\ell \star \beta_2^\ell$ is a uniform quasigeodesic in $\hat{Y}$. By applying Lemma 1.101, we conclude that $\bar{y}$ is uniformly close to $P_{Y^\ell, q(Z)}(q(y))$.

Suppose now that $\bar{y}$ lies in some H_i^h; let $a_i = a(H_i^h)$ denote the apex of the cone in Y^ℓ corresponding to H_i, and $\beta_0 \subset H_i^h$ be the maximal subsegment of β contained in H_i^h and connecting points $x_i, y_i \in H_i$. Then β is the concatenation $\beta_1 \star \beta_0 \star \beta_2$. Accordingly, β^ℓ is the concatenation $\beta_1^\ell \star [x_i a_i] \star [a_i y_i] \star \beta_2^\ell$. Thus, a_i is uniformly close to $P_{Y^\ell, q(Z)}(q(y))$. At the same time, $q(\bar{y}) = a_i$. □

Lastly, we discuss relatively hyperbolic structures on metric spaces which are already hyperbolic; see [**Bow12**, Theorem 7.11] and, in a more limited version, [**Ger96**].

THEOREM 9.25. *Suppose that X is a hyperbolic space and $\mathcal{H} = \{H_i, i \in I\}$ is a collection of rectifiable uniformly qi embedded subsets of X (in particular, these subsets are uniformly quasiconvex in X).*

(1) *If the subsets H_i are uniformly pairwise cobounded, then the pair $(X, \mathcal{H})$ is relatively hyperbolic.*
(2) *Conversely, if X^h is hyperbolic, then the subsets $\{H_i, i \in I\}$ are uniformly pairwise cobounded in X.*

PROOF.

(1) We first equip X^h with a structure of a tree of hyperbolic spaces $\mathfrak{Y} = (\pi : Y \to T)$. We define the tree T by taking the wedge of rays $R_i, i \in I$, where each R_i is the positive half-line $[0, \infty)$ equipped with the standard simplicial structure (vertices are the nonnegative integers) and the wedge of rays is obtained by identifying 0 in all R_i with a single vertex $v_0 \in T$. For $n \in \mathbb{N} \cup \{0\} \subset R_i$ we use the notation v_{in} for the corresponding vertex in T. We define vertex-spaces of $\mathfrak{Y}$ as $Y_{v_0} := X$; $Y_{v_{in}} := H_i \times \{n\} \subset H_i^h = H_i \times [1, \infty)$. The edge-space $Y_{e_{in}}$ for the edge $e_{in} = [v_{in}, v_{in+1}]$ is $H_i \times \{n + \frac{1}{2}\}$. The incidence maps

$$f_{e_{in}, v_{in}} : Y_{e_{in}} \to Y_{v_{in}}, f_{e_{in}, v_{in+1}} : Y_{e_{in}} \to Y_{v_{in+1}}$$

are given by

$$(y, n + \frac{1}{2}) \mapsto (y, n), \quad (y, n + \frac{1}{2}) \mapsto (y, n + 1)$$

respectively. These maps clearly are uniform quasi-isometries. (The inclusion maps $H_i = Y_{e_{i0}} \to Y_{v_0} = X$ are uniform qi embeddings by the assumption.) Thus, we obtain a tree of hyperbolic spaces $\mathfrak{Y} = (\pi : Y \to T)$ satisfying Axiom **H**.

Since the hyperbolic cones H_i^h are uniformly hyperbolic (see Proposition 1.68), the restrictions of the tree of spaces $\mathfrak{Y}$ to the rays R_i satisfy the uniform flaring condition with flaring constants independent of i; see Lemma 2.50. Suppose that $[\![v, w]\!] \subset T$ is an interval containing vertices (different from v_0) of different rays R_i, R_j. Let γ_0, γ_1 be K-qi sections of $\pi : Y \to T$ over $[\![v, w]\!]$. The assumption that peripheral subspaces H_i, H_j in X are uniformly cobounded implies that

$$d_{Y_{v_{i1}}}(\gamma_0(v_{i1}), \gamma_1(v_{i1})) \le C(K), \quad d_{Y_{v_{j1}}}(\gamma_0(v_{j1}), \gamma_1(v_{j1})) \le C(K).$$

In view of the uniform flaring of $\mathfrak{Y}$ over the rays R_i, R_j, it then follows that

$$d_{Y_{v_{in}}}(\gamma_0(v_{i1}), \gamma_1(v_{in}))$$

is either uniformly bounded (in terms of K) or grows at a linear rate (as a function of n). The same applies to

$$d_{Y_{v_{jn}}}(\gamma_0(v_{j1}), \gamma_1(v_{jn})).$$

Hence $\mathfrak{Y}$ satisfies the uniform flaring condition and, thus, Y is uniformly hyperbolic by Theorem 2.62.

(2) This part is a consequence of Lemma 9.11. □

This theorem has a useful addendum, relating quasigeodesics in X and in X^h. In the setting of the theorem, let β be a k-quasigeodesic in X^h connecting points in X. For each maximal subsegment β_i contained in some H_i^h, we replace β_i with a geodesic in X connecting the endpoints of β_i. We let β_X denote the resulting path in X.

LEMMA 9.26. *The paths β_X in X are $K_{9.26}(k)$-quasigeodesic.*

PROOF. Thinking of X^h as the total space Y of a tree of spaces $\mathfrak{Y}$ as above, we note that the path β_X obtained via the above procedure of converting β to β_X is exactly the cut-and-replace procedure in Definition 8.14. Now, the result follows from Theorem 8.21 (actually, it follows already from Proposition 8.22 proven earlier by Mitra in [**Mit98**]). □

9.2.1. Equivalence of the two definitions of relative hyperbolicity. We start by reviewing Farb's definition of relative hyperbolicity. Given a metric space Y and a collection of its rectifiably connected, uniformly properly embedded subspaces $\mathcal{H} = \{H_i : i \in I\}$, we get the associated electric space $Y^\ell = \mathcal{E}(Y, \mathcal{H})$ with the metric d^ℓ as described in Section 9.1.1.

Definition 9.27. The pair $(Y, \mathcal{H})$ is said to be *weakly relatively hyperbolic* if the metric space (Y^ℓ, d^ℓ) is hyperbolic.

Every two points in Y^ℓ are connected by a tight L-quasigeodesic in Y^ℓ for some uniform constant L (for instance, one can use geodesics in Y^ℓ). We refer to the tight quasigeodesics in Y^ℓ connecting points in Y as *electric quasigeodesics* in Y^ℓ.

In Section 9.1.1, given an electric quasigeodesic β in Y^ℓ connecting points $x, y \in Y$, we defined a path β^h in Y^h (the hyperbolization of β) connecting x and y. We now extend this definition to connect arbitrary pairs of points x, y in Y^h. We let $\hat{x} = q(x), \hat{y} = q(y)$ denote the projections of x, y to Y^ℓ. Let β be an electric quasigeodesic connecting $\hat{x}$ to $\hat{y}$. In the case when $x \in \mathring{H}^h, \hat{x} = a(H^\ell)$ is the apex of the cone H^ℓ, we let $x_H \in H$ denote the exit point of β from the cone H^ℓ; similarly, if $y \in H^h$, we let y_H denote the entry point of β into the cone H^ℓ. We set $x' = x$ if $x \in Y$ and $x' := x_H$ if $x \notin Y$; similarly, we define the point y'. Then for the subpath $\beta(x', y')$ of β between x', y', we define its hyperbolization $\beta(x', y')^h$ as before, and connect x to x', y to y' by geodesics in the corresponding horoballs H^h in the case when $x \neq x'$ or $y \neq y'$.

We, thus, obtain a family of paths connecting points of Y^h. As it turns out (see Theorem 9.30), under suitable assumptions, the resulting paths define a slim combing of Y^h. In the next definition, $D = D(K)$.

Definition 9.28. Two electric K-quasigeodesics α_1, α_2 with the same endpoints are said to have *the same D-intersection pattern* with respect to the collection of cones $H_i^\ell, i \in I$, or, simply, *D-track each other*, if the following conditions hold for all $i \in I$:

(1) If one path, say, α_1, contains $a(H_i^\ell)$ but then other (namely, α_2) does not, then $d_{H_i}(\alpha_1(s), \alpha_1(t)) \leq D$, where $\alpha_1^{-1}(\mathring{H}_i^\ell)$ is the open interval (s, t).
(2) Suppose that, for some $i \in I$, $\alpha_j^{-1}(\mathring{H}_i^\ell) = (s_i, t_i) \neq \emptyset, j = 1, 2$. Then
$$\max\{d_{H_i}(\alpha_1(s_1), \alpha_2(s_2)), d_{H_i}(\alpha_1(t_1), \alpha_2(t_2))\} \leq D.$$

The function $D(K)$ is the *tracking function* of $(Y, \mathcal{H})$.

Definition 9.29. A pair $(Y, \mathcal{H})$ is *relatively hyperbolic in Farb's sense (FRH)* if it is weakly relatively hyperbolic and for every $K \geq 1$ there exists $D = D(K)$ such that any two electric K-quasigeodesics with the same endpoints D-track each other.

The next theorem relating the two definitions was proven by A. Pal and A. Kumar Singh in [**PKS15**] (for relatively hyperbolic groups the equivalence of two definitions was known earlier, cf. [**Bow12**]).

Theorem 9.30. *If $(Y, \mathcal{H})$ is FRH, then Y^h is uniformly*[2] *hyperbolic and the hyperbolization of uniform electric quasigeodesics yields uniform quasigeodesics in Y^h. In particular, $(Y, \mathcal{H})$ is GRH.*

Remark 9.31. A. Pal and A. Kumar Singh in [**PKS15**] assume that the subsets $H_i, i \in I$, are uniformly separated in Y. In our setting one achieves uniform separation by replacing

[2] With respect to the tracking function and the hyperbolicity constant of Y^ℓ.

Y with the space Y' obtained by attaching the products $H_i \times [0,1]$, $i \in I$, along the subsets H_i, and replacing the subsets $H_i \subset Y$ with $H_i \times \{1\}$, $i \in I$.

To conclude the discussion, we note a relation between uniform tight quasigeodesics in Y^ℓ and uniform quasigeodesics in the space Y itself. (This is not needed for any proofs in the book, but clarifies the overall picture.) The following result is proven in [**Hru10**, Lemma 8.8] in the context of relatively hyperbolic groups, but the proof works for general relatively hyperbolic spaces:

THEOREM 9.32. *There are functions $D = D_{9.32}(K,\delta)$ and $L = L_{9.32}(K,\delta)$ such that if α is an electric K-quasigeodesic in Y^ℓ (connecting points x, y of Y), then for every L-quasigeodesic β in Y also connecting x and y, we have that $\alpha \cap Y$ is contained in the D-neighborhood of β, with respect to the metric d_Y of Y.*

9.2.2. Morphisms of relatively hyperbolic spaces. Given a pair of relatively hyperbolic spaces $(Y,\mathcal{H})$, $(Y',\mathcal{H}')$, a *relative morphism* of these pairs is a uniformly proper map $f : Y \to Y'$ such that:

(a) For each $H' \in \mathcal{H}'$, $f^{-1}(H')$ is either empty or equals some $H \in \mathcal{H}$.
(b) For each $H \in \mathcal{H}$, there exists $H' \in \mathcal{H}'$ satisfying $f(H) \subset H'$.

REMARK 9.33.
(1) The first condition implies that if $H_i, H_j \in \mathcal{H}$ are distinct, then $f(H_i), f(H_j)$ are not contained in the same $H' \in \mathcal{H}'$.
(2) One can relax a bit the above conditions by requiring existence of a uniform constant D such that:
  (a') For each $H' \in \mathcal{H}'$, $f^{-1}(H')$ either has diameter $\le D$ or is D-Hausdorff-close to some $H \in \mathcal{H}$.
  (b') For each $H \in \mathcal{H}$, there exists $H' \in \mathcal{H}'$ satisfying $f(H) \subset N_D(H')$.

This is the approach taken in [**MS20**]. However, the two definitions are easily seen to be effectively equivalent since one can replace the peripheral horoballs $H^h, H \in \mathcal{H}$ and $(H')^h, H' \in \mathcal{H}'$, by suitable smaller subsets.

A relative morphism is said to be a *relative qi embedding* of the pairs if, additionally:
(c) $f : Y \to Y'$ is a qi embedding.

The qi constants of f are called the *parameters* of the relative qi embedding f.

Given a relative morphism $f : (Y,\mathcal{H}) \to (Y',\mathcal{H}')$, the *coned-off map* $f^\ell : Y^\ell \to (Y')^\ell$ is defined as follows:
(i) The restriction of f^ℓ to Y equals f.
(ii) Consider a peripheral subset $H \in \mathcal{H}$ such that $f(H) \subset H' \in \mathcal{H}'$ and let $a = a(H^\ell)$, $a' = a((H')^\ell)$ denote the respective apexes of the cones $H^\ell, (H')^\ell$ over these peripheral subsets in $Y^\ell, (Y')^\ell$. Then $f^\ell(a) = a'$ and for each $x \in H$, $x' := f(x) \in H'$, the map $\hat{f}$ sends the radial segment xa to $x'a'$ isometrically. This defines f^ℓ on H^ℓ for each $H \in \mathcal{H}$.

Similarly, we define the *hyperbolic extension* f^h of f, $f^h : Y^h \to (Y')^h$; this extension construction goes back to the work of Mostow on strong rigidity of nonuniform lattices in rank one Lie groups.

For $x \in H \in \mathcal{H}$, $x' = f(x) \in H' \in \mathcal{H}'$, we consider the vertical geodesic rays $\rho_x : [0,\infty) \to H^h, \rho_{x'} : [0,\infty) \to (H')^h$, emanating from x, x' and asymptotic to the centers of the horoball $H^h, (H')^h$. Then for $t \in [0, \infty)$, we set

$$f^h(\rho_x(t)) := \rho_{x'}(t), \quad x \in H.$$

LEMMA 9.34.

(1) *The hyperbolic extension f^h of a proper coarse Lipschitz map is again a proper coarse Lipschitz map.*

(2) *If f, f^ℓ are qi embeddings, then so is f^h and the constants of f^h depend only on the parameters of $(Y, \mathcal{H}), (Y', \mathcal{H}')$ and qi constants of f, f^ℓ.*

(3) *Conversely, if f and f^h are qi embeddings, so is f^ℓ.*

PROOF.

(1) We will only show that f^h is coarse Lipschitz and leave it to the reader to check properness. Since Y^h is a path-metric space, the problem is local and we have to address it only in the horoballs H^h, $H \in \mathcal{H}$. Clearly, f^h is isometric along the vertical geodesic rays in H^h. Suppose, therefore, that y_1, y_2 are points within unit distance on the same horosphere $H \times \{t\} \subset H^h, t \geq 0$ (where the distance is computed in the intrinsic metric of the horosphere). Thus, $y_i = \rho_{x_i}(t)$, $x_i \in H$, $i = 1, 2$, and, by the definition of the metric on H^h,

$$d_H(x_1, x_2) = e^t.$$

Since the horospheres H, H' are uniformly qi embedded in Y, Y', we conclude that the restriction map $f : (H, d_H) \to (H', d_{H'})$ is (L, A)-coarse Lipschitz, where L, A depend only on the parameters of $(Y, \mathcal{H})$ and f. Hence,

$$d_{H'}(f(x_1), f(x_2)) \leq L d_H(x_1, x_2) + A \leq Le^t + A.$$

Again, by the definition of the metric on the horoball $(H')^h$,

$$d(\rho_{x'_1}(t), \rho_{x'_2}(t)) \leq e^{-t}(Le^t + A) = L + e^{-t}A \leq L + A.$$

This verifies that f^h is uniformly coarse Lipschitz.

(2) Note that given an ambient geodesic γ in Y^h, its electrification γ^ℓ is a uniform electric quasigeodesic in Y^ℓ, hence, f^ℓ carries it to a uniform electric quasigeodesic $\hat{\gamma}'$ in $(Y')^\ell$, coarsely preserving its arc-length. Then, applying Lemma 9.17, we obtain that the hyperbolization $\gamma' = (\hat{\gamma}')^h$ of $\hat{\gamma}'$ is a uniform quasigeodesic in $(Y')^h$. By the construction, the map f^h coarsely preserves arc-lengths of paths in X^h, cf. Section 1.11; and, thus, f^h sends geodesics to uniform quasigeodesics coarsely preserving arc-length, hence, is a qi embedding.

(3) The proof of this part is similar to the proof of Part (2), except one uses Theorem 9.19(2). □

REMARK 9.35. This lemma is a weak form of a more general recent result by J. Mackay and A. Sisto [**MS20**, Corollary 1.2], who proved Part (2) without assuming that f^ℓ is a qi embedding (in fact, they also relax the condition (c) in the definition of a relative qi embedding). Part (3) of the lemma then implies that f^ℓ is a uniform qi embedding provided that f is. See also the recent paper [**HH20**] by Healy and Hruska with a similar result for relatively hyperbolic groups.

9.2.3. CT maps. Suppose that $(X, \mathcal{F})$ and $(Y, \mathcal{H})$ are relatively hyperbolic spaces and $f : (Y, \mathcal{H}) \to (X, \mathcal{F})$ is a morphism of pairs. We then have the hyperbolic extension of f, which is a coarse Lipschitz uniformly proper map $f^h : Y^h \to X^h$ (see Lemma 9.34), as well as the coned-off map $f^\ell : Y^\ell \to X^\ell$.

The next proposition is a relative analogue of 8.3 (see the formulation in terms of distances to geodesics, Theorem 8.4). The main difference with the absolute case is that the properness condition (the existence of a function $C(y_0, K, D)$) in the relative case is formulated in terms of the map of electric spaces (instead of $f^h : Y^h \to X^h$, which would have made the result too easy and not so useful) and geodesics are replaced by uniform tight quasigeodesics.

PROPOSITION 9.36. *Assume that for some (equivalently, every) $y_0 \in Y$ and every $K \geq 1$, there is a function $C = C(y_0, K, D)$ such that:*

For all pairs of points $y_1, y_2 \in Y$, all tight K-quasigeodesic β^ℓ in Y^ℓ connecting y_1, y_2, and all tight K-quasigeodesics α^ℓ in X^ℓ connecting $x_1 = f(y_1), x_2 = f(y_2)$,

$$d_X(f(y_0), \alpha^\ell \cap X) \leq D \Rightarrow d_Y(y_0, \beta^\ell \cap Y) \leq C.$$

PROOF. Note that, in view of Corollary 9.20, the implication in the proposition can be rewritten as

$$d_X(f(y_0), [x_1x_2]_{X^h} \cap X) \leq D \Rightarrow d_Y(y_0, [y_1y_2]_{Y^h} \cap Y) \leq C := \phi(D),$$

for some function ϕ. We will be verifying that f^h satisfies Mitra's Criterion, Theorem 8.3. Suppose that $y_1, y_2 \in Y^h$ are such that

$$d_{X^h}(x_0, [x_1x_2]_{X^h}) \leq D,$$

where $f(y_i) = x_i$, $i = 0, 1, 2$. Our goal is to get a uniform bound on the distance from y_0 to $[y_1y_2]_{Y^h}$ in Y^h.

CASE 1. We first suppose that y_1, y_2 are both in Y. Then, x_1, x_2 are both in X as well. Since X is uniformly properly embedded in X^h and x_0 is in X, it follows that there is $x' \in [x_1x_2]_{X^h} \cap X$ within distance D' from x_0, with respect to the metric of X, where D' depends only on D (and geometry of $(X, \mathcal{H})$, of course), cf. Lemma 1.70. Thus, the assumptions of the proposition imply that

$$d_Y(y_0, [y_1y_2]_{Y^h} \cap Y) \leq C',$$

where C' is a function of D. This, of course, implies that

$$d_{Y^h}(y_0, [y_1y_2]_{Y^h}) \leq C',$$

as required by Mitra's criterion.

CASE 2. Suppose that both y_1, y_2 belong to the same H_i^h. In view of Lemma 1.70, we obtain that for $k = 1$ or $k = 2$, $d(x_0, x_k) \leq D'$, for some D' depending only on D. Uniform properness of f^h then implies that $d_{Y^h}(y_0, y_k)$ is also uniformly bounded, as required.

CASE 3. Suppose that $y_1 \in H_{i_1}^h, y_2 \in H_{i_2}^h$ and $H_{i_1}^h \neq H_{i_2}^h$. Thus, $x_i \in F_{j_i}^h$, $F_{j_i} \in \mathcal{F}$, are distinct, $i = 1, 2$. Our assumption that

$$d_{X^h}(x_0, [x_1x_2]_{X^h}) \leq D$$

implies that there is a point $p \in x_1x_2 = [x_1x_2]_{X^h} \cap X$ within distance $\leq D' = D'(D)$ from $[x_1x_2]_{X^h}$ (cf. the proof in Case 2). Let $x_i' \in F_{j_i}$, $i = 1, 2$, be points realizing the minimal distance in X^h between F_{i_1}, F_{i_2}. Thus, the R-neighborhood of the segment x_1x_2 in X^h contains $x_1'x_2'$, where R is a uniform constant; see Lemma 1.147. Let $x_i'' \in x_1x_2$ be points within distance R from $x_i', i = 1, 2$. By the uniform quasiconvexity of $H_{i_k}^h$ in X^h, $k = 1, 2$, if p is not in $x_1''x_2'' = [x_1''x_2'']_{X^h}$ then its distance to x_1'' or x_2'' is uniformly bounded (cf. Corollary 9.6). Thus, it suffices to consider the case when $p \in x_1''x_2''$.

Let $w_i, i = 1, 2$, be intersection points of $[y_1y_2]_{Y^h} = y_1y_2$ with $H^h_{i_1}, H^h_{i_2}$ respectively, and $z_i := f(w_i), t = 1, 2$. The points z_1, z_2 belong to the peripheral subsets F_{j_1}, F_{j_2} respectively. Thus, again, the R-neighborhood of the segment z_1z_2 contains $x'_1x'_2$. Since we are assuming that $p \in x''_1x''_2$, it follows that

$$p \in N_{2R+\delta}(z_1z_2)$$

with respect to the metric of X^h. Hence, $d_{X^h}(x_0, [z_1z_2]_{X^h}) \leq 2R + \delta + D'$. Now, we are in the setting of Case 1 and it follows that

$$d_Y(y_0, [w_1w_2]_{Y^h}) \leq \phi(2R + \delta + D').$$

Therefore,

$$d_{Y^h}(y_0, [y_1y_2]_{Y^h}) \leq d_Y(y_0, [w_1w_2]_{Y^h}) \leq \phi(2R + \delta + D'),$$

as required.

CASE 4. Suppose that $y_1 \notin Y$ and $y_2 \in Y$. This case is similar to Case 3 and we leave it to the reader. □

9.3. Trees of relatively hyperbolic spaces

We can now describe axioms of trees of relatively hyperbolic spaces; our definition follows [**MP11**].

DEFINITION 9.37. A tree of relatively hyperbolic spaces is an (abstract) tree of spaces $\mathfrak{X} = (\pi : X \to T)$ where all vertex and edge spaces have structures of uniformly[3] relatively hyperbolic spaces $(X_v, \mathcal{H}_v)$, $(X_e, \mathcal{H}_e)$, and the incidence maps $f_{ev} : X_e \to X_v$ are uniform relative qi embeddings.[4]

REMARK 9.38. The definition of a tree of relatively hyperbolic spaces given in [**MR08**] also requires the coned-off maps $f^\ell_{ev} : X^\ell_e \to X^\ell_v$ to be uniform qi embeddings. In view of [**MS20**], this is a consequence of the assumption that the incidence maps are uniform relative qi embeddings (cf. Remark 9.35). The reader not willing to rely upon the results of [**MS20**], can simply assume that the maps f^ℓ_{ev} are uniform qi embeddings.

Next, given such a tree of spaces $\mathfrak{X} = (\pi : X \to T)$, we define an equivalence relation on X generated by the following:

(1) For each edge $e = [v_1, v_2]$ of T, every point in a peripheral subspace $H_e \subset X_e$ is equivalent to all the points of the mapping cylinders of the incidence maps $f_{ev_i} : H_e \to H_v \subset X_{v_i}, i = 1, 2$, in X.
(2) All points of each peripheral subspace $H_v \subset X_v, v \in V(T)$, belong to the same equivalence class.

Each equivalence class of this equivalence relation is called a *peripheral subspace* P of X. Observe that for each peripheral subspace P of X and each vertex (resp., edge) space X_v (resp., X_e) of $\mathfrak{X}$, the intersection $P \cap X_v$ (resp, $P \cap X_e$) is either empty or equals one of the peripheral subspaces of X_v (resp., X_e). Given these peripheral subspaces, one defines the new space $X^P = \mathcal{G}(X, \mathcal{P})$, by attaching hyperbolic horoballs P^h_j to X along all peripheral subspaces $P_j, j \in J$, of X. Here and in what follows, $\mathcal{P} = \{P_j : j \in J\}$.

REMARK 9.39. The definition of X^P should not be confused with the one, X^h, of the total space of a tree of spaces over T, with the vertex/edge spaces X^h_v, X^h_e.

---

[3]I.e., each X^h_v, X^h_e is δ-hyperbolic for a uniform constant δ.

[4]Recall that this condition also requires the incidence maps to be morphisms of these relatively hyperbolic spaces, $f_{ev} : (X_e, \mathcal{H}_e) \to (X_v, \mathcal{H}_v)$.

Similarly, for each subtree $S \subset T$, we define the space X^P_S obtained by first restricting the tree of spaces $\mathfrak{X}$ to S and, thus, obtaining a tree of spaces $X_S \to S$, and then applying the above procedure, so that $X^P_S = (X_S)^P$. The peripheral structure of X_S is denoted $\mathcal{P}_S$.

Additionally, we have the space $\widehat{X} = \mathcal{E}(X, \mathcal{P})$ obtained by coning-off the peripheral subspaces $P \in \mathcal{P}$ in X.

Furthermore, we define the tree of coned-off-spaces $\mathfrak{X}^\ell = (\pi : X^\ell \to T)$ with vertex/edge spaces X^ℓ_v, X^ℓ_e and coned-off incidence maps $f^\ell_{ev} : X^\ell_e \to X^\ell_v$, which are uniform qi embeddings (see Remark 9.38).

REMARK 9.40. In the terminology of [**MR08**], X^ℓ is a *partially electrocuted space*.

The total space X^ℓ of $\mathfrak{X}^\ell$ contains a *forest* $\mathcal{F}$ which is a collection of pairwise disjoint trees F_P corresponding to the peripheral subspaces $P \in \mathcal{P}$:

(i) The vertices v_{iv} of F_P are the apexes $a(H^\ell_{iv})$ of the cones over the peripheral subspaces $H_{iv} = P \cap X_v$ of the vertex-spaces $X_v, v \in V(T)$.
(ii) Similarly, the edges ϵ_{ie} of F_P are labeled by the apexes $a(H^\ell_{ie})$ of the cones over the peripheral subspaces of the edge-spaces $X_e, e \in E(T)$, where $H_{ie} = P \cap X_e$.
(iii) A vertex v_{iv} (corresponding to $a(H^\ell_{iv})$) of the tree F_P is incident to the edge ϵ_{ie} (corresponding to $a(H^\ell_{ie})$) if and only if the incidence map f_{ev} of $\mathfrak{X}$ sends H_{ie} to H_{iv}.

The fact that each graph F_P is a tree is a consequence of Lemma 9.41.

LEMMA 9.41. *Under the projection $X^\ell \to T$, each F_P maps isomorphically to a subtree in T.*

PROOF. Connectivity of F_P is clear, we need to verify the injectivity of the map. The problem is local and reduces to analyzing vertex-spaces of $\mathfrak{X}$. The only way the map can fail to be injective is if there is an edge $e = [v, w]$ of T and two distinct peripheral subspaces H_{ei}, H_{ej} of X_e which map to the same peripheral subspace H_{vk} of X_v. But this contradicts the condition that $(X_e, \mathcal{H}_e) \to (X_v, \mathcal{H}_v)$ is a morphism of relatively hyperbolic spaces. □

In particular, the trees F_P are uniformly qi embedded in X^ℓ.

REMARK 9.42. There are natural projections $\theta_P : P \to F_P$ sending each nonempty intersection $P \cap X_v = H_{iv}$ to the corresponding vertex v_{iv} of F_P and, for $H_{ie} = P \cap X_e, e = [v, w]$, sending the (double) mapping cylinder of $(f_{ev} \cup f_{ew})|_{H_{ie}}$ to the edge ϵ_{ie} in such a way that each interval $\{x\} \times [v, w], x \in X_e$ maps linearly onto the edge ϵ_{ie}. The mapping cylinder of θ_P is then homeomorphic to the closure in X^ℓ of the component of $X^\ell \setminus X$ containing F_P. We, therefore, will use the notation $Cyl(F_P, P)$ for this closure. This viewpoint of identifying X^ℓ with the space obtained from X by attaching mapping cylinders of peripheral subsets $P \in \mathcal{P}$ to trees, is adapted from [**MR08**] and [**MP11**]. In line with the notion of radial segments in cones, we will refer to the projections of intervals $\{x\} \times [0, 1] \subset P \times [0, 1]$ to $Cyl(F_P, P)$ as *radial segments* in $Cyl(F_P, P)$.

The spaces X^ℓ and $\widehat{X}$ are related by a quotient map $\tau : X^\ell \to \widehat{X}$ which collapses each tree $F_P \in \mathcal{F}$ to a single point, the apex $a(P^\ell) \in \widehat{X}$ of the cone over the peripheral subspace P. See also the diagram in Section 9.3.1.

We next describe a *relative flaring condition* for trees of relatively hyperbolic spaces. This condition consists of two parts.

Part 1. The first part of the relative flaring condition requires that the tree of hyperbolic spaces $\mathfrak{X}^\ell$ satisfies one of the equivalent flaring conditions, equivalently, the space X^ℓ is hyperbolic (see Section 2.5 for the detailed discussion).

Part 2. The (λ, D)-flaring:

Let $\Pi = (\gamma_0, \gamma_1)$ be a pair of 1-sections[5] of $\mathfrak{X}^\ell$ over an interval $[\![u, w]\!]$ of length $\geq D$ in T, such that for each vertex $v \in V([\![u, w]\!])$, $\gamma_i(v) \in a(X_v^\ell)$, $i = 0, 1$. Then

$$\Pi_{max} \geq \lambda \Pi_0,$$

where the distances are computed in electrified vertex-spaces.

Mj and Reeves in **[MR08]** prove the following relative form of the Bestvina–Feighn combination theorem.

THEOREM 9.43. *For every $\lambda > 1, D \geq 2$, if $\mathfrak{X}$ is a tree of relatively hyperbolic spaces satisfying the relative (λ, D)-flaring condition (both parts), the space X^P is hyperbolic.*

We refer the reader to the papers by Sisto **[Sis13]**, Gautero **[Gau16]**, and Dahmani **[Dah03]** for related results.

LEMMA 9.44. *Assuming that $\mathfrak{X}$ is a tree of relatively hyperbolic spaces satisfying the relative flaring condition, the pair $(X^\ell, \mathcal{F})$ is a relatively hyperbolic space.*

PROOF. We will be using Theorem 9.25. Part 1 of the relative flaring condition (together with Theorem 2.62) implies that X^ℓ is hyperbolic. In view of Lemma 9.41, the peripheral subspaces $F_P \in \mathcal{F}$ are trees uniformly qi embedded in X^ℓ. Lastly, Part 2 of the relative flaring condition implies that distinct peripheral subspaces in $\mathcal{F}$ are uniformly pairwise cobounded. □

Since the pair $(X^\ell, \mathcal{F})$ is relatively hyperbolic, we may perform the secondary cone-off construction, coning-off the peripheral subspaces $F_P \in \mathcal{F}$ of X^ℓ. The result is a hyperbolic metric space $\widetilde{X} := \mathcal{E}(X^\ell, \mathcal{F})$. There is a natural map

$$\theta : \widehat{X} = \mathcal{E}(X, \mathcal{P}) \to \widetilde{X} = \mathcal{E}(X^\ell, \mathcal{F})$$

which is the identity on X and sends the cone over each $P \in \mathcal{P}$ to the union of $Cyl(F_P, P)$ and $C(a_F, F)$ (cone over $F = F_P$) so that each radial line segment connecting the apex $a = a(P^\ell)$ of $P^\ell = C(a, P)$ to a point $x \in P$, maps homeomorphically to the concatenation of the radial line segment in $C(a, F)$ connecting a to a point $y \in F$ with the line radial segment in $C(F, P)$ connecting y to x. The next lemma is immediate from the fact that each cone P^ℓ has unit diameter and for each $F \in \mathcal{F}$, the union of $Cyl(F_P, P)$ with the cone $C(a_F, F)$ has diameter 2:

LEMMA 9.45. *The map θ is a quasi-isometry.*

In the follow-up (to **[MR08]**) paper **[MP11]**, Mj and Pal prove the following result regarding existence of CT maps:

THEOREM 9.46. *For each vertex $v \in V(T)$, the inclusion map $X_v^h \to X^P$ admits a CT extension.*

The main result of this chapter is to generalize this theorem to subtrees $S \subset T$:

THEOREM 9.47. *For each subtree $S \subset T$, the inclusion map $X_S^P \to X^P$ admits a CT extension.*

[5]For each edge $e_j = [v_j, v_{j+1}]$ in $[\![u, w]\!]$, $\gamma_i(e_j)$ maps to $\gamma_i(v_j), \gamma_i(v_{j+1})$ under the maps $a(X_{e_j}^\ell) \to a(X_{v_j}^\ell)$ and $a(X_{e_j}^\ell) \to a(X_{v_{j+1}}^\ell)$.

Note that the main tools in the proof of Theorem 9.46 in **[MP11]** were:

(a) A construction of *derived ladders* in the induced tree of coned-off space $X^\ell \to T$ and

(b) a construction of qi sections in X lying inside these ladders.

We include below proofs of these results for the sake of completeness. (We will also need these results in order to prove Theorem 9.47.) However, we provide simplified proofs modulo the corresponding results for trees of hyperbolic metric spaces.

9.3.1. Comparison of quasigeodesics. Let $\mathfrak{X} = (\pi : X \to T)$ be a tree of relatively hyperbolic spaces. There are several spaces and, accordingly, several types of quasigeodesics associated with $\mathfrak{X}$. In this section we discuss the relation between different types of quasigeodesics. We begin, however, with a diagram describing spaces where these quasigeodesics live in. Recall that the tree of spaces $\mathfrak{X} = (\pi : X \to T)$ gives rise to two other trees of spaces:

$$\mathfrak{X}^h = (\pi : X^h \to T) \leftsquigarrow \mathfrak{X} \rightsquigarrow \mathfrak{X}^\ell = (\pi : X^\ell \to T).$$

Both $\mathfrak{X}^h$ and $\mathfrak{X}^\ell$ are trees of hyperbolic spaces (satisfying Axiom **H**) but only $\mathfrak{X}^\ell$ satisfies the flaring condition. The following diagram describes the relation between five different spaces associated with X; the arrow θ is the quasi-isometry described in Lemma 9.45 and the map τ is the collapsing map from Section 9.3:

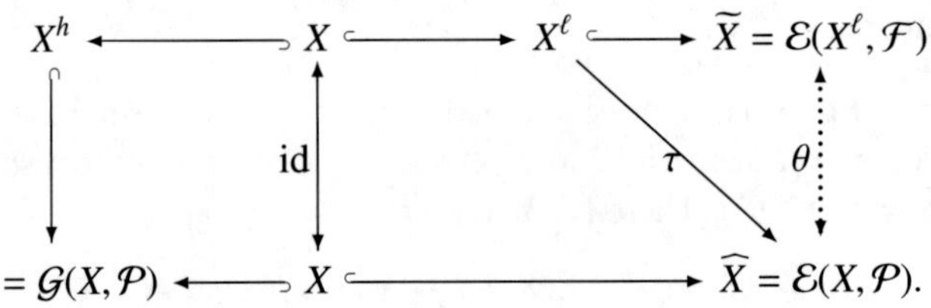

We will discuss the following classes of quasigeodesics in these spaces:

- Quasigeodesics in $X^P = \mathcal{G}(X, \mathcal{P})$;
- Tight quasigeodesics in $\widehat{X} = \mathcal{E}(X, \mathcal{P})$;
- Tight quasigeodesics in X^ℓ;
- Tight quasigeodesics in $\widetilde{X} = \widehat{X^\ell} = \mathcal{E}(X^\ell, \mathcal{F})$.

Our goal is to relate these quasigeodesics. We already know that the first two types of quasigeodesics are uniformly Hausdorff-close to each other (when intersected with X, where the distance is computed via the metric d_X); see Theorem 9.19 and Corollary 9.20. In this section we prove that the same holds for the remaining types of quasigeodesics under a suitable tightness assumption on the quasigeodesics in X^ℓ.

Definition 9.48. We say that a a continuous quasigeodesic in X^ℓ is *tight* if its intersection with every $\mathring{C}(F_P, P), P \in \mathcal{P}$, is either empty or is a concatenation of two radial geodesics with a geodesic in F_P.

By analogy with electrification of paths β in X^P (where the result is a tight path in $\widehat{X}$), we define the *partial electrification* β^ℓ of β as Definition 9.49.

Definition 9.49 (Partial electrification). Suppose that β is a path in X^P which is tight with respect to $\mathcal{P}$. For each pair of points x, y in β which belong to some $P \in \mathcal{P}$ and such that the subpath $\beta(x, y)$ between points x, y is contained in $\mathring{P}^h$, except for the points x, y, we replace $\beta(x, y)$ with a geodesic in $\mathring{C}(F_P, P)$ connecting x and y. The resulting path β^ℓ is the partial electrification of β.

It follows from the definition that the path β^ℓ is tight in X^ℓ.

Given a tight quasigeodesic β in X^ℓ, we can electrify it with respect to the collection of peripheral subsets $\mathcal{F}$ and obtain a tight quasigeodesic $\widetilde{\beta} = \mathcal{E}_{\mathcal{F}}(\beta)$ in $\widetilde{X}$; see Section 9.1.3. Clearly, this defines an injective map $\beta \mapsto \widetilde{\beta}$ from the set of tight quasigeodesics in X^ℓ to those in $\widetilde{X}$, under which uniform quasigeodesics correspond to uniform quasigeodesics.

We also have the quasi-isometry

$$\theta : \mathcal{E}(X, \mathcal{P}) = \widehat{X} \to \widetilde{X} = \mathcal{E}(X^\ell, \mathcal{F});$$

see Lemma 9.45. This map induces a bijection Θ between the sets of tight quasigeodesics in $\widehat{X}$ and quasigeodesics of the form $\widetilde{\beta}$ in $\widetilde{X}$. Uniform quasigeodesics again correspond to uniform quasigeodesics. Combining the two maps, we obtain a bijection

$$\Phi : \alpha \mapsto \Theta(\alpha) = \widetilde{\beta} \mapsto \beta,$$

between the sets of tight quasigeodesics in $\mathcal{E}(X, \mathcal{P})$ and those in X^ℓ. Moreover, the paths α and β agree in X.

We record these observations as in Lemma 9.50.

Lemma 9.50. *There exists a bijection Φ between the sets of tight quasigeodesics in $\mathcal{E}(X, \mathcal{P})$ and those in X^ℓ. Moreover, a tight path in $\mathcal{E}(X, \mathcal{P})$ is a uniform quasigeodesic if and only if the path $\beta = \Phi(\alpha)$ is in X^ℓ.*

As an application, we obtain a result which appears as [**MP11**, Lemma 1.21].

Lemma 9.51. *Let x, y be in X and let α be a continuous L-quasigeodesic in X^P between these points. Then there exists a tight $L_{9.51}(L)$-quasigeodesic β in X^ℓ connecting x and y such that $\alpha \cap X = \beta \cap X$.*

Proof. As described in Section 9.1.3, we convert α to a uniform tight quasigeodesic $\widehat{\alpha}$ in $\widehat{X}$, the electrification of α. Then applying the map Φ as above, we obtain $\beta = \Phi(\alpha)$, a tight quasigeodesic in X^ℓ whose qi constant depends only on that of α. By the construction. $\alpha \cap X = \beta \cap X$. □

Corollary 9.52. *Any two points in $X \subset X^\ell$ are connected by a uniform tight quasigeodesic in X^ℓ.*

Suppose that β_1, β_2 are tight L-quasigeodesics in X^ℓ with the same endpoints in X. Applying the inverse bijection Φ^{-1} to β_1, β_2 we obtain tight L'-quasigeodesics α_1, α_2 in $\widehat{X}$, again connecting the same points in X and such that $\alpha_i \cap X = \beta_i \cap X, i = 1, 2$. Since, by Theorem 9.30 the paths α_1, α_2 uniformly track each other in X, we obtain Corollary 9.53.

Corollary 9.53. *Any two tight quasigeodesics in X^ℓ connecting points of X uniformly track each other in X.*

9.3.2. Ladders in trees of relatively hyperbolic spaces. By Lemma 9.34, we have a tree of hyperbolic spaces $\mathfrak{X}^h = (\pi^h : X^h \to T)$ such that edge-spaces are uniformly quasi-isometrically embedded in vertex-spaces.

Hence, given $x, y \in X_u \subset X_u^h$ we construct a (K, D, E)-ladder $\mathcal{L}^h \subset \mathfrak{X}^h$ of the geodesic segment $L_u^h = [xy]_{X_u^h}$, as we have done in Chapter 3. (Note that flaring conditions were not used in this construction.)

For each vertex v (resp., edge e) in $\pi(L^h) \subset T$ we have the oriented vertical geodesic segment $L_v^h = [x_v y_v]_{X_v^h}$ (resp., $L_e^h = [x_e y_e]_{X_e^h}$) of the ladder $\mathcal{L}^h$. Coning-off the subsegments of L_v (resp., L_e) contained in the peripheral horoballs of X_v^h (resp., X_e^h) we obtain a collection of uniform quasigeodesics in the electrified vertex/edge spaces X_v^ℓ (resp., X_e^ℓ); see

Theorem 9.19(2). We let L_v^ℓ (resp., L_e^ℓ) denote the geodesics in X_v^ℓ (resp., X_e^ℓ) connecting the endpoints of the above quasigeodesics.

LEMMA 9.54. *The collection of segments L_v^ℓ and L_e^ℓ defines a $(\hat{K}, \hat{D}, \hat{E})$-ladder $\mathcal{L}^\ell$ in $\mathfrak{X}^\ell$.*

PROOF. The proof is based on Lemma 3.17. The collection of segments L_v^ℓ and L_e^ℓ satisfies the assumptions of Lemma 3.17 because $\mathcal{L}^h$ is a ladder and because coning-off of quasigeodesics coarsely commutes with the nearest-point projections; see Lemma 9.24 as well as Remark 3.2(viii). □

We let L^ℓ, L^h denote the total spaces of the ladders $\mathcal{L}^\ell$ and $\mathcal{L}^h$ respectively. We will say that the ladder $\mathcal{L}^\ell$ is *derived* from the ladder $\mathcal{L}^h$.

REMARK 9.55. The ladder construction given in [**MP11**] is a bit more complicated.

By the construction, each ladder is a tree of relatively hyperbolic spaces with the total space L and the peripheral structure $\mathcal{P}_L$ (given by the pull-back of the peripheral structure $\mathcal{P}$ of X), and the inclusion map $(L, \mathcal{P}_L) \to (X, \mathcal{P})$ is a morphism of relatively hyperbolic spaces (we will need only that distinct peripheral subsets map to distinct ones).

Corollary 3.13 (the existence of coarse Lipschitz retractions to ladders in hyperbolic trees of spaces), combined with Lemma 9.54 implies:

COROLLARY 9.56. *There is a coarsely Lipschitz retraction $X^\ell \to L^\ell$ with Lipschitz constant depending only on the parameters K, D, E.*

Applying Corollary 9.52, we obtain Corollary 9.57.

COROLLARY 9.57. *Any two points in $L \subset L^\ell$ are connected by a uniform tight quasi-geodesic in L^ℓ.*

QI sections in ladders. We next discuss the construction in [**MP11**] of *vertical quasi-geodesic rays* contained in ladders. We assume that

$$\mathcal{L}^h = \mathcal{L}^h([x_u y_u]_{X_u^h})$$

is a (K, D, E)-ladder in $\mathfrak{X}^h$ and $\mathcal{L}^\ell$ is the ladder in $\mathfrak{X}^\ell$ derived from it.

LEMMA 9.58. *Given a vertex $v \in \pi(L^h)$ and a point $z_v \in X_v \cap L_v^\ell$, there is a k-qi section $\sigma : [\![u, v]\!] \to X$ of $\pi : X \to T$ such that $\sigma([\![u, v]\!]) \subset L^\ell$, $\sigma(v) = z_v$. Here, k depends only on K and E.*

PROOF. Arguing inductively, it is clear that it suffices to prove the lemma in the case when u, v span an edge $e = [u, v]$ of T. We will use the fact that $\mathcal{L}^\ell$ is derived from the K-ladder $\mathcal{L}^h$. Thus, there exists a point $z \in L_u^h$ within distance K from z_v (the distance is measured in X_{uv}^h). Since z_v is in X_v, the incidence maps of $\mathfrak{X}$ are morphisms of relatively hyperbolic spaces, there exists a point $z_u \in L_u^h \cap X_u$ within uniformly bounded distance from z_v and, hence, within uniformly bounded distance from z in X_{uv}. Hence, we set $\sigma(u) := z_u$. □

9.4. Cannon–Thurston maps for trees of relatively hyperbolic spaces

Proof of Theorem 9.47: In the proof we shall use the following criterion for the existence of CT maps, which appears as [**MP11**, Lemma 1.29]. Recall that we have a subtree $\mathfrak{X}_S^\ell = (X^\ell \to S)$ in the tree of hyperbolic spaces $\mathfrak{X}^\ell = (X^\ell \to T)$. We fix a point $x_0 \in X_{v_0}$, where $v_0 \in V(S)$.

LEMMA 9.59. *A CT map for* $X_S^P \to X^P$ *exists provided that the following condition holds:*

For each $k \geq 1$ *and* $M \geq 0$, *there is* $N \geq 0$ *such that for all* $x, y \in X_S$, *if* $\gamma^\ell \subset X^\ell, \gamma_S^\ell \subset X_S^\ell$ *are tight k-quasigeodesics joining* x, y, *then* $d_X(x_0, \gamma^\ell \cap X) \leq M$ *implies* $d_{X_S}(x_0, \gamma_S^\ell \cap X_S) \leq N$.

PROOF. Our proof closely follows the one of [**MP11**, Lemma 1.29]. We will verify that the conditions of Proposition 9.36 are satisfied. The main difference with the setting of the lemma is that Proposition 9.36 was stated in terms of coned-off quasigeodesics $\widehat{\gamma}, \widehat{\gamma}_S$ in $\widehat{X} = \mathcal{E}(X, \mathcal{P})$, $\widehat{X}_S = \mathcal{E}(X_S, \mathcal{P}_S)$ respectively, connecting x, y. Here $\mathcal{P}_S$ is the collection of intersections $P \cap X_S$, $P \in \mathcal{P}$.

However, according to Lemma 9.51, the Hausdorff distances (computed in X and X_S respectively) between $\gamma^\ell \cap X, \widehat{\gamma} \cap X$ and between $\gamma_S^\ell \cap X_S, \widehat{\gamma}_S \cap X_S$ are uniformly bounded, with bounds depending only on K. With this in mind, Proposition 9.36 applies and the lemma follows. □

We now proceed to proving the theorem. Let $x, y \in X_S$ be arbitrary points and let $\gamma^\ell = [xy]_{X^\ell}$ be a uniform tight quasigeodesic in X^ℓ joining them (see Corollary 9.52). Suppose $z \in \gamma^\ell \cap X$ is a point such that $d_X(x_0, z) \leq D$ for some $D \geq 0$.

We apply the cut-and-replace operation (see Definition 8.14) to the tree of spaces $\mathfrak{X}^\ell$ and the quasigeodesic γ^ℓ, transforming it to a path $\widehat{\gamma^\ell}$ in X_S^ℓ. By Theorem 8.21, the path $\widehat{\gamma^\ell}$ in X_S^ℓ is a uniform quasigeodesic. Tightness of γ^ℓ implies that of $\widehat{\gamma^\ell}$.

CASE 1. Suppose that $z \in X_S$. By the construction of $\widehat{\gamma^\ell}$, the point z lies on $\widehat{\gamma^\ell}$. By Corollary 9.53, any two tight uniform quasigeodesics in X_S^ℓ uniformly track each other in X_S. Hence, each tight uniform quasigeodesic γ_S^ℓ in X_S^ℓ (connecting x and y) passes within uniformly bounded distance (in terms of the metric of X_S) from the point z. Thus, the implication required by Proposition 9.59 holds and we are done in this case.

CASE 2. Suppose that $z \notin X_S$. Then there is a vertex $v \in V(S)$ and a component γ_1^ℓ of $\gamma^\ell \setminus X_S^\ell$, such that $z \in \gamma_1^\ell$ and the endpoints x_1, y_1 of γ_1^ℓ belong to X_v^ℓ.

Subcase 2.1. We first consider the subcase when x_1, y_1 both belong to X_v. Let T_1 be the smallest subtree of T such that γ_1^ℓ is contained in $X_{T_1}^\ell$. In other words, T_1 is the span of a component of $T \setminus S$ and the vertex v. In the tree of spaces $\mathfrak{X}_{T_1}^\ell$ we construct a ladder

$$\mathcal{L}^\ell = \mathcal{L}^\ell([x_1 y_1]_{X_v^\ell})$$

with the total space L^ℓ, as described in Section 9.3.2.

Let $\bar{\gamma}_1^\ell$ be a uniform tight quasigeodesic joining x_1, y_1 in L^ℓ. Since the ladder L^ℓ is uniformly qi embedded in $X_{T_1}^\ell$ (see Corollary 3.13), $\bar{\gamma}_1^\ell$ is also a uniform quasigeodesic in $X_{T_1}^\ell$. Moreover, since the inclusion $\mathcal{L}^\ell \to \mathfrak{X}_{T_1}^\ell$ corresponds to a morphism of relatively hyperbolic spaces

$$(L, \mathcal{P}_L) \to (X_{T_1}, \mathcal{P}_{T_1}),$$

the path $\bar{\gamma}_1^\ell$ is tight in $X_{T_1}^\ell$.

Applying Corollary 9.53 again, we obtain that $z \in \gamma_1^\ell \cap X_{T_1}$ is C-close (in terms of the metric of X_{T_1}) to a point $\bar{z} \in \bar{\gamma}_1^\ell \cap X_w$ for some vertex $w \in V(T_1)$, where C depends only on qi constants of the original quasigeodesic γ_1^ℓ. Recall that $d_X(x_0, z) \leq D$. Then $d_X(x_0, \bar{z}) \leq D_1 := D + C$.

According to Lemma 9.58, there exists a k-qi section σ over $[\![v,w]\!]$ with image in $L^\ell \cap X$, such that $\sigma(w) = \bar{z}$ and

$$\sigma(v) = z_1 \in [x_1 y_1]_{X_v^\ell}.$$

Since z_1 belongs to X_S, $d_T(v,w) \le D_1$ and $d_X(x_0,\bar{z}) \le D_1$, it follows that

$$d_X(x_0, z_1) \le kD_1 + D_1.$$

Thus, we can apply the same reasoning as in the case when z is in X_S, to conclude that for every uniform tight quasigeodesic γ_S^ℓ in X_S^ℓ connecting x, y such that the minimal distance (in X_S) from $\gamma_S^\ell \cap X_S$ to z is uniformly bounded.

Subcase 2.2. Suppose that one of the two points x_1, y_1 is in X_v and the other is not. After relabeling, we can assume that $x_1 \in X_v$ and $y_1 \notin X_v$.

The (tight) geodesic $[x_1 y_1]_{X_v^\ell}$ contains a maximal subpath $[y_1' y_1]_{X_{v_1}^\ell}$ of length ≤ 1 contained in $X_v^\ell \setminus X_v$. The concatenation γ_1' of this path and γ_1^ℓ is still tight, γ_1' is again a uniform quasigeodesic in $\mathfrak{X}_{T_1}^\ell$, and both of its endpoints are in X_{v_1}. Now, we repeat the argument in Subcase 2.1 with respect to the pair of points $x_1, y_1' \in X_v$ and the uniform tight quasigeodesic γ_1' connecting them.

Subcase 2.3. The subcase when both points x_1, y_1 are not in X_v but $[x_1 y_1]_{X_v^\ell} \cap X_v \neq \emptyset$ is similar to Subcase 2.2 and we leave it to the reader.

Subcase 2.4. $[x_1 y_1]_{X_v^\ell} \cap X_v = \emptyset$, which implies that the points x_1, y_1 are within unit distance from each other in X_v^ℓ. Tightness of the path γ_1^ℓ then implies that γ_1^ℓ is disjoint from X, which contradicts the assumption that $z \in \gamma_1^\ell \cap X$.

9.5. Cannon–Thurston laminations for trees of relatively hyperbolic spaces

Since the inclusion map $X_S^P \to X^P$ admits a CT-map, one can also define its Cannon–Thurston lamination. Unlike Chapter 8 (specifically, Sections 8.7, 8.8 and 8.9) where these laminations were discussed in great detail in the absolute case, here we limit our discussion to a (weak) analogue of Theorem 8.52(1), relating these CT-laminations to that of the inclusion maps $X_v^h \to X^P$ (Theorem 9.60 below).

From now on, we assume, as we did in Theorem 8.52, that X is a proper metric space. Let $S \subset T$ be a subtree. We then have the inclusion map $X_S^P \to X^P$ that, according to Theorem 9.47, admits a CT-map $\partial_{X_S^P, X^P}$.

THEOREM 9.60. *Suppose that $\xi^\pm$ are distinct points in $\partial_\infty X_S^P$ such that*

$$\eta = \partial_{X_S^P, X^P}(\xi^-) = \partial_{X_S^P, X^P}(\xi^+).$$

(1) *Then there exists a vertex-space $X_v \subset X_S$ and a complete geodesic $\alpha : \mathbb{R} \to X_v^h$, asymptotic to $\xi_\pm$ in X_S^P and asymptotic (in X_v^h) to points $\xi_v^\pm \in \partial_\infty X_v^h$, such that*

$$\partial_{X_v^h, X^P}(\xi_v^+) = \partial_{X_v^h, X^P}(\xi_v^-) = \eta.$$

(2) *Moreover, α is a uniform quasigeodesic in X_S^P.*

REMARK 9.61. In this situation we necessarily have

$$\partial_{X_v^h, X_S^P}(\xi_v^\pm) = \xi_\pm.$$

PROOF. The proof follows the arguments of Proposition 8.49. Our first task is to modify the *cut-and-replace* procedure (Definition 8.14), used in the proof of Proposition 8.49. As before, we will be identifying spaces X_v^h with their images in X^P, $v \in V(T)$; ditto X_S^P. The modification of a path β in X^P to a path $\widehat{\beta}$ in X_S^P will be done in two steps.

STEP 1. The first step simply repeats what is done in Definition 8.14 (except that X^P and X^P_S are not exactly total spaces of trees of spaces!): We identify *primary detour subpaths* ζ_v of β, connecting points x_v, y_v of vertex-spaces X^h_v, $v \in V(S)$, and, apart from these endpoints, lying outside of X^P_S. We then replace each detour subpath with a geodesic $[x_v y_v]_{X^h_v}$ in X^h_v. These geodesic segments are the *primary replacement subpaths* for β. As in Definition 8.14, we refer to the resulting path as β_S: Its image entirely lies in X^P_S. The trouble is that, unlike the absolute case, even if β is uniformly quasigeodesic, the paths β_S *are priori* are not even uniformly proper in X^P_S.

STEP 2. We then modify β_S with respect to the collection of peripheral subsets $\mathcal{P}_S$ of X_S as follows. We define *secondary* detour subpaths of β_S as subpaths ζ_P in β_S connecting points of peripheral subsets $P \in \mathcal{P}_S$ of X_S and, besides those points, lying entirely inside the open peripheral horoballs $\mathring{P}^h \subset X^P_S$. The path

$$\widehat{\beta} = \mathcal{G}_{X_S,\mathcal{P}_S}(\beta_S)$$

the *hyperbolization* of β_S, is obtained by replacing each secondary detour subpath ζ_P with a geodesic in P^h connecting the endpoints of ζ_P.

Our next goal is to relate quasigeodesic properties of β to that of $\widehat{\beta}$. We will be using the notion of *tight paths*, Definition 9.15. The following lemma is clear:

LEMMA 9.62. *Suppose that β in X^P is tight with respect to $\mathcal{P}$.*

(1) *For each subtree $S \subset T$, the cut-and-replace path β_S is tight with respect to $\mathcal{P}_S$, and so is the path $\widehat{\beta}$.*
(2) *$(\beta_S)^\ell = (\beta^\ell)_S$, where β_S is the result of application Step 1 to β as above and $(\beta_S)^\ell$ is its partial electrification (a tight path in X^ℓ_S; see Definition 9.49), while $(\beta^\ell)_S$ is the result of application of the cut-and-replace procedure from Definition 8.14 to the path β^ℓ (the partial electrification of β) in the tree of spaces $\mathfrak{X}^\ell = (X^\ell \to T)$.*

In view of Part 2 of this lemma, we will use the notation β^ℓ_S for $(\beta_S)^\ell = (\beta^\ell)_S$.

Recall that the map Φ defined in Section 9.3.1 establishes a bijection between tight paths in $\mathcal{E}(X,\mathcal{P})$ and those in X^ℓ; same for the corresponding map Φ_S for the electric space $\mathcal{E}(X_S,\mathcal{P}_S)$ and X^ℓ_S. The following lemma is again a direct consequence of the definitions:

LEMMA 9.63. *For every path β in X^P, tight with respect to $\mathcal{P}$, the path $\widehat{\beta}$ can be described as follows. Set $\hat{\beta}_S := \Phi_S^{-1}(\beta^\ell_S)$. Then $\widehat{\beta}$ equals the path obtained via hyperbolization of $\hat{\beta}_S$ with respect to $\mathcal{P}_S$, i.e.,*

$$\widehat{\beta} = \mathcal{G}_{X_S,\mathcal{P}_S}(\hat{\beta}_S).$$

This lemma has an important consequence.

COROLLARY 9.64. *If β is a tight L-quasigeodesic in X, then $\widehat{\beta}$ is an $L_{9.64}(L)$- quasigeodesic in X^P_S.*

PROOF. First of all, according to Theorem 9.19(2), the path β^ℓ is a uniform (in terms of L) quasigeodesic in X^ℓ. Thus, by Theorem 8.21, so is the path $\beta^\ell_S = (\beta^\ell)_S$, in X^ℓ_S. By Lemma 9.50, the path $\hat{\beta}_S$ is also. Lastly, by Lemma 9.17, it follows that the path $\widehat{\beta} = \mathcal{G}_{X_S,\mathcal{P}_S}(\hat{\beta}_S)$ is also a uniform quasigeodesic in X^P_S. □

With these preliminaries out of the way, we can now proceed with the proof of Theorem 9.60. Fix a base-point $x_0 \in X_S$. Let $(x^\pm_n)$ denote sequences in X_S converging in $X^P_S \cup \partial_\infty X^P_S$ to the points $\xi^\pm$ respectively; see Lemma 9.12. We let β_n denote a sequence of tight (with respect to $\mathcal{P}$) uniformly quasigeodesic paths in X^P connecting the points x_n^-, x_n^+. According to Corollary 9.64, the paths $\widehat{\beta}_n$ are uniformly quasigeodesic. Thus, since, by the assumption, $\xi^+ \neq \xi^-$, it follows that there is a constant C such that $d(x_0, \widehat{\beta}_n) \le C$. We let z_n denote a point in $\widehat{\beta}_n$ within distance C from x_0. Without loss of generality, we may assume that these points lie in the images in X_S^P of spaces $X_{v_n}^h$, $v_n \in S$. Since X is assumed to be proper, so is X_S. Therefore, after extraction, we may assume that all points z_n lie in X_v^h for some vertex $v \in V(S)$.

CASE 1. All points z_n lie in X_v. Since the sequence of distances $d(x_0, \beta_n)$ diverges to ∞, for all but finitely many n, the points z_n lie in secondary replacement segments

$$[z_n^+ z_n^-]_{X_v} = \widehat{\zeta}_{vn} \subset \widehat{\beta}_n \cap X_v.$$

Recall that, by the construction of $\widehat{\beta}_n$, for each n we also have a primary replacement segment $\zeta_{vn} = [y_n^+ y_n^-]_{X_v^h}$ of β_n, containing $[z_n^+ z_n^-]_{X_v}$, where $y_n^\pm$ lie on β_n. Since $d(x_0, \beta_n) \to \infty$,

$$\lim_{n\to\infty} d(x_0, y_n^\pm) = \infty.$$

By the properness of X_v, since z_n is in ζ_{vn} and $d(x_0, z_n) \le C$, after further extraction, the sequence of geodesics $[y_n^+ y_n^-]_{X_v^h}$ converges to a bi-infinite geodesic α in X_v^h connecting points $\xi_v^\pm \in \partial_\infty X_v^h$. Using again the assumption that $d(x_0, \beta_n) \to \infty$, and the points $x_n^\pm, y_n^\pm$ all lie on the uniform quasigeodesic β_n, we see that

$$\partial_{X_v^h, X^P}(\xi_v^+) = \partial_{X_v^h, X^P}(\xi_v^-) = \partial_{X_S^P, X^P}(\xi^+) = \partial_{X_S^P, X^P}(\xi^-).$$

This proves the first claim of the theorem in Case 1.

It remains to show that α (or, for this matter, any uniform quasigeodesic in X_v^h asymptotic to $\xi_v^\pm$) is a uniform quasigeodesic in X_S^P. Note that, since paths $\widehat{\beta}_n$ are uniform quasigeodesics in X_S^P, the same holds for the subpaths $[z_n^- z_n^+]_{X_v} \subset \widehat{\beta}_n$.

Subcase 1a: $\lim_{n\to\infty} d(x_0, z_n^\pm) = \infty$. Then the sequence of subsegments $[z_n^- z_n^+]_{X_v}$ (uniformly quasigeodesic in X_S^P) subconverges to a geodesic α_v in X_v also asymptotic to $\xi_v^\pm$; it follows that α_v is a uniform quasigeodesic in X_S^P.

Subcase 1b: There exists a constant D such that $\lim_{n\to\infty} d(x_0, z_n^+) = \infty$ and $d(x_0, z_n^-) \le D$ for all n. By properness, after further extraction, we can assume that all points z_n^- lie on a fixed peripheral subset $P_v \in \mathcal{P}_{X_v}$ of X_v. We then consider the sequence of segments $[y_n^- z_n^+]_{X_v}$ instead of the segments $[z_n^- z_n^+]_{X_v}$ used in Case 1.

Each segment $[y_n^- z_n^+]_{X_v}$ is a concatenation of a vertical geodesic segment

$$[y_n^- z_n^-]_{X_v} \subset P_v^h$$

and the geodesic segment $[z_n^- z_n^+]_{X_v}$. Since the path β_{nS} is tight (with respect to $\mathcal{P}_S$), its subsegment $[y_n^- z_n^+]_{X_v}$ is also tight in X_v^h with respect to its peripheral structure $\mathcal{P}_v$. In particular, the subsegments $[z_n^- z_n^+]_{X_v}$ are all disjoint from the open peripheral horoball $\mathring{P}_v^h$. In particular, $[z_n^- z_n^+]_{X_v}$ is disjoint from the open peripheral horoball $\mathring{P}_S^h \subset X_S^P$ containing $\mathring{P}_v^h$. Using the fact that $[z_n^- z_n^+]_{X_v}$ and $[y_n^- z_n^+]_{X_v}$ are both uniform quasigeodesic in X_S^P, we apply Lemma 9.9 to conclude that the segments $[y_n^- z_n^+]_{X_v}$ are uniformly quasigeodesic in X_S^P as well. Thus, their limit α_v, a uniform quasigeodesic in X_v^h asymptotic to $\xi_v^\pm$, is also a uniform quasigeodesic in X_S^P.

Subcase 1c: There exists a constant D such that $d(x_0, z_n^\pm) \le D$ for all n. The proof is similar to the subcase 1b and we give only a sketch. We consider the sequence of geodesic segments $[y_n^- y_n^+]_{X_v}$ and break each of these as a concatenation of three geodesic segments, two of which are contained in distinct horoballs $P_{v\pm}^h$ and one has uniformly bounded length. We again apply Lemma 9.9 to conclude that the segments $[y_n^- y_n^+]_{X_v}$ are uniformly quasigeodesic in X_S^P.

CASE 2. We now assume that none of the points z_n belongs to X_S. Then, after further extraction, each z_n lies in a peripheral horoball P^h for some $P \in \mathcal{P}_S$. Hence, z_n belongs to a geodesic $[p_n q_n]_{P^h} \subset \widehat{\beta}_n$, where $p_n \in P_{v_n}$, $q_n \in P_{w_n} \in \mathcal{P}_{X_{w_n}}$ and

$$P_{v_n} = X_{v_n} \cap P, \quad P_{w_n} = X_{w_n} \cap P.$$

By the description of such geodesics $[p_n q_n]_{P^h}$, up to a uniformly bounded error, the path $[p_n q_n]_{P^h}$ is the concatenation of two vertical geodesics segments $[p_n p_n']_{X_{v_n}^h}$, $[q_n q_n']_{X_{w_n}^h}$ and a unit horizontal segment in P^h connecting p_n', q_n'. Thus, either

$$d_{X_S^P}(z_n, p_n) \leq C,$$

or

$$d_{X_S^P}(z_n, q_n) \leq C.$$

Accordingly, $d_{X_S^P}(x_0, p_n) \leq 2C$ or $d_{X_S^P}(x_0, q_n) \leq 2C$. Since p_n, q_n are in X_S, Case 2 is reduced to Case 1.

This concludes the proof of the theorem. □

Bibliography

[ABC+91] J. M. Alonso, T. Brady, D. Cooper, V. Ferlini, M. Lustig, M. Mihalik, M. Shapiro, and H. Short, *Notes on word hyperbolic groups*, Group theory from a geometrical viewpoint (Trieste, 1990), World Sci. Publ., River Edge, NJ, 1991, pp. 3–63. Edited by Short. MR1170363

[Ali05] E. Alibegović, *A combination theorem for relatively hyperbolic groups*, Bull. London Math. Soc. **37** (2005), no. 3, 459–466, DOI 10.1112/S0024609304004059. MR2131400

[Bas93] H. Bass, *Covering theory for graphs of groups*, J. Pure Appl. Algebra **89** (1993), no. 1-2, 3–47, DOI 10.1016/0022-4049(93)90085-8. MR1239551

[BBI01] D. Burago, Y. Burago, and S. Ivanov, *A course in metric geometry*, Graduate Studies in Mathematics, vol. 33, American Mathematical Society, Providence, RI, 2001, DOI 10.1090/gsm/033. MR1835418

[BCM12] J. F. Brock, R. D. Canary, and Y. N. Minsky, *The classification of Kleinian surface groups, II: The ending lamination conjecture*, Ann. of Math. (2) **176** (2012), no. 1, 1–149, DOI 10.4007/annals.2012.176.1.1. MR2925381

[BF92] M. Bestvina and M. Feighn, *A combination theorem for negatively curved groups*, J. Differential Geom. **35** (1992), no. 1, 85–101. MR1152226

[BF96] M. Bestvina and M. Feighn, *Addendum and correction to: "A combination theorem for negatively curved groups" [J. Differential Geom.* **35** *(1992), no. 1, 85–101; MR1152226 (93d:53053)]*, J. Differential Geom. **43** (1996), no. 4, 783–788. MR1412684

[BFH97] M. Bestvina, M. Feighn, and M. Handel, *Laminations, trees, and irreducible automorphisms of free groups*, Geom. Funct. Anal. **7** (1997), no. 2, 215–244, DOI 10.1007/PL00001618. MR1445386

[BH99] M. R. Bridson and A. Haefliger, *Metric spaces of non-positive curvature*, Grundlehren der mathematischen Wissenschaften [Fundamental Principles of Mathematical Sciences], vol. 319, Springer-Verlag, Berlin, 1999, DOI 10.1007/978-3-662-12494-9. MR1744486

[Bon86] F. Bonahon, *Bouts des variétés hyperboliques de dimension* 3 (French), Ann. of Math. (2) **124** (1986), no. 1, 71–158, DOI 10.2307/1971388. MR847953

[Bow95] B. H. Bowditch, *Geometrical finiteness with variable negative curvature*, Duke Math. J. **77** (1995), no. 1, 229–274, DOI 10.1215/S0012-7094-95-07709-6. MR1317633

[Bow99] B. H. Bowditch, *Convergence groups and configuration spaces*, Geometric group theory down under (Canberra, 1996), de Gruyter, Berlin, 1999, pp. 23–54. MR1714838

[Bow08] B. H. Bowditch, *Tight geodesics in the curve complex*, Invent. Math. **171** (2008), no. 2, 281–300, DOI 10.1007/s00222-007-0081-y. MR2367021

[Bow12] B. H. Bowditch, *Relatively hyperbolic groups*, Internat. J. Algebra Comput. **22** (2012), no. 3, 1250016, 66, DOI 10.1142/S0218196712500166. MR2922380

[Bow13] B. H. Bowditch, *Stacks of hyperbolic spaces and ends of 3-manifolds*, Geometry and topology down under, Contemp. Math., vol. 597, Amer. Math. Soc., Providence, RI, 2013, pp. 65–138, DOI 10.1090/conm/597/11769. MR3186670

[Bow14] B. H. Bowditch, *Uniform hyperbolicity of the curve graphs*, Pacific J. Math. **269** (2014), no. 2, 269–280, DOI 10.2140/pjm.2014.269.269. MR3238474

[BR13] O. Baker and T. R. Riley, *Cannon-Thurston maps do not always exist*, Forum Math. Sigma **1** (2013), Paper No. e3, 11, DOI 10.1017/fms.2013.4. MR3143716

[BR20] O. Baker and T. Riley, *Cannon-Thurston maps, subgroup distortion, and hyperbolic hydra*, Groups Geom. Dyn. **14** (2020), no. 1, 255–282, DOI 10.4171/ggd/543. MR4077662

[Bri00] P. Brinkmann, *Hyperbolic automorphisms of free groups*, Geom. Funct. Anal. **10** (2000), no. 5, 1071–1089, DOI 10.1007/PL00001647. MR1800064

[BS00] M. Bonk and O. Schramm, *Embeddings of Gromov hyperbolic spaces*, Geom. Funct. Anal. **10** (2000), no. 2, 266–306, DOI 10.1007/s000390050009. MR1771428

[Bum05] I. Bumagin, *On definitions of relatively hyperbolic groups*, Geometric methods in group theory, Contemp. Math., vol. 372, Amer. Math. Soc., Providence, RI, 2005, pp. 189–196, DOI 10.1090/conm/372/06884. MR2139687

[CB88] A. J. Casson and S. A. Bleiler, *Automorphisms of surfaces after Nielsen and Thurston*, London Mathematical Society Student Texts, vol. 9, Cambridge University Press, Cambridge, 1988, DOI 10.1017/CBO9780511623912. MR964685

[CDP90] M. Coornaert, T. Delzant, and A. Papadopoulos, *Géométrie et théorie des groupes* (French), Lecture Notes in Mathematics, vol. 1441, Springer-Verlag, Berlin, 1990. Les groupes hyperboliques de Gromov. [Gromov hyperbolic groups]; With an English summary. MR1075994

[CM17] C. H. Cashen and A. Martin, *Quasi-isometries between groups with two-ended splittings*, Math. Proc. Cambridge Philos. Soc. **162** (2017), no. 2, 249–291, DOI 10.1017/S0305004116000530. MR3604915

[CT07] J. W. Cannon and W. P. Thurston, *Group invariant Peano curves*, Geom. Topol. **11** (2007), 1315–1355, DOI 10.2140/gt.2007.11.1315. MR2326947

[Dah03] F. Dahmani, *Combination of convergence groups*, Geom. Topol. **7** (2003), 933–963, DOI 10.2140/gt.2003.7.933. MR2026551

[DDLS23] S. Dowdall, M. G. Durham, C. J. Leininger, and A. Sisto, *Extensions of Veech groups I: A hyperbolic action*, J. Topol. **16** (2023), no. 2, 757–805, DOI 10.1112/topo.12296. MR4637976

[DK18] C. Druţu and M. Kapovich, *Geometric group theory: With an appendix by Bogdan Nica*, American Mathematical Society Colloquium Publications, vol. 63, American Mathematical Society, Providence, RI, 2018, DOI 10.1090/coll/063. MR3753580

[DKT16] S. Dowdall, I. Kapovich, and S. J. Taylor, *Cannon-Thurston maps for hyperbolic free group extensions*, Israel J. Math. **216** (2016), no. 2, 753–797, DOI 10.1007/s11856-016-1426-2. MR3557464

[DM17] F. Dahmani and M. Mj, *Height, graded relative hyperbolicity and quasiconvexity* (English, with English and French summaries), J. Éc. polytech. Math. **4** (2017), 515–556, DOI 10.5802/jep.50. MR3646028

[DS05] C. Druţu and M. Sapir, *Tree-graded spaces and asymptotic cones of groups (English summary): With an appendix by Denis Osin and Mark Sapir*, Topology **44** (2005), no. 5, 959–1058, DOI 10.1016/j.top.2005.03.003. MR2153979

[DT18] S. Dowdall and S. J. Taylor, *Hyperbolic extensions of free groups*, Geom. Topol. **22** (2018), no. 1, 517–570, DOI 10.2140/gt.2018.22.517. MR3720349

[Far98] B. Farb, *Relatively hyperbolic groups*, Geom. Funct. Anal. **8** (1998), no. 5, 810–840, DOI 10.1007/s000390050075. MR1650094

[FHS83] M. Freedman, J. Hass, and P. Scott, *Least area incompressible surfaces in 3-manifolds*, Invent. Math. **71** (1983), no. 3, 609–642, DOI 10.1007/BF02095997. MR695910

[Gau03] F. Gautero, *Hyperbolicity of mapping-torus groups and spaces*, Enseign. Math. (2) **49** (2003), no. 3-4, 263–305. MR2026897

[Gau16] F. Gautero, *Geodesics in trees of hyperbolic and relatively hyperbolic spaces*, Proc. Edinb. Math. Soc. (2) **59** (2016), no. 3, 701–740, DOI 10.1017/S0013091515000450. MR3572767

[Gd90] É. Ghys and P. de la Harpe, *La propriété de Markov pour les groupes hyperboliques* (French), Sur les groupes hyperboliques d'après Mikhael Gromov (Bern, 1988), Progr. Math., vol. 83, Birkhäuser Boston, Boston, MA, 1990, pp. 165–187, DOI 10.1007/978-1-4684-9167-8_9. MR1086657

[Ger96] S. M. Gersten, *Subgroups of word hyperbolic groups in dimension 2*, J. London Math. Soc. (2) **54** (1996), no. 2, 261–283, DOI 10.1112/jlms/54.2.261. MR1405055

[Ger98] S. M. Gersten, *Cohomological lower bounds for isoperimetric functions on groups*, Topology **37** (1998), no. 5, 1031–1072, DOI 10.1016/S0040-9383(97)00070-0. MR1650363

[Ger09] V. Gerasimov, *Expansive convergence groups are relatively hyperbolic*, Geom. Funct. Anal. **19** (2009), no. 1, 137–169, DOI 10.1007/s00039-009-0718-7. MR2507221

[GGKW17] F. Guéritaud, O. Guichard, F. Kassel, and A. Wienhard, *Anosov representations and proper actions*, Geom. Topol. **21** (2017), no. 1, 485–584, DOI 10.2140/gt.2017.21.485. MR3608719

[GL04] F. Gautero and M. Lustig, *Relative hyperbolization of (one-ended hyperbolic)-by-cyclic groups*, Math. Proc. Cambridge Philos. Soc. **137** (2004), no. 3, 595–611, DOI 10.1017/S0305004104007881. MR2103918

[GL07] F. Gautero, *Combinatorial mapping-torus, branched surfaces and free group automorphisms*, Ann. Sc. Norm. Super. Pisa Cl. Sci. (5) **6** (2007), no. 3, 405–440. MR2370267

[GMRS98] R. Gitik, M. Mitra, E. Rips, and M. Sageev, *Widths of subgroups*, Trans. Amer. Math. Soc. **350** (1998), no. 1, 321–329, DOI 10.1090/S0002-9947-98-01792-9. MR1389776

[GP15] V. Gerasimov and L. Potyagailo, *Non-finitely generated relatively hyperbolic groups and Floyd quasiconvexity*, Groups Geom. Dyn. **9** (2015), no. 2, 369–434, DOI 10.4171/GGD/317. MR3356972

[GP16] V. Gerasimov and L. Potyagailo, *Similar relatively hyperbolic actions of a group*, Int. Math. Res. Not. IMRN **7** (2016), 2068–2103, DOI 10.1093/imrn/rnv170. MR3509947

[Gro87] M. Gromov, *Hyperbolic groups*, Essays in group theory, Math. Sci. Res. Inst. Publ., vol. 8, Springer, New York, 1987, pp. 75–263, DOI 10.1007/978-1-4613-9586-7_3. MR919829

[Gro93] M. Gromov, *Asymptotic invariants of infinite groups*, Geometric group theory, Vol. 2 (Sussex, 1991), London Math. Soc. Lecture Note Ser., vol. 182, Cambridge Univ. Press, Cambridge, 1993, pp. 1–295. MR1253544

[GS19] S. Gouëzel and V. Shchur, *Corrigendum: A corrected quantitative version of the Morse lemma [MR3003738]*, J. Funct. Anal. **277** (2019), no. 4, 1258–1268, DOI 10.1016/j.jfa.2019.02.021. MR3959731

[Gui04] O. Guichard. Déformation de sous-groupes discrets de groupes de rang un. PhD thesis, Université Paris 7, 2004.

[GW12] O. Guichard and A. Wienhard, *Anosov representations: domains of discontinuity and applications*, Invent. Math. **190** (2012), no. 2, 357–438, DOI 10.1007/s00222-012-0382-7. MR2981818

[Ham05] U. Hamenstädt. Word hyperbolic extension of surface groups. Preprint, `arXiv:math/0505244`, 2005.

[Hat02] A. Hatcher, *Algebraic topology*, Cambridge University Press, Cambridge, 2002. MR1867354

[Hem04] J. Hempel, *3-manifolds: Reprint of the 1976 original*, AMS Chelsea Publishing, Providence, RI, 2004, DOI 10.1090/chel/349. MR2098385

[HH20] B. Healy and G. C. Hruska. Cusped spaces and quasi-isometries of relatively hyperbolic groups. Arxiv, `arXiv:2010.09876`, 2020.

[Hru10] G. C. Hruska, *Relative hyperbolicity and relative quasiconvexity for countable groups*, Algebr. Geom. Topol. **10** (2010), no. 3, 1807–1856, DOI 10.2140/agt.2010.10.1807. MR2684983

[JKLO16] W. Jeon, I. Kapovich, C. Leininger, and K. Ohshika, *Conical limit points and the Cannon-Thurston map*, Conform. Geom. Dyn. **20** (2016), 58–80, DOI 10.1090/ecgd/294. MR3488025

[Kap01] I. Kapovich, *The combination theorem and quasiconvexity*, Internat. J. Algebra Comput. **11** (2001), no. 2, 185–216, DOI 10.1142/S0218196701000553. MR1829050

[Kap09] M. Kapovich, *Hyperbolic manifolds and discrete groups*, Modern Birkhäuser Classics, Birkhäuser Boston, Ltd., Boston, MA, 2009. Reprint of the 2001 edition, DOI 10.1007/978-0-8176-4913-5. MR2553578

[Kap23] M. Kapovich. A note on properly discontinuous actions. *San Paulo Journal of Math Sciences*, 2023. DOI 10.1007/s40863-023-00353-z.

[KL18a] M. Kapovich and B. Leeb, *Discrete isometry groups of symmetric spaces*, Handbook of group actions. Vol. IV, Adv. Lect. Math. (ALM), vol. 41, Int. Press, Somerville, MA, 2018, pp. 191–290. MR3888689

[KL18b] M. Kapovich and B. Leeb, *Finsler bordifications of symmetric and certain locally symmetric spaces*, Geom. Topol. **22** (2018), no. 5, 2533–2646, DOI 10.2140/gt.2018.22.2533. MR3811766

[KL19] M. Kapovich and B. Liu, *Geometric finiteness in negatively pinched Hadamard manifolds*, Ann. Acad. Sci. Fenn. Math. **44** (2019), no. 2, 841–875, DOI 10.5186/aasfm.2019.4444. MR3973544

[Kla99] E. Klarreich, *Semiconjugacies between Kleinian group actions on the Riemann sphere*, Amer. J. Math. **121** (1999), no. 5, 1031–1078. MR1713300

[Kla18] E. Klarreich, *The boundary at infinity of the curve complex and the relative Teichmüller space*, Groups Geom. Dyn. **16** (2022), no. 2, 705–723, DOI 10.4171/ggd/662. MR4502619

[KLP17] M. Kapovich, B. Leeb, and J. Porti, *Anosov subgroups: dynamical and geometric characterizations*, Eur. J. Math. **3** (2017), no. 4, 808–898, DOI 10.1007/s40879-017-0192-y. MR3736790

[KM98] O. Kharlampovich and A. Myasnikov, *Hyperbolic groups and free constructions*, Trans. Amer. Math. Soc. **350** (1998), no. 2, 571–613, DOI 10.1090/S0002-9947-98-01773-5. MR1390041

[KS20] M. Mj and P. Sardar, *A combination theorem for metric bundles*, Geom. Funct. Anal. **22** (2012), no. 6, 1636–1707, DOI 10.1007/s00039-012-0196-1. MR3000500

[Man05] J. F. Manning, *Geometry of pseudocharacters*, Geom. Topol. **9** (2005), 1147–1185, DOI 10.2140/gt.2005.9.1147. MR2174263

[Min94] Y. N. Minsky, *On rigidity, limit sets, and end invariants of hyperbolic* 3*-manifolds*, J. Amer. Math. Soc. **7** (1994), no. 3, 539–588, DOI 10.2307/2152785. MR1257060

[Min03a] Y. N. Minsky, *Combinatorial and geometrical aspects of hyperbolic 3-manifolds*, Kleinian groups and hyperbolic 3-manifolds (Warwick, 2001), London Math. Soc. Lecture Note Ser., vol. 299,

Cambridge Univ. Press, Cambridge, 2003, pp. 3–40, DOI 10.1017/CBO9780511542817.002. MR2044543

[Min03b] Y. N. Minsky, *End invariants and the classification of hyperbolic 3-manifolds*, Current developments in mathematics, 2002, Int. Press, Somerville, MA, 2003, pp. 181–217. MR2062319

[Mit97] M. Mitra, *Ending laminations for hyperbolic group extensions*, Geom. Funct. Anal. **7** (1997), no. 2, 379–402, DOI 10.1007/PL00001624. MR1445392

[Mit98] M. Mitra, *Cannon-Thurston maps for trees of hyperbolic metric spaces*, J. Differential Geom. **48** (1998), no. 1, 135–164. MR1622603

[Mit99] M. Mitra, *On a theorem of Scott and Swarup*, Proc. Amer. Math. Soc. **127** (1999), no. 6, 1625–1631, DOI 10.1090/S0002-9939-99-04935-7. MR1610757

[Mit04] M. Mitra, *Height in splittings of hyperbolic groups*, Proc. Indian Acad. Sci. Math. Sci. **114** (2004), no. 1, 39–54, DOI 10.1007/BF02829670. MR2040599

[Mj14a] M. Mj, *Cannon-Thurston maps for surface groups*, Ann. of Math. (2) **179** (2014), no. 1, 1–80, DOI 10.4007/annals.2014.179.1.1. MR3126566

[Mj14b] M. Mj, *Ending laminations and Cannon-Thurston maps: With an appendix by Shubhabrata Das and Mj*, Geom. Funct. Anal. **24** (2014), no. 1, 297–321, DOI 10.1007/s00039-014-0263-x. MR3177384

[Mj16] M. Mj, *Cannon-Thurston maps for surface groups: an exposition of amalgamation geometry and split geometry*, Geometry, topology, and dynamics in negative curvature, London Math. Soc. Lecture Note Ser., vol. 425, Cambridge Univ. Press, Cambridge, 2016, pp. 221–271. MR3497262

[Mj17] M. Mj, *Cannon-Thurston maps for Kleinian groups*, Forum Math. Pi **5** (2017), e1, 49, DOI 10.1017/fmp.2017.2. MR3652816

[MMS85] D. McCullough, A. Miller, and G. A. Swarup, *Uniqueness of cores of noncompact 3-manifolds*, J. London Math. Soc. (2) **32** (1985), no. 3, 548–556, DOI 10.1112/jlms/s2-32.3.548. MR825931

[Mor24] H. M. Morse, *A fundamental class of geodesics on any closed surface of genus greater than one*, Trans. Amer. Math. Soc. **26** (1924), no. 1, 25–60, DOI 10.2307/1989225. MR1501263

[Mos97] L. Mosher, *A hyperbolic-by-hyperbolic hyperbolic group*, Proc. Amer. Math. Soc. **125** (1997), no. 12, 3447–3455, DOI 10.1090/S0002-9939-97-04249-4. MR1443845

[MP11] M. Mj and A. Pal, *Relative hyperbolicity, trees of spaces and Cannon-Thurston maps*, Geom. Dedicata **151** (2011), 59–78, DOI 10.1007/s10711-010-9519-2. MR2780738

[MR08] M. Mj and L. Reeves, *A combination theorem for strong relative hyperbolicity*, Geom. Topol. **12** (2008), no. 3, 1777–1798, DOI 10.2140/gt.2008.12.1777. MR2421140

[MR18] M. Mj and K. Rafi, *Algebraic ending laminations and quasiconvexity*, Algebr. Geom. Topol. **18** (2018), no. 4, 1883–1916, DOI 10.2140/agt.2018.18.1883. MR3797060

[MS09] M. Mj and P. Sardar, *A combination theorem for metric bundles*, Geom. Funct. Anal. **22** (2012), no. 6, 1636–1707, DOI 10.1007/s00039-012-0196-1. MR3000500

[MS13] H. Masur and S. Schleimer, *The geometry of the disk complex*, J. Amer. Math. Soc. **26** (2013), no. 1, 1–62, DOI 10.1090/S0894-0347-2012-00742-5. MR2983005

[MS20] J. M. Mackay and A. Sisto, *Maps between relatively hyperbolic spaces and between their boundaries*, Trans. Amer. Math. Soc. **377** (2024), no. 2, 1409–1454, DOI 10.1090/tran/9063. MR4688555

[Mut21] J. P. Mutanguha, *The dynamics and geometry of free group endomorphisms*, Adv. Math. **384** (2021), Paper No. 107714, 60, DOI 10.1016/j.aim.2021.107714. MR4237417

[Ohs09] K. Ohshika, *Constructing geometrically infinite groups on boundaries of deformation spaces*, J. Math. Soc. Japan **61** (2009), no. 4, 1261–1291. MR2588511

[Osi06] D. V. Osin, *Relatively hyperbolic groups: intrinsic geometry, algebraic properties, and algorithmic problems*, Mem. Amer. Math. Soc. **179** (2006), no. 843, vi+100, DOI 10.1090/memo/0843. MR2182268

[Ota01] J.-P. Otal, *The hyperbolization theorem for fibered 3-manifolds: Translated from the 1996 French original by Leslie D. Kay*, SMF/AMS Texts and Monographs, vol. 7, American Mathematical Society, Providence, RI; Société Mathématique de France, Paris, 2001. MR1855976

[PKS15] A. Pal and A. Kumar Singh, *Relatively hyperbolic spaces*, Geometry, groups and dynamics, Contemp. Math., vol. 639, Amer. Math. Soc., Providence, RI, 2015, pp. 307–325, DOI 10.1090/conm/639/12792. MR3379836

[Roe03] J. Roe, *Lectures on coarse geometry*, University Lecture Series, vol. 31, American Mathematical Society, Providence, RI, 2003, DOI 10.1090/ulect/031. MR2007488

[RS94] E. Rips and Z. Sela, *Structure and rigidity in hyperbolic groups. I*, Geom. Funct. Anal. **4** (1994), no. 3, 337–371, DOI 10.1007/BF01896245. MR1274119

[Sar18] P. Sardar, *Graphs of hyperbolic groups and a limit set intersection theorem*, Proc. Amer. Math. Soc. **146** (2018), no. 5, 1859–1871, DOI 10.1090/proc/13871. MR3767341

[Sco73a] G. P. Scott, *Finitely generated 3-manifold groups are finitely presented*, J. London Math. Soc. (2) **6** (1973), 437–440, DOI 10.1112/jlms/s2-6.3.437. MR380763

[Sco73b] G. P. Scott, *Compact submanifolds of 3-manifolds*, J. London Math. Soc. (2) **7** (1973), 246–250, DOI 10.1112/jlms/s2-7.2.246. MR326737

[Sel97] Z. Sela, *Acylindrical accessibility for groups*, Invent. Math. **129** (1997), no. 3, 527–565, DOI 10.1007/s002220050172. MR1465334

[Ser03] J.-P. Serre, *Trees*: *Translated from the French original by John Stillwell; Corrected 2nd printing of the 1980 English translation*, Springer Monographs in Mathematics, Springer-Verlag, Berlin, 2003. MR1954121

[Sho91] H. Short, *Quasiconvexity and a theorem of Howson's*, Group theory from a geometrical viewpoint (Trieste, 1990), World Sci. Publ., River Edge, NJ, 1991, pp. 168–176. MR1170365

[Sis13] A. Sisto, *Projections and relative hyperbolicity*, Enseign. Math. (2) **59** (2013), no. 1-2, 165–181, DOI 10.4171/LEM/59-1-6. MR3113603

[SW79] P. Scott and T. Wall, *Topological methods in group theory*, Homological group theory (Proc. Sympos., Durham, 1977), London Math. Soc. Lecture Note Ser., vol. 36, Cambridge Univ. Press, Cambridge-New York, 1979, pp. 137–203. MR564422

[Swe01] E. L. Swenson, *Quasi-convex groups of isometries of negatively curved spaces (English translation)*: *Geometric topology and geometric group theory (Milwaukee, WI, 1997)*, Topology Appl. **110** (2001), no. 1, 119–129, DOI 10.1016/S0166-8641(99)00166-2. MR1804703

[Thu80] W. P. Thurston. The Geometry and Topology of 3-Manifolds. *Princeton University Notes*, 1980.

[Tuk98] P. Tukia, *Conical limit points and uniform convergence groups*, J. Reine Angew. Math. **501** (1998), 71–98, DOI 10.1515/crll.1998.081. MR1637829

[V05] J. Väisälä, *Gromov hyperbolic spaces*, Expo. Math. **23** (2005), no. 3, 187–231, DOI 10.1016/j.exmath.2005.01.010. MR2164775

[Yam04] A. Yaman, *A topological characterisation of relatively hyperbolic groups*, J. Reine Angew. Math. **566** (2004), 41–89, DOI 10.1515/crll.2004.007. MR2039323

List of symbols

Δ	geodesic triangles in metric spaces
L, ϵ	coarse Lipschitz constants
D, E, ϵ, R, r	bounds on distances
R, r	radii of tubular neighborhoods
$\ell(\alpha)$, length(α)	length of a path α
$c, \alpha, \beta, \gamma, \phi, \psi$	a path
β^h	hyperbolization of a path β
$\widehat{\beta}$	hyperbolization of a path β_S
β^ℓ	electrification of a path β
$\hat{c} = c_S$	path obtained from c via the cut-and-replace procedure, Definition 8.14
$\overleftarrow{c}$	reversed path of c
$c_1 \star c_2$	concatenation of paths c_1, c_2
D_0	Equation (3.5)
K_0	Notation 2.6.4
κ, k, K	quasi-isometry and flow constants
Function K'	Lemma 3.17
Functions $K^\vee$ and $K^\wedge$	Equation (3.11)
λ	quasiconvexity constant
ρ, μ, ν	retractions
f, g, h	maps
P, $P_{X,Y}$	nearest-point projection
$\bar{P}$	the modified projection to a tripod
$\bar{x}$	nearest-point projection of x
η	a distortion function
η_0	the distortion function of vertex-spaces in a tree of spaces $\mathfrak{X}$
$V(T)$	the vertex set of a tree T
$e = [u, v]$	the edge of T connecting vertices u, v
$\dot{e}$	an edge in a tree minus its vertices
$E(T)$	the edge set of a tree T
γ_x	qi sections over an interval or a subtree, passing through x
Hd	Hausdorff distance

Symbol	Meaning
$[\![u,v]\!]$ or uv	the geodesic segment (interval) connecting vertices $u, v \in V(T)$ in a tree
$]\!]u,v]\!]$, $]\!]u,v[\![$, $[\![u,v[\![$	open and half-open intervals in T
$S(u,r)$	the sphere of radius r and center u
$B(u,r)$	the closed ball of radius r and center u
$N_R(A)$	closed R-neighborhood of A
$b = \mathrm{center}(\Delta)$	center of a tripod in a tree
$\mathcal{X} := \coprod_{v\in V(T)} X_v$	union of vertex-spaces of a tree of spaces
$\mathcal{F}l_K(\cdot)$, $Fl_K(\cdot)$, $\mathfrak{F}l_K(\cdot)$	flow-space (as a tree of spaces), the total space and the intersection with $\mathcal{X}$ respectively
$Fl(X_v)$	ideal boundary flow
$\mathfrak{L}$, $\mathcal{L}_{K,D,E}$	a ladder, as a tree of spaces and intersection with $\mathcal{X}$ respectively
$\mathfrak{A}$, $\mathcal{A}_{K,C}$	carpets
$\Lambda(f)$, $\Lambda(Y,X)$	CT-lamination
$\Lambda^{\xi}(X_v, X)$	ξ-ending lamination, Definition 8.55
$\partial_{Y,X}$, $\partial_{H,G}$	Cannon–Thurston maps for the inclusion maps $Y \to X$ and $H \hookrightarrow G$
$\partial_\infty(Y,X)$	relative ideal boundary of Y in X
$\partial^{\xi}_\infty(X_v, X)$	ξ-relative ideal boundary, Definition 8.55
δ_0	hyperbolicity constant of vertex spaces (total space and intersection with $\mathcal{X}$)
δ'_0	hyperbolicity constant of spaces X_{uv}, $[u,v] \in E(T)$
λ_0	the quasiconvexity constant of $X_{eu} \subset X_u$, $e = [u,v] \in E(T)$
λ'_0	the quasiconvexity constant for the inclusion maps $X_u \to X_{uv}$, $[u,v] \in E(T)$
L_0	the quasi-isometry constant for the incidence maps $f_{eu} : X_e \to X_v$
L'_0	the quasi-isometry constant for the inclusion maps $X_v \to X_{uv}$, $e = [u,v] \in E(T)$
N^{fib}_R	closed fiberwise R-neighborhood
N^e_R	closed neighborhood in X_{uv}, where $[u,v] = e \in E(T)$
$\Pi = (\gamma_0, \gamma_1)$	a pair of K-qi sections over the same interval in T
Π_0	the girth of Π
$\Pi_{\max}$	the maximal separation of the ends of Π
$\mathfrak{X} = (\pi : X \to T)$	a tree of spaces
$X_S := \pi^{-1}(S)$	the total space of a subtree of spaces over $S \subset T$
$(Y^\ell, d^\ell) = \mathcal{E}(Y, \mathcal{H})$	the electrified space
$(Y^h, d^h) = \mathcal{G}(Y, \mathcal{H})$	the horoballification
$a(Y^\ell)$	cone-locus
$\mathring{H}^h$	open horoball, §9.1.1
$\mathring{H}^\ell$	open cone, §9.1.1

Index

Selected Published Titles in This Series

282 **Michael Kapovich and Pranab Sardar,** Trees of Hyperbolic Spaces, 2024
281 **Horst R. Thieme,** Discrete-Time Dynamics of Structured Populations and Homogeneous Order-Preserving Operators, 2024
280 **Tadashi Ochiai,** Iwasawa Theory and Its Perspective, Volume 2, 2024
279 **Ellen Elizabeth Eischen, Wee Teck Gan, Aaron Pollack, and Zhiwei Yun,** Automorphic Forms Beyond GL_2, 2024
278 **Delaram Kahrobaei, Ramón Flores, Marialaura Noce, Maggie E. Habeeb, and Christopher Battarbee,** Applications of Group Theory in Cryptography, 2024
277 **George Metcalfe, Francesco Paoli, and Constantine Tsinakis,** Residuated Structures in Algebra and Logic, 2023
276 **Balázs Bárány, Károly Simon, and Boris Solomyak,** Self-similar and Self-affine Sets and Measures, 2023
275 **Alekos Vidras and Alain Yger,** Multidimensional Residue Theory and Applications, 2023
274 **Willi Freeden and M. Zuhair Nashed,** Recovery Methodologies: Regularization and Sampling, 2023
273 **Gennadiy Feldman,** Characterization of Probability Distributions on Locally Compact Abelian Groups, 2023
272 **Tadashi Ochiai,** Iwasawa Theory and Its Perspective, Volume 1, 2023
271 **Hideto Asashiba,** Categories and Representation Theory, 2022
270 **Leslie Hogben, Jephian C.-H. Lin, and Bryan L. Shader,** Inverse Problems and Zero Forcing for Graphs, 2022
269 **Ralph S. Freese, Ralph N. McKenzie, George F. McNulty, and Walter F. Taylor,** Algebras, Lattices, Varieties, 2022
268 **Ralph S. Freese, Ralph N. McKenzie, George F. McNulty, and Walter F. Taylor,** Algebras, Lattices, Varieties, 2022
267 **Dragana S. Cvetković Ilić,** Completion Problems on Operator Matrices, 2022
266 **Kate Juschenko,** Amenability of Discrete Groups by Examples, 2022
265 **Nabil H. Mustafa,** Sampling in Combinatorial and Geometric Set Systems, 2022
264 **J. Scott Carter and Seiichi Kamada,** Diagrammatic Algebra, 2021
263 **Vugar E. Ismailov,** Ridge Functions and Applications in Neural Networks, 2021
262 **Ragnar-Olaf Buchweitz,** Maximal Cohen–Macaulay Modules and Tate Cohomology, 2021
261 **Shiri Artstein-Avidan, Apostolos Giannopoulos, and Vitali D. Milman,** Asymptotic Geometric Analysis, Part II, 2021
260 **Lindsay N. Childs, Cornelius Greither, Kevin P. Keating, Alan Koch, Timothy Kohl, Paul J. Truman, and Robert G. Underwood,** Hopf Algebras and Galois Module Theory, 2021
259 **William Heinzer, Christel Rotthaus, and Sylvia Wiegand,** Integral Domains Inside Noetherian Power Series Rings, 2021
258 **Pramod N. Achar,** Perverse Sheaves and Applications to Representation Theory, 2021
257 **Juha Kinnunen, Juha Lehrbäck, and Antti Vähäkangas,** Maximal Function Methods for Sobolev Spaces, 2021
256 **Michio Jimbo, Tetsuji Miwa, and Fedor Smirnov,** Local Operators in Integrable Models I, 2021
255 **Alexandre Boritchev and Sergei Kuksin,** One-Dimensional Turbulence and the Stochastic Burgers Equation, 2021
254 **Karim Belabas and Henri Cohen,** Numerical Algorithms for Number Theory, 2021

For a complete list of titles in this series, visit the AMS Bookstore at **www.ams.org/bookstore/survseries/**.